Probability at Saint-Flour

Editorial Committee: Jean Bertoin, Erwin Bolthausen, K. David Elworthy

T0241150

For further volumes:
http://www.springer.com/series/10212

Saint-Flour Probability Summer School

Founded in 1971, the Saint-Flour Probability Summer School is organised every year by the mathematics department of the Université Blaise Pascal at Clermont-Ferrand, France, and held in the pleasant surroundings of an 18th century seminary building in the city of Saint-Flour, located in the French Massif Central, at an altitude of 900 m.

It attracts a mixed audience of up to 70 PhD students, instructors and researchers interested in probability theory, statistics, and their applications, and lasts 2 weeks. Each summer it provides, in three high-level courses presented by international specialists, a comprehensive study of some subfields in probability theory or statistics. The participants thus have the opportunity to interact with these specialists and also to present their own research work in short lectures.

The lecture courses are written up by their authors for publication in the LNM series.

The Saint-Flour Probability Summer School is supported by:

– Université Blaise Pascal
– Centre National de la Recherche Scientifique (C.N.R.S.)
– Ministère délégué à l'Enseignement supérieur et à la Recherche

For more information, see back pages of the book and
http://math.univ-bpclermont.fr/stflour/

Jean Picard
Summer School Chairman
Laboratoire de Mathématiques
Université Blaise Pascal
63177 Aubière Cedex
France

Philippe Biane • Alice Guionnet
Dan-Virgil Voiculescu

Noncommutative Probability
and Random Matrices
at Saint-Flour

 Springer

Philippe Biane
Inst. Gaspard Monge
Université Paris-Est
Champs-Sur-Marne
France

Alice Guionnet
Ecole Normale Supérieure de Lyon
Lyon Cedex 07
France

Dan-Virgil Voiculescu
Department of Mathematics
University of California, Berkeley
Berkeley, CA
USA

Reprint of lectures originally published in the Lecture Notes in Mathematics volumes 1608 (1995), 1738 (2000) and 1957 (2009)

ISBN 978-3-642-32798-8
Springer Heidelberg New York Dordrecht London

Library of Congress Control Number: 2012949819

Mathematics Subject Classification (2010): 46L54; 46L53; 60H99; 60G60; 60H07; 60-02; 82-02; 15-02

Printed on acid-free paper

Springer is part of Springer Science+Business Media (www.springer.com)

Preface for Saint Flour collections

The *École d'Été de Saint-Flour*, founded in 1971, is organised every year by the *Laboratoire de Mathématiques* of the *Université Blaise Pascal* (Clermont-Ferrand II) and the *CNRS*. It is intended for PhD students, teachers and researchers who are interested in probability theory, statistics, and in applications of stochastic techniques. The summer school has been so successful in its 40 years of existence that it has long since become one of the institutions of probability as a field of scholarship.

The school has always had three main simultaneous goals:

1. to provide, in three high-level courses, a comprehensive study of 3 fields of probability theory or statistics;
2. to facilitate exchange and interaction between junior and senior participants;
3. to enable the participants to explain their own work in lectures.

The lecturers and topics of each year are chosen by the Scientific Board of the school. Further information may be found at http://math.univ-bpclermont.fr/stflour/

The published courses of Saint-Flour have, since the school's beginnings, been published in the *Lecture Notes in Mathematics* series, originally and for many years in a single annual volume, collecting 3 courses. More recently, as lecturers chose to write up their courses at greater length, they were published as individual, single-author volumes. See www.springer.com/series/7098. These books have become standard references in many subjects and are cited frequently in the literature.

As probability and statistics evolve over time, and as generations of mathematicians succeed each other, some important subtopics have been revisited more than once at Saint-Flour, at intervals of 10 years or so.

On the occasion of the 40th anniversary of the *École d'Été de Saint-Flour*, a small ad hoc committee was formed to create selections of some courses on related topics from different decades of the school's existence that would seem interesting viewed and read together. As a result Springer is releasing a number of such theme volumes under the collective name "Probability at Saint-Flour".

Jean Bertoin, Erwin Bolthausen and K. David Elworthy

Jean Picard, Pierre Bernard, Paul-Louis Hennequin
  (current and past Directors of the *École d'Été de Saint-Flour*)

September 2012

Table of Contents

Calcul Stochastique Non-Commutatif..1
Philippe Biane

Introduction...4
1. Espaces de probabilité non-commutatifs finis...11
2. Théorie spectrale 22
3. Variables de Gauss et de Poisson et relations de commutation 28
4. Espaces de Fock 41
5. Intégration stochastique non-commutative 52
6. Applications du calcul stochastique non-commutatif 72
Commentaires et bibliographie...89
Références..93

Lectures on Free Probability Theory..97
Dan Voiculescu

0. Introduction 101
1. Noncommutative Probability and Operator Algebra Background...............102
2. Addition of Freely Independent Noncommutative Random Variables.........112
3. Multiplication of freely Independent Noncommutative
   Random Variables 126
4. Generalized Canonical Form, Noncrossing Partitions 131
5. Free Independence with Amalgamation...134
6. Some Basic Free Processes 137
7. Random Matrices in the Large N Limit..143
8. Free Entropy 150
References..164

Large Random Matrices: Lectures on Macroscopic Asymptotics................169
Alice Guionnet

Introduction...179
1. Wigner's Theorem...185
2. Wigner's Matrices; More Moments Estimates...207
3. Words in Several Independent Wigner Matrices 219
4. Concentration Inequalities and Logarithmic Sobolev Inequalities 227
5. Generalizations 237
6. Concentration Inequalities for Random Matrices 243
7. Maps and Gaussian Calculus...271

8.   First-order Expansion 287
9.   Second-order Expansion for the Free Energy 299
10. Large Deviations for the Law of the Spectral Measure of Gaussian
     Wigner's Matrices 327
11. Large Deviations of the Maximum Eigenvalue 337
12. Stochastic Analysis for Random Matrices 345
13. Large Deviation Principle for the Law of the Spectral Measure
     of Shifted Wigner Matrices 361
14. Asymptotics of Harish–Chandra–Itzykson–Zuber Integrals
     and of Schur Polynomials..389
15. Asymptotics of Some Matrix Integrals 395
16. Free Probability Setting...405
17. Freeness 409
18. Free Entropy 423
19. Basics of Matrices 441
20. Basics of Probability Theory 445
References..453
Index 465
List of Participants of the Summer School...467
List of Short Lectures Given at the Summer School....................................471

CALCUL STOCHASTIQUE

NON-COMMUTATIF

Philippe BIANE

Originally published in: *École d'Été de Probabilités de Saint-Flour XXIII – 1993*,
Lecture Notes in Mathematics, Vol. **1608**, 1–96, DOI: 10.1007/ BFb0095746,
© Springer-Verlag Berlin Heidelberg 1995, Reprint by Springer-Verlag Berlin Heidelberg 2012

Sommaire

Introduction   4

1. Espaces de probabilités non-commutatifs finis   11

    1.1. Espaces de probabilités   11
    1.2. Variables aléatoires, lois   12
    1.3. Espace de bernoulli non-commutatif   13
    1.4. Addition de variables de Bernoulli quantiques indépendantes   15
    1.5. Le processus de spin   17
    1.6. Un théorème limite pour la marche de Bernoulli quantique   19

2. Théorie spectrale   22

    2.1. Quelques définitions   22
    2.2. Résolutions de l'identité et théorème spectral   24
    2.3. Opérateurs à trace et espaces de probabilités non-commutatifs   25
    2.5. Théorèmes de Stone et de Nelson   26

3. Variables de Gauss et de Poisson et relations de commutation   28

    3.1. Relations de commutation d'Heisenberg   28
    3.2. Modèle gaussien des relations de commutation   30
    3.3 Opérateurs de Weyl   31
    3.4. Vecteurs exponentiels   32
    3.5. Opérateur de nombre et loi de Poisson   33
    3.6. Modèle poissonien des relations de commutation   35
    3.7. Théorème de Stone-von Neumann   36
    3.8. Conséquences du théorème de Stone-von Neumann   39

4. Espaces de Fock   41

    4.1. Puissances symétriques d'un espace de Hilbert   41
    4.2. Opérateurs de création et d'annihiltion   43
    4.3. Seconde quantification   44
    4.4. Opérateurs de Weyl sur l'espace de Fock   45
    4.5. Interprétations gaussienne et poissonienne de l'espace de Fock   47
    4.6. Mouvement brownien et processus de Poisson   49
    4.7. Martingales normales et repésentations chaotiques   51

5. Intégration stochastique non-commutative   52

5.1. Filtration et processus adaptés sur l'espace de Fock   52
5.2. Intégration de processus simples   53
5.3. Divergence et gradient sur l'espace $\Gamma(L^2_{\mathbf{C}}(\mathbf{R}_+))$ 56
5.4. Intégrale stochastique et formule d'Itô   58
5.5. Exemples d'intégrales stochastiques   68
5.6. Extensions de l'intégrale stochastique   68

6. Applications du calcul stochastique non-commutatif   72

6.1. Processus de Markov non-commutatif   72
6.2. Une autre construction de la marche de Bernoulli quantique   75
6.3. Equations différentielles linéaires   78
6.4. Dilatations de semi-groupes complètement positifs   82
6.5. Application aux processus de Markov   84

Commentaires et bibliographie   89

Références   93

Introduction

Le calcul stochastique non-commutatif s'est développé depuis quelques années à la suite des travaux de Hudson et Parthasarathy [40], qui ont défini des intégrales stochastiques par rapport à trois "martingales non-commutatives" a_t^+, a_t^-, et a_t^0. Ces trois processus ne sont pas composés de variables aléatoires au sens classique, mais sont des familles d'opérateurs sur l'espace L^2 de la mesure de Wiener. Le théorème spectral permet d'interpréter les combinaisons auto-adjointes de ces opérateurs comme des variables aléatoires, et le fait qu'ils ne commutent pas entre eux, leur permet d'avoir des propriétés remarquables; par exemple, le processus $(a_t^+ + a_t^-)_{t \geq 0}$ s'interprète comme un mouvement brownien ou, plus exactement, chaque opérateur $a_t^+ + a_t^-$ est l'opérateur de multiplication par la variable aléatoire B_t, où $(B_t)_{t \geq 0}$ est le mouvement brownien canonique sur l'espace de Wiener. D'autre part, pour tout $z \in \mathbb{C}$ $(a_t^0 + z a_t^+ + \bar{z} a_t^- + |z|^2 t)_{t \geq 0}$ est un processus de Poisson d'intensité $|z|^2$, au sens où ces opérateurs s'interprètent comme les opérateurs de multiplication définis par un processus de Poisson sur son espace L^2, lorsqu'on a identifié l'espace L^2 de la mesure de Wiener et celui d'un processus de Poisson au moyen des décompositions en chaos. Il est clairement impossible d'obtenir de telles propriétés en utilisant des familles de variables aléatoires au sens habituel du terme. Toutes les notions évoquées ci-dessus seront expliquées en détails dans la suite du cours.

La motivation initiale du calcul stochastique stochastique non-commutatif était d'utiliser les techniques d'équations différentielles stochastiques pour résoudre des problèmes de mécanique quantique, mais il s'est avéré que la théorie ainsi développée est intéressante par elle-même et donne un point de vue nouveau sur certains aspects des probabilités classiques. C'est ainsi que la formule d'Itô y apparaît comme intimement liée aux relations de commutation d'Heisenberg qui jouent un rôle fondamental en mécanique quantique. Un sous-produit remarquable de ce calcul stochastique est la possibilité d'obtenir, en principe, tout processus de Markov comme solution d'une équation différentielle stochastique qui a la même structure formelle que celle d'une diffusion, les mouvements browniens qui dirigent l'équation étant remplacés par une version multidimensionnelle des "processus non-commutatifs" a_t^+, a_t^-, et a_t^0 (en fait pour avoir ce résultat en toute généralité il faut affronter des problèmes analytiques qui ne sont pas encore complètement résolus). On verra à la fin du cours des exemples explicites de telles constructions, dont celui des chaînes de Markov en temps continu sur un espace d'état fini, pour lequel la théorie est complète. Il apparaît que ce sont les sauts des processus de Markov qui nécessitent l'introduction de martingales non-commutatives dans l'équation différentielle stochastique.

Le but de ce cours est de présenter, à un public de probabilistes, les bases de cette théorie qui est actuellement en plein développement. L'expérience de plusieurs exposés devant des probabilistes "classiques" m'a appris que l'évocation des "variables aléatoires non-commutatives" avait tendance à plonger l'auditoire dans la perplexité. Ma première tâche va donc être de tenter de démythifier cette notion, et pour cela je vais commencer par évoquer son origine, qui est au coeur de la mécanique quantique.

C'est à von Neumann [56] que l'on doit d'avoir dégagé le formalisme mathématique de la mécanique quantique, après les travaux de nombreux physiciens dont W.Heisenberg, E.Schrödinger, et M.Born. Le postulat de base est que tout système physique peut être décrit par un espace de Hilbert complexe H et un vecteur $\psi \in H$ de norme 1 (que l'on apelle l'état du système). Souvent, le système considéré est très simple, par exemple il peut consister en une seule particule dans un espace vide, auquel cas un choix plausible de H est l'espace $L^2(\mathbf{R}^3)$, (si on ne tient pas compte du spin) et ψ s'appelle alors la fonction d'onde de la particule. Mais rien n'empêche en théorie de traiter des système complexes composés d'un grand nombre de particules; on a alors affaire à de "gros" espaces de Hilbert, et à des fonctions d'onde compliquées que l'on ne peut pas, en général, expliciter.

A toute quantité physique du système que l'on peut mesurer au moyen d'une expérience correspond un opérateur auto-adjoint A sur H. D'après le théorème spectral il existe une mesure μ_ψ sur le spectre $\sigma(A)$ telle que $\mu_\psi(f) = < f(A)\psi, \psi >$ pour toute fonction borélienne bornée sur $\sigma(A)$, $f(A)$ étant défini par le cacul fonctionnel des opérateurs auto-adjoints. Lorsque A a un spectre discret formé de valeurs propres λ_i de multiplicité 1, cette mesure de probabilité est simple à décrire, on a $\mu_\psi(\{\lambda_i\}) = | < \psi, \phi_i > |^2$, ϕ_i étant un vecteur propre de norme 1 associé à λ_i. On postule que le résultat d'une mesure de la quantité physique correspondant à A est une variable aléatoire de loi μ_ψ lorsque le système est dans l'état ψ. En particulier, cette loi n'est une mesure de Dirac que lorsque le vecteur ψ est un vecteur propre de A. Un autre postulat ("réduction du paquet d'ondes") énonce qu'après la mesure de la quantité correspondant à l'opérateur A, si cette mesure a donné comme résultat λ (où λ est une valeur propre de A de vecteur propre correspondant ϕ), le système se trouve dans l'état ϕ.

La mécanique quantique est une théorie intrinsèquement probabiliste, car il existe des systèmes physiques pour lesquels la connaissance exacte de toutes les données (i.e. de la fonction d'onde du système) ne permet pas de prédire avec certitude le résultat d'une expérience.

Pour les lecteurs qui ne sont pas familiers avec la mécanique quantique, je vais illustrer les postulats ci-dessus par la description d'expériences de physique (comme dans Feynman [35]) inspirées de celle de Stern et Gerlach, et en donner une interprétation avec les postulats de la mécanique quantique. Cette description nous permettra de rencontrer un premier exemple de variable aléatoire "non-commutative", exemple sur lequel on reviendra au chapitre 1.

On suppose donné un dispositif produisant un faisceau linéaire horizontal de particules de même nature (par exemple des atomes d'hydrogène, ou des molécules d'azote, ...), que l'on envoie à travers un appareil qui crée un champ magnétique ayant un fort gradient vertical. A la sortie de l'appareil, on constate que le faisceau de particules s'est scindé en plusieurs sous-faisceaux présentant des déviations verticales par rapport à la direction initiale (voir *fig.* 1).

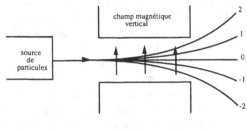

fig. 1

Les déviations de ces faisceaux par rapport à la direction initiale sont des multiples entiers d'une même quantité. Cette quantité ainsi que le nombre de faisceaux sont des caractéristiques du type des particules utilisées, elles ne dépendent pas de la direction du champ magnétique, si celui-ci reste orthogonal au faisceau. Si le faisceau initial se scinde en N faisceaux, la particule est dite de spin J où J est le demi-entier $J = \frac{N-1}{2}$ (dans le cas de la figure 1, le spin est 2). On interprète cette expérience en postulant que chaque particule possède un "moment magnétique" dont la composante suivant l'axe vertical est un multiple demi-entier d'une certaine quantité, ce demi-entier étant de la forme $-J, -J+1, \ldots, J$. (C'est cette propriété de quantification de certaines quantités physiques associées aux particules élémentaires qui a donné son nom à la mécanique quantique).

On peut modifier cette expérience en faisant passer les faisceaux à travers un second appareil, placé juste à la sortie du premier, et créant un champ magnétique opposé, de façon à remettre les particules dans leur direction initiale. En disposant judicieusement des caches entre les deux champs magnétiques, on peut filtrer les faisceaux de manière à n'en garder qu'un.

fig. 2

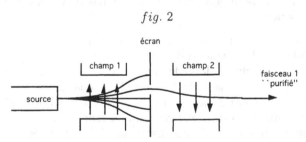

En faisant passer un tel faisceau "purifié" à travers un troisième champ vertical, on constate qu'il ne se sépare plus et qu'il dévie de la même valeur qu'auparavant. Ce procédé permet donc de sélectionner les particules suivant la valeur de leur moment magnétique vertical (*fig.* 3).

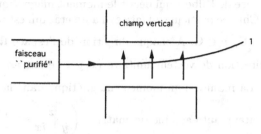

fig. 3

Afin de simplifier la discussion, je vais maintenant supposer que les particules ont un spin $J = \frac{1}{2}$, c'est à dire que le moment magnétique d'une telle particule suivant une direction donnée ne peut prendre que les valeurs $+1$ ou -1 dans une unité de mesure convenable. Au moyen du procédé décrit ci-dessus on se donne un faisceau de telles particules de moment magnétique vertical $+1$, que l'on fait passer à travers un champ magnétique faisant un angle Δ avec la verticale (*fig.* 4, $\Delta \in [0, \frac{\pi}{2}]$).

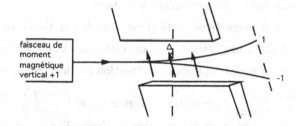

fig. 4

On constate que le faisceau s'est séparé en deux. Si l'on mesure l'intensité de chaque faisceau (c'est à dire la proportion de particules du faisceau initial dans chacun des deux faisceaux de sortie), on obtient respectivement les valeurs $\cos^2 \frac{\Delta}{2}$ et $\sin^2 \frac{\Delta}{2}$.

Utilisons un dispositif analogue à celui de la figure 2 pour sélectionner les particules de la figure 4 qui ont un moment magnétique $+1$ dans la direction Δ. Si maintenant on fait passer à nouveau ces particules dans un champ magnétique vertical, on constate que le faisceau se sépare en 2, autrement dit, après être passées dans le champ magnétique non vertical les particules ont "oublié" leur moment magnétique vertical. On peut recommencer l'expérience avec des directions arbitraires, à chaque fois le résultat est le suivant: si on a sélectionné un faisceau de particules ayant un moment magnétique $+1$ dans la direction D, et que l'on mesure leurs moments magnétiques dans la direction D' faisant un angle Δ avec D, on trouve $+1$ et -1 avec des proportions respectives de $\cos^2 \frac{\Delta}{2}$ et $\sin^2 \frac{\Delta}{2}$.

Pour essayer d'expliquer ces expériences de manière classique il faudrait pour chaque particule introduire une famille de variables aléatoires indexées par les directions de l'espace, et prenant les valeurs $\{-1, +1\}$, qui changeraient de façon compliquée par un passage au travers d'un champ magnétique. L'interprétation quantique, elle, ne fait intervenir que des vecteurs de dimension 2 et des matrices 2×2. Voici cette interprétation. L'espace de Hilbert qui décrit le moment magnétique d'une particule est de dimension 2. Chaque particule du faisceau a un état qui est un vecteur colonne $\psi = \begin{pmatrix} u \\ v \end{pmatrix}$ avec $u \in \mathbb{C}$, $v \in \mathbb{C}$. A chaque direction de l'espace $\mathbb{R}^3$, on associe une matrice 2×2, la direction de vecteur unitaire (x, y, z) étant associée à la matrice $\begin{pmatrix} z & y + ix \\ y - ix & -z \end{pmatrix}$. La mesure d'un moment magnétique dans la direction (x, y, z) correspond à l'opérateur auto-adjoint de matrice $\begin{pmatrix} z & y + ix \\ y - ix & -z \end{pmatrix}$. D'après les postulats de la mécanique quantique, si une particule est dans l'état $\psi = \begin{pmatrix} u \\ v \end{pmatrix}$, la mesure du moment magnétique dans la direction (x, y, z) est une variable aléatoire qui vaut 1 avec probabilité $| < \psi, \alpha_+ > |^2$ et -1 avec probabilité $| < \psi, \alpha_- > |^2$, où α_+ et α_- sont des vecteurs propres unitaires de $\begin{pmatrix} z & y + ix \\ y - ix & -z \end{pmatrix}$ de valeurs propres respectives $+1$ et -1. D'autre part, si le résultat de la mesure a donné $+1$ (ou -1), après cette mesure, la particule se trouve dans l'état α_+ (ou α_-).

Reprenons l'exemple des expériences décrites ci-dessus.

La mesure du moment magnétique vertical est donnée par l'opérateur $\begin{pmatrix} 1 & 0 \\ 0 & -1 \end{pmatrix}$. Les particules de moment magnétique $+1$ dans cette direction sont donc dans l'état $e_1 = \begin{pmatrix} 1 \\ 0 \end{pmatrix}$. Le moment magnétique dans la direction $(0, \sin\Delta, \cos\Delta)$ est mesuré par l'opérateur $\begin{pmatrix} \cos\Delta & \sin\Delta \\ \sin\Delta & -\cos\Delta \end{pmatrix}$, de vecteurs propres $\alpha_+ = \begin{pmatrix} \cos\frac{\Delta}{2} \\ \sin\frac{\Delta}{2} \end{pmatrix}$ $\alpha_- = \begin{pmatrix} \sin\frac{\Delta}{2} \\ \cos\frac{\Delta}{2} \end{pmatrix}$.

Le résultat de la mesure est donc une variable de Bernoulli de probabilités: $P(1) = | < e_1, \alpha_+ > |^2 = \cos^2\frac{\Delta}{2}$ et $P(-1) = | < e_1, \alpha_- > |^2 = \sin^2\frac{\Delta}{2}$, comme le montre l'expérience. Un calcul semblable avec des directions arbitraires dans l'espace permet de retrouver tous les résultats de l'expérience.

Cet exemple nous a permis de rencontrer pour la première fois des "variables de Bernoulli quantiques". Nous les étudierons plus en détails au chapitre 1 du cours.

Après cette brève incursion dans la physique des particules, nous allons nous concentrer sur l'aspect purement mathématique des probabilités quantiques. Le plan du cours suit un chemin parallèle à celui d'un cours de probabilités classique, partant des probabilités élémentaires sur un ensemble fini, passant aux variables "continues" avec le théorème central limite, puis aux processus stochastiques, (mouvement brownien, processus de Poisson), et enfin à l'intégration stochastique.

On commence au chapitre 1 par définir la notion d'espace de probabilité non-commutatif fini, et les variables aléatoires sur un tel espace. Il s'agit en fait simplement d'algèbre linéaire de dimension finie, mais l'interprétation probabiliste que l'on en fait amène à des considérations intéressantes. Alors que la théorie élémentaire des probabilités utilise essentiellement la combinatoire et le dénombrement (voir Feller [34]), ces méthodes sont remplacées ici par l'algèbre linéaire. Même dans une situation aussi simple on voit rapidement apparaître des effets "quantiques" non-triviaux comme on le montre en étudiant la marche de Bernoulli quantique, un analogue non-commutatif du jeu de pile ou face, et le "processus de spin" associé. A la fin du chapitre 1, on démontre un théorème limite qui contient les deux théorèmes limite en loi sur les sommes de variables de Bernoulli, d'une part le théorème de Moivre-Laplace, d'autre part le théorème de convergence vers la loi de Poisson. On verra que le formalisme non-commutatif permet de réunir ces deux résultats en un seul. Les variables limites obtenues joueront un rôle fondamental par la suite.

L'énoncé du théorème limite montre la nécessité de sortir du cadre de la dimension finie, pour traiter aussi bien des variables continues, comme les gaussiennes, que des variables discrètes prenant une infinité de valeurs, comme les variables de Poisson. Cela est rendu possible grâce à la théorie spectrale des opérateurs auto-adjoints dont les principaux résultats sont rappelés au chapitre 2.

Avec le chapitre 3 on rentre dans le vif du sujet en étudiant de façon approfondie les objets introduits à la fin du chapitre 1, qui sont connus en physique quantique sous le nom d'opérateurs de création, d'annihilation, et de nombre. Le chapitre 4 est consacré aux espaces de Fock qui fournissent un moyen canonique de construire des familles d'opérateurs de création, annihilation et nombre. Ce chapitre contient des notions qui sont bien connues des spécialistes du calcul de Malliavin (cf par exemple Watanabe [87]), seul le langage dans lequel elles sont exprimées est un peu différent. C'est dans ce chapitre que l'on introduit les processus a_t^+, a_t^-, et a_t^0 de Hudson et Parthasarathy.

Dans le chapitre 5, on aborde le calcul stochastique non-commutatif proprement dit, inspiré du calcul stochastique d'Itô. On verra que ce nouveau calcul stochastique est très lié à l'intégrale de Skorokhod (voir Skorokhod [80]), une extension de l'intégrale d'Itô, qui est en fait un objet purement Hilbertien, défini comme une divergence, comme l'on remarqué Gaveau et Trauber [36]. Dans le chapitre 6 on montre comment ce calcul permet de construire explicitement des processus de Markov non-commutatifs, cette notion étant une extension de la notion usuelle de processus de Markov. Dans le cas particulier des processus de Markov au sens habituel, cette construction permet d'étendre à des processus de Markov non-nécessairement continus la construction des diffusions à l'aide d'équation différentielles stochastiques. On décrira comment fonctionne ce formalisme dans le cas simple des processus de naissance et des chaînes de Markov sur un espace d'états fini.

Un dernier chapitre est consacré à des commentaires et des compléments, ainsi qu'aux références bibliographiques.

Les connaissances requises pour lire le cours se limitent à une bonne familiarité avec les notions de base de probabilité et de théorie des opérateurs (théorie spectrale), mais en pratique il serait utile d'avoir une bonne connaissance du calcul stochastique usuel comme exposé par exemple dans [17], [21] ou [69]. En particulier, les sujets suivants, populaires parmi les probabilistes, utilisent des notions très proches: calcul de Malliavin et analyse sur l'espace de Wiener, intégrale de Skorokhod, équations différentielles stochastiques. La théorie des représentations des groupes est présente implicitement dans une grande partie du cours, mais on ne l'utilise pas vraiment (sauf au paragraphe 6.2). Le point de vue des groupes est beaucoup plus présent dans le livre de Parthasarathy [60].

Je remercie vivement les organisateurs de l'Ecole d'été de Saint-Flour de m'avoir donné la possibilité de faire ce cours. Je suis également reconnaissant à tous ceux qui ont bien voulu faire des commentaires sur la première version du cours, et je remercie tout particulièrement Monique Pontier dont les suggestions ainsi que les corrections qu'elle m'a indiquées m'ont considérablement aidé, et Jacques Azéma pour sa lecture critique et méticuleuse de la première version du cours.

Chapitre 1

Espaces de probabilité non-commutatifs finis

1.1 Espaces de probabilités

Ce chapitre est consacré à la notion d'espace de probabilité non-commutatif, en commençant par l'exemple le plus simple, qui est l'analogue d'un espace de probabilité usuel fini. Comme indiqué dans l'introduction l'idée de base, qui a son origine dans la mécanique quantique, est qu'un opérateur auto-adjoint sur un espace de Hilbert peut s'interpréter comme une variable aléatoire. Pour préciser cette correspondance, commençons par montrer comment associer à une variable aléatoire un opérateur auto-adjoint.

Soit $(\Omega, \mathcal{F}, P)$ un espace de probabilité fini, avec la tribu formée de toutes les parties de Ω. L'espace des variables aléatoires complexes $L_{\mathbf{C}}^2(\Omega, \mathcal{F}, P)$ est muni de sa structure d'espace de Hilbert complexe de dimension finie. Soit X une variable aléatoire réelle sur $(\Omega, \mathcal{F}, P)$, on définit un opérateur auto-adjoint

$$T_X : L_{\mathbf{C}}^2(\Omega, \mathcal{F}, P) \to L_{\mathbf{C}}^2(\Omega, \mathcal{F}, P)$$

$$f \to Xf$$

$L_{\mathbf{C}}^2(\Omega, \mathcal{F}, P)$ a une base orthonormale $(\delta_\omega = \frac{\chi_\omega}{\sqrt{P(\{\omega\})}})_{\omega \in \Omega}$ où χ_ω désigne la fonction indicatrice de ω (on suppose que $P(\{\omega\}) > 0$ pour tout $\omega \in \Omega$). Dans cette base, l'opérateur T_X est diagonal, ses valeurs propres sont les valeurs prises par la variable aléatoire X, puisque $T_X(\delta_\omega) = X(\omega)\delta_\omega$. De plus, l'espérance de la variable aléatoire X est donnée par la formule $E[X] = <X1, 1>$ où 1 désigne la variable identiquement égale à 1.

Réciproquement, donnons nous un opérateur auto-adjoint A sur un espace de Hilbert complexe E de dimension finie , et $(\varphi_i)_{i \in I}$, $I = \{1, \ldots, n\}$ une base orthonormale de vecteurs propres de A, correspondant aux valeurs propres $\{\lambda_i\}_{i \in I}$ (avec éventuellement des répétitions en cas de valeurs propres de multiplicité > 1). Soit $u \in E$ un vecteur unitaire, il détermine une loi de probabilité μ sur $(I, \mathcal{I})$ (I est muni de la tribu maximale $\mathcal{I}$) en posant $\mu(\{i\}) = |<u, \varphi_i>|^2$, ainsi qu'une isométrie $\iota : l^2(I, \mathcal{I}, \mu) \to E$ qui envoie χ_i sur $<u, \varphi_i> \varphi_i$. Soit X la variable aléatoire sur $(I, \mathcal{I})$ définie par $X(i) = \lambda_i$, alors $A = \iota \circ T_X \circ \iota^*$, donc l'opérateur A s'interprète comme un opérateur de multiplication par la variable aléatoire X. De plus, l'espérance de X s'exprime à l'aide de A par la formule $E[X] = <Au, u>$.

Les deux exemples ci-dessus montrent comment on peut passer de la donnée de (E, u, A), E étant un espace de Hilbert u un vecteur unitaire de E et A un opérateur auto-adjoint sur E à celle de $(\Omega, \mathcal{F}, P, X)$ où $(\Omega, \mathcal{F}, P)$ est un espace de probabilités et X une variable aléatoire. Le principe des probabilités quantiques consiste à considérer la première donnée comme fondamentale, et à remplacer l'espace de probabilité usuel par un couple (E, u) et les variables aléatoires par les opérateurs auto-adjoints sur E.

Prendre l'espérance d'une variable aléatoire correspond alors à appliquer la forme linéaire $A \mapsto < Au, u >$. Comme en probabilités classiques, où l'on a parfois besoin de considérer plusieurs probabilités sur le même espace, on veut pouvoir considérer simultanément des vecteurs unitaires $u_1, \ldots, u_n$. Lorsqu'on fait une combinaison convexe des fonctionnelles $A \mapsto < Au_k, u_k >$ définies sur $\mathcal{L}(E)$ l'espace des opérateurs sur E, on n'obtient pas une fonctionnelle du type $A \mapsto < Au, u >$. Observons que $< Au, u >= tr(A\pi_u)$, tr étant la forme linéaire donnant la trace d'un opérateur, et π_u le projecteur orthogonal sur la droite engendrée par u. On voit donc que si $\sum_k p_k = 1$, $\sum_k p_k < Au_k, u_k >= tr(AS)$ où $S = \sum_k p_k \pi_{u_k}$ est un opérateur positif de trace 1. En fait, tout opérateur positif de trace 1 s'obtient de cette façon.

Donnons une première définition.

DÉFINITION 1 – Un espace de probabilité non-commutatif fini est la donnée d'un espace de Hilbert de dimension finie, E et d'un opérateur positif S sur E, de trace 1. La forme linéaire sur $\mathcal{L}(E)$ (l'espace des opérateurs sur E) donnée par

$$A \mapsto tr(AS)$$

est appelée l'état correspondant à S.

Par abus de langage, on confondra parfois l'opérateur S avec l'état qu'il définit. L'ensemble des états est une partie convexe compacte du dual de $\mathcal{L}(E)$, dont les points extrémaux sont les états de la forme $A \mapsto < Au, u >$ (appelés aussi états purs), où u est un vecteur de norme 1. Un état pur est associé à un opérateur π_u qui est le projecteur orthogonal sur la droite engendrée par u; en particulier, deux vecteurs unitaires déterminent le même état si et seulement s'ils sont colinéaires. Remarquons tout de suite que tout état est combinaison convexe d'un nombre fini d'états purs mais qu'il n'y pas unicité de la décomposition (par exemple l'état associé à $\frac{1}{dim(E)} Id$ est égal à $\frac{1}{dim(E)} \sum_i \pi_{u_i}$ pour toute base orthonormale (u_i) de E).

1.2 Variables aléatoires, lois

Après avoir introduit les espaces de probabilité, passons aux variables aléatoires.

DÉFINITION 2 – Une variable aléatoire non-commutative sur l'espace de probabilité non-commutatif fini (E, S) est un opérateur auto-adjoint sur E.

Lorsqu'aucune confusion n'est possible, on parlera de variable aléatoire tout court.

L'exemple du paragraphe 1.1 montre que toute variable aléatoire réelle au sens usuel peut être considérée comme une variable aléatoire non-commutative, en la faisant agir par multiplication sur un espace L^2. Soit A une variable aléatoire non-commutative, appelons $\sigma(A)$ son spectre, et soit f une fonction réelle sur $\sigma(A)$. On définit un opérateur autoadjoint en posant $f(A)x = f(\lambda)x$ pour tout $\lambda \in \sigma(A)$ et x vecteur propre de A de valeur propre λ.

DÉFINITION 3 – La loi de la variable A dans l'état S est la mesure de probabilité sur $\sigma(A)$ donnée par la formule:

$$\mu(f) = tr(f(A)S)$$

pour toute fonction f sur $\sigma(A)$.

Soit $\lambda \in \sigma(A)$ on a $\mu(\{\lambda\}) = tr(S\pi_\lambda)$ où π_λ est le projecteur orthogonal sur le sous-espace propre associé à λ. On vérifie immédiatement en reprenant l'exemple du paragraphe 1.1 que $T_{f(X)} = f(T_X)$ et que la loi de T_X dans l'état pur 1 (ici 1 désigne la fonction constante égale à 1) est bien la loi de la variable aléatoire X. Les notions de variable aléatoire et de loi que nous venons de définir sont donc des extensions des notions usuelles. La première différence avec la théorie des probabilités classiques apparait quand on essaie d'appliquer les opérations algébriques élémentaires. En effet, alors qu'on peut toujours faire le produit de deux variables aléatoires usuelles, le produit de deux opérateurs auto-adjoints n'est auto-adjoint que s'ils commutent. Nous allons examiner plus en détail ce dernier cas. Soient A et B deux opérateurs auto-adjoints qui commutent, on peut décomposer E en somme directe de sous-espaces propres communs à A et B. Soit $(E_{\lambda,\nu})_{\lambda \in \sigma(A), \nu \in \sigma(B)}$ une telle décomposition on peut définir $f(A, B)$ pour une fonction f sur $\sigma(A) \times \sigma(B)$ par $f(A, B)x = f(\lambda, \nu)x$ si $x \in E_{\lambda,\nu}$. Ce calcul fonctionnel respecte l'addition et la multiplication des fonctions, ce qui fait qu'on peut définir la loi jointe de A et B qui est la mesure sur $\sigma(A) \times \sigma(B)$ déterminée par

$$\mu_{A,B}(f) = tr(f(A, B)S)$$

Plus généralement, une famille $(X_t)_{t \in T}$ d'opérateurs auto-adjoints qui commutent admet un calcul fonctionnel qui permet de définir leur loi jointe.

DÉFINITION 4 – *Une famille d'opérateurs sur un espace de Hilbert, indexée par un ensemble T (le temps) est appelée un processus stochastique non-commutatif. Si ces opérateurs sont auto-adjoints et commutent, le processus est dit classique.*

DÉFINITION 5 – *La loi d'un processus classique $(X_t)_{t \in T}$ est la loi jointe des variables $(X_t)_{t \in T}$, c'est à dire la mesure de probabilité sur $\prod_{t \in T} \sigma(X_t)$ déterminée par la forme linéaire*

$$f \mapsto tr(Sf((X_t)_{t \in \mathbb{R}}))$$

sur l'ensemble des fonctions cylindriques bornées sur $\prod_{t \in T} \sigma(X_t)$.

Lorsqu'on considère des opérateurs non auto-adjoints, ou qui ne commutent pas entre eux, la notion de loi jointe de ces opérateurs n'a pas de sens.

Dans la suite du cours, nous allons rencontrer de nombreuses familles d'opérateurs auto-adjoints qui commutent, et calculer les lois des processus classiques correspondants. Un des aspects remarquables de ces processus sera leur construction à partir d'opérateurs qui ne commutent pas entre eux, et pour lesquels la notion de "loi jointe" n'a pas de sens.

1.3 Espace de Bernoulli non-commutatif

Nous allons maintenant étudier plus en détail l'exemple le plus simple d'espace de probabilité non-commutatif, en considérant l'espace de Hilbert $E = \mathbb{C}^2$ avec sa base canonique (e_1, e_2), et le produit hermitien usuel.

Les opérateurs linéaires sur E sont représentés dans la base canonique par des matrices 2×2. Un état se représente par une matrice de la forme:

$$S = \begin{pmatrix} \alpha & \bar{z} \\ z & 1 - \alpha \end{pmatrix} \quad \text{avec } \alpha \in \mathbb{R} \ z \in \mathbb{C} \ \alpha(1 - \alpha) \geq z\bar{z}, \ 0 \leq \alpha \leq 1 \tag{1}$$

Lorsque $\alpha(1 - \alpha) = z\bar{z}$, S est un projecteur orthogonal sur une droite complexe, l'état associé est donc un état pur.

Les variables aléatoires forment un espace vectoriel réel de dimension 4, noté V. La formule $< A, B >= \frac{1}{2} tr(AB^*)$ définit un produit hermitien sur $M_2(\mathbb{C})$ dont la restriction à V est un produit euclidien. Une base orthogonale de V est donnée par les matrices

$$I = \begin{pmatrix} 1 & 0 \\ 0 & 1 \end{pmatrix}, \ \sigma_x = \begin{pmatrix} 0 & 1 \\ 1 & 0 \end{pmatrix}, \ \sigma_y = \begin{pmatrix} 0 & -i \\ i & 0 \end{pmatrix}, \ \sigma_z = \begin{pmatrix} 1 & 0 \\ 0 & -1 \end{pmatrix} \quad (2)$$

Les matrices σ_x, σ_y, σ_z sont connues en physique sous le nom de matrices de Pauli, et nous les avons rencontrées dans l'introduction, elles servent à décrire les particules de spin $\frac{1}{2}$. Ces matrices ne commutent pas entre elles mais vérifient les relations suivantes

$$[\sigma_x, \sigma_y] = 2i\sigma_z \quad [\sigma_z, \sigma_x] = 2i\sigma_y \quad [\sigma_y, \sigma_z] = 2i\sigma_x \quad (3)$$

(Dans la suite, la notation $[X, Y]$ désigne le commutateur $XY - YX$).

Nous aurons également besoin des matrices adjointes l'une de l'autre

$$\delta^- = \begin{pmatrix} 0 & 1 \\ 0 & 0 \end{pmatrix} \ \delta^+ = \begin{pmatrix} 0 & 0 \\ 1 & 0 \end{pmatrix} \quad (4)$$

grâce auxquelles ont peut représenter les autres

$$I = \delta^+\delta^- + \delta^-\delta^+, \ \ \sigma_x = \delta^+ + \delta^-, \ \ \sigma_y = i(\delta^+ - \delta^-) \ \ \sigma_z = \delta^-\delta^+ - \delta^+\delta^- \quad (5)$$

Le groupe des matrices 2×2 unitaires agit par conjugaison sur l'espace V. Pour toute matrice U unitaire, l'application $A \mapsto UAU^*$ est $\mathbb{R}$-linéaire sur V et préserve le produit des matrices, le passage à l'adjoint, et donc la forme hermitienne scalaire et les relations de commutation. Quitte à conjuguer par une telle matrice, on pourra supposer que l'état S est donné par la matrice

$$\begin{pmatrix} p & 0 \\ 0 & q \end{pmatrix} \ \text{avec } p + q = 1, \ p \geq q \geq 0$$

ce que nous ferons par la suite.

Les deux cas extrémaux $p = 1$ et $p = \frac{1}{2}$ sont particulièrement intéressants. Le premier correspond à l'état pur associé au vecteur e_0, le second est une trace, c'est à dire que l'on a $tr(SAB) = tr(SBA)$ pour tout couple d'opérateurs (A, B) (c'est d'ailleurs le seul état qui vérifie cette propriété).

Terminons ce chapitre en calculant la loi d'une variable aléatoire

$$A = x\sigma_x + y\sigma_y + z\sigma_z + t\, Id \ \ x, y, z, t \in \mathbb{R}$$

dans l'état S. Le spectre de A est composé de deux points,

$$\sigma(A) = \{t \pm \sqrt{x^2 + y^2 + z^2}\}$$

et la loi de A est la mesure μ telle que

$$\mu(t + \sqrt{x^2 + y^2 + z^2}) = \frac{1}{2}(1 + \frac{z}{\sqrt{x^2 + y^2 + z^2}}(p - q))$$

$$\mu(t - \sqrt{x^2 + y^2 + z^2}) = \frac{1}{2}(1 - \frac{z}{\sqrt{x^2 + y^2 + z^2}}(p - q))$$

On remarque en particulier que lorsque $p = 1$, en faisant varier (x, y, z, t) on obtient toutes les lois sur $\mathbb{R}$ portées par (au plus) deux points. Lorsque $p = \frac{1}{2} = q$, les lois obtenues sont toutes les lois uniformes portées par deux points.

L'espace (E, S) est l'analogue pour les probabilités quantiques de l'espace de probabilité d'une variable de Bernoulli en probabilités classiques. Dans tous les cours de probabilités élémentaires, une des premières choses que l'on fait avec des variables de Bernoulli c'est d'en additioner des copies indépendantes. Nous allons faire la même chose dans le paragraphe suivant, en introduisant les marches de Bernoulli quantiques.

1.4 Addition de variables de Bernoulli quantiques indépendantes

Nous allons étudier le premier exemple de processus stochastique non-commutatif qui est un analogue quantique du jeu de pile ou face. Cela nous permettra de nous familiariser avec les notions introduites jusqu'ici en faisant des calculs explicites sur un exemple non-trivial.

En probabilités classiques, il est facile de construire des copies indépendantes $X_1, \ldots, X_N$ d'une même variable aléatoire X sur $(\Omega, \mathcal{F}, P)$, en considérant l'espace produit $(\Omega^N, \mathcal{F}^{\otimes N}, P^{\otimes N})$ et en posant $X_j(\omega_1, \ldots, \omega_N) = X(\omega_j)$.

L'espaces L^2 obtenu est un produit tensoriel $L^2(\Omega^N, \mathcal{F}^{\otimes N}, P^{\otimes N}) = (L^2(\Omega, \mathcal{F}, P))^{\otimes N}$, et l'opérateur de multiplication par la variable X_j se représente sous la forme

$$T_{X_j} = I \otimes \ldots \otimes T_X \otimes \ldots \otimes I$$

l'opérateur T_X agissant sur le j^e facteur du produit tensoriel.

On va faire la même chose dans le cas non-commutatif, c'est à dire considérer l'espace de Hilbert $E^{\otimes N}$, et l'état $S^{\otimes N}$. Afin de rester avec des espaces de dimension finie, on se restreint à N fini. Si A est une variable aléatoire sur E, les opérateurs

$$A^{(k)} = I \otimes \ldots \otimes I \underset{\underset{k^e \text{ place}}{\uparrow}}{\otimes A \otimes} I \ldots I \otimes I$$

sont des variables aléatoires non-commutatives.

Si $l \neq k$, et A, B sont des variables, alors $A^{(k)}$ et $B^{(l)}$ commutent. De plus dans l'état $S^{\otimes N}$ leur loi jointe est celle de deux variables indépendantes. En effet si f et g sont deux fonctions sur $\sigma(A)$ et $\sigma(B)$ on a

$$tr\left(S^{\otimes N} f(A^{(k)}) g(B^{(l)})\right) = tr\left(S^{\otimes N}(I \otimes \ldots \otimes f(A) \otimes \ldots \otimes g(B) \otimes \ldots \otimes I)\right)$$
$$= tr(Sf(A)) \, tr(Sg(B))$$

16

d'où l'on déduit que la loi jointe de $A^{(k)}$ et $B^{(l)}$ est une loi produit.

En particulier, pour toute variable A, la famille $(A^{(k)})_{1 \leq k \leq N}$ est un processus classique dont la loi est celle d'une famille de variables indépendantes équidistribuées avec comme loi commune la loi de A.

Reprenons les notations du paragraphe 1.3 et posons

$$x_k = \sigma_x^{(k)}, y_k = \sigma_y^{(k)}, z_k = \sigma_z^{(k)}$$

$$X_n = \sum_{k=1}^n x_k \quad Y_n = \sum_{k=1}^n y_k \quad Z_n = \sum_{k=1}^n z_k \tag{6}$$

Les variables $(x_k)_{1 \leq k \leq N}$ (resp. y_k, z_k) commutent, sont indépendantes et suivent des lois de Bernoulli portées par $\{\pm 1\}$.

Les trois processus $(X_n)_{n \leq N}, (Y_n)_{n \leq N}, (Z_n)_{n \leq N}$, sont donc des processus classiques, dont les lois sont celles de marches de Bernoulli. En revanche, ces trois processus ne commutent pas entre eux, en fait d'après les relations de commutation (3) on a

$$[X_k, Y_l] = 2i Z_{k \wedge l} \tag{7}$$

ainsi que les relations qui s'en déduisent par permutation circulaire de X, Y, Z.

DÉFINITION 6 – On appelle le processus non-commutatif $(X_n, Y_n, Z_n)_{n \leq N}$ une marche de Bernoulli quantique.

Afin de mieux comprendre comment les trois processus X, Y, Z sont reliés, nous allons examiner plus en détails leur action sur l'espace $E^{\otimes N}$. Pour cela, introduisons les opérateurs

$$A_n^+ = \frac{1}{2}(X_n - iY_n), \quad A_n^- = \frac{1}{2}(X_n + iY_n), \quad B_n = \frac{1}{2}(nI - Z_n) \tag{8}$$

On a

$$A_n^+ = \sum_{k=0}^n \delta^{+(k)} \quad A_n^- = \sum_{k=0}^n \delta^{-(k)} \quad B_n = \sum_{k=0}^n \delta^{+(k)} \delta^{-(k)} \tag{9}$$

L'espace $E^{\otimes N}$ est muni d'une base orthonormale naturelle $(e_U)_{U \subset \{1,\ldots,N\}}$ avec $e_U = e_{i_1} \otimes \ldots \otimes e_{i_N}$ où $i_k = 1 \Leftrightarrow k \in U$ et $i_k = 0 \Leftrightarrow k \notin U$.

Dans cette base les opérateurs A_n^-, A_n^+, B_n sont donnés par les formules suivantes

$$A_n^+ e_U = \sum_{k \notin U, k \leq n} e_{U \cup \{k\}}$$

$$A_n^- e_U = \sum_{k \in U, k \leq n} e_{U \setminus \{k\}} \tag{10}$$

$$B_n e_U = |U \cap \{1, \ldots, n\}| e_U$$

Considérons le sous-espace de $E^{\otimes N}$ engendré par les vecteurs orthogonaux ε_j^n définis par

$$\varepsilon_j^n = (C_n^j)^{-\frac{1}{2}} \sum_{|U \cap \{1,\ldots,n\}| = j} e_U \quad \text{pour } j = 0, 1, \ldots n \tag{11}$$

Ce sous-espace est laissé invariant par les opérateur A_n^+, A_n^-, B_n et on a

$$A_n^+ \varepsilon_j^n = \sqrt{(j+1)(n-j)}\varepsilon_{j+1}^n \tag{12}$$

$$A_n^- \varepsilon_j^n = \sqrt{j(n-j+1)}\varepsilon_{j-1}^n \tag{13}$$

$$B_n \varepsilon_j^n = j\varepsilon_j^n \tag{14}$$

$$Z_n \varepsilon_j^n = (n-2j)\varepsilon_j^n \tag{15}$$

pour $j = 0, 1, \ldots, n$ en posant $\varepsilon_{n+1}^n = \varepsilon_{-1}^n = 0$.

1.5 Le processus de spin

On a vu que la conjugaison par une matrice unitaire induit une rotation sur l'espace V. De même, si U est une matrice unitaire, l'opérateur $U^{\otimes N}$ agissant par conjugaison produit une rotation sur l'espace vectoriel réel engendré par les opérateurs X_n, Y_n, Z_n, pour $n \leq N$ et laisse invariantes les relations de commutation (3) entre ces opérateurs. Cela suggère d'étudier la "norme" de la marche de Bernoulli quantique, définie comme étant la racine carrée de $X_n^2 + Y_n^2 + Z_n^2$. Pour des raisons qui apparaitront plus bas, il est plus judicieux de considérer la racine carrée de $X_n^2 + Y_n^2 + Z_n^2 + I$.

DÉFINITION 7 – *On définit les variables Σ_n, $0 \leq n \leq N$ par les conditions*
$$\Sigma_n \text{ est positif}$$

$$\Sigma_n^2 = X_n^2 + Y_n^2 + Z_n^2 + I = 2(A_n^+ A_n^- + A_n^- A_n^+) + (2B_n - n)^2 + I \tag{16}$$

Le processus $(\Sigma_n)_{n \leq N}$ est appelé processus de spin.

PROPOSITION 1 – *Si $n \leq m$ on a*

$$0 = [\Sigma_n, X_m] = [\Sigma_n, Y_m] = [\Sigma_n, Z_m] = [\Sigma_n, \Sigma_m]$$

Démonstration
Des relations (7) on tire pour $n \leq m$: $[Y_n^2, X_m] = -2i(Y_n Z_n + Z_n Y_n)$ et $[Z_n^2, X_m] = 2i(Y_n Z_n + Z_n Y_n)$ d'où $[\Sigma_n^2, X_m] = 0$ et $[\Sigma_n, X_m] = 0$. Les autres relations en découlent par permutation circulaire de X_m, Y_m, Z_m.

Le processus de spin est un processus classique, on peut donc se poser le problème de déterminer sa loi. C'est ce que nous allons faire dans la suite du paragraphe, et pour cela on va déterminer une base de vecteurs propres communs à tous les opérateurs $(\Sigma_n)_{1 \leq n \leq N}$.
Remarquons tout de suite que l'on a, d'après (12), (13), (14) et (16)

$$\Sigma_n \varepsilon_j^n = (n+1)\varepsilon_j^n$$

Les vecteurs ε_j^n sont donc des vecteurs propres de Σ_n. Nous allons généraliser ce résultat dans la proposition suivante.

PROPOSITION 2 – *Soit $J = (j_1, \ldots, j_N)$ une suite d'entiers telle que*
a) *$j_1 = 2$*
b) *$j_i \geq 1$ pour tout $i \leq N$.*
c) *$j_{i+1} - j_i \in \{+1, -1\}$ $\forall i < N$*

Alors il existe un sous-espace H_J de $E^{\otimes N}$ de dimension $j_N = l + 1$, espace propre commun à $\Sigma_1, \ldots, \Sigma_N$ de valeurs propres correspondantes $(j_1, \ldots, j_N)$, avec une base orthonormale $(\phi_0^J, \phi_1^J, \ldots, \phi_l^J)$ telle que

$$A_N^+ \phi_j^J = \sqrt{(j+1)(l-j)}\, \phi_{j+1}^J$$
$$A_N^- \phi_j^J = \sqrt{j(l-j+1)}\, \phi_{j-1}^J \qquad (17)$$
$$Z_N \phi_j^J = (2j - l)\, \phi_j^J$$

(avec $\phi_{l+1}^J = \phi_{-1}^J = 0$). De plus, les espace H_J sont orthogonaux et $E^{\otimes N} = \oplus_J H_J$.

Démonstration

On le démontre par récurrence sur N.

C'est vrai pour N=1, avec $J = (2)$, $H_J = E$ et $e_0 = \phi_0^J$, $e_1 = \phi_1^J$.

Soit $J' = (j_1, \ldots, j_{N+1})$ une suite de longueur N+1 satisfaisant les hypothèses a) et b), et $J = (j_1, \ldots, j_N)$. Par hypothèse, l'espace $H_J \subset E^{\otimes N}$ a une base $\phi_0^J, \ldots \phi_l^J$ avec $j_N = l + 1$ vérifiant (17). Le sous-espace $H_J \otimes E \subset E^{\otimes(N+1)}$ est un sous-espace propre commun à $\Sigma_1, \ldots, \Sigma_N$ sur lequel ces opérateurs admettent les valeurs propres $(j_1, \ldots j_N)$. Nous allons le décomposer sous l'action de Σ_{N+1}.

Posons

$$\eta_j = \sqrt{\frac{l-j+1}{l+1}}\, \phi_j^J \otimes e_0 + \sqrt{\frac{j}{l+1}}\, \phi_{j-1}^J \otimes e_1 \quad 0 \le j \le l+1$$
$$\xi_j = \sqrt{\frac{j+1}{l+1}}\, \phi_j^J \otimes e_0 - \sqrt{\frac{l-j}{l+1}}\, \phi_{j-1}^J \otimes e_1 \quad 0 \le j \le l-1$$

Les vecteurs η_j, ξ_j forment une base orthonormale de $H_J \otimes E$.

Comme $A_{N+1}^+ = A_N^+ + \delta^{+(N+1)}$ et $A_{N+1}^- = A_N^- + \delta^{-(N+1)}$ un calcul simple utilisant les formules pour A_N^+, A_N^- et Z_N donne

$$A_{N+1}^+ \eta_j = \sqrt{(j+1)(l+1-j)}\, \eta_{j+1}$$
$$A_{N+1}^- \eta_j = \sqrt{j(l-j+2)}\, \eta_{j-1}$$
$$Z_{N+1} \eta_j = (l+1-2j)\, \eta_j$$
$$\Sigma_{N+1} \eta_j = (l+1)\, \eta_j$$

pour $0 \le j \le l+1$ et

$$A_{N+1}^+ \xi_j = \sqrt{(j+1)(l-1-j)}\, \xi_{j+1}$$
$$A_{N+1}^- \xi_j = \sqrt{j(l-j)}\, \xi_{j-1}$$
$$Z_{N+1} \xi_j = (l-1-2j)\, \xi_j$$
$$\Sigma_{N+1} \xi_j = (l-1)\, \xi_j$$

pour $0 \le j \le l-1$, en posant $\xi_l = \eta_{l+2} = \xi_{-1} = \eta_{-1} = 0$. On en déduit qu'en posant

$H_{J'} = vect(\eta_j)_{0 \le j \le l+1}, \phi_j^{J'} = \eta_j$ si $j_{N+1} = J_N + 1$

$H_{J'} = vect(\xi_j)_{0 \le j \le l-1}, \phi_j^{J'} = \xi_j$ si $j_{N+1} = J_N - 1$

alors $H_{J'}$ est un sous-espace propre commun aux Σ_j, de suite de valeurs propres J', et on a le résultat de la proposition au rang $N + 1$.

Nous pouvons maintenant calculer la loi du processus Σ.

THÉORÈME 1 – Si $p = 1$ la loi de Σ est celle d'un processus déterministe, la trajectoire $(2, 3, \ldots, N + 1)$ ayant une probabilité 1.

Si $p = \frac{1}{2}$, la loi de Σ est celle d'une chaîne de Markov sur $\mathbb{N}^*$ de probabilités de transition:

$$p(n, n + 1) = \frac{n + 1}{2n} \quad p(n, n - 1) = \frac{n - 1}{2n} \tag{18}$$

Démonstration

Tout d'abord, si $p = 1$, l'état $\Sigma^{\otimes N}$ est l'état pur associé au vecteur $e_0^{\otimes N}$, donc c'est un vecteur propre commun à $\Sigma_1, \ldots, \Sigma_N$, de valeurs propres $2, 3, \ldots, N + 1$, et la loi de Σ est celle d'un processus déterministe, qui met la probabilité 1 sur la trajectoire $(2, 3, \ldots, N + 1)$.

Si $p = \frac{1}{2}$, la probabilité d'une trajectoire $J = (j_1, \ldots, j_N)$ est égale à $\frac{1}{2^N} tr(\Pi_J)$ où Π_J est le projecteur orthogonal sur le sous espace H_J. Comme $dim(H_J) = j_N$ cette probabilité est $\frac{j_N}{2^N}$, et il est facile d'en déduire que la loi de Σ est celle d'une chaîne de Markov sur $\mathbb{N}^*$ dont les probabilités de transition sont données par celles du théorème.

REMARQUES– Cette dernière chaîne de Markov est bien connue en probabilités classiques, il s'agit d'un analogue discret du processus de Bessel de dimension 3 (cf J.Pitman [68]). En fait, si $(B_t)_{t \geq 0}$ est un mouvement brownien de dimension 3 issu de 0, et les instants T_n sont les instants successifs de passage de $|B_t|$ par des valeurs entières, (i.e. $T_0 = 0$ et $T_{n+1} = \inf\{t \geq T_n \mid |B_t| - |B_{T_n}| = \pm 1\}$) alors le processus $|B_{T_n}|_{n \geq 0}$ est une chaîne de Markov sur $\mathbb{N}$ dont les probabilités de transition sont données par la formule (18)

Si U, V, W sont trois marches de Bernoulli symétriques indépendantes, leur norme n'est pas un processus de Markov, et n'est pas à valeurs entières. Les propriétés du processus Σ reflètent l'invariance par rotation du triplet X, Y, Z (ou plus fondamentalement, des relations de commutation (3)). La marche de Bernoulli quantique peut être considérée comme une approximation discrète du mouvement brownien de dimension 3, invariante par rotation.

Nous verrons au chapitre 6 que l'on peut interpréter ce processus comme un processus de Markov à valeurs dans un espace non-commutatif. Cela proviendra de résultats de la théorie des représentations des groupes. Les calculs de la proposition 2 qui proviennent de la théorie des représentations de l'algèbre de Lie $sl(2, \mathbb{C})$ en sont déjà un exemple (comparer à [78]).

Exercice: Calculer la loi de $(\Sigma_n)_{n \geq 1}$ pour $p \neq 1, \frac{1}{2}$.

1.6 Un théorème limite pour la marche de Bernoulli quantique

Il y a essentiellement deux théorèmes limite pour l'addition de variables de Bernoulli indépendantes. Le premier est le théorème de Moivre-Laplace (cas particulier du théorème central limite), qui concerne l'addition de variables de Bernoulli symétriques et qui donne à la limite une loi normale. Le second est le théorème de convergence des lois binomiales de paramètres (N, α_N) vers la loi de Poisson de paramètre α lorsque $N \to +\infty$ et $N\alpha_N \to \alpha$. On peut réaliser la loi binomiale (N, α_N) comme loi d'une somme de N variables de Bernoulli indépendantes de même loi sur $\{0, 1\}$ donnée par $P(\{0\}) = 1 - \alpha_N$, $P(\{1\}) = \alpha_N$.

Nous avons vu au paragraphe 1.3 que pour $p = 1$, l'espace de probabilité non-commutatif fini (E, S) permet de réaliser toutes les lois de Bernoulli sur $\mathbb{R}$. Cela va nous permettre de donner un "théorème limite non-commutatif" qui englobe les deux résultats cités ci-dessus. Plus précisément, pour tout $\theta \in \mathbb{R}$ la variable $e^{i\theta}\delta^+ + e^{-i\theta}\delta^-$ suit une loi de Bernoulli symétrique sur $\{+1, -1\}$, donc d'après le théorème de Moivre-Laplace, dans l'état $S^{\otimes N}$, la loi de $e^{i\theta}\frac{A_N^+}{\sqrt{N}} + e^{-i\theta}\frac{A_N^-}{\sqrt{N}}$ converge vers la loi normale centrée réduite.

De même, on peut construire une variable de Bernoulli sur $\{0, 1\}$ en utilisant l'opérateur

$$(1+\frac{|z|^2}{N})^{-1} \begin{pmatrix} \frac{|z|^2}{N} & \frac{\bar{z}}{\sqrt{N}} \\ \frac{z}{\sqrt{N}} & 1 \end{pmatrix} = (1+\frac{|z|^2}{N})^{-1}((1-\frac{|z|^2}{N})\delta^+\delta^- + \frac{z}{\sqrt{N}}\delta^+ + \frac{\bar{z}}{\sqrt{N}}\delta^- + \frac{|z|^2}{N}Id)$$

Sa loi est portée par $\{0, 1\}$, et $P(\{0\}) = \frac{1}{1+\frac{|z|^2}{N}}$, $P(\{1\}) = \frac{\frac{|z|^2}{N}}{1+\frac{|z|^2}{N}}$, par conséquent, dans l'état $S^{\otimes N}$, l'opérateur

$$(1 + \frac{|z|^2}{N})^{-1}((1 - \frac{|z|^2}{N})B_N + \frac{zA_N^+}{\sqrt{N}} + \frac{\bar{z}A_N^-}{\sqrt{N}} + |z|^2 Id)$$

suit une loi binomiale de paramètres $(N, \frac{\frac{|z|^2}{N}}{1+\frac{|z|^2}{N}})$ qui converge vers la loi de Poisson de paramètre $|z|^2$ quand $N \to +\infty$.

Ces deux résultats classiques suggèrent d'étudier la convergence "en loi" du triplet d'opérateurs $(B_N, \frac{A_N^+}{\sqrt{N}}, \frac{A_N^-}{\sqrt{N}})$. Comme ces opérateurs ne sont pas auto-adjoints et ne commutent pas, nous ne pouvons pas considérer leur loi jointe, mais néanmoins nous pouvons calculer leurs "moments" c'est à dire les expressions de la forme $< P(B_N, \frac{A_N^+}{\sqrt{N}}, \frac{A_N^-}{\sqrt{N}})e_0^{\otimes N}, e_0^{\otimes N} >$ où P est un polynome en trois indéterminées non-commutatives. Le théorème suivant montre que ces moments convergent quand $N \to +\infty$, vers les moments d'un triplet (a^0, a^+, a^-) d'opérateurs sur un espace de dimension infinie.

Les opérateurs (a^0, a^+, a^-) vont jouer un rôle fondamental par la suite, et nous les étudierons en détail au chapitre 3, après avoir revu les notions nécessaires de théorie spectrale au chapitre 2.

Soit H_0 un espace vectoriel complexe, pré-hilbertien de dimension infinie avec une base orthonormale, $(\varepsilon_0 = \Omega, \varepsilon_1, \ldots, \varepsilon_n, \ldots)$. On définit des applications linéaires (a^+, a^-, a^0) sur H_0 par les formules

$$\begin{aligned} a^+\varepsilon_j &= \sqrt{j+1}\,\varepsilon_{j+1} \\ a^-\varepsilon_j &= \sqrt{j}\,\varepsilon_{j-1} \\ a^0\varepsilon_j &= j\varepsilon_j \end{aligned} \tag{19}$$

pour $j \geq 0$, avec $\varepsilon_{-1} = 0$.

THÉORÈME 2 – Pour tout polynome P en trois indéterminées non-commutatives, on a

$$\lim_{N\to\infty} < P(B_N, \frac{A_N^+}{\sqrt{N}}, \frac{A_N^-}{\sqrt{N}}) e_0^{\otimes N}, e_0^{\otimes N} > = < P(a^0, a^+, a^-)\Omega, \Omega >$$

Démonstration

Il suffit de le montrer pour un monome de degré d, pour tout entier $d \geq 0$. Le vecteur $e_0^{\otimes N}$ est égal à ε_0^N d'après (11). Le sous-espace de $E^{\otimes N}$ engendré par les vecteurs ε_j^N est stable par application des opérateurs B_N, A_N^+, A_N^-, par conséquent dans le calcul de $< P(B_N, \frac{A_N^+}{\sqrt{N}}, \frac{A_N^-}{\sqrt{N}}) \varepsilon_0^N, \varepsilon_0^N >$ seules interviennent les valeurs des opérateurs $B_N, \frac{A_N^+}{\sqrt{N}}, \frac{A_N^-}{\sqrt{N}}$ évalués sur ε_j^N pour $j \leq d$. Or d'après (12)(13)(14) on a

$$\frac{A_N^+}{\sqrt{N}} \varepsilon_j^N = \sqrt{\frac{(j+1)(N-j)}{N}} \varepsilon_{j+1}^N$$

$$\frac{A_N^-}{\sqrt{N}} \varepsilon_j^N = \sqrt{\frac{j(N-j+1)}{N}} \varepsilon_{j-1}^N$$

$$B_N \varepsilon_j^N = j \varepsilon_j^N$$

le théorème suit en comparant ces formules avec (19) quand $N \to \infty$.

Comme les opérateurs $(B_N, \frac{A_N^+}{\sqrt{N}}, \frac{A_N^-}{\sqrt{N}})$ ne commutent pas, ils n'ont pas de loi jointe au sens classique et ce théorème est donc purement non-commutatif. Cependant, si on considère des polynômes $P(B_N, \frac{A_N^+}{\sqrt{N}}, \frac{A_N^-}{\sqrt{N}})$ qui s'expriment sous la forme d'un polynôme en $(e^{i\theta} \frac{A_N^+}{\sqrt{N}} + e^{-i\theta} \frac{A_N^-}{\sqrt{N}})$ ou en $((1+\frac{|z|^2}{N})^{-1}((1-\frac{|z|^2}{N})B_N + \frac{zA_N^+}{\sqrt{N}} + \frac{\bar{z}A_N^-}{\sqrt{N}} + |z|^2 Id)$ les théorèmes limite cités plus haut montrent qu'il y a convergence de ces moments vers les moments correspondants des lois de Gauss ou de Poisson. Cela suggère que les opérateurs $e^{i\theta} a^+ + e^{-i\theta} a^-$ "sont" des variables normales, tandis que les opérateurs $a^0 + za^+ + \bar{z}a^- + |z|^2$ "sont" des variables de Poisson. C'est ce que nous vérifierons dans le chapitre 3, après avoir vu les notions nécessaires de théorie spectrale au chapitre 2.

Chapitre 2

Théorie spectrale

2.1 Quelques définitions

Après les espaces de probabilité non-commutatifs finis nous allons passer au cadre plus général des espaces de Hilbert de dimension infinie, qui nous permettra de considérer des variables non-commutatives dont les lois ne seront pas à support fini. Pour cela, nous allons devoir faire des rappels d'analyse fonctionnelle, plus précisément de théorie des opérateurs auto-adjoints. Cette théorie classique, due entre autres à von Neumann et Stone est exposée par exemple dans les livres de Reed et Simon [70], Rudin [72], Halmos [39].

Dans la suite, H désigne un espace de Hilbert complexe séparable.
Commençons par quelques définitions.

DÉFINITION 8 – Un opérateur T sur H est la donnée d'un couple $(D(T), T)$ où $D(T)$, le domaine de l'opérateur, est un sous-espace vectoriel de H et T une application linéaire de $D(T)$ dans H.
L'opérateur T est dit borné si $D(T) = H$ et $\|T\| = \sup_{x \in H, x \neq 0} \frac{|Tx|}{|x|} < +\infty$.

Dans la suite, quand on parlera d'un opérateur, son domaine sera souvent sous-entendu. Les opérateurs bornés sur H forment une algèbre notée $\mathcal{B}(H)$, qui est une algèbre de Banach pour la norme $\|\ \|$. Il y a d'autres topologies intéressantes sur $\mathcal{B}(H)$, citons entre autres la topologie forte, qui est celle de la convergence simple sur H, et qui est (malgré son nom) plus faible que celle de la norme.

EXEMPLE FONDAMENTAL– Soient $(\Omega, \mathcal{F}, P)$ un espace de probabilités,
$H = L^2_{\mathbf{C}}(\Omega, \mathcal{F}, P)$, et $X : \Omega \to \mathbb{C}$ une variable aléatoire.
Posons $D(T) = \{f \in L^2_{\mathbf{C}}(\Omega, \mathcal{F}, P) | Xf \in L^2_{\mathbf{C}}(\Omega, \mathcal{F}, P)\}$, et $T_X(f) = Xf$, pour $f \in D(T)$, alors T_X est un opérateur. L'opérateur T_X ne dépend que de la classe de X modulo les ensembles de mesure nulle, et il est borné si et seulement si $X \in L^\infty(\Omega, \mathcal{F}, P)$. Cet exemple est une généralisation immédiate de l'exemple du début du chapitre 1.

DÉFINITION 9 – Le graphe de l'opérateur T est l'ensemble

$$\Gamma(T) = \{(x, Tx) \in H \times H | x \in D(T)\}$$

T est dit fermé si $\Gamma(T)$ est fermé dans $H \times H$.
Un opérateur S est une extension de T (on note $T \subset S$) si $\Gamma(T) \subset \Gamma(S)$ (ou encore $D(T) \subset D(S)$ et $T(x) = S(x)$, $\forall x \in D(T)$).
Un opérateur T est dit fermable s'il possède une extension fermée, et sa fermeture est alors l'opérateur dont le graphe est $\overline{\Gamma(T)}$.

DÉFINITION 10 – On suppose $D(T)$ dense dans H. L'adjoint de T est l'opérateur T^* de domaine

$D(T^*) = \{y \in H | x \mapsto < Tx, y >$ *est continue sur* $D(T)\}$ *tel que pour* $y \in D(T^*)$ *on a* $< Tx, y >=< x, T^*y > \forall x \in D(T)$ $(D(T)$ *étant dense dans* H *cette dernière égalité détermine* T^*y).

Un opérateur de la forme T^* *est toujours fermé.*

T *est dit symétrique si* $T \subset T^*$ *(ou encore* $< Tx, y >=< x, Ty > \forall x, y \in D(T))$
L'opérateur T *est dit auto-adjoint si* $T = T^*$, *unitaire s'il est borné, inversible, et* $T^* = T^{-1}$.

Revenons à l'exemple fondamental: si X est une variable aléatoire réelle, l'opérateur T_X associé est auto-adjoint. Réciproquement, le théorème spectral (énoncé ci-dessous) permet de réaliser tout opérateur auto-adjoint comme un opérateur de multiplication sur un espace L^2.

DÉFINITION 11 − *Le spectre de l'opérateur* T *est l'ensemble*

$$\sigma(T) = \{\lambda \in \mathbb{C} \mid T - \lambda Id \text{ n'a pas d'inverse borné}\}$$

Le spectre d'un opérateur auto-adjoint est un sous-ensemble fermé de $\mathbb{R}$.
L'opérateur auto-adjoint T *est dit positif si son spectre est inclus dans* $\mathbb{R}_+$.
Les valeurs propres de T *sont les* $\lambda \in \mathbb{C}$ *tels que* $ker(T - \lambda Id) \neq \{0\}$.

DÉFINITION 12 − *Un projecteur est un opérateur auto-adjoint borné* P *tel que* $P^2 = P$.
C'est l'opérateur de projection orthogonale sur son image.

2.2 Résolutions de l'identité et théorème spectral

Soient $(A, \mathcal{A})$ un espace mesurable et H un espace de Hilbert.

DÉFINITION 13 − *Une résolution de l'identité sur* $(A, \mathcal{A})$ *est une famille* $(E_X)_{X \in \mathcal{A}}$ *de projecteurs orthogonaux de* H *telle que*

$$i)\ E_A = id, \quad E_\emptyset = 0$$
$$ii) \forall X_1, X_2 \in \mathcal{A}\ E_{X_1} E_{X_2} = E_{X_1 \cap X_2} \qquad (20)$$
$$iii)\ \textit{Pour toute famille dénombrable } (X_j)_{j \in J}$$
$$\textit{telle que } X_j \cap X_k = \emptyset \textit{ si } k \neq j \textit{ et tout } x \in H \textit{ on a}$$
$$E_{\cup_{j \in J} X_j}(x) = \sum_{j \in J} E_{X_j}(x)$$

Une résolution de l'identité est portée par $X \in \mathcal{A}$ *si* $Y \cap X = \emptyset \Rightarrow E_Y = 0$.

Soit $f : (A, \mathcal{A}) \to (B, \mathcal{B})$ une application mesurable, on définit une résolution de l'identité $f(E)$ sur $(B, \mathcal{B})$ en posant $f(E)_Y = E_{f^{-1}(Y)}$.
Soit E une résolution de l'identité sur $(A, \mathcal{A})$, pour tout couple de vecteurs $x, y \in H$ d'après i) ii) et iii), on définit une mesure complexe bornée $E_{x,y}$ sur $(A, \mathcal{A})$ par $E_{x,y}(B) =< E_B(x), y >$.
On notera $E_x = E_{x,x}$, si $x \in H$.
Soit E une résolution de l'identité sur $\mathbb{R}$, posons

$$D(T) = \{x \in H | \int_{\mathbb{R}} \lambda^2 \, dE_x(\lambda) < +\infty\} \qquad (21)$$

$$< Tx, y > = \int_{\mathbb{R}} \lambda dE_{x,y}(\lambda) \tag{22}$$

Ces formules définissent ainsi un opérateur auto-adjoint T. Réciproquement, le théorème spectral montre qu'à tout opérateur auto-adjoint T on peut associer une résolution de l'identité pour laquelle T est définie par (21) et (22).

THÉORÈME 3 – *Soit T un opérateur auto-adjoint sur H, il existe une unique résolution de l'identité E sur $\mathbb{R}$, portée par $\sigma(T)$, telle que*

$$D(T) = \{x \in H| \int_{\mathbb{R}} \lambda^2 dE_x(\lambda) < +\infty\}$$

$$\forall x, y \in D(T) \quad < Tx, y > = \int_{\mathbb{R}} \lambda dE_{x,y}(\lambda)$$

Dans le cas où H est de dimension finie, la résolution de l'identité est facile à décrire. Le spectre de l'opérateur T est l'ensemble de ses valeur propres, et si λ est une valeur propre de T, $E_{\{\lambda\}}$ est le projecteur orthogonal sur le sous-espace propre correspondant.

On déduit du théorème spectral le résultat suivant. Si T est un opérateur auto-adjoint sur H, il existe un espace de probabilités $(\Omega, \mathcal{F}, P)$, une isométrie $\iota : H \to L^2(\Omega, \mathcal{F}, P)$ et une variable aléatoire réelle X telle que l'opérateur de multiplication par X soit égal à $\iota \circ T \circ \iota^*$ (voir par exemple Reed et Simon [70]).

Une autre conséquence du théorème spectral est la possibilité d'un calcul borélien sur les opérateurs auto-adjoints. Soient T un opérateur auto-adjoint de résolution de l'identité E, et f une fonction borélienne sur $\sigma(T)$, on peut définir un opérateur auto-adjoint $f(T)$ par sa résolution de l'identité $f(E)$. En particulier, si f est bornée, $f(T)$ est borné. Ce calcul fonctionnel prolonge le calcul polynomial usuel sur les opérateurs bornés et il respecte les opérations algébriques:

$$(f + g)(T) = f(T) + g(T), fg(T) = f(T)g(T), (\lambda f)(T) = \lambda \, f(T)$$

pour des fonctions f et g bornées, et $\lambda \in \mathbb{R}$.
De plus, si $f_n \to f$ simplement, et les f_n sont uniformément bornées, alors $f_n(T) \to f(T)$ fortement (cette dernière propriété est une conséquence du théorème de convergence dominée de Lebesgue).

DÉFINITION 14 – *Deux résolutions de l'identité E et E' sur $(A, \mathcal{A})$ et $(A', \mathcal{A}')$ commutent si*

$$\forall X \in \mathcal{A}, Y \in \mathcal{A}' \quad E_X E'_Y = E'_Y E_X$$

PROPOSITION 3 – *Soient $(A_j, \mathcal{A}_j)_{j \in J}$ des espaces lusiniens. Soit $(E^j)_{j \in J}$ une famille de résolutions de l'identité sur $(A_j, \mathcal{A}_j)_{j \in J}$, commutant deux à deux, il existe une unique résolution de l'identité E sur $(\prod_{j \in J} A_j, \otimes_{j \in J} \mathcal{A}_j)$ telle que pour tout ensemble cylindrique $\prod_{j \in J} X_j$ (où $X_j = A_j$ sauf pour un nombre fini d'indices) on ait*

$$E_{\prod_{j \in J} X_j} = \prod_{j \in J} E^j_{X_j},$$

(le produit dans le membre de droite est bien défini car $E^j_{X_j} = Id$ sauf pour un nombre fini de $j \in J$ et ces projecteurs commutent).

Démonstration

Soit $\phi \in H$, posons $\mu_{\phi,\phi}(\prod_{j \in J} X_j) = \prod_{j \in J} < E^j_{X_j}(\phi), \phi >$ pour un ensemble cylindrique $\prod_{j \in J} X_j$. On définit ainsi une application additive sur la sous-algèbre de Boole de $\otimes_{j \in J} \mathcal{A}_j$ engendrée par les ensembles cylindriques $\prod_{j \in J} X_j$. Soit U_n une suite décroissante dans cette algèbre de Boole telle que $\mu_{\phi,\phi}(U_n) \geq \varepsilon > 0$ pour tout n. Pour tout n il existe un compact $V_n \subset U_n$ tel que $\mu_{\phi,\phi}(V_n) \geq \varepsilon(1 - 3^{-n})$. Il s'ensuit que $\cap_{n \leq N} V_n$ est une suite décroissante de compacts non vides donc $\cap_{n \geq 0} V_n \neq \emptyset$ et $\cap_{n \geq 0} U_n \neq \emptyset$. D'après le théorème de Carathéodory, $\mu_{\phi,\phi}$ se prolonge en une mesure positive sur $\otimes_{j \in J} \mathcal{A}_j$.

La résolution de l'identité E sur $\otimes_{j \in J} \mathcal{A}_j$ est obtenue en posant $< E_X(\phi), \phi >= \mu_{\phi,\phi}(X)$, ce qui détermine E_X de façon unique par polarisation.

On dit que deux opérateurs auto-adjoints commutent si leurs résolutions de l'identité commutent (pour des opérateurs bornés, cela revient au même que la commutation usuelle).

On peut alors définir un calcul fonctionnel mesurable pour une famille $(T_j)_{j \in J}$ d'opérateurs auto-adjoints qui commutent en posant que $f((T_j))$ est l'opérateur auto-adjoint dont la résolution de l'identité est $f(E)$, où f est une fonction mesurable sur $\prod_{j \in J} \sigma(X_j)$ et E la résolution de l'identité sur $\prod_{j \in J} \sigma(X_j)$ obtenue en appliquent la proposition précédente aux résolutions des opérateurs X_j. Ce calcul vérifie les mêmes propriétés que celui décrit plus haut pour un seul opérateur.

Comme dans le cas d'un seul opérateur, on montre qu'il existe un espace de probabilité $(\Omega, \mathcal{F}, P)$, une isométrie: $\iota : H \to L^2(\Omega, \mathcal{F}, P)$ et des variable aléatoire réelle X_j telles que les opérateur de multiplication par X_j soient $\iota \circ T_j \circ \iota^*$.

2.3 Opérateurs à trace et espaces de probabilité non-commutatifs

Soit $\phi \in H$ un vecteur de norme 1, et E une résolution de l'identité sur $(A, \mathcal{A})$, on obtient une loi de probabilité sur $(A, \mathcal{A})$ en posant $\mu(X) =< E_X(\phi), \phi >$. Lorsque E est la résolution de l'identité d'un opérateur auto-adjoint T, la mesure μ est une loi de probabilité sur $\sigma(T)$. De même, si E est la résolution de l'identité associée à une famille d'opérateurs qui commutent, on a ainsi une loi sur le produit des spectres de ces opérateurs.

DÉFINITION 15 – *La forme linéaire sur $\mathcal{B}(H)$ définie par $T \to< T\phi, \phi >$ est appelée l'état pur associé à ϕ.*

Comme dans le cas de la dimension finie, on considèrera des états obtenus comme combinaison convexe d'états purs. Ces combinaisons sont décrites au moyen de la notion d'opérateur à trace.

DÉFINITION 16 – *Soit T un opérateur borné, on dit que T est à trace si $\sum_{n=0}^{+\infty} | < Te_n, e_n > | < +\infty$ pour une base orthonormée $(e_n)_{n \in \mathbb{N}}$ de H. (Cette propriété ne dépend pas de la base choisie). La trace de T est la quantité $tr(T) = \sum_{n=0}^{+\infty} < Te_n, e_n >$, qui ne dépend pas de la base utilisée pour la calculer.*

Les opérateurs à trace forment un idéal bilatère de l'algèbre $\mathcal{B}(H)$.

DÉFINITION 17 – *Si S est un opérateur à trace, positif, de trace 1, la forme linéaire sur $\mathcal{B}(H)$ définie par $T \mapsto tr(TS)$ est appelée l'état associé à S (par abus de langage on appelle aussi parfois état l'opérateur S lui-même).*

Soit π_ϕ le projecteur orthogonal sur le sous-espace engendré par un vecteur ϕ de norme 1, alors $tr(\pi_\phi T) = <T\phi, \phi>$ est l'état pur associé à ϕ. Comme en dimension finie, l'ensemble des états est l'enveloppe convexe faiblement fermée, dans le dual de $\mathcal{B}(H)$, de l'ensemble des états purs, qui en sont les points extrémaux.

On peut maintenant définir la notion d'espace de probabilité non-commutatif.

DÉFINITION 18 – *Un espace de probabilité non-commutatif est la donnée d'un couple (H, S) où H est un espace de Hilbert et S un opérateur positif à trace, de trace 1.*

DÉFINITION 19 – *Une variable aléatoire non-commutative sur l'espace (H, S) est un opérateur auto-adjoint sur H.*
Une famille de $(T_j)_{j \in J}$ d'opérateurs sur H est appelée un processus non-commutatif, et si ces opérateurs sont auto-adjoints et commutent, le processus est dit classique.

Soit $(T_j)_{j \in J}$ un processus classique, et E la résolution de l'identité sur $\prod_{j \in J} \sigma(T_j)$ associée, la loi jointe des T_j dans l'état S, est la mesure sur $\prod_{j \in J} \sigma(T_j)$
$X \mapsto tr(E_X S)$, c'est la loi du processus classique $(T_j) j \in J$.

Les définitions que nous venons de voir sont les généralisations en dimension infinie des définitions du chapitre 1. Une famille de variables aléatoires réelles au sens usuel peut s'interpréter comme une famille d'opérateurs auto-adjoints sur leur espace L^2, ce qui fait que la théorie des probabilités non-commutative contient la théorie usuelle. Chaque fois que l'on rencontre, en probabilités non-commutatives un processus classique, on est en situation d'appliquer les résultat usuels de probabilités. La situation se complique (mais devient aussi plus intéressante) lorsque l'on considère *simultanément* des variables qui ne commutent pas entre elles. Une des conséquences est que l'on ne peut pas parler des trajectoires d'un processus non-commutatif, tout ce qu'on en connait sont des propriétés en loi. En un sens, on est ramené à la situation des probabilités classiques avant Kolmogorov, lorsqu'on ne possédait pas d'espace de probabilité, et que l'on ne savait parler des processus qu'en loi.

2.4 Théorèmes de Stone et de Nelson

Dans la suite nous aurons à travailler avec des opérateurs symétriques définis sur un domaine dense d'un espace de Hilbert H. Afin de leur appliquer la théorie spectrale, il faudra leur trouver une extension auto-adjointe. Le but de ce paragraphe est de fournir des critères pour l'existence et l'unicité d'une telle extension.

DÉFINITION 20 – *Un opérateur symétrique est dit essentiellement auto-adjoint s'il admet une seule extension auto-adjointe (ou, ce qui est équivalent, si sa fermeture est auto-adjointe).*

DÉFINITION 21 – *Un groupe à un paramètre fortement continu d'opérateurs unitaires est une famille $(U_t)_{t \in \mathbb{R}}$ d'opérateurs unitaires telle que $\forall s, t \in \mathbb{R}$ $U_t U_s = U_{t+s}$, et l'application $t \mapsto U_t x$ est continue pour tout $x \in H$.*

THÉORÈME 4 – Soit $(U_t)_{t\in\mathbf{R}}$ un groupe fortement continu d'opérateurs unitaires, et

$$D(A) = \{\phi \in H \mid \lim_{t\to 0} \frac{1}{i}\frac{U_t(\phi) - \phi}{t} \text{ existe}\}$$

alors $D(A)$ est dense dans H, et l'opérateur A défini sur le domaine $D(A)$ par $A\phi = \lim_{t\to 0}\frac{1}{i}\frac{U_t(\phi)-\phi}{t}$ est un opérateur auto-adjoint, de plus on a $U_t = e^{itA}$.

Ce théorème est dû à Stone (cf [70] ou [72]).

DÉFINITION 22 – Soit A un opérateur, $\phi \in D(A)$ est dit analytique pour A si $A\phi \in D(A)$, $A^2\phi \in D(A)$, ... $A^n\phi \in D(A)$ pour tout entier n, et il existe $t \in \mathbf{R}^*_+$ tel que $\sum_{n=0}^{+\infty} \frac{t^n}{n!}|A^n\phi| < +\infty$.

THÉORÈME 5 – Soit A un opérateur symétrique, possédant un sous-espace dense de vecteurs analytiques, alors A est essentiellement auto-adjoint.

Ce théorème est dû à Nelson ([55] voir Reed et Simon [70]).

28

Chapitre 3

Variables de Gauss et de Poisson et relations de commutation

3.1 Relations de commutation d'Heisenberg

Après ces points de théorie spectrale nous allons reprendre le fil, en étudiant les opérateurs introduits à la fin du chapitre 1 dans le théorème limite pour les marches de Bernoulli quantiques.

Soit H un espace de Hilbert complexe muni d'une base orthonormale $(e_n)_{n \in \mathbb{N}}$. On notera parfois Ω le vecteur e_0, et on appellera "état vide" l'état pur associé. On considère les opérateurs a^0, a^+, a^- de domaine $D = vect(e_n)_{n \in \mathbb{N}}$, définis par

$$a^0 e_n = n\, e_n$$
$$a^+ e_n = \sqrt{n+1}\, e_{n+1} \tag{23}$$
$$a^- e_n = \sqrt{n}\, e_{n-1}$$

(avec la convention $e_{-1} = 0$).

Ce chapitre est consacré à l'étude approfondie de ces opérateurs. Tout d'abord, d'un point de vue analytique, nous allons montrer que les combinaisons linéaires symétriques de ces opérateurs sont essentiellement auto-adjointes, puis nous calculerons les lois dans l'état vide de ces variables aléatoires. On vérifiera que l'on obtient bien des lois normales et des lois de Poisson comme le laisse présager le théorème limite (2). Le triplet (a^0, a^+, a^-) joue en probabilités quantiques le rôle qui est tenu par les variables aléatoires de Gauss et de Poisson en probabilités classiques. Ce qui est remarquable ici, ce sont les relations algébriques simples entre ces opérateurs.

Les opérateurs a^0, a^+, a^- laissent stable le domaine D, ce qui permet de les composer et de donner les relations suivantes

$$a^0 = a^+ a^- \tag{24}$$
$$[a^+, a^-] = -Id \tag{25}$$
$$[a^0, a^+] = a^+ \tag{26}$$
$$[a^0, a^-] = -a^- \tag{27}$$

Les opérateurs a^+ et a^- sont formellement adjoints l'un de l'autre, c'est-à-dire que

$$< a^+ x, y > = < x, a^- y > \quad \forall x, y \in D \tag{28}$$

Ceci entraine que $a^+ \subset (a^-)^*$ et $a^- \subset (a^+)^*$, et en particulier les opérateurs a^+ et a^- sont fermables.

Les combinaisons linéaires

$$Q = a^+ + a^- \text{ et } P = \frac{1}{i}(a^+ - a^-)$$

sont des opérateurs symétriques. En terme de ces opérateurs, les relations (25) s'écrivent

$$a^0 = \frac{1}{2}(P^2 + Q^2) \tag{29}$$

$$[P, Q] = 2iId \tag{30}$$

$$[a^0, P] = -iQ \tag{31}$$

$$[a^0, Q] = iP \tag{32}$$

Les opérateurs a^+ et a^- sont connus en physique sous le nom d'opérateurs de création et d'annihilation, et les deux relation équivalentes (25) et (30) sont les "relations de commutation d'Heisenberg".
L'opérateur a^0 quant à lui porte le nom d'opérateur de nombre.

On va maintenant construire toutes les combinaisons linéaires symétriques de ces opérateurs et montrer que ce sont des opérateurs essentiellement auto-adjoints.

PROPOSITION 4 – Soient $\tau \in \mathbb{R}$, $z \in \mathbb{C}$, l'opérateur $\tau a^0 + za^+ + \bar{z}a^-$ est essentiellement auto-adjoint.

Démonstration
On commence par montrer par récurrence sur n que pour tout $k \in \mathbb{N}$

$$|(\tau a^0 + za^+ + \bar{z}a^-)^n e_k| \le (|\tau| + 2|z|)^n \frac{(n+k)!}{k!}$$

Cette relation étant vraie pour $n = 0$, supposons la vraie pour n, et calculons

$$|(\tau a^0 + za^+ + \bar{z}a^-)^{n+1} e_k|$$
$$= |(\tau a^0 + za^+ + \bar{z}a^-)^n (\tau k e_k + z\sqrt{k+1}\, e_{k+1} + \bar{z}\sqrt{k}\, e_{k-1})|$$
$$\le |\tau| k |(\tau a^0 + za^+ + \bar{z}a^-)^n (e_k)|$$
$$+ |z|\sqrt{k+1}|(\tau a^0 + za^+ + \bar{z}a^-)^n (e_{k+1})|$$
$$+ |z|\sqrt{k}|(\tau a^0 + za^+ + \bar{z}a^-)^n (e_{k-1})|$$
$$\le (|\tau| + 2|z|)^n \Big(|\tau| k \frac{(n+k)!}{k!}$$
$$+ |z|\sqrt{k+1}\frac{(n+k+1)!}{(k+1)!} + |z|\sqrt{k}\frac{(n+k-1)!}{(k-1)!} \Big)$$
$$\le (|\tau| + 2|z|)^{n+1} \frac{(n+k+1)!}{k!}$$

ce qui démontre l'inégalité. On en déduit que les vecteur e_k sont analytiques pour $\tau a^0 + za^+ + \bar{z}a^-$ car

$$\sum_{n=0}^{\infty} |(\tau a^0 + za^+ + \bar{z}a^-)^n e_k| \frac{t^n}{n!} \le \sum_{n=0}^{\infty} (|\tau| + 2|z|)^n \frac{(n+k)!}{k!n!} t^n$$
$$= (1 - t(|\tau| + 2|z|))^{-(k+1)} < +\infty$$
$$\text{pour } t \ge 0 \text{ assez petit}$$

D'après le théorème de Nelson l'opérateur $\tau a^0 + za^+ + \bar{z}a^-$ est essentiellement auto-adjoint.

Dans la suite, par abus de langage on désignera par $\tau a^0 + za^+ + \bar{z}a^-$ sa fermeture.
REMARQUE– Le calcul ci-dessus nous dit en fait un peu plus. Si on a un espace vectoriel dense dans H formé de vecteurs de la forme $\phi = \sum_k \phi_k e_k$ tels qu'il existe un $\beta > 1$ avec $\sum_k |\phi_k| \beta^k < +\infty$, alors les vecteurs de ce sous-espace sont dans le domaine de la fermeture de $\tau a^0 + za^+ + \bar{z}a^-$ et sont analytiques, donc la restriction de $\tau a^0 + za^+ + \bar{z}a^-$ à ce sous-espace est essentiellement auto-adjointe.

3.2 Modèle gaussien des relations de commutation

On va maintenant calculer les lois, dans l'état vide, des opérateurs auto-adjoints que l'on vient de construire, en commençant par le cas $\tau = 0$. La façon dont les opérateurs a^+, a^- ont été obtenus suggère que les lois de P et Q sont des lois normales centrées réduites. C'est ce que nous allons vérifier en calculant leurs transformées de Fourier au moyen des groupes unitaires à un paramètre e^{itP} et e^{itQ}. Ces deux groupes engendrent un groupe d'opérateurs unitaires qui s'expriment à l'aide des opérateurs de Weyl définis ci-dessous. Ces opérateurs sont obtenus au moyen d'une représentation des opérateurs a^+ et a^- sur l'espace L^2 d'une mesure gaussienne, dans laquelle, $Q = a^+ + a^-$ correspond à l'opérateur de multiplication par la fonction x.
Remarquons que les vecteurs e_k de H sont obtenus à partir de la suite de vecteurs $(a^+ + a^-)^k(e_0)$ par le procédé d'orthogonalisation de Gramm-Schmidt. Dans l'espace $L^2(\mathbf{R}, \nu)$, en itérant l'opérateur de multiplication par la fonction x appliqué à la fonction 1 on obtient la suite des monômes $1, x, x^2, \ldots, x^n, \ldots$. Les fonctions obtenues à partir de cette suite par le procédé de Gramm-Schmidt sont les polynômes d'Hermite, dont on rappelle ci-dessous les principales propriétés.

Les polynômes d'Hermite peuvent être définis par la série génératrice

$$e^{tx - \frac{t^2}{2}} = \sum_{n=0}^{+\infty} \frac{t^n}{n!} \mathcal{H}_n(x) \tag{33}$$

La suite des polynômes d'Hermite $(\mathcal{H}_n)_{n \in \mathbb{N}}$ vérifie

$$\int_{-\infty}^{+\infty} \mathcal{H}_n(x) \mathcal{H}_m(x) \frac{e^{-\frac{x^2}{2}}}{\sqrt{2\pi}} \, dx = \delta_{n,m} \, n! \tag{34}$$

On en déduit la suite de polynômes orthonormale dans $L^2(\mathbf{R}, \nu)$

$$h_n(x) = \frac{\mathcal{H}_n(x)}{\sqrt{n!}} \tag{35}$$

A partir de la série génératrice, on obtient les relations de récurrence

$$x \, h_n(x) = \sqrt{n+1} \, h_{n+1}(x) + \sqrt{n} \, h_{n-1}(x) \tag{36}$$

$$\frac{d}{dx} \, h_n(x) = \sqrt{n} \, h_{n-1}(x) \tag{37}$$

Ces relations de récurrence fournissent un modèle pour les opérateurs a^+ et a^-. En effet, soit U l'isomorphisme de H dans $L^2(\mathbf{R}, \nu)$ qui envoie e_k sur h_k alors $a^+ = U^{-1} \circ (x - \frac{d}{dx}) \circ U$ et $a^- = U^{-1} \circ \frac{d}{dx} \circ U$ sur le domaine D. En particulier, l'opérateur Q est conjugué à l'opérateur de multiplication par x, alors que P est conjugué à $\frac{1}{i}(x - 2\frac{d}{dx})$. De même, U transforme l'opérateur $a^0 = a^+ a^-$ en $(-\frac{d^2}{dx^2} + x\frac{d}{dx})$. Cet opérateur est l'opérateur d'Ornstein-Uhlenbeck.

Ce modèle des opérateurs a^+ et a^- sur un espace Gaussien va nous permettre de définir des groupes unitaires à un paramètre.

3.3 Opérateurs de Weyl

DÉFINITION 23 – Soit $z = u + iv \in \mathbf{C}$, on définit un opérateur W_z sur $L^2(\mathbf{R}, \nu)$ par

$$W_z f(x) = f(x - 2u)\,e^{zx - u^2 - iuv}$$

Les opérateurs W_z sont appelés opérateurs de Weyl.

PROPOSITION 5 – Soient $z = u + iv, z' = u' + iv' \in \mathbf{C}$
i) W_z est un opérateur unitaire, $W_0 = Id$.
ii) on a

$$\forall z, z' \in \mathbf{C} \quad W_z W_{z'} = W_{z+z'}e^{i(u'v - uv')} = W_{z+z'}e^{i\Im m(z\bar{z}')} \tag{38}$$

Démonstration
Vérifions ii)

$$W_z W_{z'} f(x) = f(x - 2u - 2u')e^{zx - u^2 - iuv}e^{z'(x-2u)-u'^2-iu'v'}$$

$$= f(x - 2(u + u'))e^{(z+z')x - (u+u')^2 - i(u+u')(v+v') + i(u'v - uv')}$$

$$= e^{i(u'v - uv')}W_{z+z'}f(x)$$

Pour i) on vérifie tout d'abord que $W_0 = Id$, puis on calcule

$$< W_z(f), W_z(g) > = \int_{-\infty}^{+\infty} f(x - 2u)\bar{g}(x - 2u)e^{zx - u^2 - iuv}e^{\bar{z}x - u^2 + iuv}\frac{e^{-\frac{x^2}{2}}}{\sqrt{2\pi}}\,dx$$

$$= \int_{-\infty}^{+\infty} f(x - 2u)\bar{g}(x - 2u)\frac{e^{-\frac{(x-2u)^2}{2}}}{\sqrt{2\pi}}\,dx$$

$$= < f, g >$$

ce qui montre que W_z est isométrique, et comme d'après ii) $W_z W_{-z} = W_{-z} W_z = Id$ on voit que W_z est inversible donc unitaire.

COROLLAIRE– Pour tout $z \in \mathbf{C}$, $(W_{tz})_{t \in \mathbf{R}}$ est un groupe d'opérateurs unitaires, fortement continu, dont le générateur est $\frac{1}{i}(za^+ - \bar{z}a^-)$.

Démonstration
La continuité de $t \to W_{tz}f$ se vérifie aisément pour f continue à support compact par le théorème de Lebesgue. Ces fonctions formant un sous-espace dense dans $L^2(\mathbf{R}, \nu)$ on en déduit la continuité forte de $(W_{tz})_{t \in \mathbf{R}}$.

Calculons le générateur infinitésimal de ce groupe unitaire. Pour tout polynôme p on a

$$\frac{W_{tz}p(x) - p(x)}{it} \rightarrow_{t \to 0} \left(-2u\frac{d}{dx}p(x) + zxp(x)\right) = \frac{1}{i}(za^+ - \bar{z}a^-)(p)(x)$$

comme $\frac{1}{i}(za^+ - \bar{z}a^-)$ est essentiellement auto-adjoint sur l'espace des polynômes, il coïncide avec le générateur de $(W_{tz})_{t \in \mathbf{R}}$.

COROLLAIRE– La variable $\frac{1}{i}(za^+ - \bar{z}a^-)$ suit une loi gaussienne centrée de variance $|z|^2$ dans l'état vide.

Démonstration

Calculons la transformée de Fourier de cette loi

$$< e^{it(za^+ + \bar{z}a^-)}\Omega, \Omega > =< W_{tz}h_0, h_0 >$$

$$= \int_{-\infty}^{+\infty} e^{tzx - t^2(u^2 + iuv)} \frac{e^{-\frac{x^2}{2}}}{\sqrt{2\pi}} dx$$

$$= e^{\frac{1}{2}t^2 z^2} e^{-t^2(u^2 + iuv)}$$

$$= e^{-\frac{1}{2}t^2 |z|^2}$$

d'où le résultat.

La relation (38) entraine la relation suivante entre les groupes à un paramètre engendrés par les opérateurs P et Q

$$e^{itP}e^{isQ} = e^{-2ist}e^{isQ}e^{itP}$$

(remarquer que $W_t = e^{itP}$ et $W_{is} = e^{isQ}$).

Cette dernière relation permet de retrouver la relation (30), en effet en dérivant on obtient

$$\frac{d}{dt}\frac{d}{ds}_{s=t=0} e^{-itP}e^{isQ}e^{itP} = [P, Q] = 2iId$$

Les relations de commutations (38) entre opérateurs de Weyl sont donc une version "intégrée" des relations de commutation (25) et (30) d'Heisenberg.

Rappelons que le *groupe d'Heisenberg* est $\mathbf{C} \times \mathbf{R}$ muni de la loi de groupe

$$(z, t) \star (z', t') = (z + z', t + t' + \Im m(z\bar{z}'))$$

La relation de commutation de Weyl (38) signifie que $(z, t) \mapsto W_z e^{it}$ est une représentation du groupe d'Heisenberg par des opérateurs unitaires sur $L^2(\mathbf{R}, \nu)$.

3.4 Vecteurs exponentiels

Avant de calculer la loi des opérateurs $\tau a^0 + za^+ + \bar{z}a^-$ pour $\tau \neq 0$ (on peut toujours se ramener à $\tau = 1$), introduisons une définition qui sera très utile quand nous étudierons les espaces de Fock.

DÉFINITION 24 – *Soit* $\alpha \in \mathbf{C}$, *on pose*

$$\xi(\alpha) = \sum_{k=0}^{\infty} \frac{\alpha^k}{\sqrt{k!}} h_k = e^{\alpha x - \frac{\alpha^2}{2}} = W_\alpha(h_0)e^{\frac{1}{2}|\alpha|^2} \tag{39}$$

Les vecteurs $\xi(\alpha)$ sont appelés vecteurs exponentiels, et on note Ξ le sous-espace vectoriel qu'ils engendrent.

PROPOSITION 6 – *Les vecteurs exponentiels vérifient les propriétés suivantes*
i) $< \xi(\alpha), \xi(\alpha') > = e^{\alpha \bar{\alpha}'}$
ii) Ξ *est dense dans $L^2(\mathbf{R}, \nu)$, et les vecteurs $\xi(\alpha)$ sont linéairement indépendants.*
iii) $W_z(\xi(\alpha)) = \xi(\alpha + z) e^{-\alpha \bar{z} - \frac{1}{2}|z|^2}$

Démonstration
i) $< \xi(\alpha), \xi(\alpha') > = \sum_{n=0}^{\infty} \frac{\alpha^k \bar{\alpha}'^k}{k!} = e^{\alpha \bar{\alpha}'}$
ii) Soient $\alpha_0, \ldots, \alpha_n \in \mathbf{C}$. Les projections orthogonales de $\xi(\alpha_0), \ldots, \xi(\alpha_n)$ sur le sous-espace engendré par $h_0, \ldots, h_n$ ont pour déterminant dans la base $h_0, \ldots, h_n$

$$
\begin{vmatrix}
1 & 1 & \cdots & 1 \\
\alpha_0 & \alpha_1 & \cdots & \alpha_n \\
& & \cdots & \\
\frac{\alpha_0^n}{\sqrt{n!}} & \frac{\alpha_1^n}{\sqrt{n!}} & \cdots & \frac{\alpha_n^n}{\sqrt{n!}}
\end{vmatrix}
$$

Ce déterminant est un multiple d'un déterminant de Vandermonde. Il ne s'annule que si deux des α_k sont égaux, par conséquent les vecteurs $\xi(\alpha_0), \ldots, \xi(\alpha_n)$ forment un système libre si les α_k sont distincts.

La densité de Ξ provient de l'égalité $h_k = \frac{1}{\sqrt{k!}} \frac{d^k}{d\alpha^k} \xi(\alpha)_{|\alpha=0}$ qui montre que les vecteurs h_k sont dans la fermeture de Ξ.
iii) $W_z(e^{\alpha x - \frac{\alpha^2}{2}}) = e^{\alpha(x-2u) - \frac{\alpha^2}{2}} e^{zx - u^2 - iuv}$
$= e^{(\alpha+z)x - \frac{(\alpha+z)^2}{2}} e^{-\alpha \bar{z} - \frac{1}{2}|z|^2} = \xi(\alpha + z) e^{-\alpha \bar{z} - \frac{1}{2}|z|^2}$

D'après la remarque qui suit la proposition (4), les vecteurs de Ξ sont analytiques pour les opérateurs de la forme $\tau a^0 + z a^+ + \bar{z} a^-$, par conséquent, les restrictions de ces opérateurs à Ξ sont essentiellement auto-adjointes.
L'action de a^+ et a^- sur les vecteurs exponentiels prend une forme simple.

$$a^+(\xi(\alpha)) = \frac{d}{d\alpha} \xi(\alpha) \tag{40}$$

$$a^-(\xi(a)) = \alpha \xi(\alpha) \tag{41}$$

3.5 Opérateur de nombre et loi de Poisson

Nous allons maintenant examiner les variables $\tau a^0 + z a^+ + \bar{z} a^-$ avec $\tau \neq 0$. L'opérateur de nombre a^0 est donné par sa décomposition spectrale, les vecteurs e_k forment une base de vecteurs propres de a^0. Le groupe unitaire à un paramétre qu'il engendre est donc donné par la formule:

$$e^{ita^0} h_k = e^{itk} h_k$$

L'action sur les vecteurs exponentiels est obtenue facilement, de $a^0 = a^+ a^-$ on tire

$$a^0 \xi(\alpha) = \alpha \frac{d}{d\alpha} \xi(\alpha) \tag{42}$$

et

$$e^{ita^0}\xi(\alpha) = \xi(e^{it}\alpha) \tag{43}$$

On va en déduire les relations de commutation entre ce groupe unitaire et les opérateurs de Weyl.

PROPOSITION 7 –

$$e^{ita^0}W_z e^{-ita^0} = W_{e^{it}z} \tag{44}$$

$$W_{-z}e^{ita^0}W_z = e^{it(a^0+za^++\bar{z}a^-+|z|^2)} \tag{45}$$

Démonstration
Pour la première formule, considérons un vecteur exponentiel $\xi(\alpha)$, et calculons

$$
\begin{aligned}
e^{ita^0}W_z e^{-ita^0}\big(\xi(\alpha)\big) &= e^{ita^0}W_z\big(\xi(e^{-it}\alpha)\big) \\
&= e^{ita^0}\Big(\xi(e^{-it}\alpha + z)e^{-\bar{z}\alpha e^{-it}-\frac{|z|^2}{2}}\Big) \\
&= \xi(\alpha + ze^{it})e^{-\bar{z}\alpha e^{-it}-\frac{|z|^2}{2}} \\
&= W_{e^{it}z}\xi(\alpha)
\end{aligned}
$$

les vecteurs exponentiels forment une partie totale, donc les deux opérateurs unitaires coïncident ce qui montre la première partie.
La famille $(W_{-z}e^{ita^0}W_z)_{t\in\mathbf{R}}$ est un groupe à un paramètre d'opérateurs unitaires. Calculons son générateur.

$$
\begin{aligned}
\frac{1}{i}\frac{d}{dt}\Big|_{t=0}(W_{-z}e^{ita^0}W_z(\xi(\alpha)) &= \frac{1}{i}\frac{d}{dt}\Big|_{t=0}\big(\xi(e^{it}(\alpha+z)-z)e^{-\alpha\bar{z}-\frac{1}{2}|z|^2}e^{e^{it}(\alpha+z)\bar{z}-\frac{1}{2}|z|^2}\big) \\
&= (\alpha+z)\frac{d}{d\alpha}(\xi(\alpha)) + (\bar{z}\alpha+|z|^2)\xi(\alpha) \\
&= (a^0+za^++\bar{z}a^-+|z|^2)(\xi(\alpha))
\end{aligned}
$$

Le générateur de $(W_{-z}e^{ita^0}W_z)_{t\in\mathbf{R}}$ coincide avec $a^0+za^++\bar{z}a^-+|z|^2$ sur le domaine Ξ, or $a^0+za^++\bar{z}a^-+|z|^2$ est essentiellement auto-adjoint sur ce domaine, d'après la remarque qui suit la proposition (4) donc ces deux opérateurs sont égaux, et en appliquant le théorème de Stone on a la seconde formule.

La proposition précédente nous permet d'obtenir la loi de la variable $a^0+za^+ +\bar{z}a^-+|z|^2$ dans l'état vide:

PROPOSITION 8 – La variable $a^0+za^++\bar{z}a^-+|z|^2$ suit une loi de Poisson de paramètre $|z|^2$.

Démonstration
je vais donner deux façons de calculer cette loi, afin d'illustrer les principes de base des probabilités non-commutatives. La première consiste à calculer la transformée

de Fourier

$$< e^{it(a^0+za^++\bar{z}a^-+|z|^2)}\Omega, \Omega > \; = < W_{-z}e^{ita^0}W_z h_0, h_0 >$$
$$= < e^{ita^0}W_z h_0, W_z h_0 >$$
$$= e^{-|z|^2} < e^{ita^0}\xi(z), \xi(z) >$$
$$= e^{-|z|^2} < \xi(e^{it}z), \xi(z) >$$
$$= e^{|z|^2(e^{it}-1)}$$

Pour la seconde, on remarque que $a^0 + za^+ + \bar{z}a^- + |z|^2 = W_{-z}a^0 W_z$, et on applique le lemme suivant.

LEMME – Soient A un opérateur auto-adjoint sur H, U un opérateur unitaire et $x \in H$ de norme 1, la loi de $U^{-1}AU$ dans l'état pur x est égale à la loi de A dans l'état pur Ux.

La loi de a^0 dans l'état $W_z(h_0)$ est facile à calculer, le spectre de a^0 est $\mathbb{N}$ et le vecteur h_k est un vecteur propre de a^0 de valeur propre k, donc la loi est portée par les entiers positifs et donne comme probabilité à k

$$| < W_z(h_0), h_k > |^2 = \frac{|z|^{2k}}{k!}e^{-|z|^2}$$

on reconnait la loi de Poisson de paramètre $|z|^2$.

Les relations de commutation de Weyl fournissent une représentation du groupe d'Heisenberg. Si on tient compte de la relation (44) on obtient une représentation d'un groupe qui contient le groupe d'Heisenberg. Ce groupe est $U(1) \times \mathbb{C} \times \mathbb{R}$ avec la loi de groupe

$$(e^{i\theta}, z, t) \star (e^{i\theta'}, z', t') = (e^{i(\theta+\theta')}, z + e^{i\theta}z', t + t' + \Im m\,(z\bar{z}'))$$

La relation (44) montre que l'application

$$(e^{i\theta}, z, t) \mapsto W_z e^{i\theta a^0}e^{it}$$

définit une représentation unitaire de ce groupe.
Remarquons également que la relation (44) entraine

$$e^{i\frac{\pi}{2}a^0}e^{itP}e^{-i\frac{\pi}{2}a^0} = e^{itQ}$$

autrement dit les variables P et Q sont unitairement équivalente par la transformation $e^{i\frac{\pi}{2}a^0}$. Cette transformation unitaire de $L^2(\mathbb{R}, \nu)$ n'est autre que la transformation de Fourier.

3.6 Modèle poissonien des relations de commutation

Au cours de ce chapitre, j'ai utilisé l'espace L^2 de la mesure de Gauss pour donner une représentation des relations de commutation d'Heisenberg. On aurait pu également utiliser l'espace L^2 de la mesure de Poisson bien que cela soit moins habituel. Je vais indiquer brièvement dans ce paragraphe comment s'interprètent alors les opérateurs de création, d'annihilation et de nombre.

On considère la mesure de Poisson de paramètre 1, sur $\mathbb{N}$, donnée par la formule $\mu(\{k\}) = \frac{e^{-1}}{k!}$.

Les polynômes orthogonaux par rapport à cette mesure sont les polynômes de Charlier (voir Chihara [16]), $(C_n)_{n \in \mathbb{N}}$ définis par la série génératrice

$$\sum_{n=0}^{\infty} C_n(x) \frac{w^n}{n!} = e^{-x}(1+w)^x \tag{46}$$

Ils vérifient les relations de récurrence

$$C_{n+1}(x) = (x - n)C_n(x) - nC_{n-1}(x) \tag{47}$$

$$C_n(x+1) - C_n(x) = nC_{n-1}(x) \tag{48}$$

Les polynômes $c_n = \sqrt{n!}\, C_n$ forment une base orthonormale de $L^2(\mathbb{N}, \mu)$. Introduisons l'opérateur de différence finie Δ défini par

$$\Delta f(x) = f(x+1) - f(x)$$

Il possède un adjoint Δ^* donné par $\Delta^* f(x) = x f(x-1) - f(x)$. Les relations de récurrence des polynomes de Charlier entrainent alors que

$$\Delta^* c_n = \sqrt{n+1}\, c_{n+1} \tag{49}$$

$$\Delta c_n = \sqrt{n}\, c_{n-1} \tag{50}$$

autrement dit, les opérateurs Δ^* et Δ donnent un modèle des opérateurs de création et d'annihilation. L'opérateur de multiplication par la fonction x est donné par la formule $(\Delta^* + Id)(\Delta + Id)$ et par construction, il suit une loi de Poisson de paramètre 1 dans l'état vide.

3.7 Théorème de Stone-von Neumann

La fin du chapitre 3 est constituée de compléments qui ne sont pas indispensables pour la suite du cours.

Les opérateurs de Weyl vérifient les relations de commutation (38), qui sont une version intégrée des relations d'Heisenberg, et définissent ainsi une représentation du groupe d'Heisenberg. Nous allons voir que cette représentation est irréductible, et que toute autre représentation en est un multiple, c'est le contenu du théorème de Stone-von Neumann démontré dans ce paragraphe. Ensuite, on exploitera cette propriété des relations de commutation pour introduire de façon naturelle de nouvelles variables aléatoires et faire quelques calculs de lois. Ces variables jouent un rôle important en optique dans la théorie des états dits "comprimés" de la lumière (cf S.Reynaud [71]).

Voici l'énoncé du théorème de Stone-von Neumann.

THÉORÈME 6 – Soit $(\tilde{W}_z)_{z \in \mathbb{C}}$ une famille fortement continue d'opérateurs unitaires sur un espace de Hilbert $\tilde{H}$ telle que

$$\forall z, z' \in \mathbb{C} \quad \tilde{W}_z \tilde{W}'_z = \tilde{W}_{z+z'} e^{i\Im m (z\bar{z}')} \tag{51}$$

alors il existe un espace de Hilbert $\tilde{H}_0$ et isomorphisme d'espaces de Hilbert

$$U : \tilde{H} \to L^2(\mathbf{R}, \nu) \otimes \tilde{H}_0$$

tel que $\tilde{W}_z = U^*(W_z \otimes Id)U$.
D'autre part, les opérateurs W_z opèrent de façon irréductible sur $L^2(\mathbf{R}, \nu)$, c'est à dire que les seuls sous-espaces fermés invariants par tous ces opérateurs sont $\{0\}$ et $L^2(\mathbf{R}, \nu)$.

Démonstration

Il existe de nombreuses démonstrations du théorème de Stone-von Neumann (cf [73], [48], [70]). En voici une de nature algébrique.
Commençons par remarquer que la relation (51) entraine que $\tilde{W}_0 = Id$ et pour tout $z \in \mathbf{C}$, $\tilde{W}_z^* = \tilde{W}_{-z}$.
On va établir les deux lemmes suivants:

LEMME 1 – L'opérateur $\tilde{P} = \frac{1}{2\pi} \int_{\mathbf{C}} \tilde{W}_z e^{-\frac{1}{2}|z|^2} dz$ est auto-adjoint, et pour tout $\zeta \in \mathbf{C}$ on a

$$\tilde{P}\tilde{W}_\zeta\tilde{P} = e^{-\frac{1}{2}|\zeta|^2} \tilde{P}$$

En particulier (en prenant $\zeta = 0$), $\tilde{P}$ est un projecteur.

Démonstration

Cet opérateur est bien défini et borné car $z \to \tilde{W}_z$ est fortement continue bornée. Il est auto-adjoint car $\tilde{P}^* = \frac{1}{2\pi} \int_{\mathbf{C}} \tilde{W}_{-z} e^{-\frac{1}{2}|z|^2} dz = \frac{1}{2\pi} \int_{\mathbf{C}} \tilde{W}_z e^{-\frac{1}{2}|z|^2} dz = \tilde{P}$. Calculons

$$\tilde{P}\tilde{W}_\zeta\tilde{P} = (\frac{1}{2\pi})^2 \int\int_{\mathbf{C}^2} \tilde{W}_z\tilde{W}_\zeta\tilde{W}_{z'} e^{-\frac{1}{2}(|z|^2+|z'|^2)} dz\,dz'$$

$$= (\frac{1}{2\pi})^2 \int\int_{\mathbf{C}^2} \tilde{W}_{z+\zeta+z'} e^{-\frac{1}{2}(|z|^2+|z'|^2)+\bar{z}\zeta-z\bar{\zeta}+\bar{z}z'-z\bar{z}'+\bar{\zeta}z'-\zeta\bar{z}'} dz\,dz'$$

$$= (\frac{1}{2\pi})^2 \int\int_{\mathbf{C}^2} \tilde{W}_{z+\zeta+z'} e^{-\frac{1}{2}|z+\zeta+z'|^2+(\bar{z}+\bar{\zeta})(z'+\zeta)-\frac{1}{2}|\zeta|^2} dz\,dz'$$

$$= (\frac{1}{2\pi})^2 \int\int_{\mathbf{C}^2} \tilde{W}_{z''} e^{-\frac{1}{2}|z''|^2+(\bar{z}''-\bar{\zeta}-\bar{z}''')z'''-\frac{1}{2}|\zeta|^2} dz''\,dz'''$$

en faisant le changement de variables: $z'' = z + \zeta + z'$, $z''' = z' + \zeta$

$$= (\frac{1}{2\pi}) \int_{\mathbf{C}} \tilde{W}_{z''} e^{-\frac{1}{2}|z''|^2-\frac{1}{2}|\zeta|^2} dz''$$

en intégrant par rapport à z''' et en utilisant la formule $\frac{1}{2\pi} \int_{\mathbf{C}} e^{tz-\bar{z}z} dz = 1$

$$= e^{-\frac{1}{2}|\zeta|^2} \tilde{P}$$

ce qui termine la démonstration du lemme.

LEMME 2 – $P = \frac{1}{2\pi} \int_{\mathbf{C}} W_z e^{-\frac{1}{2}|z|^2} dz$ est le projecteur orthogonal sur la droite engendrée par h_0 dans $L^2(\mathbf{R}, \nu)$.

Démonstration
Soit $f \in L^2(\mathbf{R}, \nu)$, on a

38

$$Pf(x) = \frac{1}{2\pi} \int_{-\infty}^{+\infty} \int_{-\infty}^{+\infty} f(x - 2u)e^{x(u+iv)-u(u+iv)-\frac{1}{2}(u^2+v^2)} \, du \, dv$$

$$= \frac{1}{\sqrt{2\pi}} \int_{-\infty}^{+\infty} f(x - 2u)e^{-\frac{1}{2}(x-u)^2+xu-\frac{3}{2}u^2} \, du$$

$$= \frac{1}{\sqrt{2\pi}} \int_{-\infty}^{+\infty} f(x - 2u)e^{-\frac{1}{2}(x-2u)^2} \, du$$

$$= <f, h_0 > h_0(x)$$

COROLLAIRE – Les opérateurs W_z opèrent de façon irréductible.

Démonstration

Soit K un sous-espace fermé invariant par les W_z. Le projecteur orthogonal sur K commute avec les opérateurs W_z, donc également avec les opérateurs $W_z P W_{-z}$ pour $z \in \mathbf{C}$. Or, d'après le lemme 2, $W_z P W_{-z}$ est le projecteur orthogonal sur la droite engendrée par le vecteur $W_z(h_0)$. Ceci entraine que pour tout $z \in \mathbf{C}$ on a soit $W_z(h_0) \in K$ soit $W_z(h_0) \in K^{\perp}$. Comme $< W_z(h_0), W_{z'}(h_0) > \neq 0$ on voit que soit K soit son orthogonal contient tous les vecteurs $W_z(h_0)$ donc tous les vecteurs $\xi(z)$. Ces vecteurs forment une partie totale de $L^2(\mathbf{R}, \nu)$, et K étant fermé on a $K = H$ ou $K = \{0\}$.

LEMME 3 – Pour tous $z_1, z_2 \in \mathbf{C}$, $x_1, x_2 \in \tilde{H}$ on a

$$< \tilde{W}_{z_1} \tilde{P} x_1, \tilde{W}_{z_2} \tilde{P} x_2 > = < \tilde{P} x_1, \tilde{P} x_2 > < W_{z_1} h_0, W_{z_2} h_0 >$$

Démonstration

$$< \tilde{W}_{z_1} \tilde{P} x_1, \tilde{W}_{z_2} \tilde{P} x_2 > = < \tilde{P} \tilde{W}_{-z_2} \tilde{W}_{z_1} \tilde{P} x_1, x_2 >$$

$$= e^{-i\Im m(z_2 \bar{z}_1)} < \tilde{P} \tilde{W}_{z_1 - z_2} \tilde{P} x_1, x_2 >$$

$$= e^{-i\Im m(z_2 \bar{z}_1) - \frac{1}{2}|z_1 - z_2|^2} < \tilde{P} x_1, x_2 > \quad \text{d'après le lemme 1}$$

$$= < \tilde{P} x_1, \tilde{P} x_2 > < W_{z_1} h_0, W_{z_2} h_0 >$$

Le lemme 3 entraine que l'application

$$U : \Xi \otimes Im(\tilde{P}) \to \tilde{H}$$

$$W_z h_0 \otimes x \mapsto \tilde{W}_z x$$

se prolonge en une isométrie de $L^2(\mathbf{R}, \nu) \otimes Im(\tilde{P})$ dans $\tilde{H}$.
Montrons que son image est $\tilde{H}$ tout entier. Supposons que $y \in \tilde{H}$ est orthogonal à $\tilde{W}_z(Px) \; \forall x \in \tilde{H} \; z \in \mathbf{C}$. Pour tout $\zeta \in \mathbf{C}$ on a:

$$< \tilde{W}_{\zeta} \tilde{P} \tilde{W}_{-\zeta} x, y > = 0$$

$$= \int_{\mathbf{C}} < W_z x, y > e^{-\frac{1}{2}|z|^2 + 2i\Im m(z\bar{\zeta})} dz$$

La fonction $z \mapsto < W_z x, y >$ est continue, donc elle est nulle (injectivité de la transformée de Fourier) et $< x, y > = 0 \; \forall x \in \tilde{H}$, d'où $y = 0$.

L'application U est bien l'isomorphisme cherché, on a $\tilde{W}_z = U^*(W_z \otimes Id)U$ par construction, ce qui termine la démonstration.

3.8 Conséquences du théorème de Stone-von Neumann

Supposons, avec les notations du théorème 6, que les opérateurs $\tilde{W}_z$ agissent de façon irréductible, alors l'espace $\tilde{H}_0$ est de dimension 1, et U est un isomorphisme entre $\tilde{H}$ et $L^2(\mathbf{R}, \nu)$.

Soit $\tau : \mathbb{C} \to \mathbb{C}$ une application $\mathbf{R}$-linéaire, alors $\Im m\,(\tau(z)\tau(\bar{z}')) = \det(\tau)\Im m\,(z\bar{z}')$ donc l'application τ vérifie

$$\forall z, z' \in \mathbb{C}\;\; \Im m\,(\tau(z)\tau(\bar{z}')) = \Im m\,(z\bar{z}')$$

si et seulement si $\det(\tau) = 1$, et les opérateurs $\tilde{W}_z = W_{\tau(z)}$ vérifient alors les hypothèses du théorème de Stone-von Neumann. Ils agissent de façon irréductible sur $L^2(\mathbf{R}, \nu)$ ce qui entraine l'existence d'un opérateur unitaire U_τ tel que $W_{\tau(z)} = U_\tau^* W_z U_\tau$ pour tout $z \in \mathbb{C}$. Si V_τ est un autre opérateur unitaire vérifiant cette égalité, alors $U_\tau V_\tau^*$ commute avec W_z pour tout $z \in \mathbb{C}$ et donc c'est un multiple de l'identité, par irréductibilité de la représentation. L'opérateur U_τ est donc déterminé à un nombre complexe de module 1 près.

Je vais donner ci-dessous deux exemples de groupes à un paramètre de telles transformations.

Le premier exemple est celui des applications $\tau_t(z) = ze^{it}$. D'après les relations (44) on a

$$e^{ita^0}W_z e^{-ita^0} = W_{ze^{it}}$$

ce qui montre que l'opérateur unitaire U_{τ_t} correspondant est e^{ita^0}.

Le second exemple est donné par les applications $\tau_t(u+iv) = e^t u + ie^{-t}v$. L'opérateur U_{τ_t} donné par la formule suivante

$$U_{\tau_t} f(x) = f(e^t x)e^{\frac{t}{2} - \frac{1}{4}(e^{2t}-1)x^2}$$

vérifie pour $z = u + iv$

$$U_{\tau_t}^* W_{u+iv} U_{\tau_t} f(x) = (W_{u+iv} U_{\tau_t} f)(e^{-t}x)e^{-\frac{t}{2} - \frac{1}{4}(e^{-2t}-1)x^2}$$

$$= U_{\tau_t} f(e^{-t}(x - 2u))e^{ze^{-t}x - u^2 - iuv}e^{-\frac{t}{2} - \frac{1}{4}(e^{-2t}-1)x^2}$$

$$= f(x - 2e^t u)e^{ze^{-t}x - u^2 - iuv}e^{-\frac{t}{2} - \frac{1}{4}(e^{-2t}-1)x^2}e^{\frac{t}{2} - \frac{1}{4}(e^{2t}-1)(e^{-t}x - 2u)^2}$$

$$= f(x - 2e^t u)e^{(e^t u + ie^{-t}v)x - e^{2t}u^2 - iuv}$$

$$= W_{e^t u + ie^{-t}v} f(x)$$

Les opérateurs U_{τ_t} forment un groupe à un paramètre, dont nous allons calculer le générateur.

Pour un polynome p en la variable x on a

$$\lim_{t \to 0} \frac{U_{\tau_t} p(x) - p(x)}{it} = \frac{1}{i}(x\frac{d}{dx}p(x) - (\frac{x^2}{2} + 1)p(x))$$

On en déduit que, sur le domaine des polynômes, ce générateur coïncide avec $\frac{1}{2}(PQ + QP)$.

Comme dans la proposition 4 on montre facilement que $\frac{1}{2}(PQ+QP)$ est essentielle-ment auto-adjoint sur le domaine formé des polynômes, et sa fermeture auto-adjointe a donc une loi dans l'état vide dont la transformée de Fourier est

$$< e^{\frac{it}{2}(PQ+QP)}\Omega, \Omega > = < U_{\tau_t}\Omega, \Omega >$$

$$= \int_{-\infty}^{+\infty} e^{\frac{t}{2} - \frac{1}{4}(e^{2t}-1)x^2} \frac{e^{-\frac{1}{2}x^2}}{\sqrt{2\pi}} dx$$

$$= \frac{1}{\sqrt{\operatorname{ch}t}}$$

Terminons ce paragraphe par le calcul de la loi de $\alpha P^2 + \beta Q^2$ ($\alpha, \beta \in \mathbb{R}_+^*$). Les opérateurs $P^2 + Q^2$ et $PQ + QP$ vérifient la relation

$$[P^2 + Q^2, PQ + QP] = 8i(P^2 - Q^2)$$

on en déduit que

$$e^{-i\frac{t}{2}(PQ+QP)}(P^2 + Q^2)e^{i\frac{t}{2}(PQ+QP)} = e^{-4t}P^2 + e^{4t}Q^2$$

(pour vérifier cette égalité, dériver deux fois par rapport à t et utiliser la relation de commutation précédente). On a donc $\alpha P^2 + \beta Q^2 = \sqrt{\alpha\beta}U_{\tau_{-t}}(P^2 + Q^2)U_{\tau_t}$ avec $\beta = \alpha e^{8t}$. Le calcul de la loi de $\alpha P^2 + \beta Q^2$ dans l'état vide est ramené à celui de loi de $P^2 + Q^2$ dans l'état $U_{\tau_t}\Omega$, or $P^2 + Q^2 = 4a^0 + 2Id$ donc il suffit de calculer la loi de a^0 dans l'état $U_{\tau_t}\Omega$. L'opérateur a^0 est donné par sa décomposition spectrale, le vecteur h_n étant vecteur propre de valeur propre n. La loi de a^0 dans l'état pur $U_{\tau_t}\Omega$ est donc la loi sur $\mathbb{N}$ donnant la masse $| < h_n, U_{\tau_t}\Omega > |^2$ au point n. On a

$$\sum_0^\infty \frac{u^n}{\sqrt{n!}} < h_n, U_{\tau_t}\Omega > = (2\pi)^{-\frac{1}{2}} \int_{-\infty}^{+\infty} e^{ux - \frac{u^2}{2} - \frac{x^2}{4}} e^{\frac{t}{2} - (e^{2t}-1)\frac{x^2}{4}} e^{-\frac{x^2}{2}} dx$$

$$= \frac{1}{\sqrt{\operatorname{ch}t}} e^{\frac{u^2}{2}\operatorname{th}t}$$

d'où l'on tire la valeur de la probabilité de n qui est 0 si n est impair, ct $\frac{1}{\operatorname{ch}t}(\frac{\operatorname{th}t}{2})^n \frac{n!}{(\frac{n}{2}!)^2}$ sinon.

Chapitre 4

Espaces de Fock

4.1 Puissances symétriques d'un espace de Hilbert

Nous avons construit dans le chapitre précédent des variables de Gauss et de Poisson à l'aide d'opérateurs de création, annihilation et nombre. Afin de pouvoir étudier non plus seulement des variables mais des processus, nous allons introduire la notion d'espace de Fock qui est un moyen naturel de considérer un produit de copies indépendantes de l'espace du chapitre 3. Nous allons retrouver dans ce contexte des vecteurs exponentiels, opérateurs d'annihilation, de création, de Weyl etc ....

On commence par des préliminaires algébriques au paragraphes 4.1 et 4.2, puis on aborde les probabilités avec les paragraphes 4.3 et suivants.

Soit K un espace de Hilbert, et $n \in \mathbb{N}^*$, le groupe $\mathfrak{S}_n$ des permutations de $[1, \ldots, n]$ agit sur $K^{\otimes n}$ par

$$\sigma(x_1 \otimes \ldots \otimes x_n) = x_{\sigma(1)} \otimes \ldots \otimes x_{\sigma(n)}$$

DÉFINITION 25 – On note $K^{\circ n}$ le sous-espace de $K^{\otimes n}$ des éléments invariants par $\mathfrak{S}_n$. C'est un sous-espace de Hilbert de $K^{\otimes n}$ appelé puissance symétrique n^e de K et l'opérateur $\mathfrak{s} = \frac{1}{n!} \sum_{\sigma \in \mathfrak{S}_n} \sigma$ est le projecteur orthogonal sur ce sous-espace.

Soient $h_1, \ldots, h_n, \ k_1, \ldots, k_n \in K$ on pose

$$k_1 \circ \ldots \circ k_n = \sqrt{n!}\, \mathfrak{s}(k_1 \otimes \ldots \otimes k_n) \tag{52}$$

On a

$$< h_1 \circ \ldots \circ h_n, k_1 \circ \ldots \circ k_n > = \sum_{\sigma \in \mathfrak{S}_n} \prod_{i=1}^{n} < h_i, k_{\sigma(i)} > \tag{53}$$

L'espace vectoriel engendré par les vecteurs $k^{\circ n}$ est $K^{\circ n}$.

DÉFINITION 26 – Soit K un espace de Hilbert on appelle espace de Fock construit sur K l'espace de Hilbert

$$\Gamma(K) = \bigoplus_{n=0}^{\infty} K^{\circ n}$$

où la somme directe est une somme Hilbertienne, et par convention $K^{\circ 0}$ est un espace de Hilbert de dimension 1, engendré par un vecteur unitaire noté Ω et appelé "vecteur vide".

On notera $\Gamma_0(K)$ la somme directe non complétée $\bigoplus_{n=0}^{\infty} K^{\circ n}$.

DÉFINITION 27 – Soit $k \in K$ on pose $k^{\circ 0} = \Omega$ et

$$\xi(k) = \sum_{n=0}^{\infty} \frac{k^{\circ n}}{n!}$$

Les vecteurs de la forme $\xi(k)$ pour $k \in K$ sont appelés vecteurs exponentiels, et on note $\Xi(K)$ l'espace vectoriel qu'ils engendrent.

Si $K = \mathbb{C}.u$ est de dimension 1 avec $|u| = 1$, alors en identifiant $u^{\circ n}$ avec $\sqrt{n!}e_n$ on obtient une isométrie entre $\Gamma(K)$ et l'espace de Hilbert H du chapitre 3. Les vecteurs exponentiels $\xi(\alpha)$ du chapitre 3 correspondent aux vecteurs $\xi(\alpha.u)$ ci-dessus.

PROPOSITION 9 – *Les vecteurs exponentiels forment un système libre et total dans $\Gamma(K)$, et on a*

$$\forall h, k \in K \quad < \xi(h), \xi(k) >= e^{<h,k>} \tag{54}$$

Démonstration
La formule résulte de la définition des vecteurs exponentiels. Pour tout $k \in K$ on a $\frac{d^n}{d\varepsilon^n}\xi(\varepsilon k)_{\varepsilon=0} = k^{\circ n}$, donc les vecteurs $k^{\circ n}$ sont dans l'adhérence de $\Xi(K)$, et ces vecteurs engendrent Γ_0 tout entier.
La liberté du système $\xi(k)$ se démontre de manière analogue à celle des vecteurs exponentiels du chapitre 3.

On déduit de la proposition 9 que ξ est une application continue, en effet, on a

$$|\xi(h) - \xi(k)|^2 = e^{<h,h>} - e^{<h,k>} - e^{<k,h>} + e^{<k,k>}$$

qui tend vers 0 quand $h \to k$. En particulier, si $S \subset K$ est une partie dense, l'espace $\Xi(S)$ engendré par les vecteurs exponentiels d'éléments de S est dense dans $\Gamma(K)$.

PROPOSITION 10 – *Soit $K = K_1 \oplus K_2$ une décomposition orthogonale, l'application*

$$\Xi(K) \to \Xi(K_1) \otimes \Xi(K_2)$$
$$\xi(k) \mapsto \xi(k_1) \otimes \xi(k_2)$$

(où $k = k_1 + k_2$ est la décomposition de k suivant $K_1 \oplus K_2$) se prolonge en une isométrie de $\Gamma(K)$ sur $\Gamma(K_1) \otimes \Gamma(K_2)$.

Démonstration
D'après la formule (54) on a

$$< \xi(k), \xi(l) > = e^{<k,l>} = e^{<k_1,l_1>+<k_2,l_2>}$$
$$= < \xi(k_1), \xi(l_1) > < \xi(k_2), \xi(l_2) > = < \xi(k_1) \otimes \xi(k_2), \xi(l_1) \otimes \xi(l_2) >$$

qui montre que l'application est une isométrie surjective. Comme $\Xi(K)$ est dense dans $\Gamma(K)$, on conclut.

En particulier, si $K_1 = \mathbb{C}k$ et $K_2 = (\mathbb{C}k)^\perp$ alors $\Gamma(K) \sim H \otimes \Gamma(K_2)$.
La proposition (10) s'étend immédiatement au cas d'une décomposition finie

$$\Gamma(\bigoplus_{i=1}^{n} K_i) \sim \bigotimes_{i=1}^{n} \Gamma(K_i)$$

Si on a une décomposition orthogonale infinie, $K = \bigoplus_{i=1}^{\infty} K_i$ l'isomorphisme

$$\Gamma(\bigoplus_{i=1}^{\infty} K_i) \sim \bigotimes_{i=1}^{\infty} \Gamma(K_i)$$

reste valable, le produit infini $\bigotimes_{i=1}^{\infty} \Gamma(K_i)$, étant défini de la façon suivante.
On considère l'espace vectoriel engendré par les vecteurs de la forme
$u_1 \otimes \ldots \otimes u_n \otimes \ldots$ où $u_n \in \Gamma(K_n)$ est égal à Ω_n (le vecteur vide de $\Gamma(K_n)$) pour
tous les n en dehors d'un nombre fini. C'est un espace préhilbertien pour le produit

$$< h_1 \otimes \ldots \otimes h_n \otimes \ldots, k_1 \otimes \ldots \otimes k_n \otimes \ldots >=< h_1, k_1 > \ldots < h_n, k_n > \ldots$$

(le produit converge car les termes valent 1 à partir d'un certain rang) et sa
complétion est l'espace de Hilbert $\bigotimes_{i=1}^{\infty} \Gamma(K_i)$.
Si (k_i) est une base orthonormale de K, on a l'isomorphisme $\Gamma(K) = \bigotimes_i \Gamma(\mathbb{C}k_i)$.

4.2 Opérateurs de création et d'annihilation

DÉFINITION 28 – Soit $k \in K$ on définit deux opérateurs de domaine $\Gamma_0(K)$ par:

$$a_k^+(k_1 \circ \ldots \circ k_n) = k \circ k_1 \circ \ldots \circ k_n \tag{55}$$

$$a_k^-(k_1 \circ \ldots \circ k_n) = \sum_{i=1}^{n} < k_i, k > k_1 \circ \ldots \circ \widehat{k_i} \circ \ldots \circ k_n \tag{56}$$

Les opérateurs a_k^+ sont appelés opérateurs de création, les a_k^- opérateurs d'annihilation.

a_k^+ dépend linéairement de k alors que a_k^- en dépend antilinéairement.
Ils vérifient la relation d'adjonction

$$\forall u, v \in \Gamma_0(K) \quad < a_k^+(u), v >=< u, a_k^-(v) > \tag{57}$$

en particulier, ils sont fermables et $a_k^+ \subset (a_k^-)^*$, $a_k^- \subset (a_k^+)^*$.
Le domaine $\Gamma_0(K)$ est invariant par a_k^+ et a_k^-.
On vérifie aisément que les vecteurs exponentiels sont dans le domaine de la
fermeture des opérateurs de création et d'annihilation et on a

$$a_k^+(\xi(h)) = \frac{d}{dt}\xi(h + tk)_{t=0} \tag{58}$$

$$a_k^-(\xi(h)) =< h, k > \xi(h) \tag{59}$$

Lorsque $K = \mathbb{C}.u$ est de dimension 1, en utilisant l'identification ci-dessus de $\Gamma(K)$
avec H, on voit que l'opérateur a_{zu}^+ correspond à za^+, et a_{zu}^- à $\bar{z}a^-$. Si $k \in K$, et
$|k| = 1$ dans la décomposition $\Gamma(K) \sim \Gamma(\mathbb{C}k) \otimes \Gamma((\mathbb{C}k)^{\perp})$, on a $a_k^{\pm} \sim a^{\pm} \otimes Id$. On en
déduit que l'opérateur $a_k^+ + a_k^-$ est essentiellement auto-adjoint sur $\Gamma_0(K)$ et qu'il
suit une loi normale centrée réduite dans l'état Ω. En particulier, si on dispose d'une
base orthonormée $(k_i)_{i \in I}$ de K, les opérateurs $a_{k_i}^+$, $a_{k_i}^-$, $i \in I$ forment une famille de
"copies indépendantes" (dans un sens analogue à celui du chapitre 1) des opérateurs
a^+, a^-.

Les relations de commutation (25) deviennent

$$[a_h^+, a_k^-] = - < h, k > Id \tag{60}$$

Si nous utilisons les opérateurs $Q_u = a_u^+ + a_u^-$ et $P_u = \frac{1}{i}(a_u^+ - a_u^-) = Q_{\frac{1}{i}u}$ ces
relations prennent la forme

$$[Q_u, Q_v] = -2i\Im m < u, v > Id$$

$$[P_u, P_v] = -2i\Im < u, v > Id \tag{61}$$

Terminons ce paragraphe en introduisant deux opérateurs que nous utiliserons dans le chapitre suivant.

DÉFINITION 29 – *L'opérateur gradient sur l'espace de Fock est la fermeture de l'opérateur de* $\Gamma_0(K)$ *dans* $K \otimes \Gamma(K)$ *défini par*

$$\forall k \in K \ \forall X, Y \in \Gamma_0(K) \quad < \nabla(X), k \otimes Y > = < a_k^-(X), Y > \tag{62}$$

La divergence δ *est l'adjoint de* ∇, *c'est donc un opérateur fermé de* $Dom(\delta) \subset K \otimes \Gamma(K)$ *dans* $\Gamma(K)$, *et on a*

$$\delta(k \otimes X) = a_k^+(X)$$

4.3 Seconde quantification

Soit A un opérateur sur K, l'opérateur $A^{\circ n}$ est défini sur le sous-espace $D(A)^{\circ n}$ de $K^{\circ n}$ engendré par les vecteurs $k_1 \circ \ldots \circ k_n$, $k_i \in D(A)$ par

$$A^{\circ n} k_1 \circ \ldots \circ k_n = A k_1 \circ \ldots \circ A k_n$$

Si A est borné, alors $A^{\circ n}$ est borné de norme $\|A\|^n$.

DÉFINITION 30 – *La seconde quantification de* A, *notée* $\Gamma(A)$ *est l'opérateur de domaine* $\bigoplus_{n \in \mathbb{N}} D(A)^{\circ n}$ *(somme algébrique) défini par* $\Gamma(A) = \bigoplus_{n \in \mathbb{N}} A^{\circ n}$.

Si A est borné de norme ≤ 1, alors $\Gamma(A)$ a une fermeture bornée, de norme ≤ 1, mais si $\|A\| > 1$, $\Gamma(A)$ n'est pas borné.

Si U est unitaire sur K alors $\Gamma(U)$ est unitaire, et si $(U_t)_{t \in \mathbb{R}}$ est un groupe à un paramètre de tels opérateurs, fortement continu, alors $(\Gamma(U_t))_{t \in \mathbb{R}}$ est un groupe unitaire fortement continu sur $\Gamma(K)$. Le générateur de ce groupe s'exprime à l'aide du générateur V de U_t. Si $k_1, \ldots k_n \in D(V)$ alors

$$\lim_{t \to 0} \frac{1}{it} (\Gamma(U_t) - Id)(k_1 \circ \ldots \circ k_n) = \sum_{j=1}^{n} k_1 \circ \ldots \circ V k_j \circ \ldots \circ k_n$$

par conséquent, $k_1 \circ \ldots \circ k_n$ est dans le domaine du générateur de $\Gamma(U_t)$.

DÉFINITION 31 – *Soit* A *un opérateur sur* K, *la seconde quantification différentielle de* A, *notée* $d\Gamma(A)$ *est l'opérateur de domaine* $\bigoplus_{n \in \mathbb{N}} D(A)^{\circ n}$ *(somme algébrique) donné par*

$$d\Gamma(A)(k_1 \circ \ldots \circ k_n) = \sum_{i=1}^{n} k_1 \circ \ldots \circ A k_i \circ \ldots \circ k_n \tag{63}$$

Lorsque V est auto-adjoint, $d\Gamma(V)$ est essentiellement auto-adjoint, en effet les vecteurs de la forme $k_1 \circ \ldots \circ k_n$ où les k_i sont analytiques pour V forment un ensemble total de vecteurs analytiques pour $d\Gamma(V)$. En particulier, si $(U_t)_{t \in \mathbb{R}}$ est un groupe unitaire fortement continu, le générateur du groupe $(\Gamma(U_t))_{t \in \mathbb{R}}$ est la fermeture de $d\Gamma(V)$. Dans la suite, si V est auto-adjoint, on désignera encore par $d\Gamma(V)$ la fermeture de $d\Gamma(V)$.

La seconde quantification différentielle de l'identité est appelée l'opérateur de nombre. C'est un opérateur auto-adjoint, le sous-espace $K^{\circ n}$ de $\Gamma(K)$ étant un espace propre de valeur propre n.

L'action des opérateurs de seconde quantification sur les vecteurs exponentiels a la forme suivante, soit $k \in D(A)$ (avec A auto-adjoint), alors k est dans le domaine de $d\Gamma(A)$ et

$$\Gamma(A)\xi(k) = \xi(Ak) \tag{64}$$

$$d\Gamma(A)\xi(k) = a_{Ak}^+\xi(k) \tag{65}$$

Il y a des relations de commutation entre les opérateurs de seconde quantification et les opérateurs de création et d'annihilation.

PROPOSITION 11 – *Soient U un opérateur unitaire, A un opérateur auto-adjoint borné et $u \in K$ on a les relations suivantes*

$$\Gamma(U)a_u^{\pm}\Gamma(U^{-1}) = a_{Uu}^{\pm} \tag{66}$$

$$[d\Gamma(A), a_u^{\pm}] = a_{Au}^{\pm} \tag{67}$$

sur le domaine $\Gamma_0(K)$.

Démonstration
Cela résulte d'un calcul simple utilisant les définitions de ces opérateurs.

On peut résumer les définitions des opérateurs de création, annihilation et seconde quantification en donnant les trois formules suivantes qui caractérisent leurs restrictions à l'espace $\Xi(K)$. Soient $u \in K$, A borné, pour tous $h, k \in K$ on a

$$< a_u^+\xi(k), \xi(h) > = < u, h >< \xi(k), \xi(h) > \tag{68}$$

$$< a_u^-\xi(k), \xi(h) > = < k, u >< \xi(k), \xi(h) > \tag{69}$$

$$< d\Gamma(A)\xi(k), \xi(h) > = < Ak, h >< \xi(k), \xi(h) > \tag{70}$$

4.4 Opérateurs de Weyl sur l'espace de Fock

Comme dans la proposition 4, on montre facilement que si A est borné, l'opérateur $d\Gamma(A) + a_u^+ + a_u^-$ est essentiellement auto-adjoint. Nous allons introduire des opérateurs qui réalisent les groupes unitaires engendrés par les extensions auto-adjointes de ces opérateurs, en les définissant sur les vecteurs exponentiels.

PROPOSITION 12 – *Soient U un opérateur unitaire sur K et $u \in K$ il existe un unique opérateur unitaire $W_{(U,u)}$ sur $\Gamma(K)$ tel que*

$$\forall k \in K \ W_{(U,u)}\xi(k) = \xi(Uk + u)e^{-<Uk,u>-\frac{1}{2}|u|^2} \tag{71}$$

de plus, on a la relation

$$W_{(U,u)}W_{(V,v)} = W_{(UV,u+Uv)}e^{-i\Im <u,Uv>} \tag{72}$$

Démonstration

Les vecteurs exponentiels forment un système libre, donc la formule (71) détermine un opérateur de domaine $\Xi(K)$. On vérifie facilement que c'est une isométrie, il suffit de voir que $< W_{(U,u)}\xi(k), W_{(U,u)}\xi(h) >=< \xi(k), \xi(h) > \quad \forall k, h \in K$. Comme $\Xi(K)$ est dense dans $\Gamma(K)$, on peut étendre $W_{(U,u)}$ par continuité en une isométrie de $\Gamma(K)$. La formule (72) se vérifie sur les vecteurs exponentiels, et de là sur tout $\Gamma(K)$ par continuité. Elle entraine en particulier que les $W_{(U,u)}$ sont inversibles donc unitaires.

Remarquons que $W_{U,0} = \Gamma(U)$. La formule (72) montre que les opérateurs de la forme $W_{(U,u)}e^{i\theta}$ où U est unitaire, $u \in K$ et $\theta \in \mathbf{R}$, forment un groupe d'opérateurs unitaires. En fait si nous faisons de $U(K) \times K \times \mathbf{R}$ (où $U(K)$ est le groupe unitaire de K) un groupe avec le produit

$$(U, u, t) \star (U', u', t') = (UU', u + Uu', t + t' + \Im m < u, Uu' >)$$

alors

$$(U, u, t) \mapsto W_{(U,u)}e^{it}$$

est une représentation unitaire de ce groupe.

Nous allons en extraire certains sous-groupes à un paramètre et calculer les lois de leurs générateurs dans l'état vide.

Le premier exemple est celui des groupes $(W_{(Id,tu)})_{t\in\mathbf{R}}$. Son générateur s'obtient en calculant

$$\lim_{t \to 0} \frac{W_{(Id,tu)})\xi(h) - \xi(h)}{it} = \frac{1}{i}(a_u^+ - a_u^-)\xi(h)$$

Le générateur de ce groupe est donc $P_u = \frac{1}{i}(a_u^+ - a_u^-)$. Dans l'état vide, la variable aléatoire correspondante suit une loi de Gauss centrée de variance $|u|^2$, comme on peut le vérifier en calculant $< W_{(Id,tu)}\xi(0), \xi(0) >= e^{-|u|^2\frac{t^2}{2}}$. D'après les relations de commutation (61) et (72), si on a un sous-espace vectoriel réel $\mathfrak{K}$ de K tel que $\forall u, v \in \mathfrak{K} < u, v >\in \mathbf{R}$ alors les opérateurs $(P_u)_{u\in\mathfrak{K}}$ (respectivement $(Q_u)_{u\in\mathfrak{K}}$) forment un processus classique dont la loi est celle d'un processus gaussien de covariance $\mathrm{cov}(P_u, P_v) =< u, v >$.

Considérons maintenant des opérateurs de seconde quantification. Soient $U_t = e^{itA}$ un groupe fortement continu à un paramètre d'opérateurs unitaires, et $u \in K$. On sait que $(\Gamma(U_t))_{t\in\mathbf{R}}$ est un groupe unitaire fortement continu ainsi que son conjugué par l'opérateur $W_{(Id,u)}$

$$\left(W_{(Id,-u)}\Gamma(U_t)W_{(Id,u)} = W_{(U_t, U_t u - u)}e^{i\Im m < u, U_t u>}\right)_{t\in\mathbf{R}}$$

On peut caculer son générateur si $u \in D(A)$, c'est l'opérateur dont la restriction à $\Xi(K)$ est $d\Gamma(A) + a_{Au}^+ + a_{Au}^- + < u, Au >$, qui est essentiellement auto-adjoint. Pour calculer la loi de cette variable dans l'état vide, calculons sa transformée de Fourier.

On trouve

$$< W_{(Id,-u)}\Gamma(U_t)W_{(Id,u)}\Omega, \Omega > = < \Gamma(U_t)W_{(Id,u)}\Omega, W_{(Id,u)}\Omega >$$
$$= < \xi(U_t u)\ e^{-\frac{1}{2}|u|^2}, \xi(u)\ e^{-\frac{1}{2}|u|^2} >$$
$$= e^{<U_t u, u> - <u,u>}$$
$$= e^{\int_{-\infty}^{+\infty}(e^{itx}-1)\ d\mu_u(x)}$$

où μ_u est la mesure spectrale de A associée au vecteur u.

D'après la formule de Lévy-Khintchine, cette loi est indéfiniment divisible, de mesure de Lévy μ_u.

Que devient le théorème_de Stone von Neumann sur l'espace de Fock? Il s'agit de savoir si des opérateurs $\widetilde{W_u}$, $u \in K$ satifaisant aux relations de commutation

$$\widetilde{W_u}\widetilde{W_v} = \widetilde{W_{u+v}}e^{-i\Im m <u,v>} \tag{73}$$

se mettent sous la forme d'une somme de copies des opérateurs $W_{(I,u)}$. La réponse est oui si la dimension de K est finie (et la démonstration est essentiellement la même que celle donnée au chapitre 3 voir [48]). Par contre lorsque la dimension de K est infinie, il existe de nombreuses représentations irréductibles des relations (73) qui ne sont pas unitairement équivalentes à la représentation sur l'espace de Fock. Des exemples en sont fournis par le théorème de Shale [79]. Soit $\tau : K \to K$ une application $\mathbf{R}$-linéaire qui préserve la forme symplectique $\Im m < u,v >$ sur K considéré comme espace réel, alors les opérateurs $W_{(I,\tau(u))}, u \in K$ forment une représentation des relations de commutation (73). Le théorème de Shale énonce que cette représentation est unitairement équivalente à la représentation $W_{(I,u)}$ (i.e. il existe U_τ unitaire tel que $U_\tau^* W_{(I,u)} U_\tau = W_{(I,\tau(u))}\ \forall u \in K$) si et seulement si $\tau^*\tau - Id$ est un opérateur de Hilbert-Schmidt. Ce théorème est à rapprocher du résultat classique suivant sur les mesures gaussiennes. Les lois de deux processus gaussiens linéaires $(X_u^1)_{u\in K}$ et $(X_u^2)_{u\in K}$ de covariances respectives q_1 et q_2 sont absolument continues si et seulement si $q_1 - q_2$ est de Hilbert-Schmidt (cf Neveu [57]). En fait la démonstration du théorème de Shale se ramène à cette propriété des mesures gaussiennes.

Le théorème de Shale est démontré dans [79] et [60].

4.5 Interprétations gaussienne et poissonnienne de l'espace de Fock

Lorsque K est de la forme $L_{\mathbf{C}}^2(E, \mathcal{E}, m)$ pour un espace mesurable $(E, \mathcal{E}, m)$ il y a deux interprétations probabilistes classiques de l'espace de Fock $\Gamma(K)$, comme espace L^2 d'un processus stochastique, que l'on va rappeler ci-dessous.

La première de ces interprétations est l'interprétation gaussienne. Donnons nous, sur un espace de probabilité $(\Omega, \mathcal{F}, P)$, une famille gaussienne X_B indexée par $\mathcal{E}$ de covariance $E(X_B X_C) = m(B \cap C)$. Le sous-espace fermé de $L^2(\Omega, \mathcal{F}, P)$ engendré par cette famille gaussienne est une famille gaussienne indexée par $L_{\mathbf{R}}^2(E, \mathcal{E}, m)$ de covariance $E(X_f X_g) = < f,g >$. Appelons $\mathcal{X}$ la tribu engendrée par les variables $X_B, B \in \mathcal{E}$.

PROPOSITION 13 – *L'application linéaire de* $\Xi(L^2_{\mathbb{C}}(E,\mathcal{E},m))$ *dans* $L^2_{\mathbb{C}}(\Omega,\mathcal{X},P)$ *qui à* $\xi(k)$ *associe la variable aléatoire*

$$e^{X_k - \frac{1}{2}|u|^2 + \frac{1}{2}|v|^2 + i<u,v>}$$

(avec $k = u + iv$*) se prolonge en un isomorphisme d'espaces de Hilbert de* $\Gamma(L^2_{\mathbb{C}}(E,\mathcal{E},m))$ *dans* $L^2_{\mathbb{C}}(\Omega,\mathcal{X},P)$.
Dans cet isomorphisme, l'opérateur de multiplication par la variable X_u *est transformé en l'opérateur* Q_u *sur* $\Gamma(L^2_{\mathbb{C}}(E,\mathcal{E},m))$.

Démonstration
On vérifie que l'application en question est une isométrie grâce à la formule gaussienne:

$$E(e^{X_u + iX_v}) = e^{\frac{1}{2}<u,u> - <v,v> + i<u,v> + i<v,u>}$$

L'application se prolonge donc de façon unique par continuité à $\Gamma(L^2_{\mathbb{C}}(E,\mathcal{E},m))$ en une isométrie. Pour voir qu'elle est surjective, il suffit de voir que les variables de la forme e^{X_k} forment une partie totale de $L^2_{\mathbb{C}}(\Omega,\mathcal{X},P)$ (cf Neveu [57]).
Pour montrer l'assertion relative à Q_u on peut procéder de la manière suivante. On a $K = \mathbb{C}u + (\mathbb{C}u)^\perp$ d'où $\Gamma(K) = \Gamma(\mathbb{C}u) \otimes \Gamma((\mathbb{C}u)^\perp)$. Pour les espaces L^2 cela correspond à la décomposition $L^2_{\mathbb{C}}(\Omega,\mathcal{X},P) = L^2_{\mathbb{C}}(\Omega,\mathcal{X}_u,P) \otimes L^2_{\mathbb{C}}(\Omega,\mathcal{X}u^\perp,P)$ ($\mathcal{X}_u$ est la tribu engendrée par X_u et $\mathcal{X}u^\perp$ celle engendrée par les variables gaussiennes orthogonales à u). Dans cette décomposition on a $Q_u = Q_u \otimes Id$, et on utilise alors l'identification du chapitre 3 entre l'espace H et l'espace L^2 de la mesure de Gauss.

Les images des espaces $K^{\circ n}$ dans l'isométrie de la proposition 13 sont les chaos du processus gaussien, qui sont définis inductivement comme étant les sous-espaces C_n tels que $C_0 \oplus \ldots \oplus C_n$ est le sous-espace fermé engendré par les polynômes (complexes) de degrés $\leq n$ en les variables X_f.

La seconde interprétation de l'espace de Fock fait intervenir un processus de Poisson ponctuel de mesure caractéristique m. C'est par définition une famille $(M(B))_{B \in \mathcal{E}}$ de variables aléatoires telle que pour toute famille $B_1, \ldots, B_n$ d'éléments disjoints de $\mathcal{E}$ de mesures $m(B_i)$ finies, les variables $M(B_1), \ldots, M(B_n)$ sont des variables de Poisson indépendantes de paramètres $m(B_i)$. On peut trouver une version de ce processus définie sur l'espace canonique $(\mathfrak{M},\mathcal{M})$ des mesures ponctuelles sur E (c'est à dire des mesures qui sont des sommes dénombrables de masses de Dirac). Un élément générique de cet espace est une mesure ponctuelle M. La loi P_m du processus ponctuel est donnée par sa fonction caractéristique

$$\int_{\mathfrak{M}} e^{\int_E f(x)dM(x)} dP_m(M) = e^{\int_E (e^{f(x)}-1)dm(x)} \tag{74}$$

pour toute fonction f, telle que $e^f - 1$ soit m-intégrable.

PROPOSITION 14 – *Il existe un isomorphisme de* $\Gamma(L^2_{\mathbb{C}}(E,\mathcal{E},m))$ *sur* $L^2_{\mathbb{C}}(\mathfrak{M},\mathcal{M},P_m)$ *tel que* $\xi(f)$ *pour* $f \in L^1 \cap L^2(E,\mathcal{E},m))$ *corresponde à la variable* $\eta(f)$ *donnée par la formule*

$$\eta(f)(M) = \prod_{x \in \text{supp}(M)} (1 + f(x))e^{-\int_E f(x)dm(x)} = e^{\int_E \text{Log}(1+f(x))dM(x) - \int_E f(x)dm(x)} \tag{75}$$

Démonstration

Les variables de la forme $\eta(f)$ forment une partie totale dans $L^2(\mathfrak{M}, \mathcal{M}, P_m)$ (cf Neveu [58]), donc il suffit de montrer la formule $E(\eta(f)\eta(\bar{g})) = e^{\int_E f(x)\bar{g}(x)dm(x)}$ pour deux fonctions f et g de $L^1 \cap L^2(E, m)$, or d'après la formule caractéristique (74) on a

$$E(\eta(f)\eta(\bar{g})) = e^{\int_E (e^{\mathrm{Log}(1+f(x))(1+\bar{g}(x))} - 1)\ dm(x) - \int_E f(x)+\bar{g}(x)dm(x)} = e^{\int_E f(x)\bar{g}(x)dm(x)}$$

Soit u une fonction réelle dans $L^2 \cap L^\infty(E, \mathcal{E}, m)$ la variable aléatoire correspondant à $a_u^+\xi(f)$ est

$$\lim_{t\to 0} \frac{\eta(f + tu) - \eta(f)}{t} = \left(\int_E \frac{\cdot u(x)}{1 + f(x)} dM(x) - \int_E u(x)dm(x) \right)\eta(f)$$

La fonction u peut également être considérée comme un opérateur de multiplication sur $L^2(E, \mathcal{E}, m)$, et alors on a

$$d\Gamma(u)\eta(f) = a_{uf}^+\eta(f) = \left[\int_E \frac{u(x)f(x)}{1 + f(x)} dM(x) - \int_E u(x)f(x)dm(x) \right]\eta(f)$$

On en déduit que l'opérateur $d\Gamma(u) + a_u^+ + a_u^- + \int_E u(x)dm(x)$, agissant sur l'espace de Fock s'interprète comme l'opérateur de multiplication par la variable aléatoire $\int_E u(x)dM(x)$. En particulier, prenant $u = 1_B$ avec $B \in \mathcal{E}$ et $m(B) < \infty$ on a $M(B) = d\Gamma(1_B) + a_{1_B}^+ + a_{1_B}^- + m(B)$.

4.6 Mouvement brownien et processus de Poisson

Nous allons maintenant nous intéresser plus particulièrement au cas où $K = L^2_{\mathbb{C}}(\mathbb{R}_+)$, et introduire ainsi les trois processus de Hudson et Parthasarathy. On définira au chapitre suivant des intégrales stochastiques par rapport à ces trois processus.

Il y a des identifications naturelles

$$L^2(\mathbb{R}_+)^{\circ n} \sim L^2(\mathbb{R}_+^n)_s \sim L^2(\Delta_n)$$

où $\Delta_n = \{(s_1, \ldots, s_n) | s_1 < \ldots < s_n\} \subset \mathbb{R}_+^n$ et $L^2_{\mathbb{C}}(\mathbb{R}_+^n)_s$ est le sous espace de $L^2_{\mathbb{C}}(\mathbb{R}_+^n)$ formé des fonction symétriques. Une fonction sur Δ_n est la restriction d'une unique fonction symétrique dans $\mathbb{R}_+^n$ ce qui donne la seconde identification, et pour la première, on peut identifier $h^{\circ n}$ avec la fonction $(s_1, \ldots, s_n) \mapsto \sqrt{n!}h(s_1)\ldots h(s_n)$. D'après le paragraphe précédent, l'espace de Fock peut s'interpréter comme l'espace L^2 d'un processus gaussien, indexé par $L^2(\mathbb{R}_+)$. Un tel processus gaussien peut se réaliser à l'aide d'un mouvement brownien B par $X_f = \int f(s)dB_s$. Le mouvement brownien se retrouve par la formule $B_t = \int 1_{[0,t]}(s)dB_s$. Sur l'espace de Fock $\Gamma(L^2(\mathbb{R}_+))$, $Q_t = Q_{1_{[0,t]}}$ est l'opérateur de multiplication par la variable B_t. Le processus $(Q_t)_{t\in\mathbb{R}_+}$ est donc un mouvement brownien réel dans l'état vide. On dispose en fait de toute une famille de mouvements browniens, qui ne commutent pas entre eux, qui sont $P_t = P_{1_{[0,t]}}$, et toutes les combinaisons linéaires $\cos\theta\, Q_t + \sin\theta\, P_t$.

Nous allons maintenant interpréter les notions que nous avons introduites sur l'espace de Fock abstrait, au moyen du mouvement brownien B_t.

Soit $f \in L^2_{\mathbf{R}}(\mathbf{R}_+)$ l'opérateur Q_f est l'opérateur de multiplication par la variable aléatoire $\int_0^\infty f(s)dB_s$. L'opérateur a_f^- est l'opérateur de dérivation de Malliavin (gradient) dans la direction f et $L^2_{\mathbf{R}}(\mathbf{R}_+)$ est l'espace de Cameron-Martin du mouvement brownien.

Les chaos du mouvement brownien s'expriment à l'aide d'intégrales stochastiques itérées. Le n^e chaos est formé des intégrales stochastiques

$$\int \ldots \int_{\Delta_n} h(s_1, \ldots, s_n)dB_{s_1} \ldots dB_{s_n}$$

Le vecteur exponentiel $\xi(f)$ correspond à la variable aléatoire

$$e^{\int_0^\infty f(s)dB_s - \frac{1}{2}\int_0^\infty f^2(s)ds}$$

qui est la valeur terminale de la martingale exponentielle

$$t \mapsto e^{\int_0^t f(s)dB_s - \frac{1}{2}\int_0^t f^2(s)ds}$$

Passons maintenant à l'interprétation poissonienne de l'espace de Fock de $L^2_{\mathbf{C}}(\mathbf{R}_+)$. Un processus de Poisson ponctuel sur $\mathbf{R}_+$ ayant la mesure de Lebesgue comme mesure caractéristique donne un processus de Poisson $(N_t)_{t \in \mathbf{R}_+} = M([0,t])$ d'intensité 1. L'espace L^2 du processus de Poisson se décompose en chaos, le n^e chaos étant l'espace des intégrales stochastiques multiples

$$\int \ldots \int_{\Delta_n} f(s_1, \ldots, s_n)d\tilde{N}_{s_1} \ldots d\tilde{N}_{s_n}$$

$\tilde{N}_t = N_t - t$ étant le processus de Poisson compensé. Comme pour le mouvement brownien on obtient ainsi l'isomorphisme avec l'espace de Fock.
Sur l'espace de Fock, l'opérateur correspondant à la multiplication par N_t est $d\Gamma(1_{[0,t]}) + a^+_{1_{[0,t]}} + a^-_{1_{[0,t]}} + t$.
La variable N_t suit bien une loi de Poisson de paramètre t.
L'exponentielle $\xi(f)$ correspond à la variable $e^{\int_0^\infty \text{Log}(1+f(s))dN_s - \int_0^\infty f(s)ds}$ qui est la valeur terminale de la martingale exponentielle $t \mapsto e^{\int_0^t \text{Log}(1+f(s))dN_s - \int_0^t f(s)ds}$.

Nous avons maintenant rencontré les trois processus d'opérateurs fondamentaux de Hudson et Parthasarathy. Ce sont les processus a^-, a^+, a^0 sur l'espace $\Gamma(L^2_{\mathbf{C}}(\mathbf{R}_+))$ définis par

$$a^0_t = d\Gamma(1_{[0,t]})$$
$$a^-_t = a^-_{1_{[0,t]}}$$
$$a^+_t = a^+_{1_{[0,t]}}$$

Les processus $a^+_t + a^-_t$ et $a^0_t + a^+_t + a^-_t + t$ sont des processus classiques, dans l'état vide leurs lois respectives sont celles d'un mouvement brownien, et d'un processus de Poisson de paramètre 1. La théorie classique du calcul stochastique par rapport à une martingale de carré intégrable montre que l'on dispose d'une notion d'intégrale stochastique par rapport à chacune des martingales $a^+_t + a^-_t$ et $a^0_t + a^+_t + a^-_t$. Nous allons généraliser ce calcul stochastique en développant une théorie de l'intégrale stochastique par rapport aux trois processus non-commutatifs a_t^0, a_t^+, a_t^-. Dans cette théorie, les processus que nous chercherons à intégrer seront des processus d'opérateurs, c'est à dire des fonctions sur $\mathbf{R}_+$ à valeurs dans les opérateurs sur $\Gamma(L^2(\mathbf{R}_+))$.

4.7 Martingales normales et représentations chaotiques

En dehors des interprétations brownienne et poissonienne de l'espace de Fock $\Gamma(L_{\mathbf{C}}^2(\mathbf{R}_+))$ il existe d'autres représentations probabilistes, obtenues en considérant des martingales normales.

DÉFINITION 32 – *Une martingale de carré intégrable M est dite normale si* $< M, M >_t = t.$
(Ici $< M, M >$ désigne le crochet de la martingale M, cf [21] *).*

Pour une telle martingale, On peut envoyer isométriquement l'espace de Fock $\Gamma(L_{\mathbf{C}}^2(\mathbf{R}_+))$ dans l'espace L^2 d'une telle martingale en posant

$$\phi(f_n) = \int_{\Delta_n} f_n(s_1, \ldots, s_n) dM_{s_1} \ldots dM_{s_n}$$

où $f_n \in L^2(\mathbf{R}_+)^{\circ n}$ est identifiée à une fonction de carré intégrable sur Δ_n.

DÉFINITION 33 – *On dit que M possède la propriété de représentation chaotique si l'image de ϕ est L^2 tout entier.*

Ainsi, dès que l'on dispose d'une martingale normale ayant la propriété de représentation chaotique, on a une interprétation probabiliste de l'espace de Fock. Alors que les propriétés de représentation chaotique du mouvement brownien et du processus de poisson compensé sont connues depuis longtemps, ce n'est que récemment que de nouvelles martingales possédant cette propriété ont été découvertes (voir [23]). Nous reviendrons plus en détails sur certains de ces exemples dans la suite.

Chapitre 5

Intégration stochastique non-commutative

5.1 Filtration et processus adaptés sur l'espace de Fock

L'espace de Hilbert $L^2_{\mathbf{C}}(\mathbf{R}_+)$ possède une structure de somme continue d'espaces de Hilbert, ce qui signifie que pour tout $t \in \mathbf{R}_+$ on a une décomposition orthogonale

$$L^2_{\mathbf{C}}(\mathbf{R}_+) = L^2_{\mathbf{C}}([0,t]) \oplus L^2_{\mathbf{C}}([t,\infty[)$$

et plus généralement si $t_1 < \ldots < t_n$

$$L^2_{\mathbf{C}}(\mathbf{R}_+) = L^2_{\mathbf{C}}([0,t_1]) \oplus L^2_{\mathbf{C}}([t_1,t_2]) \oplus \ldots \oplus L^2_{\mathbf{C}}([t_n,\infty[)$$

Au niveau des espaces de Fock, cela se traduit par une structure de *produit tensoriel continu*

$$\Gamma = \Gamma_{t_1]} \otimes \Gamma_{t_1,t_2} \otimes \ldots \otimes \Gamma_{[t_n}$$

où on a posé

$$\Gamma = \Gamma(L^2_{\mathbf{C}}(\mathbf{R}_+)), \quad \Gamma_{t]} = \Gamma(L^2_{\mathbf{C}}([0,t])), \quad \Gamma_{s,t} = \Gamma(L^2_{\mathbf{C}}([s,t])) \quad \Gamma_{[t} = \Gamma(L^2_{\mathbf{C}}([t,+\infty]))$$

De même, pour les espaces engendrés par les vecteurs exponentiels on a, avec des notations semblables

$$\Xi = \Xi_{t_1]} \otimes \Xi_{[t_1,t_2]} \otimes \ldots \otimes \Xi_{[t_n}$$

(ici le produit tensoriel est pris au sens algébrique).

Dans la suite nous utiliserons souvent les notations $h_{t]} = h1_{[0,t]}$ et $h_{[t} = h1_{[t,+\infty[}$

Lorsqu'on utilise l'interprétation brownienne (ou poissonienne) de l'espace de Fock, en notant $\Omega_{s]}$ le vecteur vide de $\Gamma_{s]}$, l'espace L^2 engendré par les variables aléatoires mesurables par rapport à la tribu $\mathcal{F}_{[s,t]}$, engendrée par les accroissements du mouvement brownien (resp du processus de poisson) entre les instants s et t, est $\Omega_{s]} \otimes \Gamma_{[s,t]} \otimes \Omega_{[t}$. L'opérateur de multiplication correspondant à une variable X, $\mathcal{F}_{[s,t]}$-mesurable est de la forme $Id \otimes T_X \otimes Id$ sur la décomposition $\Gamma_{s]} \otimes \Gamma_{[s,t]} \otimes \Gamma_{[t}$ de l'espace de Fock. Ces remarques justifient la définition, donnée plus bas, des processus d'opérateurs adaptés sur l'espace de Fock.

Nous allons imiter la théorie des intégrales stochastiques par rapport à une martingale et définir des intégrales stochastiques par rapport aux trois processus $(a_t^+)_{t \in \mathbf{R}_+}$, $(a_t^-)_{t \in \mathbf{R}_+}$, et $(a_t^0)_{t \in \mathbf{R}_+}$. Comme on l'a vu au chapitre précédent, le mouvement brownien et le processus de Poisson s'expriment par des combinaisons linéaires de ces processus, il est donc naturel d'essayer de définir l'intégration stochastique par rapport à chacun de ces trois processus séparément. Les processus que nous allons intégrer seront des familles d'opérateurs $(X_t)_{t \in \mathbf{R}_+}$ sur l'espace Γ, adaptés au sens donné plus bas à ce mot. Pour définir ces intégrales stochastiques non-commutatives, la méthode usuelle (celle utilisée par Hudson et Parthasarathy dans leur article fondamental [40]) consiste à commencer par définir les intégrales stochastiques de processus simples (i.e. étagés) par la formule à laquelle on s'attend, puis à étendre par

continuité l'application linéaire ainsi obtenue à une classe de processus plus générale, définie par une condition simple d'intégrabilité par raport à la mesure de Lebesgue, en utilisant une majoration de la norme de l'intégrale stochastique d'un processus simple. Cette méthode suppose que l'on sache approcher convenablement un processus "intégrable" par un processus simple, ce qui n'est pas si facile que cela. Dans ce cours, nous allons suivre une autre voie, inspirée de travaux de Belavkin et Lindsay. Nous allons commencer par définir les intégrales stochastiques de processus simples adaptés, puis obtenir une formule fondamentale qui caractérise ces intégrales stochastiques et qui a un sens pour des processus seulement intégrables. Ensuite nous utiliserons cette formule pour définir l'intégrale stochastique d'un processus adapté en général, la difficulté consistant à montrer l'existence de l'opérateur qui est caractérisé par la formule fondamentale. Pour cela nous allons utiliser de façon essentielle les propriétés de l'opérateur de divergence (définition 29).
Passons maintenant à la définition des processus adaptés.

DÉFINITION 34 – *Soit S un sous-espace dense de $L^2_{\mathbb{C}}(\mathbb{R}_+)$, stable par multiplication par les fonctions $1_{[s,t]}$ pour tous les $s < t \in \mathbb{R}_+$, une famille d'opérateurs $(X_t)_{t \in \mathbb{R}_+}$ sur Γ est dite S−adaptée (ou plus simplement adaptée si elle est S-adaptée pour un certain S) si pour tout $t \in \mathbb{R}_+$*
i) on a $D(X_t) \supset \Xi(S_{t]}) \otimes \Gamma_{[t}$, où $\Xi(S_{t]})$ est l'espace engendré par les vecteurs exponentiels de la forme $\xi(h1_{[0,t]}), h \in S$.
ii) X_t est de la forme $\widetilde{X_t} \otimes Id$ sur l'espace $\Xi(S_{t]}) \otimes \Gamma_{[t}$, (on note Φ_t l'ensemble des opérateurs de cette forme).

EXEMPLE – Les processus a^+, a^-, a^0 vérifient les relations $a^\varepsilon_t = \widetilde{a^\varepsilon_t} \otimes Id$ sur l'espace $\Xi_{t]} \otimes \Gamma_{[t}$, ils sont par conséquent $L^2_{\mathbb{C}}(\mathbb{R}_+)$-adaptés.

5.2 Intégration de processus simples

Dans ce paragraphe, on fixe S une partie de $L^2(\mathbb{R}_+)$ comme dans la définition 34.

DÉFINITION 35 – *Un processus adapté $(X_t)_{t \in \mathbb{R}_+}$ est dit simple s'il existe une suite de réels $0 = t_0 < t_1 < \ldots < t_n$ telle que $X_t = \sum_{j=0}^{n-1} 1_{[t_j, t_{j+1}[}(t) X^{(j)}$ où chaque $X^{(j)} \in \Phi_{t_j}$.*

Soient $(X_t)_{t \in \mathbb{R}_+}$ un processus adapté simple, $T_1 \leq T_2 \in \mathbb{R}_+ \cup \{\infty\}$ on définit trois opérateurs sur le domaine $\Xi(S_{T_2]}) \otimes \Gamma_{[T_2}$ ($\Xi(S)$ si $T_2 = \infty$) par la formule

$$\int_{T_1}^{T_2} X_s da^\varepsilon_s = \sum_{j=0}^{n-1} X^{(j)}(a^\varepsilon_{t_{j+1} \wedge T_2 \vee T_1} - a^\varepsilon_{t_j \wedge T_2 \vee T_1}) \tag{76}$$

où $\varepsilon = +, -$ ou 0.

DÉFINITION 36 – *Les opérateurs $\int_{T_1}^{T_2} X_s da^\varepsilon_s$ sont appelés les intégrales stochastiques de X par rapport aux trois processus a^ε, sur l'intervalle $[T_1, T_2]$.*

Vérifions immédiatement que la formule (76) définit bien un opérateur. Cette formule contenant des produits d'opérateurs non bornés, il faut vérifier qu'ils sont bien définis, or si $T_1 \leq s_1 \leq s_2 \leq T_2$ et X est dans Φ_{s_1}, sur la décomposition

$$\Xi(S_{T_2]}) \otimes \Gamma_{[T_2} = \Xi(S_{s_1]}) \otimes \Xi(S_{[s_1, T_2]}) \otimes \Gamma_{[T_2}$$

l'opérateur X est de la forme $\widetilde{X} \otimes Id \otimes Id$, et $a_{s_2}^\varepsilon - a_{s_1}^\varepsilon$ de la forme $Id \otimes (\widetilde{a_{s_2}^\varepsilon - a_{s_1}^\varepsilon}) \otimes Id$, par conséquent leur produit est bien défini sur ce domaine.

La proposition suivante va nous donner la caractérisation cherchée de l'intégrale stochastique d'un processus simple.

PROPOSITION 15 – Soit X un processus simple

$$\forall h, k \in S \quad < \int_{T_1}^{T_2} X_s da_s^\varepsilon(\xi(h)), \xi(k) > = \int_{T_1}^{T_2} < X_s(\xi(h)), \xi(k) > \kappa^\varepsilon(s)\, ds \quad (77)$$

avec $\kappa^- = h$, $\kappa^+ = \bar{k}$, $\kappa^0 = h\bar{k}$.

Démonstration

Par linéarité il suffit de le montrer pour un processus de la forme $X_t = X^1 1_{[t_1,t_2[}(t)$ avec $X^1 \in \Phi_{t_1}$, $T_1 \leq t_1 \leq t_2 \leq T_2$. Dans ce cas $\int_{T_1}^{T_2} X_s da_s^\varepsilon = X^1(a_{t_2}^\varepsilon - a_{t_1}^\varepsilon)$.

On a $X^1 = \widetilde{X^1} \otimes Id$ sur $\Xi(S_{t_1]}) \otimes \Xi_{[t_1}$ et $(a_{t_2}^\varepsilon - a_{t_1}^\varepsilon) = Id \otimes (\widetilde{a_{t_2}^\varepsilon - a_{t_1}^\varepsilon})$ sur le produit tensoriel $\Xi(S_{t_1]}) \otimes \Xi_{[t_1}$, donc $\int_{T_1}^{T_2} X_s da_s^\varepsilon = \widetilde{X^1} \otimes (\widetilde{a_{t_2}^\varepsilon - a_{t_1}^\varepsilon})$ et pour $h, k \in S$

$$< \int_{T_1}^{T_2} X_s da_s^\varepsilon(\xi(h)), \xi(k) > =$$

$$= < \widetilde{X^1}(\xi(h_{t_1]})) \otimes (\widetilde{a_{t_2}^\varepsilon - a_{t_1}^\varepsilon})\xi(h_{[t_1}), \xi(k) >$$

$$= < \widetilde{X^1}(\xi(h_{t_1]})), \xi(k_{t_1]}) > < (\widetilde{a_{t_2}^\varepsilon - a_{t_1}^\varepsilon})\xi(h_{[t_1}), \xi(k_{[t_1}) >$$

$$= < \widetilde{X^1}(\xi(h_{t_1]})), \xi(k_{t_1]}) > \int_{t_1}^{t_2} \kappa^\varepsilon(s)ds < \xi(h_{[t_1}), \xi(k_{[t_1}) >$$

d'après les formules (68)(69)(70)

$$= \int_{t_1}^{t_2} < X^1(\xi(h)), \xi(k) > \kappa^\varepsilon(s)\, ds$$

$$= \int_{T_1}^{T_2} < X_s(\xi(h)), \xi(k) > \kappa^\varepsilon(s)\, ds$$

Ce qui montre la proposition.

Comme $\Xi(S)$ est dense dans Γ la formule (77) détermine entièrement l'opérateur $\int_0^T X_s da_s^\varepsilon$ sur le domaine $\Xi(S)$.

Lorsque X est un processus qui n'est pas nécessairement simple ou adapté, si le membre de droite est bien défini pour tous les $h, k \in S$ (avec éventuellement $T_2 = +\infty$), comme les vecteurs exponentiels d'éléments de S forment une partie libre et totale dans $\Gamma(L_{\mathbf{C}}^2(\mathbf{R}_+))$ la formule (77) détermine au plus un opérateur $\int_{T_1}^{T_2} X_s da_s^\varepsilon$ sur le domaine $\Xi(S)$, et c'est cet opérateur que nous appellerons l'intégrale stochastique de X. Il reste à trouver un critère qui permet d'établir l'existence de cet opérateur.

Tout d'abord, nous allons énoncer quelques propriétés élémentaires de l'intégrale stochastique. Soit X (resp Y) un processus tel que

i) le membre de droite de (77) ait un sens pour tous les $h, k \in S$

ii) il existe un opérateur $\int_{T_1}^{T_2} X_s$ (resp Y_s)da_s^ε vérifiant (77)

alors les propriétés suivantes se vérifient immédiatement.

i) $\int_{T_1}^{T_2} X_s da_s^\varepsilon + \int_{T_1}^{T_2} Y_s da_s^\varepsilon = \int_{T_1}^{T_2} (X_s + Y_s) da_s^\varepsilon$

ii) $\int_{T_1}^{T_2} X_s da_s^\varepsilon + \int_{T_2}^{T_3} X_s da_s^\varepsilon = \int_{T_1}^{T_3} X_s da_s^\varepsilon$

iii) $\int_0^T X_s da_s^\varepsilon$ est de la forme $U \otimes Id$ sur $\Xi(S_{T]}) \otimes \Xi(S_{[T})$

iv) Posons $X_t^{S,T} = X_t$ pour $S < t < T$
$\qquad\qquad\quad = 0$ sinon

alors

$$\int_0^\infty X_s^{S,T} da_s^\varepsilon = \int_S^T X_s da_s^\varepsilon$$

Il reste à trouver des conditions sur X pour qu'un opérateur vérifiant (77) existe. Pour cela, nous allons nous limiter aux intégrales de 0 à ∞ (grâce à iv) et considérer les trois cas $\varepsilon = -, 0, +$ séparément.

Le plus simple est celui de $\varepsilon = -$ car

$$\int_0^\infty < X_s(\xi(h)), \xi(k) > \kappa^-(s)\ ds = < \int_0^\infty h(s) X_s(\xi(h)) ds, \xi(k) >$$

par conséquent on peut poser

$$\left(\int_0^\infty X_s da_s^-\right) \xi(h)) = \int_0^\infty h(s) X_s(\xi(h)) ds \qquad (78)$$

qui existe dès que la fonction de $\mathbf{R}_+$ dans Γ définie par $t \mapsto X_t(\xi(h))$ est faiblement mesurable, de carré intégrable.

Pour $\varepsilon = +$, on a

$$\int_0^\infty < X_s(\xi(h)), \xi(k) > \kappa^+(s)\ ds = \int_0^\infty < X_s(\xi(h)), k(s)\xi(k) > ds \qquad (79)$$

Supposons de nouveau que la fonction $t \mapsto X_t(\xi(h))$ soit faiblement mesurable de carré intégrable. On peut la considérer comme un élément de $L_{\mathbf{C}}^2(\mathbf{R}_+) \otimes \Gamma(L_{\mathbf{C}}^2(\mathbf{R}_+))$, noté $X(\xi(h))$. De même, la fonction $t \mapsto k(t)\xi(k)$ définit un élément de $L_{\mathbf{C}}^2(\mathbf{R}_+) \otimes \Gamma(L_{\mathbf{C}}^2(\mathbf{R}_+))$. La définition (29) de l'opérateur gradient montre que cet élément est égal à $\nabla(\xi(k))$.

L'expression (79) vaut donc $< X(\xi(h)), \nabla(\xi(k)) >$ le produit scalaire étant pris dans $L_{\mathbf{C}}^2(\mathbf{R}_+) \otimes \Gamma(L_{\mathbf{C}}^2(\mathbf{R}_+))$. On voit donc qu'il existe un opérateur $\int_0^\infty X_s da_s^+ (\xi(h))$ vérifiant (77) dès que $X(\xi(h))$ (i.e. la fonction $t \mapsto X_t(\xi(h))$ est dans le domaine de l'opérateur de divergence δ et alors, $\int_0^\infty X_s da_s^+ (\xi(h)) = \delta(X(\xi(h)))$.

Un calcul du même type pour $\varepsilon = 0$ montre que $\int_0^\infty X_s da_s^0(\xi(h))$ existe dès que la fonction $t \mapsto h(t) X_t(\xi(h))$ considérée comme élément de $L_{\mathbf{C}}^2(\mathbf{R}_+) \otimes \Gamma(L_{\mathbf{C}}^2(\mathbf{R}_+))$ est dans le domaine de δ, et alors $\int_0^\infty X_s da_s^0(\xi(h)) = \delta(hX(\xi(h)))$.

Nous voyons donc que le problème d'existence de l'opérateur $\int X_s da_s^\varepsilon$ pour $\varepsilon = +, 0$ se ramène à celui de montrer qu'une fonction est dans le domaine de l'opérateur divergence. Nous allons étudier plus en détail les opérateurs gradient et divergence dans le paragraphe suivant. On sait en fait, depuis Gaveau et Trauber [36] que l'opérateur de divergence sur l'espace de Fock peut s'interpréter comme une intégrale stochastique, l'intégrale de Skorokhod [80]. Je renvoie à l'article de Nualart [59] pour une discussion plus détaillée de l'intégrale de Skorokhod et de ses liens avec l'opérateur de divergence.

5.3 Divergence et gradient sur l'espace $\Gamma(L^2_{\mathbf{C}}(\mathbf{R}_+))$

Rappelons que dans l'espace de Fock $\Gamma(L^2_{\mathbf{C}}(\mathbf{R}_+))$, le produit tensoriel symétrique $L^2_{\mathbf{C}}(\mathbf{R}_+)^{\circ m}$ est isomorphe à l'espace $L^2_{\mathbf{C}}(\mathbf{R}^m_+)_s$ des fonctions symétriques de carré intégrable sur $\mathbf{R}^m_+$, le produit scalaire étant multiplié par $m!$. Un élément de $\Gamma(L^2_{\mathbf{C}}(\mathbf{R}_+))$ admet donc un développement convergent $a = \sum_{m=0}^{\infty} a_m$ où $a_m \in L^2_{\mathbf{C}}(\mathbf{R}^m_+)$ est une fonction symétrique.

Un élément u de $L^2_{\mathbf{C}}(\mathbf{R}_+) \otimes \Gamma(L^2_{\mathbf{C}}(\mathbf{R}_+))$ est donné par une fonction $t \mapsto u(t)$ de $\mathbf{R}_+$ dans Γ.

Si $t \in \mathbf{R}_+$ et $a = \sum_{m=0}^{\infty} a_m$ est un élément de $\Gamma(L^2_{\mathbf{C}}(\mathbf{R}_+))$ on pose $D_t a = \sum_{m=1}^{\infty} m a_m(.,t)$.

PROPOSITION 16 – *Soit* $a = \sum_{m=0}^{\infty} a_m \in \Gamma$, *alors* $a \in Dom(\nabla)$ *si et seulement si* $\int_0^{\infty} |D_t a|^2 dt < +\infty$ *et alors* $\nabla(a)$ *est la fonction* $t \mapsto D_t a \in L^2_{\mathbf{C}}(\mathbf{R}_+) \otimes \Gamma(L^2_{\mathbf{C}}(\mathbf{R}_+))$.

Démonstration

Il suffit de montrer que pour toute $h \in L^2(\mathbf{R}_+)$ $\int_0^{\infty} \bar{h}(s) D_s a ds = a_h^-(a)$, or par définition de a_h^- on a $a_h^-(a) = \sum_{m=0}^{\infty} m \int_0^{\infty} a_m(.,s) \bar{h}(s) ds$.

On note $\mathcal{L}^2$ la classe des fonctions $u \in L^2_{\mathbf{C}}(\mathbf{R}_+) \otimes \Gamma(L^2_{\mathbf{C}}(\mathbf{R}_+))$ telles que

$$\int_0^{\infty} \int_0^{\infty} |D_s u(t)|^2 ds dt < +\infty$$

Il est clair que l'espace $\mathcal{L}^2$ est dense dans $L^2_{\mathbf{C}}(\mathbf{R}_+) \otimes \Gamma(L^2_{\mathbf{C}}(\mathbf{R}_+))$ et que c'est un espace de Hilbert pour le produit scalaire

$$< u, v >_{\mathcal{L}^2} = < u, v > + \int_0^{\infty} \int_0^{\infty} < D_s u(t), D_s v(t) > ds dt$$

Pour cette norme, l'espace vectoriel engendré par les éléments de la forme $h \otimes a$ où $h \in L^2_{\mathbf{C}}(\mathbf{R}_+)$ et a est dans la somme algébrique des espaces $L^2_{\mathbf{C}}(\mathbf{R}_+)^{\circ k}$, est dense dans $\mathcal{L}^2$.

PROPOSITION 17 – *On a* $\mathcal{L}^2 \subset Dom(\delta)$ *et pour tous* $u, v \in \mathcal{L}^2$

$$< \delta(u), \delta(v) > = < u, v > + \int_0^{\infty} \int_0^{\infty} < D_s u(t), D_t v(s) > ds dt \qquad (80)$$

Démonstration

Montrons d'abord la formule de la proposition pour u et v de la forme $u = h \otimes a$, $v = k \otimes b$ avec $h, k \in L^2_{\mathbf{C}}(\mathbf{R}_+)$ et $a, b \in \Gamma$ appartenant à la somme algébrique des sous-espaces $L^2_{\mathbf{C}}(\mathbf{R}_+)^{\otimes n}$. On a $D_s u(t) = h(t) D_s(a)$, et $D_t v(s) = k(s) D_t(b)$ et on calcule

$$
\begin{aligned}
< \delta(u), \delta(v) > &= < a_h^+(a), a_k^+(b) > \\
&= < a_k^- a_h^+(a), b > \\
&= < a_h^+ a_k^-(a).b > + < h, k >< a, b > \qquad \text{d'après (60)} \\
&= < a_k^-(a), a_h^-(b) > + < u, v >
\end{aligned}
$$

$$= \int_0^\infty \int_0^\infty < D_s u(t), D_t v(s) > dsdt + < u, v >$$

Cette égalité s'étend par linéarité au sous-espace engendré par les fonctions u comme ci-dessus, et en particulier pour u dans cet espace on a

$$|\delta(u)|^2 = |u|^2 + \int_0^\infty \int_0^\infty < D_s u(t), D_t u(s) > dsdt \leq |u|_{\mathcal{L}^2}^2$$

On en déduit la proposition par densité de ce sous-espace dans $\mathcal{L}^2$ et par le fait que δ est un opérateur fermé.

Nous avons vu au paragraphe précédent que l'intégrale stochastique par rapport à a^+ ou a^0 d'un processus adapté X_t appliquée à un vecteur exponentiel, si elle existe, s'exprime à l'aide de la divergence.
Si X est adapté, on a $X_t(\xi(h)) = X_t(\xi(h_{t]})) \otimes \xi(h_{[t})$ dans $\Gamma_{t]} \otimes \Xi_{[t}$, nous allons donc nous intéresser à l'appartenance au domaine de la divergence d'éléments de $L_{\mathbb{C}}^2(\mathbb{R}_+) \otimes \Gamma(L_{\mathbb{C}}^2(\mathbb{R}_+))$ de la forme $t \mapsto u(t) = x(t) \otimes \xi(h_{[t})$ avec $x(t) \in \Gamma_{t]}$.
On a $\nabla \xi(h) = h \otimes \xi(h)$ on en déduit que pour une telle fonction on a $D_s u(t) = h(s)x(t) \otimes \xi(h_{[t})$ si $s > t$ et $D_s u(t) = D_s x(t) \otimes \xi(h_{[t})$ si $s < t$. On en déduit le corollaire suivant. (Rappelons que pour une fonction u sur $\mathbb{R}_+$ on note $u_{s]}$ la fonction égale à u sur $[0, t]$ et à 0 sur $]t, +\infty[$).

COROLLAIRE– On suppose que $(u(t) \equiv x(t) \otimes \xi(h_{[t}))_{t \in \mathbb{R}_+}$ *et* $(v(t) \equiv y(t) \otimes \xi(k_{[t}))_{t \in \mathbb{R}_+} \in L_{\mathbb{C}}^2(\mathbb{R}_+) \otimes \Gamma(L_{\mathbb{C}}^2(\mathbb{R}_+))$. *Alors* u, v *sont dans le domaine de* δ, *et de plus*

$$< \delta(u), \delta(v) > = \int_0^\infty < u(s), \delta(v_{s]}) > \bar{k}(s)ds + \int_0^\infty < \delta(u_{s]}), v(s) > h(s)ds$$
$$+ < u, v > \tag{81}$$

La divergence vérifie de plus l'inégalité suivante

$$|\delta(u)| \leq (|h| + \sqrt{|h|^2 + 1})(\int_0^\infty |u_s|^2)^{1/2} \tag{82}$$

Démonstration
Supposons tout d'abord que les fonctions x et y sont dans $\mathcal{L}^2$. Il s'ensuit que

$$\int_0^\infty \int_0^\infty |D_s u(t)|^2 dsdt = \int\int_{s<t} |D_s x(t)|^2 \otimes \xi(h_{[t})|^2 dsdt$$
$$+ \int\int_{s>t} |h(s)x(t) \otimes \xi(h_{[t})|^2 dsdt$$
$$\leq \int_0^\infty \int_0^\infty |D_s x(t)|^2 dsdt |\xi(h)|^2 + \int_0^\infty |u(t)|^2 dt \int_0^\infty |h(s)|^2 ds$$
$$< +\infty$$

donc la fonction u est dans $\mathcal{L}^2$ (ainsi que v pour la même raison). En appliquant la proposition précédente à ces deux fonctions on obtient

$$< \delta(u), \delta(v) > =< u, v > + \int_0^\infty \int_0^\infty < D_s u(t), D_t v(s) > ds dt$$

$$=< u, v > + \int_0^\infty \Big[\int_0^t < D_s u(t), v(s) > ds \Big] \bar{k}(t) dt$$

$$+ \int_0^\infty \Big[\int_0^s h(t) < u(t), D_t v(s) > dt \Big] ds$$

$$=< u, v > + \int_0^\infty < u(t), \delta(v_{t]}) > \bar{k}(t) dt$$

$$+ \int_0^\infty h(s) < \delta(u_{s]}), v(s) > ds$$

d'après la proposition 16, car δ est l'adjoint de ∇

ce qui montre la première identité dans le cas où $x, y \in \mathcal{L}^2$.

Nous allons en déduire l'inégalité (82) sur la norme de $\delta(u)$. D'après (81)

$$\left| \delta(u_{t]}) \right|^2 = \int_0^t |u(s)|^2 ds + 2 \int_0^t \Re e \left(< u(s), h(s) \delta(u_{s]}) > \right) ds$$

d'où, en posant $\phi(t) = \sup_{s \le t} |\delta(u_{s]})|$ et en appliquant l'inégalité de Cauchy-Schwarz

$$\phi(t)^2 \le 2 \Big(\int_0^t |h(s)|^2 ds \Big)^{\frac{1}{2}} \Big(\int_0^t | < u(s), \delta(u_{s]}) > |^2 ds \Big)^{1/2} + \int_0^t |u(s)|^2 ds$$

$$\le 2\phi(t) |h| \Big(\int_0^t |u(s)|^2 ds \Big)^{\frac{1}{2}} + \int_0^t |u(s)|^2 ds$$

de cette dernière inégalité on tire

$$\left| \delta(u_{t]}) \right| \le \left(|h| + \sqrt{|h|^2 + 1} \right) \Big(\int_0^t |u(s)|^2 ds \Big)^{1/2}$$

Soit maintenant une fonction $u(t) \equiv x(t) \otimes \xi(h_{[t})$ telle que pour tout t on ait $x(t) \in \Gamma_{t]}$ et $\int_0^\infty |u(t)|^2 dt < \infty$, on peut trouver une suite de fonctions $u^{(n)}$ de la forme $u^{(n)}(t) = x^{(n)}(t) \otimes \xi(h_{[t})$ telle que $x^{(n)} \in \mathcal{L}^2$ et $u^{(n)} \to_{n \to \infty} u$ dans $L^2_{\mathbf{C}}(\mathbf{R}_+) \otimes \Gamma(L^2_{\mathbf{C}}(\mathbf{R}_+))$ (par exemple on peut prendre pour $x^{(n)}(t)$ la projection de $x(t)$ sur l'espace engendré par les $n + 1$ premiers chaos $\sum_{m \le n} L^2_{\mathbf{C}}(\mathbf{R}_+)^{\circ m}$). En utilisant l'inégalité (82) pour les $u^{(n)}$ et le fait que δ est un opérateur fermé, on voit que $u \in Dom(\delta)$ et les formules (81) et (82) s'obtiennent par passage à la limite.

5.4 Intégrale stochastique non-commutative et formule d'Itô

En appliquant le corollaire précédent à une fonction de la forme $u_t = X_t \xi(h)$, où X est un processus adapté, on obtient le résultat suivant, dû à Hudson et Parthasarathy.

THÉORÈME 7 – Soient S une partie de $L_{\mathbf{C}}^2(\mathbf{R}_+)$ comme dans la définition (34), et X un processus S-adapté tel que $(t \mapsto \alpha^\varepsilon(t) X_t(\xi(h))) \in L_{\mathbf{C}}^2(\mathbf{R}_+) \otimes \Gamma(L_{\mathbf{C}}^2(\mathbf{R}_+))$ (où $\alpha^\pm(t) = 1, \alpha^0(t) = h(t)$) pour tout $h \in S$. Il existe un unique opérateur $\int_0^\infty X_s da_s^\varepsilon$ défini sur $\Xi(S)$ tel que pour tous $h, k \in S$ on ait

$$< \int_0^\infty X_s da_s^\varepsilon(\xi(h)), \xi(k) > = \int_0^\infty < X_s \xi(h), \xi(k) > \kappa^\varepsilon(s) ds$$

avec $\kappa^+ = \bar{k}$, $\kappa^- = h$, $\kappa^0 = h\bar{k}$.

De plus on a

$$\int_0^\infty X_s da_s^-(\xi(h)) = \int_0^\infty h(s) X_s(\xi(h)) ds \qquad (83)$$

$$\int_0^\infty X_s da_s^0(\xi(h)) = \delta(hX(\xi(h))) \qquad (84)$$

$$\int_0^\infty X_s da_s^+(\xi(h)) = \delta(X(\xi(h))) \qquad (85)$$

Les intégrales stochastique vérifient les inégalités de norme

$$\left| \int_0^\infty X_s da_s^- \xi(h) \right| \leq |h| \left(\int_0^\infty |X_s \xi(h)|^2 ds \right)^{\frac{1}{2}}$$

$$\left| \int_0^\infty X_s da_s^0 \xi(h) \right| \leq (|h| + \sqrt{|h|^2 + 1}) \left(\int_0^\infty |h(s) X_s \xi(h)|^2 ds \right)^{\frac{1}{2}} \qquad (86)$$

$$\left| \int_0^\infty X_s da_s^+ \xi(h) \right| \leq (|h| + \sqrt{|h|^2 + 1}) \left(\int_0^\infty |X_s \xi(h)|^2 ds \right)^{\frac{1}{2}}$$

REMARQUES

1) Intégrales itérées.

Le processus $t \mapsto \int_0^t X_s da_s^\varepsilon$ est adapté, et d'autre part, les inégalités (86) montrent que pour $h \in S$, $t \mapsto \int_0^t X_s da_s^\varepsilon(\xi(h))$ est une application continue à valeurs dans l'espace de Fock. Il s'ensuit que l'on peut calculer l'intégrale stochastique itérée $\int_0^t (\int_0^s X_u da_u^\varepsilon) da_s^\eta$, où $\eta \in \{-, 0, +\}$ ainsi que des intégrales itérées d'ordre supérieur. Ceci sera très utile quand on cherchera à résoudre des équations différentielles stochastiques par la méthode de Picard au chapitre 6.

2) Adjoint d'une intégrale stochastique.

Soit X un processus S-adapté vérifiant les hypothèses du théorème (7) (X est dit ε-intégrable), supposons qu'il existe un processus S-adapté $*\varepsilon$-intégrable (où $*- = +, *0 = 0, *+ = -$) $X^\sharp$ tel que

$$\forall t \geq 0 \ \forall h, k \in S \quad < X_t \xi(h), \xi(k) > = < \xi(h), X_t^\sharp \xi(k) >$$

(c'est à dire que $X_t^\sharp \subset X_t^*$) la formule (77) montre que

$$\forall h, k \in S \quad < \int_0^\infty X_s da_s^\varepsilon(\xi(h)), \xi(k) > = < \xi(h), \int_0^\infty X_s^\sharp da_s^{*\varepsilon}(\xi(k)) > \qquad (87)$$

Autrement dit on a $\int_0^\infty X_s^\sharp da_s^{*\varepsilon} \subset \left(\int_0^\infty X_s da_s^\varepsilon \right)^*$.

Nous avons ainsi réussi à définir l'intégrale stochastique de processus adaptés à l'aide de la formule (77). L'intégrale stochastique que nous obtenons est un opérateur défini sur un domaine exponentiel dense $\Xi(S)$. La première chose que nous allons faire est de vérifier que les intégrales stochastiques que nous venons de définir coïncident avec celles de la théorie classique. On va donc se placer dans l'interprétation brownienne de l'espace de Fock, et considérer le processus classique $a_t^+ + a_t^-$ comme un mouvement brownien B_t avec sa filtration $\mathcal{F}_t$. Soit X_t un processus adapté tel que chaque X_t est un opérateur de multiplication par une variable aléatoire (notée x_t) $\mathcal{F}_t$-mesurable. On va vérifier que l'intégrale stochastique non-commutative $\int_0^\infty X_s(da_s^+ + da_s^-)$ est la restriction au domaine $\Xi(S)$ de l'opérateur de multiplication par la variable aléatoire $\int_0^\infty x_s dB_s$. Pour cela, on utilise la formule fondamentale (77) et on calcule pour $h, k \in S$

$$E[\int_0^\infty x_s dB_s \xi(h)\overline{\xi(k)}]$$

Rappelons que dans l'interprétation brownienne $\xi(h)$ s'identifie à la variable aléatoire $\exp(\int_0^\infty h(s)dB_s - \frac{1}{2}\int_0^\infty h(s)^2 ds)$.

En particulier, $\xi(h)\overline{\xi(k)} = \xi(h + \bar{k})e^{<h,k>}$ le produit étant celui des variables aléatoires. D'autre part

$$\exp(\int_0^\infty h(s)dB_s - \frac{1}{2}\int_0^\infty h(s)^2 ds) = 1 + \int_0^\infty h(t)\xi(h_{t]})dB_t$$

on a donc, en utilisant la formule d'Ito

$$\int_0^\infty x_s dB_s \xi(h)\overline{\xi(k)} = e^{<h,k>}\int_0^\infty x_t\Big(\int_0^t \xi(h_{s]} + \bar{k}_{s]})(h(s) + \bar{k}(s))dB_s\Big)dB_t$$

$$+ e^{<h,k>}\int_0^\infty \Big(\int_0^t x_s dB_s\Big)\xi(h_{t]} + \bar{k}_{t]})(h(t) + \bar{k}(t))dB_t$$

$$+ e^{<h,k>}\int_0^\infty x_t\xi(h_{t]} + \bar{k}_{t]})(h(t) + \bar{k}(t))dt$$

$$+ \int_0^\infty x_s dB_s$$

En prenant l'espérance on trouve

$$E\Big[\int_0^\infty x_s dB_s \xi(h)\overline{\xi(k)}\Big] = E\Big[\int_0^\infty x_t\xi(h_{t]} + \bar{k}_{t]})(h(t) + \bar{k}(t))dt\Big]e^{<h,k>}$$

$$= \int_0^\infty < X_t\xi(h), \xi(k) > (h(t) + \bar{k}(t))dt$$

$$=< \int_0^\infty X_t(da_t^- + da_t^+)\xi(h), \xi(k) >$$

qui est bien ce qu'il fallait vérifier.

On démontre par un calcul analogue (bien qu'un peu plus fastidieux) que dans l'interprétation poissonnienne de l'espace de Fock, l'intégrale stochastique par rapport au processus de Poisson compensé s'obtient en prenant l'intégrale par rapport à $a^0 + a^- + a^+$.

Nous allons maintenant aborder l'un des aspects les plus remarquables du travail de Hudson et Parthasarathy qui est la généralisation de la formule d'Itô pour le mouvement brownien et le processus de Poisson aux intégrales stochastiques non-commutatives.

Une des façons possibles d'énoncer la formule d'Itô classique consiste à écrire le produit de deux intégrales stochastiques comme somme de trois termes, les deux premiers étant des intégrales stochastiques, et le troisième, le terme de correction d'Itô faisant intervenir le crochet des deux intégrales stochastiques. Plus précisément, si M et N sont deux martingales, H et K deux processus prévisibles intégrables, on a

$$\int_0^t H_s dM_s \int_0^t K_s dN_s = \int_0^t H_s \left(\int_0^s K_u dN_u \right) dM_s + \int_0^t \left(\int_0^s H_u dM_u \right) K_s dN_s$$
$$+ \int_0^t H_s K_s dM_s \cdot dN_s$$

Dans cette dernière expression on a noté $dM \cdot dN$ le crochet $d[M, N]$ afin d'éviter la confusion avec les commutateurs.

La formule d'Itô pour les intégrales stochastiques non-commutatives s'écrira formellement de la même façon

$$\int_0^t X_s da_s^\varepsilon \int_0^t Y_s da_s^\eta = \int_0^t X_s \left(\int_0^s Y_u da_u^\eta \right) da_s^\varepsilon + \int_0^t \left(\int_0^s X_u da_u^\varepsilon \right) Y_s da_s^\eta$$
$$+ \int_0^t X_s Y_s da_s^\varepsilon \cdot da_s^\eta$$

où $da_s^\varepsilon \cdot da_s^\eta$ est le "crochet" des processus a^ε et a^η. Nous calculerons ces crochets et nous verrons que suivant les valeurs de ε et η, le produit $da_s^\varepsilon \cdot da_s^\eta$ est égal à ds ou da_s^τ pour $\tau \in \{-, 0, +\}$.

L'intégrale stochastique étant un opérateur défini sur un domaine $\Xi(S)$, et ce domaine n'ayant aucune raison d'être stable, on ne peut pas multiplier (i.e. composer) deux intégrales stochastiques si bien que la formule ci-dessus n'a pas de sens en général. Pour tourner cette difficulté, on utilise l'identité

$$< \int_0^t X_s^* da_s^{*\varepsilon} \int_0^t Y_s da_s^\eta (\xi(h)), \xi(k) > = < \int_0^t Y_s da_s^\eta \xi(h), \int_0^t X_s da_s^\varepsilon \xi(k) >$$

justifiée formellement par la remarque 2 ci-dessus, et on en déduit que l'égalité formelle

$$< \int_0^t X_s^* da_s^{*\varepsilon} \int_0^t Y_s da_s^\eta \xi(h), \xi(k) > = \int_0^t < X_s^* \left(\int_0^s Y_u da_u^\eta \right) da_s^{*\varepsilon} \xi(h), \xi(k) >$$
$$+ \int_0^t < \left(\int_0^s X_u^* da_u^{*\varepsilon} \right) Y_s da_s^\eta \xi(h), \xi(k) >$$
$$+ < \int_0^t X_s^* Y_s da_s^{\cdot\varepsilon} \cdot da_s^\eta \xi(h), \xi(k) >$$

est équivalente à l'égalité suivante qui, elle, a un sens pour $h, k \in S$

$$< \int_0^t Y_s da_s^\eta \xi(h), \int_0^t X_s da_s^\varepsilon \xi(k) > = \int_0^t < \int_0^s Y_u da_u^\eta \xi(h), X_s \xi(k) > \kappa^{*\varepsilon}(s) ds$$

$$+ \int_0^t < Y_s \xi(h), \int_0^s X_u da_u^\varepsilon \xi(k) > \kappa^\eta(s) ds$$

$$+ \int_0^t < Y_s \xi(h), X_s \xi(k) > \lambda^{\varepsilon \eta}(s) ds \quad (88)$$

où κ est défini comme dans (77) et $\lambda^{\varepsilon \eta} = \kappa^\tau$ si $da_s^{*\varepsilon} \cdot da_s^\eta = da_s^\tau$.
C'est cette dernière égalité que nous allons démontrer en calculant les valeurs des produits $da_s^\varepsilon \cdot da_s^\eta$ au moyen de la propriété fondamentale de la divergence (la formule (81)).
Commençons par un cas simple, $\varepsilon = \eta = -$ et calculons

$$< \int_0^t X_s da_s^- \xi(h), \int_0^t Y_s da_s^- \xi(k) > = \int_0^t \int_0^t h(s) \bar{k}(r) < X_s \xi(h), Y_r \xi(k) > ds dr$$

$$= \int_0^t < \int_0^s X_u da_u^- \xi(h), Y_s \xi(k) > \bar{k}(s) ds$$

$$+ \int_0^t h(s) < X_s \xi(h), \int_0^s Y_u da_u^- \xi(k) > ds$$

En rapprochant cette identité de (88) on obtient $da_s^+ \cdot da_s^- = 0$.
Les autres produits $da_s^\varepsilon . da_s^\eta$ se traitent de la même façon. Je vais expliciter le calcul de $da_s^- \cdot da_s^0$ et $da_s^- \cdot da_s^+$ et laisser les autres au lecteur.
D'après (81) on a

$$< \int_0^t X_s da_s^0 \xi(h), \int_0^t Y_s da_s^+ \xi(k) > = < \delta(h X(\xi(h))_{t]}), \delta(Y(\xi(k))_{t]}) >$$

$$= \int_0^t < \delta(h X(\xi(h))_{s]}), Y_s \xi(k) > h(s) ds$$

$$+ \int_0^t h(s) < X_s \xi(h), \delta(Y(\xi(k))_{s]}) > \bar{k}(s) ds$$

$$+ \int_0^t < h(s) X_s \xi(h), Y_s \xi(k) > ds$$

$$= \int_0^t < \int_0^s X_u da_u^0 \xi(h), Y_s \xi(k) > h(s) ds$$

$$+ \int_0^t h(s) < X_s \xi(h), \int_0^s Y_u da_u^+ \xi(k) > \bar{k}(s) ds$$

$$+ \int_0^t < h(s) X_s \xi(h), Y_s \xi(k) > ds$$

Par comparaison avec (88) nous obtenons donc les formules $da_s^- \cdot da_s^0 = da_s^-$ et en passant à l'adjoint $da_s^0 \cdot da_s^+ = da_s^+$.

Le calcul de $da^- \cdot da^+$ procède comme suit

$$< \int_0^t X_s da_s^+ \xi(h), \int_0^t Y_s da_s^+ \xi(k) >=< \delta(X(\xi(h))_{t]}), \delta(Y(\xi(k))_{t]}) >$$

$$= \int_0^t < \delta(X(\xi(h))_{s]}), Y_s \xi(k) > h(s)ds$$

$$+ \int_0^t < X_s \xi(h), \delta(Y(\xi(k))_{s]}) > \bar{k}(s)ds$$

$$+ \int_0^t < X_s \xi(h), Y_s \xi(k) > ds$$

$$= \int_0^t < \int_0^s X_u da_u^+ \xi(h), Y_s \xi(k) > h(s)ds$$

$$+ \int_0^t < X_s \xi(h), \int_0^s Y_u da_u^+ \xi(k) > \bar{k}(s)ds$$

$$+ \int_0^t < X_s \xi(h), Y_s \xi(k) > ds$$

et on obtient $da_s^- \cdot da_s^+ = ds$.
Finalement on trouve que la "table de multiplication d'Itô" s'écrit

$\cdot$	da_t^-	da_t^0	da_t^+
da_t^-	0	da_t^-	dt
da_t^0	0	da_t^0	da_t^+
da_t^+	0	0	0

On peut à partir de cette table retrouver les formules d'Itô pour le mouvement brownien et le processus de Poisson. Le mouvement brownien se réalise sur l'espace de Fock au moyen des opérateurs $B_t = a_t^+ + a_t^-$, et on a

$$d < B, B >_t = dB_t \cdot dB_t = (da_t^+ + da_t^-)^2 = dt$$

(ici $< B, B >$ est le crochet de la martingale B).
De même, le processus de Poisson compensé est $N_t = a_t^0 + a_t^+ + a_t^-$ et son crochet droit se calcule de la manière suivante

$$d[N, N]_t = dN_t^2 = (da_t^0 + da_t^+ + da_t^-)^2 = da_t^0 + da_t^+ + da_t^- + dt = dN_t + dt$$

On peut également incorporer à la table d'Itô un autre processus en plus des a^ε, qui est simplement le temps t. Il lui correspond les "intégrales stochastiques" $\int X_s ds$ qui sont des intégrales d'opérateurs au sens ordinaire. Il est immédiat de vérifier que la table de multiplication d'Itô se prolonge pour contenir dt, les produits $dt \cdot da_t^\varepsilon$ et $da_t^\varepsilon \cdot dt$ étant tous nuls.

A ce stade, se pose la question de savoir si on peut donner un sens directement à la formule

$$\int_0^t X_s da_s^\varepsilon \int_0^t Y_s da_s^\eta = \int_0^t X_s \left(\int_0^s Y_u da_u^\eta \right) da_s^\varepsilon + \int_0^t \left(\int_0^s X_u da_u^\varepsilon \right) Y_s da_s^\eta$$
$$+ \int_0^t X_s Y_s da_s^\varepsilon \cdot da_s^\eta$$

Par exemple, si le membre de droite de l'égalité est bien défini, on peut chercher à quelles conditions les opérateurs $\int_0^t X_s da_s^\varepsilon$ et $\int_0^t Y_s da_s^\eta$ sont prolongeables de façon à être composables.

La théorie des opérateurs de Maassen fournit une classe d'opérateurs qui possèdent un domaine dense et stable, ce qui fait qu'ils sont composables sur ce domaine. Ces opérateurs sont des sommes d'intégrales stochastiques multiples du type

$$\int f(s_1, \ldots s_n) da_{s_1}^{\varepsilon_1} \ldots da_{s_n}^{\varepsilon_n}$$

où f est une fonction déterministe. Sous des conditions convenable de croissance des fonctions f, ces opérateurs forment une classe stable par intégrale stochastique avec laquelle on peut exprimer directement la formule d'Itô. Je renvoie à [50] pour plus de détails sur ces opérateurs.

Pour terminer ce paragraphe, nous allons faire quelques calculs formels avec la table de multiplication d'Itô, pour arriver à un énoncé remarquable dû à Belavkin [6]. Soient X et Y deux processus adaptés qui s'écrivent sous la forme

$$X_t = \int_0^t H_s^- da_s^- + \int_0^t H_s^0 da_s^0 + \int_0^t H_s^+ da_s^+ + \int_0^t H_s ds$$

$$Y_t = \int_0^t K_s^- da_s^- + \int_0^t K_s^0 da_s^0 + \int_0^t K_s^+ da_s^+ + \int_0^t K_s ds$$

la formule d'Itô peut s'écrire de façon condensée

$$d(X_t Y_t) = X_t dY_t + (dX_t) Y_t + dX_t \cdot dY_t$$

avec $dX_t = \sum H_t^\varepsilon da_t^\varepsilon$, $dY_t = \sum K_t^\varepsilon da_t^\varepsilon$, les symboles da_t^ε commutant aux opérateurs K et H.

Introduisons les matrices à coefficients opérateurs suivantes

$$\mathsf{X}_t = \begin{pmatrix} X_t & 0 & 0 \\ 0 & X_t & 0 \\ 0 & 0 & X_t \end{pmatrix} \quad \mathsf{H}_t = \begin{pmatrix} 0 & H_t^- & H_t \\ 0 & H_t^0 & H_t^+ \\ 0 & 0 & 0 \end{pmatrix}$$

$$\mathsf{Y}_t = \begin{pmatrix} Y_t & 0 & 0 \\ 0 & Y_t & 0 \\ 0 & 0 & Y_t \end{pmatrix} \quad \mathsf{K}_t = \begin{pmatrix} 0 & K_t^- & K_t \\ 0 & K_t^0 & K_t^+ \\ 0 & 0 & 0 \end{pmatrix}$$

ainsi que la matrice

$$dA_t = \begin{pmatrix} 0 & 0 & 0 \\ da_t^- & da_t^0 & 0 \\ da_t^. & da_t^+ & 0 \end{pmatrix}$$

où on a posé $da_t^. = dt$. On a $dX_t = tr(H_t dA_t)$, et pour le processus adjoint $dX_t^* = tr(H_t^\star dA_t)$ où l'involution $\star$ est définie par rapport à la forme quadratique

de matrice $\begin{pmatrix} 0 & 0 & 1 \\ 0 & 1 & 0 \\ 1 & 0 & 0 \end{pmatrix}$ ce qui donne $H_t^\star = \begin{pmatrix} 0 & (H_t^+)^* & (H_t^.)^* \\ 0 & (H_t^0)^* & (H_t^-)^* \\ 0 & 0 & 0 \end{pmatrix}$.

En utilisant ces matrices on peut réécrire la formule d'Itô

$$d(X_t Y_t) = tr\left[((\mathbf{X}_t + \mathbf{H}_t)(\mathbf{Y}_t + \mathbf{K}_t) - \mathbf{X}_t \mathbf{Y}_t) dA_t \right]$$

En particulier, on peut donner la forme fonctionnelle suivante de la formule d'Itô. Soit P un polynôme en une variable, alors

$$dP(X_t) = tr\left[(P(\mathbf{X}_t + \mathbf{H}_t) - P(\mathbf{X}_t)) dA_t \right] \tag{89}$$

Cette formule est à rapprocher de la forme classique de la formule d'Itô pour le mouvement brownien. Si f est une fonction de classe C^2, alors

$$df(B_t) = f'(B_t)dB_t + \frac{1}{2}f''(B_t)dt$$

Cette formule se retrouve pour une fonction polynomiale f à partir de la formule de Belavkin (89) en considérant $X_t = a_t^+ + a_t^-$.

5.5 Exemples d'intégrales stochastiques

Commençons par l'exemple le plus simple. Pour toute $h \in L_{\mathbf{C}}^2(\mathbf{R}_+)$ on vérifie grâce à la formule (77) que
$\int_0^\infty h(s)da_s^+ = a_h^+$ et $\int_0^\infty \bar{h}(s)da_s^- = a_h^-$.
Si $h \in L^\infty(\mathbf{R}_+)$ est considéré comme un opérateur de multiplication sur $L_{\mathbf{C}}^2(\mathbf{R}_+)$ alors on a $d\Gamma(h_{t]}) = \int_0^t h(s)da_s^0$.
En prenant des produits de ces opérateurs et en appliquant la formule d'Itô on obtient

$$a_h^+ a_k^- = \int_0^\infty h(t)\left(\int_0^t \bar{k}(s)da_s^-\right)da_s^+ + \int_0^\infty \bar{k}(t)\left(\int_0^t h(s)da_s^+\right)da_s^-$$

$$a_k^- a_h^+ = \int_0^\infty h(t)\left(\int_0^t \bar{k}(s)da_s^-\right)da_s^+ + \int_0^\infty \bar{k}(t)\left(\int_0^t h(s)da_s^+\right)da_s^- + \int_0^\infty \bar{k}(t)h(t)dt$$

en particulier nous retrouvons la relation d'Heisenberg (60). Je laisse au lecteur le soin d'écrire les formules comprenant des intégrales par rapport à a^0, et de retrouver les relations (67).

Passons maintenant à une classe d'exemples plus générale, provenant de la théorie des martingales.

Soit M une martingale normale possédant la propriété de représentation chaotique (definitions 32 et 33). Nous pouvons donc identifier l'espace L^2 de cette martingale avec l'espace de Fock $\Gamma(L^2_{\mathbf{C}}(\mathbf{R}_+))$, ce que nous faisons dans la suite.

On suppose que la martingale M est dans L^4, ce qui entraine que son crochet est dans L^2 (cf [21]). Comme elle possède la propriété de représentation prévisible, il existe un processus prévisible ϕ tel que $[M,M]_t = t + \int_0^t \phi_s dM_s$. Nous allons noter par la même lettre ϕ le processus d'opérateurs de multiplication sur l'espace de Fock qu'il définit. Pour simplifier la discussion nous supposerons ϕ borné sur tout intervalle $[0,t]$.

Soit K un processus prévisible borné.

PROPOSITION 18 – *Pour tout $t \in \mathbf{R}_+$, l'opérateur de multiplication par la variable $\int_0^t K_s dM_s$ est égal à $\int_0^t K_s \left(\phi_s da_s^0 + da_s^+ + da_s^- \right)$*

Démonstration

Soient $u,v \in L^2(\mathbf{R}_+)$ et $H_t = \xi(v_{t]})$, $G_t = \xi(u_{t]})$. Quand on identifie l'espace de Fock avec l'espace L^2 de la martingale M au moyen des chaos, H et G sont deux martingales qui vérifient $H_t = 1 + \int_0^t v(s)H_s dM_s$ et $G_t = 1 + \int_0^t u(s)G_s dM_s$.

(remarque: la martingale M ayant un crochet oblique égal à $< M,M >_t = t$, les martingales H et G n'ont pas de sauts prévisibles, et donc si on en prend une version càdlàg, on a $H_{s-} = H_s$ presque surement pour tout s.) Si les processus H et G sont considérés comme des éléments de l'espace de Hilbert $L^2(\mathbf{R}_+) \otimes L^2(M)$ cette formule est la même que la formule usuelle $H_t = 1 + \int_0^t v(s)H_{s-}dM_s$.

Notons $\gamma_t = \int_0^t K_s dM_s$ Il faut montrer que

$$E[\gamma_t G_\infty \bar{H}_\infty] = \int_0^t < K_s \phi_s \xi(u), \xi(v) > u(s)\bar{v}(s)ds \tag{90}$$

$$+ \int_0^t < K_s \xi(u), \xi(v) > (u(s) + \bar{v}(s))ds$$

On a

$$E[\gamma_t G_\infty \bar{H}_\infty] = E[\gamma_t G_t \bar{H}_t]e^{\int_t^\infty u(s)\bar{v}(s)ds} \tag{91}$$

et

$$< K_t \phi_t \xi(u), \xi(v) > = < K_t \phi_t \xi(u_{t]}), \xi(v_{t]}) > e^{\int_t^\infty u(s)\bar{v}(s)ds}$$

En dérivant les deux égalités (90) et (91) par rapport à t, il suffit de vérifier que

$$E[\gamma_t G_t \bar{H}_t] = \int_0^t \left(u(s)\bar{v}(s)E[\gamma_s G_s \bar{H}_s] + u(s) < K_s \phi_s \xi(u_{s]}), \xi(v_{s]}) > \bar{v}(s) \right.$$

$$+ < K_s \xi(u_{s]}), \xi(v_{s]}) > (u(s) + \bar{v}(s))ds$$

Appliquons la formule d'Itô (pour la martingale M) au calcul de $\gamma_t G_t \bar{H}_t$. Tout d'abord on a

$$G_t \bar{H}_t = 1 + \int_0^t G_s d\bar{H}_s + \int_0^t \bar{H}_s dG_s + [G, \bar{H}]_t$$

$$= 1 + \int_0^t G_s \bar{H}_s \bar{v}(s) dM_s + \int_0^t u(s) G_s \bar{H}_s dM_s$$

$$+ \int_0^t u(s) G_s \bar{H}_s \bar{v}(s)(\phi_s dM_s + ds)$$

Si nous effectuons le produit avec γ_t, appliquons la formule d'Itô et prenons l'espérance, il suffit de garder les termes sans intégrale stochastique par rapport à M. On trouve

$$E[\gamma_t G_t \bar{H}_t] = E\Big[\int_0^t K_s G_s \bar{H}_s \bar{v}(s) ds + \int_0^t u(s) K_s G_s \bar{H}_s ds$$

$$+ \int_0^t u(s) G_s \bar{H}_s \bar{v}(s)(K_s \phi_s + \gamma_s) ds \Big]$$

Dans l'espace de Fock cela s'écrit

$$< \gamma_t \xi(u), \xi(v) > = \int_0^t < K_s \xi(u_{s]}), \xi(v_{s]}) > \bar{v}(s) ds$$

$$+ \int_0^t u(s) < K_s \xi(u_{s]}), \xi(v_{s]}) > ds$$

$$+ \int_0^t u(s) < (K_s \phi_s + \gamma_s)\xi(u_{s]}), \xi(v_{s]}) > \bar{v}(s) ds \Big]$$

c'est bien la formule qu'il fallait vérifier.

Nous allons appliquer ce dernier résultat à la construction des martingales d'Azéma, en suivant un article de Parthasarathy [61].

Les martingales d'Azéma sont des martingales de carré intégrable dont le crochet droit vérifie l'identité

$$d[X, X]_t = (c - 1) X_{t-} dX_t + dt \tag{92}$$

M.Emery [26] a montré que pour toute valeur du paramètre réel c il existe une seule solution (en loi) à cette équation (avec $X_0 = 0$). Ce processus est un processus de Markov sur $\mathbf{R}$, et de plus lorsque $c \in [-1, +1]$ la martingale correspondante a la propriété de représentation chaotique. (A l'heure actuelle on ne sait pas si les martingales d'Azéma correspondant à une valeur de c en dehors de cet intervalle possèdent la propriété de représentation chaotique).

En utilisant la proposition précédente l'équation de structure (92) se transforme sur l'espace de Fock en l'équation

$$dX_t = (c - 1) X_t da_t^0 + da_t^+ + da_t^- \tag{93}$$

(les variables aléatoires X_t et X_{t-} sont égales presque partout pour tout t). On va résoudre cette équation à l'aide d'intégrales stochastiques non-commutatives, et

68

montrer que l'on obtient ainsi un processus classique, qui est une martingale solution de (92). Je vais me contenter d'esquisser les calculs et renvoyer à l'article [61] pour plus de détails.

L'équation (93) se résout par la méthode de variation des constantes. On commence par résoudre l'équation sans second membre $dJ_t^c = (c-1)J_t^c da_t^0$, dont la solution est $J_t^c = \Gamma(c1_{[0,t[} + 1_{[t,+\infty[})$. En cherchant la solution de l'équation complète sous la forme $X_t = J_t^c(\int_0^t x_s^+ da^+ s + x_s^- da_s^-)$ et en appliquant la formule d'Itô on trouve $dX_t = (c-1)X_t da_t^0 + J_t^c(x_t^+ da_t^+ + x_t^- da_t^-) + (c-1)J_t^c x_t^+ da_t^-$ d'où les équations $J_t^c x_t^- = I$ et $cJ_t^c x_t^+ = I$. Ces équation se résolvent pour $c \neq 0$ et on obtient, pour $c \in [-1,1]\backslash\{0\}$,

$$X_t = J_t^c \int_0^t J_s^{c^{-1}} (da_s^- + c^{-1}da_s^+)$$

(Pour $c = 0$, la description de X se fait à l'aide de la notation de Guichardet pour l'espace de Fock, voir [61]).

Le processus X se décompose en somme de deux opérateurs adjoints l'un de l'autre $L_t^+ = J_t^c \int_0^t J_s^{c^{-1}} c^{-1} da_s^+$ et $L_t^- = J_t^c \int_0^t J_s^{c^{-1}} da_s^-$.

La formule d'Itô montre que pour tout $x \in \Xi(L_{\mathbf{C}}^2(\mathbf{R}_+))$

$$< L_t^+ x, L_t^+ x > -c < L_t^- x, L_t^- x >=< \int_0^t \Gamma(1_{[0,s[} + c^2 1_{[s,t[} 1_{[s,+\infty[})x, x > ds$$

On en déduit que si $c \in [-1,0[$ les opérateurs L_t^- et L_t^+ sont bornés sur Ξ et adjoints l'un de l'autre, donc X_t est auto-adjoint borné.

Montrons que le processus X est un processus classique. Fixons $s \in \mathbf{R}_+$, et posons $Y_t = Y_s$ pour $t \geq s$. D'après la formule d'Itô le commutateur $X_t Y_t - X_t Y_t$ vérifie l'équation $d(XY - YX)_t = (XY - YX)_t da_t^0$, et il est nul pour $t = s$ on en déduit par une application du lemme de Gronwall qu'il est nul pour $t \geq s$.

5.6 Extensions de l'intégrale stochastique

La première extension de la définition des intégrales stochastiques dont nous aurons besoin consiste à ajouter un espace de Hilbert à l'instant 0, c'est à dire à former le produit tensoriel $\mathcal{H}_0 \otimes \Gamma(L^2(\mathbf{R}_+))$ et à prolonger les opérateurs définis sur l'espace de Fock par l'identité sur $\mathcal{H}_0$. On peut ainsi définir l'intégrale stochastique de processus X d'opérateurs définis sur $D \otimes \Xi(S)$ où $D \subset \mathcal{H}_0$ est un domaine dense. La condition d'adaptation de X se traduit par une factorisation de X_t sur le produit $(D \otimes \Xi_{t]}) \otimes \Gamma_{[t}$. Les intégrales stochastiques sont données par des formules analogues à celles de de (77)

$$< \int_0^\infty X_s da_s^\varepsilon a \otimes \xi(h), b \otimes \xi(k) >= \int_0^\infty < X_s(a \otimes \xi(h)), b \otimes \xi(k) > \kappa^\varepsilon(s) ds$$

Il est facile en utilisant ce qui précède de vérifier que cette égalité détermine de façon unique un opérateur $\int_0^\infty X_s da_s^\varepsilon$ sur le domaine $D \otimes \Xi(S)$ dès que $\int_0^\infty |X_s a \otimes \xi(h)|^2 \alpha^\varepsilon(s) ds < +\infty$ pour tout $a \in D$ et $h \in S$.

Passons maintenant à la définition de l'intégrale stochastique en dimension supérieure c'est-à-dire sur l'espace de Fock multiple $\Gamma(L^2(\mathbf{R}_+) \otimes \mathbf{C}^d)$.

L'intégrale de Skorokhod étant un objet purement hilbertien, défini sur un espace de Fock quelconque, il est tentant d'essayer de formuler l'intégrale stochastique non-commutative en termes purement hilbertiens en se passant d'une identification de K avec $L^2(\mathbf{R}_+)$. C'est ce que nous allons faire dans ce paragraphe, en donnant une formule pour l'intégrale stochastique qui a l'avantage d'être simple et très générale, en particulier comme cette formule est la même sur tous les espaces de Fock, elle est valable sur un espace de Fock multiple $\Gamma(L^2(\mathbf{R}_+) \otimes \mathbf{C}^d)$.

Pour établir la formule en question, il faut remplacer la donnée d'un processus d'opérateurs X_t sur Γ par celle d'un seul opérateur sur un espace de Hilbert plus gros. C'est un peu l'analogue d'une idée commune en théorie générale des processus (classique!), au lieu de considérer un processus comme une famille X_t de variables aléatoires sur l'espace de probabilité Ω, on peut le considérer comme une variable aléatoire sur le gros espace $\mathbf{R}_+ \times \Omega$.

Commençons par expliciter cette formule pour l'intégrale par rapport au processus de création.

Soit $(X_t)_{t \in \mathbf{R}_+}$ un processus tel que pour tout $h \in S$, la fonction $t \mapsto X_t(\xi(h))$ soit de carré intégrable. C'est un élément de $L^2_{\mathbf{C}}(\mathbf{R}_+) \otimes \Gamma(L^2_{\mathbf{C}}(\mathbf{R}_+))$, donc $(X_t)_{t \in \mathbf{R}_+}$ définit un opérateur $X^+ : \Xi(L^2_{\mathbf{C}}(\mathbf{R}_+)) \to L^2_{\mathbf{C}}(\mathbf{R}_+) \otimes \Gamma(L^2_{\mathbf{C}}(\mathbf{R}_+))$, et on a $\int_0^\infty X_s da_s^+(\xi(h)) = \delta(X^+(\xi(h)))$ Nous voyons donc que l'intégrale stochastique de X n'est autre que la composition des opérateurs

$$\int_0^\infty X_s da_s^+ = \delta \circ X^+ \tag{94}$$

Passons maintenant au cas des intégrales d'annihilation. Soit X un processus, il définit un opérateur $X^- : L^2_{\mathbf{C}}(\mathbf{R}_+) \otimes \Xi(L^2_{\mathbf{C}}(\mathbf{R}_+)) \to L^2_{\mathbf{C}}(\mathbf{R}_+)$ en posant $X^-(k \otimes F) = \int_0^\infty k(s)X_s(F)ds$. La formule de définition de l'intégrale d'annihilation montre que

$$\int_0^\infty X_s da_s^-(\xi(h)) = \int_0^\infty X_s(\xi(h))h(s)ds = X^-(\nabla\xi(h))$$

L'intégrale stochastique est donc la composition

$$\int_0^\infty X_s da_s^- = X^- \circ \nabla \tag{95}$$

Pour l'intégrale de conservation, nous avons besoin d'une troisième interprétation d'un processus d'opérateurs. Au processus X on associe maintenant un opérateur X^0 de $L^2_{\mathbf{C}}(\mathbf{R}_+) \otimes \Xi(L^2_{\mathbf{C}}(\mathbf{R}_+))$ dans $L^2_{\mathbf{C}}(\mathbf{R}_+) \otimes \Gamma(L^2_{\mathbf{C}}(\mathbf{R}_+))$ en posant $X^0(k \otimes F) = t \mapsto k(t)X_t(F)$. L'opérateur X^0 ainsi obtenu commute avec les opérateurs de multiplication $h \otimes I : L^2_{\mathbf{C}}(\mathbf{R}_+) \otimes \Xi(L^2_{\mathbf{C}}(\mathbf{R}_+)) \to L^2_{\mathbf{C}}(\mathbf{R}_+) \otimes \Gamma(L^2_{\mathbf{C}}(\mathbf{R}_+))$. La formule (77) nous dit alors que $\int_0^\infty X_s da_s^0(\xi(h)) = \delta(X^0(\nabla\xi(h)))$ par conséquent, l'intégrale stochastique est cette fois la composition

$$\int_0^\infty X_s da_s^0 = \delta \circ X^0 \circ \nabla \tag{96}$$

Après ces considérations, nous pouvons donner l'extension multidimensionnelle de l'intégrale stochastique. On considère l'espace de Fock $\Gamma(L^2(\mathbf{R}_+) \otimes \mathbf{C}^d)$. Il est muni naturellement d'une structure de produit tensoriel continu, qui permet de définir la notion de processus adapté, et il peut s'interpréter comme l'espace L^2 d'un mouvement brownien (ou d'un processus de Poisson) de dimension d. Si on choisit une base e_j de $\mathbf{C}^d$, il y a un isomorphisme entre $\Gamma(L^2(\mathbf{R}_+) \otimes \mathbf{C}^d)$ et $\Gamma(L^2(\mathbf{R}_+))^{\otimes d}$ déduit de la proposition (10). En particulier, on a des opérateurs de création, annihilation et conservation sur la j^e composante du produit tensoriel

$$a_t^{+j} = a_{1_{[0,t]} \otimes e_j}^+ = Id \otimes \ldots \otimes a_t^+ \otimes \ldots \otimes Id,$$

$$a_t^{-j} = a_{1_{[0,t]} \otimes e_j}^- = Id \otimes \ldots \otimes a_t^- \otimes \ldots \otimes Id, \text{ et}$$

$$a_t^{0j} = d\Gamma(1_{[0,t]} \otimes E_{jj}) = Id \otimes \ldots \otimes a_t^0 \otimes \ldots \otimes Id.$$

Pour définir une intégrale de création, nous avons besoin d'un processus X^+ qui est un opérateur de $\Xi(L^2(\mathbf{R}_+) \otimes \mathbf{C}^d)$ dans $(L^2(\mathbf{R}_+) \otimes \mathbf{C}^d) \otimes \Gamma(L^2(\mathbf{R}_+) \otimes \mathbf{C}^d))$. En fixant une base e_j de $\mathbf{C}^d$, un tel opérateur est donné par une famille $(X_t^j)j = 1 \ldots d, t \in \mathbf{R}+$ d'opérateurs de $\Xi(L^2(\mathbf{R}_+) \otimes \mathbf{C}^d)$ dans $\Gamma(\Xi(L^2(\mathbf{R}_+) \otimes \mathbf{C}^d))$. L'intégrale de création de X^+ est alors la somme $\sum_j \int_0^\infty X_t^j da_t^{+j}$, chacun des termes de la somme étant défini à partir de la divergence. Une définition équivalente de cette intégrale stochastique consiste à écrire l'analogue de la formule (77)

$$< \int_0^\infty X_s da_s^+ \xi(h), \xi(k) > = \int_0^\infty < X_s \xi(h), k(s) \otimes \xi(k) > ds$$

où X_s est un opérateur de $\Xi(L^2(\mathbf{R}_+) \otimes \mathbf{C}^d)$ dans $\mathbf{C}^d \otimes \Gamma((L^2(\mathbf{R}_+) \otimes \mathbf{C}^d))$.

Comme dans le cas unidimensionnel, on vérifie que l'intégrale d'un processus adapté est toujours définie sur le domaine $\Xi(S \otimes \mathbf{C}^d)$, dès lors que les conditions $\int_0^\infty |X_s \xi(h \otimes e_j)|^2 ds < +\infty$ pour $h \in S$ sont remplies.

L'intégrale d'annihilation a une définition analogue, un processus X étant un opérateur de $L^2(\mathbf{R}_+) \otimes \mathbf{C}^d \otimes \Xi(L^2(\mathbf{R}_+) \otimes \mathbf{C}^d)$ dans $\Gamma(L^2(\mathbf{R}_+) \otimes \mathbf{C}^d)$, ou encore une famille X_s d'opérateurs de $\mathbf{C}^d \otimes \Xi(L^2(\mathbf{R}_+) \otimes \mathbf{C}^d)$ dans $\Gamma(L^2(\mathbf{R}_+) \otimes \mathbf{C}^d)$. La définition est cette fois contenue dans la formule

$$\int_0^\infty X_s da_s^- \xi(h) = \int_0^\infty X_s(h(s) \otimes \xi(h)) ds$$

En considèrant une base e_j de $\mathbf{C}^d$ nous pouvons encore décomposer cette intégrale en intégrales par rapport aux processus d'annihilation $a_t^{-j} = a_{1_{[0,t]} \otimes e_j}^-$.

La définition des intégrales de conservation va nous montrer que les processus a_t^{0j} introduits ci-dessus sont insuffisants pour décrire ces intégrales. Un processus intégrable par rapport au processus de conservation est un opérateur X^0 de $L^2(\mathbf{R}_+) \otimes \mathbf{C}^d \otimes \Xi(L^2(\mathbf{R}_+) \otimes \mathbf{C}^d)$ dans $L^2(\mathbf{R}_+) \otimes \mathbf{C}^d \otimes \Gamma(L^2(\mathbf{R}_+) \otimes \mathbf{C}^d)$ qui commute avec les opérateurs de multiplication $h \otimes I \otimes I$ sur $L^2(\mathbf{R}_+) \otimes \mathbf{C}^d \otimes \Xi(L^2(\mathbf{R}_+) \otimes \mathbf{C}^d)$. En introduisant une base e_j de $\mathbf{C}^d$ nous voyons qu'un tel processus se décompose en une famille $X_t^{jj'}$ d'opérateurs de $\Xi(L^2(\mathbf{R}_+) \otimes \mathbf{C}^d)$ dans $\Gamma(L^2(\mathbf{R}_+) \otimes \mathbf{C}^d)$. Son intégrale stochastique $\delta \circ X^0 \circ \nabla$ est alors une somme d'intégrales stochastiques $\sum_{jj'} \int_0^\infty X_t^{jj'} da_t^{0jj'}$ où les processus $a^{0jj'}$ sont définis par $a_t^{0jj'} = d\Gamma(1_{[0,t[} \otimes E_{jj'})$ ($E_{jj'}$ est l'opérateur qui envoie e_j sur $e_{j'}$ et les autres éléments de la base sur 0). Une formule analogue à (77) donne

$$< \int_0^\infty X_s^0 da_s^0 \xi(h), \xi(k) > = \int_0^\infty < X_s^0(h(s) \otimes \xi(h)), k(s) \otimes \xi(k) > ds$$

La formule d'Itô se prolonge au cas multidimensionnel nous avons cette fois d différentielles de processus de création da_t^{+j}, d d'annihilation da_t^{-j} et d^2 de conservation $da_t^{jj'}$. Il est plus agréable pour calculer la table de multiplication d'Itô d'adopter une notation plus intrinsèque en posant $a_t^{\pm u} = a_{1_{[0,t]} \otimes u}^{\pm}$ pour $u \in \mathbb{C}^d$ et $a_t^{0A} = d\Gamma(1_{[0,t]} \otimes A)$ pour $A \in M_d(\mathbb{C})$. On a alors

$$da_t^{-u} \cdot da_t^{+v} = <v, u> dt$$

$$da_t^{-u} \cdot da_t^{0A} = da^{-uA}$$

$$da_t^{0A} \cdot da_t^{+v} = da_t^{+Av}$$

(dans cette notation, u est un vecteur ligne et v un vecteur colonne).

Pour terminer signalons que la formule d'Itô sous la forme due à Belavkin (89) reste inchangée, si au lieu d'utiliser des matrices 3×3 on utilise des matrices $(d+2) \times (d+2)$. On considère des X et Y deux processus adaptés qui s'écrivent sous la forme

$$X_t = \sum_\varepsilon \int_0^t H_s^\varepsilon da_s^\varepsilon \qquad Y_t = \sum_\varepsilon \int_0^t K_s^\varepsilon da_s^\varepsilon$$

où ε parcourt les indices $-j, +j, 0jj', \cdot$ (avec toujours $da^{\cdot} = dt$). H_t^+ est un vecteur colonne d'opérateurs H_t^- un vecteur ligne et H_t^0 une matrice $d \times d$ avec lesquels on forme les matrices $(d+2) \times (d+2)$ à coefficients opérateurs

$$\mathbb{H}_t = \begin{pmatrix} 0 & H_t^- & H_t^{\cdot} \\ 0 & H_t^0 & H_t^+ \\ 0 & 0 & 0 \end{pmatrix} \quad \mathbb{K}_t = \begin{pmatrix} 0 & K_t^- & K_t^{\cdot} \\ 0 & K_t^0 & K_t^+ \\ 0 & 0 & 0 \end{pmatrix}$$

$$d\mathbb{A}_t = \begin{pmatrix} 0 & 0 & 0 \\ da_t^- & da_t^0 & 0 \\ da_t^{\cdot} & da_t^+ & 0 \end{pmatrix}$$

Le processus adjoint de X satisfait toujours à $dX_t^* = tr(\mathbb{H}_t^* d\mathbb{A}_t)$ où l'involution $\star$ est définie par rapport à la forme quadratique de matrice $\begin{pmatrix} 0 & 0 & 1 \\ 0 & Id & 0 \\ 1 & 0 & 0 \end{pmatrix}$, et la formule d'Itô est encore

$$d(X_t Y_t) = tr\Big[((X_t + \mathbb{H}_t)(Y_t + \mathbb{K}_t) - X_t Y_t)d\mathbb{A}_t\Big]$$

La forme fonctionnelle

$$dP(X_t) = tr\big[(P(X_t + \mathbb{H}_t) - P(X_t))d\mathbb{A}_t\big]$$

reste également valable.

72

Chapitre 6

Applications du calcul stochastique non-commutatif

6.1 Processus de Markov non-commutatif

Il existe une littérature importante consacrée aux applications du calcul stochastique non-commutatif à la physique quantique, qu'il ne m'est pas possible d'aborder ici faute de compétence. Les lecteurs intéressés par ces applications pourront consulter les volumes des rencontres de probabilités quantiques édités par L.Accardi et W.von Waldenfels [2] ainsi que les références qu'ils contiennent. Mentionnons juste que le calcul stochastique non-commutatif sert à construire des "évolutions stochastiques" qui décrivent le comportement d'un système quantique ouvert, en interaction avec une "source de bruit quantique" représentée par un espace de Fock. L'évolution dans le temps d'un système quantique fermé est gouvernée par un opérateur auto-adjoint, l'hamiltonien, et le groupe unitaire qu'il engendre. Lorsque le système S en considération est en interaction avec un autre système (appelé "l'extérieur") le système total formé par la réunion de S et de l'extérieur est fermé et son évolution est donc gouvernée par un hamiltonien. Lorsqu'on ne s'intéresse qu'au système lui-même, l'interaction avec l'extérieur fait que son évolution n'est plus gouvernée par un hamiltonien mais par un opérateur "dissipatif" (je renvoie au livre de Davies [18] pour une discussion précise de ces notions). Il est apparu que la théorie de l'évolution des systèmes quantiques ouverts possède une analogie formelle avec la théorie classique des processus de Markov, l'opérateur dissipatif qui décrit l'évolution du système ouvert étant un analogue non-commutatif du générateur d'un semi-groupe markovien.

Je vais essayer de donner quelques rudiments de cette théorie des processus de Markov non-commutatifs. Nous allons quitter le terrain de la physique quantique que nous n'avons fait qu'effleurer pour nous concentrer sur l'aspect probabiliste de la question.

Commençons par quelques rappels sur la théorie élémentaire des processus de Markov.

Soient E un espace lusinien (pour fixer les idées) et $(P_t)_{t \in \mathbf{R}_+}$ un semi-groupe de noyaux markoviens sur E. On sait qu'il est possible de construire un processus de Markov associé, qui consiste en la donnée d'un espace de probabilité $(\Omega, \mathcal{F}, P)$, d'une filtration $(\mathcal{F}_t)_{t \in \mathbf{R}_+}$, et d'une famille de variables aléatoires $X_t : \Omega \to E$, $t \in \mathbf{R}_+$ adaptée à la filtration $\mathcal{F}_t$ tels que

$$\forall t_1 < \ldots < t_n \ \forall f \text{ borélienne bornée sur } E^n$$
$$E[f(X_{t_1}, \ldots, X_{t_n})] =$$
$$\int_{E^{n+1}} f(x_1, \ldots, x_n) \mu(dx_0) P_{t_1}(x_0, dx_1) \ldots P_{t_n - t_{n-1}}(x_{n-1}, dx_n) \tag{97}$$

(le processus X est de loi initiale μ).

Un procédé usuel pour obtenir un tel processus consiste à utiliser le théorème de Kolmogorov pour construire la probabilité P sur l'espace canonique $\Omega = E^{\mathbb{R}_+}$, muni des applications coordonnées X_t, comme limite projective des lois marginales de rang fini données par la formule (97).

La propriété de Markov simple du processus X se traduit par l'identité

$$E[f(X_t)|\mathcal{F}_s] = P_{t-s}f(X_s) \tag{98}$$

vérifiée pour tout couple $s < t \in \mathbb{R}_+$ et toute fonction f borélienne bornée sur E.

Plusieurs auteurs ont proposé des définitions de la notion de processus de Markov non-commutatif (voir par exemple [1], [29], [9], [74], [84]). Sans entrer dans les détails, nous allons examiner les principes qui sont à la base de ces définitions.

Suivant les principes des "probabilités quantiques", nous allons commencer par traduire la situation précédente en termes algébriques. Le semi-groupe markovien P_t détermine un semi-groupe d'applications linéaires T_t sur l'algèbre $B(E)$ des fonctions boréliennes bornées sur E définies par $T_t f(x) = \int_E f(y) P_t(x, dy)$. Les applications T_t sont positives, (l'image d'une fonction positive est positive) et préservent l'identité car P_t est markovien. Réciproquement, un semi-groupe d'applications linéaires positives sur $B(E)$, préservant l'identité, provient d'un unique semi-groupe de noyaux markoviens (cf [22], on utilise le fait que E est lusinien), les données de P_t et de T_t sont donc équivalentes.

Le processus X_t définit une famille de morphismes d'algèbres $j_t : B(E) \to L^\infty(\Omega, \mathcal{F}, P)$ en posant $j_t(f) = f \circ X_t$. Ces morphismes déterminent le processus X car si Y est un autre processus tel que $f \circ Y_t = f \circ X_t$ pour tout fonction borélienne bornée, alors $X_t = Y_t$.

Si E possède une mesure m naturellement associée au semi-groupe P_t, (par exemple la mesure de comptage si E est dénombrable, ou une mesure m par rapport à laquelle les noyaux $(P_t)_{t>0}$ sont absolument continus), on peut définir les j_t sur l'algèbre $L^\infty(E, m)$, et le processus X est encore déterminé à une modification près par les morphismes j_t.

L'espérance conditionnelle sur la tribu $\mathcal{F}_t$ définit une projection

$$E_t : L^\infty(\Omega, \mathcal{F}, P) \to L^\infty(\Omega, \mathcal{F}_t, P)$$

et la propriété de Markov (98) devient alors

$$E_s \circ j_t = j_s \circ T_{t-s}$$

Décrivons maintenant des analogues non-commutatifs des objets précédents.

La donnée de base, qui est celle de l'espace E et du semi-groupe P_t est remplacée par celle d'une algèbre involutive à unité $\mathcal{A}$, qui joue le rôle d'analogue de l'algèbre $B(E)$ (ou de $L^\infty(E, m)$), et d'un semi-groupe T_t d'applications complètement positives de $\mathcal{A}$, préservant l'identité.

Expliquons ces termes.

Une algèbre involutive est une algèbre sur $\mathbb{C}$, munie d'une involution $a \mapsto a^*$ qui est une application antilinéaire, telle que $(ab)^* = b^* a^*$ (cf Dixmier [24]). Dans la pratique, $\mathcal{A}$ sera le plus souvent une C^*-algèbre.

DÉFINITION 37 – *Une C^*-algèbre est une algèbre involutive munie d'une norme qui en fait un espace de Banach telle que*
i) $\forall x, y \in \mathcal{A}$ $|xy| \leq |x||y|$ (c'est une algèbre de Banach)
ii) $|x^*| = |x|$ et $|xx^*| = |x|^2$

Un exemple immédiat est celui de $\mathcal{B}(H)$ (opérateurs bornés sur un Hilbert H) avec l'involution donnée par le passage à l'adjoint.

Un autre exemple est celui de l'algèbre $C_0(X)$ (fonctions complexes continues nulles à l'infini sur un espace topologique localement compact X, l'involution étant la conjugaison). En fait, d'après un célèbre théorème de Gelfand, toute C^*−algèbre commutative est isomorphe à une algèbre de ce type, l'espace X étant homéomorphe à l'ensemble des caractères de la C^*−algèbre.

Un élément a d'une C^*−algèbre est dit hermitien si $a^* = a$, et positif s'il est de la forme $a = x^*x$ pour un $x \in \mathcal{A}$.

Soit $\mathcal{A}$ une C^*−algèbre avec unité, un état sur $\mathcal{A}$ est une application linéaire φ continue, positive (elle envoie les éléments positifs de $\mathcal{A}$ dans $\mathbb{R}_+$), telle que $\varphi(1) = 1$. C'est l'analogue d'une mesure de probabilité, et les deux notions sont équivalentes lorsque $\mathcal{A}$ est commutative, d'après le théorème de Riesz.

Soient $\mathcal{A}$ et $\mathcal{B}$ des C^*−algèbres, une application linéaire T de $\mathcal{A}$ dans $\mathcal{B}$ est dite positive si l'image par T d'un élément positif de $\mathcal{A}$ est un élément positif de $\mathcal{B}$. La notion d'application positive entre deux C^*−algèbres a le défaut de ne pas être stable par produit tensoriel, si T est positive, il se peut que l'application $T \otimes Id : \mathcal{A} \otimes M_n(\mathbb{C}) \to \mathcal{B} \otimes M_n(\mathbb{C})$ ne soit pas positive. Pour cette raison, l'analogue non-commutatif de la notion d'application positive que nous considèrerons est celle d'application complètement positive.

DÉFINITION 38 – *On dit que l'application linéaire $T : \mathcal{A} \to \mathcal{B}$ est complètement positive si pour tout $n \geq 0$ l'application linéaire $T \otimes Id : \mathcal{A} \otimes M_n(\mathbb{C}) \to \mathcal{B} \otimes M_n(\mathbb{C})$ est positive*

Si $\mathcal{A}$ est commutative, les notions d'application positive et complètement positive coïncident.

DÉFINITION 39 – *Soient $\mathcal{A}$ une C^*−algèbre et ω un état sur $\mathcal{A}$, une espérance conditionnelle sur $\mathcal{A}$ est une application linéaire complètement positive $E : \mathcal{A} \to \mathcal{A}$, préservant l'identité, telle que*
$\omega \circ E = \omega$, et $\forall x, y \in \mathcal{A}$ $E(xE(y)) = E(x)E(y)$.

L'image d'une espérance conditionnelle est une sous-algèbre de $\mathcal{A}$ dont E laisse les éléments invariants. Lorsque $\mathcal{A} = L^\infty(\Omega, \mathcal{F}, P)$ et $\mathcal{G} \subset \mathcal{F}$ l'espérance conditionnelle usuelle sur $L^\infty(\Omega, \mathcal{G}, P)$ est une espérance conditionnelle au sens de la définition précédente.

On suppose donnés une C^*-algèbre $\mathcal{A}$ avec une identité, et un semi-groupe T_t d'applications complètement positives sur $\mathcal{A}$, préservant l'identité. Le semi-groupe T_t joue le role de semi-groupe markovien. On peut alors se poser le problème construire un "processus de Markov non-commutatif" correspondant.

Un tel processus sera la donnée d'une dilatation du semi-groupe T_t.

DÉFINITION 40 – *Une dilatation du semi-groupe T_t est la donnée de $(\mathcal{W}, \omega, \mathcal{W}_t, E_t, j_t)$ où*
i) $\mathcal{W}$ est une C^*-algèbre à unité munie d'un état ω
ii) $(\mathcal{W}_t)_{t \in \mathbb{R}_+}$ est une famille croissante de sous-algèbres de $\mathcal{W}$.

iii) Pour tout $t \in \mathbf{R}_+$, E_t est une espérance conditionnelle d'image $\mathcal{W}_t$, et $\forall s, t \in \mathbf{R}_+$, $E_t E_s = E_{s \wedge t}$

iv) Les $j_t : \mathcal{A} \to \mathcal{W}_t$ sont des morphismes qui préservent l'identité et vérifient la propriété de Markov

$$E_s \circ j_t = j_s \circ T_{t-s}$$

La définition d'une dilatation donnée ci-dessus n'est pas la seule possible, on peut demander par exemple que $\mathcal{A}$ et $\mathcal{W}$ soient des algèbres de von Neumann, avec ω, E_t, j_t normaux (cf Dixmier [24]), on peut aussi imposer des conditions moins fortes sur les E_t (comme dans Bhat et Parthasarathy [9]). Je renvoie aux articles cités pour des variantes possibles de cette définition, ainsi que pour les théorèmes d'existence correspondants.

Nous allons construire des dilatations de semi-groupes complètement positifs sur $\mathcal{B}(H_0)$ vérifiant certaines conditions analytiques en résolvant des équations différentielles stochastiques non-commutatives. Ces semi-groupes sont des analogues des semi-groupes de diffusions sur des variétés. Ensuite nous appliquerons ces idées à la construction de certains processus de Markov classiques. Le principe de ces constructions est le suivant. On se donne un espace E, avec une mesure m, et un semi-groupe markovien P_t sur E, le semi-groupe associé sur $L^\infty(E)$ étant T_t. Le semi-groupe T_t est étendu en un semi-groupe complètement positif de $\mathcal{B}(L^2(E, m))$, dont on construit une dilatation sur $\mathcal{W} = \mathcal{B}(H)$ telle que les algèbres $j_t(L^\infty(E))$ commutent. Soit $\mathcal{A}$ l'algèbre de von Neumann commutative engendrée par les $j_t(L^\infty(E))$, d'après le théorème de structure des algèbres de von Neumann commutatives (cf [24]), il existe un espace de probabilité $(\Omega, \mathcal{F}, P)$ et des variables aléatoires $X_t : \Omega \to E$ telles que $f \circ X_t = j_t(f)$ pour $f \in L^\infty(E)$. Les variables X_t forment alors un processus de Markov sur E de semi-groupe P_t.

Avant de réaliser ce programme, nous allons montrer au paragraphe suivant comment la marche de Bernoulli quantique du chapitre 1 peut être considérée comme un processus de Markov non-commutatif.

6.2 Un autre construction de la marche de Bernoulli quantique

Dans ce paragraphe nous allons donner un exemple explicite de dilatation d'un semi-groupe d'applications complètement positives en temps discret. Ce type de chaîne de Markov non-commutative est l'analogue d'une marche aléatoire sur un groupe. Nous verrons qu'il permet d'interpréter naturellement la marche de Bernoulli quantique du chapitre 1 comme une chaîne de Markov non-commutative, dont le processus de spin serait la "partie radiale".

Ce paragraphe est une parenthèse, il ne servira pas pour la suite du cours. Le lecteur peu familier avec la théorie des groupes peut passer directement au paragraphe 6.3. Commençons par quelques rappels de théorie des groupes. Je renvoie à [24] et [25] pour ce qui concerne les algèbres de groupes, et à [14] et [83] pour les représentations des groupes compacts en général et de SU(2) en particulier.

Soit G un groupe compact, muni de sa mesure de Haar normalisée. Il agit sur l'espace $L^2(G)$ par multiplication à gauche, i.e. chaque élément g du groupe définit un opérateur unitaire λ_g sur $L^2(G)$ par $\lambda_g f(h) = f(g^{-1}h)$. Appelons $vN(G)$ l'algèbre de von Neumann engendrée par ces opérateurs. Soit ν un état normal sur cette algèbre, la fonction ϕ sur G définie par $\phi(g) = \nu(\lambda_g)$ est une fonction de type positif, continue.

Notons Q l'application complètement positive déterminée par $Q(\lambda_g) = \phi(g)\lambda_g$ sur $vN(G)$.

Pour éviter les problèmes liés aux produits tensoriels infinis, et rester dans le même cadre que celui du chapitre 1, on va construire une dilatation de $(Q^n)_{n \leq N}$ pour un certain entier N. La construction d'une dilatation du semi-groupe Q^n tout entier est identique, mais avec un produit tensoriel infini.

Soit ρ une représentation unitaire de G sur un espace de Hilbert H et un état S sur H tel que $\phi(g) = tr(S\rho(g))$.

La représentation ρ étant donnée, on peut toujours trouver H et S par la construction GNS (cf [24]).

Soit $N \in \mathbb{N}$, on pose $\mathcal{W} = \mathcal{B}(H^{\otimes N})$, $\omega = S^{\otimes N}$. Pour tout $n \leq N$, $\mathcal{W}_n$ est l'algèbre des opérateurs de la forme $a \otimes Id$, avec $a \in \mathcal{B}(H^{\otimes n})$ sur la décomposition $H^{\otimes N} = H^{\otimes n} \otimes H^{\otimes(N-n)}$.

On définit des espérances conditionnelles $E_n : \mathcal{W} \to \mathcal{W}_n$ par

$$E_n(a_1 \otimes \ldots \otimes a_n \otimes a_{n+1} \otimes \ldots \otimes a_N)$$
$$= a_1 \otimes \ldots \otimes a_n \otimes tr(Sa_{n+1})Id \otimes \ldots \otimes tr(Sa_N)Id$$

On vérifie sans peine que les E_n vérifient $E_n E_m = E_{n \wedge m}$.

On construit des morphismes $j_n : vN(G) \to \mathcal{W}$ en posant $j_n(\lambda_g) = \rho(g)^{\otimes n} \otimes Id^{\otimes(N-n)}$

PROPOSITION 19 – $(\mathcal{W}, \omega, \mathcal{W}_n, E_n, j_n)_{n \leq N}$ est une dilatation de $(Q^n)_{n \leq N}$.

Démonstration
Il suffit de vérifier la propriété de Markov $E_n \circ j_{n+m} = j_n \circ Q^m$ sur un élément de la forme λ_g, or on a

$$E_n \circ j_{n+m}(\lambda_g) = \phi(g)^m \rho(g)^{\otimes n} \otimes Id^{\otimes(N-n)} = j_n \circ Q^m(\lambda_g)$$

PROPOSITION 20 – Soit $K \subset G$ un sous-groupe fermé, $vN_G(K) \subset vN(G)$ la sous algèbre de von Neumann engendrée par les éléments λ_h, $h \in K$ alors les restrictions des j_n à $vN_G(K)$ forment une dilatation de $Q_{|vN_G(K)}$. Si K est commutatif, les algèbres $j_n(vN_G(K))$ commutent.

Démonstration
La première partie est identique à celle de la proposition précédente.
Soient $m \leq n \in \mathbb{N}^*$, $h, h' \in H$ il est immédiat que $j_n(\lambda_h)$ commute avec $j_m(\lambda_{h'})$, d'où la seconde partie.

Dans le cas où G est un groupe commutatif, $vN(G)$ est isomorphe à $L^\infty(\hat{G})$ où $\hat{G}$ est le groupe dual de G, et l'état ν provient d'une mesure de probabilité sur $\hat{G}$ dont la transformée de Fourier est ϕ. La construction que nous avons faite ci-dessus se ramène alors à la construction usuelle d'une marche aléatoire sur $\hat{G}$ comme somme d'une famille de variables aléatoires indépendantes équidistribuées.

Dans le cas général, le théorème de Peter-Weyl (cf Bourbaki [14]) montre que l'espace $L^2(G)$ est une somme directe orthogonale $\oplus_{\chi \in \hat{G}} E_\chi$, où $\hat{G}$ est l'ensemble des classes d'équivalence de représentations irréductibles de G, et E_χ l'espace des fonctions coefficients de cette représentation. On en déduit que $vN(G) = \oplus_{\chi \in \hat{G}} M_\chi$ (la somme étant prise au sens des algèbres de von Neumann), où M_χ est l'algèbre de convolution à gauche par les fonctions de E_χ, qui est isomorphe à $M_d(\mathbb{C})$, d étant la dimension des représentations de classe χ.

Soit Z le centre de l'algèbre $vN(G)$, il résulte de la discussion ci-dessus que Z est isomorphe à l'algèbre $L^\infty(\hat{G})$.

PROPOSITION 21 – *Les algèbres $j_n(Z)$ commutent. Si la fonction ϕ est centrale, alors Q envoie Z dans elle-même, et les $j_{n|Z}$ forment une dilatation de $Q^n_{|Z}$.*

Démonstration

Soient $m \le n \in \mathbb{N}^*$, $c \in Z$ et $g \in G$, alors $j_n(\lambda_g) = j_m(\lambda_g)(j_m(\lambda_g)^{-1} j_n(\lambda_g))$ et $j_m(\lambda_g)^{-1} j_n(\lambda_g)$ commute avec $j_m(vN(G))$, or $j_m(c)$ commute avec $j_m(\lambda_g)$ car $c \in Z$ donc $j_m(c)$ commute avec $j_n(\lambda_g)$, on en déduit que $j_m(c)$ commute avec $j_n(vN(G))$ et donc que $j_m(Z)$ et $j_n(Z)$ commutent.

L'algèbre Z est engendrée en tant qu'algèbre de von Neumann par les opérateurs de la forme $\int_G k(g)\lambda_g dg$ où k est une fonction continue centrale. L'image par Q d'un tel opérateur est $\int_G \phi(g)k(g)\lambda_g dg$ qui est dans le centre Z. On en conclut par passage à la limite que Q préserve Z.

Pour montrer que les $j_{n|Z}$ forment une dilatation de $Q^n_{|Z}$ il suffit de montrer que pour tout m l'espérance conditionnelle E_m envoie l'algèbre engendrée par $j_{m+1}(Z) \ldots j_{m+k}(Z) \ldots$ sur celle engendrée par $j_1(Z), \ldots j_m(Z)$. Par récurrence il suffit de montrer que $E_m(j_{m+1}(Z)) \subset j_m(Z)$. Or, si $c = \int_G k(g)\lambda_g dg$ avec k centrale continue, il résulte de la définition de E_n que $E_n j_{m+1}(c) \in j_n(c)$, et par passage à la limite cela est vrai pour tout $c \in Z$.

L'algèbre de von Neumann engendrée par les $j_n(Z)$ est une algèbre commutative, donc il existe un espace de probabilité $(\Omega, \mathcal{F}, P)$ et des variables aléatoires $S_n : \Omega \to \hat{G}$, telles que $j_n(f) = f \circ S_n$. La restriction de Q à Z provient d'un noyau markovien N sur $\hat{G}$ et les variables S_n forment donc une chaîne de Markov sur $\hat{G}$ de noyau de transition N.

Nous allons retrouver la marche de Bernoulli quantique du chapitre 1, et le processus de Spin de la facon suivante. On prend $G = SU(2)$, $\phi(g) = \frac{1}{2}tr(g)$. Si ρ est une représentation de dimension finie de $SU(2)$ on peut la prolonger à son algèbre de Lie $su(2)$ en posant $\rho(\gamma) = \lim_{t \to 0} \frac{1}{t}\rho(\exp t\gamma - Id)$. Dans le cas de la fonction $\phi = \frac{1}{2}tr$, on peut prendre $H = \mathbb{C}^2$, ρ étant la représentation identique de dimension 2 de $SU(2)$, et $S = \frac{1}{2}Id$, dans la construction de la dilatation. On a alors $\mathcal{W} = \mathbb{C}^{2^N}$, et on vérifie que, avec les notations du chapitre 1, on a $j_n(i\sigma_x) = X_n$, $j_n(i\sigma_y) = Y_n$, et $j_n(i\sigma_z) = Z_n$. Les composantes de la marche de Bernoulli quantique sont donc obtenues en restreignant la dilatation j_n à des sous-algèbres engendrées par des sous-groupes à un paramètre de SU(2).

Pour ce qui est du processus de spin, nous allons faire intervenir le centre de l'algèbre de von Neumann. La théorie des représentation du groupe $SU(2)$ montre qu'il existe à isomorphisme près une unique représentation irréductible de dimension n pour tout entier $n \ge 1$. On a donc une identification naturelle $Z \sim L^\infty(\mathbb{N}^*)$. Les matrices de Pauli, en tant qu'éléments de l'algèbre de Lie complexifiée de $SU(2)$ définissent des dérivations invariantes à droite sur $L^2(SU(2))$ que nous notons X, Y et Z. L'opérateur auto-adjoint non-borné invariant à droite $\Sigma = \sqrt{X^2 + Y^2 + Z^2 + Id}$ sur $L^2(SU(2))$ commute avec tous les opérateurs λ_g. Il est diagonalisable et admet la valeur propre n sur le sous-espace de $L^2(SU(2))$ formé des fonctions coefficients de la représentation irréductible de dimension n de $SU(2)$. Cet opérateur s'interprète comme la fonction $n \mapsto n$ sur $\mathbb{N}^*$ quand on identifie $\hat{G}$ et $\mathbb{N}^*$. Remarquons que Σ n'est pas dans Z car il n'est pas borné.

On a $j_n(\Sigma) = \Sigma_n$, donc le processus de Spin est obtenu grâce à l'opérateur Σ.

Nous allons retrouver directement les probabilités de transition de cette chaîne de Markov en considérant la restriction de Q à $Z \sim L^\infty(\mathbb{N}^*)$.

Soit ρ_n une représentation irréductible de dimension n de $SU(2)$, son caractère est la fonction $\chi_n : g \mapsto tr(\rho_n(g))$ sur $SU(2)$. Cette fonction est entièrement déterminée par sa restriction aux matrices $g_\theta = \begin{pmatrix} e^{i\theta} & 0 \\ 0 & e^{-i\theta} \end{pmatrix}$ et on a $\chi_n(g_\theta) = \frac{\sin n\theta}{\sin \theta}$. De cette formule on déduit la formule de Clebsch-Gordon $\chi_n\chi_2 = \chi_{n+1} + \chi_{n-1}$. Remarquons que $\chi_2 = 2\phi$.

L'élément de Z qui correspond à l'indicatrice de n dans l'isomorphisme $Z \sim L^\infty(\mathbb{N}^*)$ est $n \int_{SU(2)} \chi_n(g) dg$. On en déduit que $Q(1_n) = \frac{1}{2}(\frac{n}{n+1}1_{n+1} + \frac{n}{n-1}1_{n-1})$ et que l'opérateur Q est bien l'opérateur markovien associé aux probabilités de transition du processus de Spin.

6.3 Equations différentielles linéaires

Nous reprenons les notations de la fin du chapitre 5. On a un espace de Fock $\Gamma(L^2(\mathbb{R}+) \otimes \mathbb{C}^d)$ sur lequel sont définis les opérateurs de création, annihilation et conservation notés collectivement a^ε. Le paramètre ε peut prendre les valeurs $+j$, $-j$, $0jk$, ou $\cdot$ avec $1 \leq j, k \leq d$. Rappelons que l'on a posé $da_t^\cdot = dt$.

On a également un espace de Hilbert initial H.

Nous allons résoudre une équation différentielle stochastique non-commutative linéaire à coefficients constants.

On se donne des opérateurs bornés L^ε sur H, et on cherche à résoudre l'équation

$$dV_t = V_t\left(\sum_\varepsilon L^\varepsilon da_t^\varepsilon\right) \tag{99}$$

Ici les opérateurs L^ε sont étendus à $H \otimes \Gamma(L^2(\mathbb{R}+) \otimes \mathbb{C}^d)$ par $L^\varepsilon \otimes Id$. Cette équation doit être interprétée comme une équation intégrale, c'est à dire que l'on doit avoir d'une part $V_t L^\varepsilon$ défini sur $H \otimes \Xi(S)$, intégrable, et d'autre part,

$$V_t = V_0 + \sum_\varepsilon \int_0^t V_s L^\varepsilon da_s^\varepsilon$$

pour tout $t \geq 0$.

Soit $S = L^2 \cap L^\infty(\mathbb{R}_+)$.

THÉORÈME 8 – Soit V_0 un opérateur borné sur H, il existe un unique processus V_t d'opérateurs de domaine $H \otimes \Xi(S)$ tel que

i) $$\forall t \geq 0 \quad V_t = V_0 + \sum_\varepsilon \int_0^t V_s L^\varepsilon da_s^\varepsilon$$

ii) $$\forall t \geq 0 \, \forall u \in S \quad \sup_{|a| \leq 1, s \leq t} |V_s(a \otimes \xi(u))| < +\infty$$

(V_0 est identifié avec l'opérateur $V_0 \otimes Id$ dans i))
Démonstration
Commençons par l'existence. On va utiliser la méthode d'itération de Picard. Posons

$$\forall t \geq 0 \quad V_t^{(0)} = V_0$$

$$V_t^{(n+1)} = V_0 + \sum_\varepsilon \int_0^t V_s^{(n)} L^\varepsilon da_s^\varepsilon$$

On vérifie par récurrence sur n en utilisant la remarque (1) après le théorème 7 que le processus $V^{(n)}$ est bien défini sur le domaine $H \otimes \Xi(S)$ et que d'après l'inégalité (86)

$$\sup_{|a| \leq 1, s \leq t} |V_s^{(n)} a \otimes \xi(u)|^2 < +\infty$$

D'autre part on a

$$(V_t^{(n+1)} - V_t^{(n)}) a \otimes \xi(u) = \sum_\varepsilon \left(\int_0^t (V_s^{(n)} - V_s^{(n-1)}) L^\varepsilon da_s^\varepsilon \right) a \otimes \xi(u)$$

d'où, d'après l'inégalité (86)

$$|(V_t^{(n+1)} - V_t^{(n)}) a \otimes \xi(u)|^2 \leq K(u) \sum_\varepsilon \int_0^t |(V_s^{(n)} - V_s^{(n-1)}) L^\varepsilon a \otimes \xi(u)|^2 ds$$

où $K(u)$ est la constante $(|u| + \sqrt{1 + |u|^2})(1 + |u|_\infty^2)|\xi(u)|^2$
Si on pose $\alpha_n(t) = \sup_{|a| \leq 1, s \leq t} |(V_s^{(n+1)} - V_s^{(n)}) a \otimes \xi(u)|^2$ alors cette inégalité montre que

$$\alpha_n(t) \leq K' \int_0^t \alpha_{n-1}(s) ds$$

avec $K' = K(u) \left(\sum_\varepsilon \|L^\varepsilon\|^2 \right)$, d'où l'on déduit par récurrence sur n que

$$\alpha_n(t) \leq \frac{K'^n}{n!} t^n |V_0 a \otimes \xi(u)|^2$$

La série $\sum_{n=0}^\infty V_t^{(n+1)} - V_t^{(n)}$ converge donc fortement sur le domaine $H \otimes \Xi(S)$.
Soit V_t la somme de cette série.
On a $\sup_{|a| \leq 1, s \leq t} |V_t a \otimes \xi(u)|^2 < +\infty$ et $\int_0^t |(V_s - V_s^{(n)}) a \otimes \xi(u)|^2 ds \to_{n \to +\infty} 0$ ce qui entraine

$$\left(\sum_\varepsilon \int_0^t V_s^{(n)} L^\varepsilon da_s^\varepsilon \right) a \otimes \xi(u) \to_{n \to +\infty} \left(\sum_\varepsilon \int_0^t V_s L^\varepsilon da_s^\varepsilon \right) a \otimes \xi(u)$$

et donc V est solution de l'équation (99).

Passons maintenant à l'unicité. Soit V' une autre solution, on pose $Y_t = V_t - V_t'$. L'inégalité (86) donne encore

$$|Y_t a \otimes \xi(u)|^2 \leq K(u) \sum_\varepsilon \int_0^t |Y_s L^\varepsilon a \otimes \xi(u)|^2 ds$$

En posant $\beta(t) = \sup_{|a| \leq 1 \ s \leq t} |Y_s a \otimes \xi(u)|^2$ on voit que $0 \leq \beta(t) \leq K' \int_0^t \beta(s) ds$ ce qui entraine que $\beta \equiv 0$.

Nous allons nous intéresser aux conditions sur les coefficients L^ε qui font que les opérateurs V_t peuvent se prolonger en des opérateurs isométriques. Afin de ne pas obscurcir la discussion par des considérations d'indices, on va traiter seulement le cas $d = 1$, le paramètre ε peut donc prendre les valeurs $-, 0, +$ ou $\cdot$. Je laisse au lecteur le soin d'énoncer les résultats analogues pour $d \geq 2$ (cf [51] et [60]).

Commençons par chercher des conditions nécessaires pour que la solution V_t de (99) soit isométrique, avec une condition initiale $V_0 = Id$.

Soient $x = a \otimes \xi(u)$ et $y = b \otimes \xi(v) \in H \otimes \Xi(S)$. Pour simplifier les notations on pose $dM_s = \sum_\varepsilon L^\varepsilon da_s^\varepsilon$.

Calculons, à l'aide de la formule d'Itô

$$< V_t x, V_t y > = < x, y > + < x, \int_0^t V_s dM_s y > + < \int_0^t V_s dM_s x, y >$$

$$+ < \int_0^t V_s dM_s x, \int_0^t V_s dM_s y > ds$$

$$= < x, y > + \int_0^t < V_s R_s x, y > ds + \int_0^t < x, V_s R_s^* y > ds$$

$$+ \int_0^t < V_s R_s x, (V_s y - y) > ds + \int_0^t < (V_s x - x), V_s R_s^* y > ds$$

$$+ \int_0^t < V_s T_s^1 x, V_s T_s^2 y > ds$$

$$= < x, y > + \int_0^t < V_s R_s x, V_s y > ds + \int_0^t < V_s x, V_s R_s^* y >$$

$$+ \int_0^t < V_s T_s^1 x, V_s T_s^2 y > ds$$

où on a posé

$$R_s = u(s) L^- + u\bar{v}(s) L^0 + \bar{v}(s) L^+ + L^\cdot$$

$$T_s^1 = L^+ + u(s) L^0 \qquad T_s^2 = L^+ + v(s) L^0$$

L'hypothèse que V est isométrique implique donc

$$\forall t \geq 0 \ \forall u, v \in S \ \forall a, b \in H \ \int_0^t < R_s x, y > + < x, R_s^* y > + < T_s^1 x, T_s^2 y > ds = 0$$

ce qui entraine que $\forall a, b \in H$

$$< L^\cdot a, b > + < a, L^\cdot b > + < L^+ a, L^+ b > = 0 \tag{100}$$

$$< L^+a, b > + < a, L^-b > + < L^+a, L^0b >= 0 \tag{101}$$

$$< L^-a, b > + < a, L^+b > + < L^0a, L^+b >= 0 \tag{102}$$

$$< L^0a, b > + < a, L^0b > + < L^0a, L^0b >= 0 \tag{103}$$

Les conditions (101) et (102) sont équivalentes, et on peut mettre (100), (101), (102), (103) sous la forme équivalente

$$L^0 + (L^0)^* + (L^0)^*L^0 = 0 \tag{104}$$

$$L^+ + (L^-)^* + (L^0)^*L^+ = 0 \tag{105}$$

$$L^{\cdot} + (L^{\cdot})^* + (L^+)^*L^+ = 0 \tag{106}$$

Ces conditions sont nécessaires. Montrons qu'elles sont suffisantes pour que V_t soit isométrique. Si ces conditions sont vérifiées, on obtient en appliquant la formule d'Itô

$$< V_tx, V_ty > - < x, y > = \int_0^t (< V_sR_sx, V_sy > - < R_sx, y >)ds$$

$$+ \int_0^t (< V_sx, V_sR_s^*y > - < x, R_s^*y >)ds$$

$$+ \int_0^t (< V_sT_s^1x, V_sT_s^2y > - < T_s^1x, T_s^2y >)ds$$

Posons $\psi(t) = \sup_{|a|,|b| \leq 1 \ s \leq t} | < V_sx, V_sy > - < x, y > |$, il existe une constante K telle que $0 \leq \psi(t) \leq K \int_0^t \psi(s)ds$ ce qui entraine que $\psi \equiv 0$.

Les conditions (104), (105), (106) sont donc nécessaires et suffisantes pour que V soit prolongeable en un processus d'isométries.

Soit L^ε une famille d'opérateurs bornés satisfaisant les conditions (104), (105), et (106), on voit facilement que ces conditions sont équivalentes à l'existence d'un opérateur isométrique W, un opérateur borné J, et un opérateur auto-adjoint borné H tels que

$$L^0 = W - I$$

$$L^+ = J^*$$

$$L^- = -JW$$

$$L^{\cdot} = iH - \frac{1}{2}JJ^*$$

L'équation différentielle (99) s'écrit alors

$$dV_t = V_t((W - I)da_t^0 - JWda_t^- + J^*da_t^+ + (iH - \frac{1}{2}JJ^*)dt) \tag{107}$$

V étant isométrique il possède un adjoint borné.

Les formules $< V_t^*x, y >=< x, V_ty >$ et (77) montrent que le processus V^* est solution de l'équation différentielle stochastique non-commutative

$$dV_t^* = ((W^* - I)da_t^0 + Jda_t^- - W^*J^*da_t^+ + (-iH - \frac{1}{2}JJ^*dt)V_t^* \tag{108}$$

(remarquons qu'il ne s'agit *pas* d'une équation du même type que (99), car les opérateurs L^e ne commutent pas a priori avec V.)

Supposons maintenant que l'opérateur W soit non seulement isométrique, mais aussi unitaire. En appliquant la formule d'Itô à l'expression

$$< V_t^* a \otimes \xi(u), V_t^* b \otimes \xi(v) >$$

à l'aide de l'équation (108) on constate que l'opérateur V_t^* est isométrique, et donc V_t est unitaire.

6.4 Dilatations de semi-groupes complètement positifs

On considère un opérateur unitaire W, un opérateur borné J, et un opérateur auto-adjoint borné H, et on note U le processus d'opérateurs unitaires solution de l'équation (107). Le processus adjoint U_t^* vérifie donc l'équation (108).
Nous allons utiliser les processus d'opérateurs U_t et U_t^* pour construire une dilatation d'un semi-groupe complètement positif sur $\mathcal{B}(H)$. Commençons par mettre en place les objets nécessaires.
Soit $\mathcal{W} = \mathcal{B}(H \otimes \Gamma(L_{\mathbb{C}}^2(\mathbb{R}_+)))$. On a une filtration naturelle en prenant pour $\mathcal{W}_t$ la sous-algèbre des opérateurs bornés de la forme $A \otimes Id$ sur l'espace $(H \otimes \Gamma_{t]}) \otimes \Gamma_{[t}$, avec $A \in \mathcal{B}(H \otimes \Gamma_{t]})$.
Donnons nous un état quelconque S sur $\mathcal{B}(H)$, et définissons l'état ω sur $\mathcal{W}$ par $\omega(A \otimes X) = tr(SA) < X\Omega, \Omega >$ pour $A \in \mathcal{B}(H)$ et $X \in \mathcal{B}(\Gamma)$.
Soit E_t le projecteur orthogonal sur le sous-espace $H \otimes \Gamma_{t]} \otimes \Omega_{[t}$.
Soient $a \otimes m_{t]} \otimes m'_{[t} \in H \otimes \Gamma_{t]} \otimes \Gamma_{[t}$, et $X \in \mathcal{W}$ alors

$$\mathsf{E}_t \circ X(a \otimes m_{t]} \otimes m'_{[t}) = n_{t]} \otimes \Omega_{[t}$$

avec $n_{[t} \in H \otimes \Gamma_{t]}$, et on pose

$$(E_t X)(a \otimes m_{t]} \otimes m'_{[t}) = n_{t]} \otimes m'_{[t}$$

On définit ainsi un opérateur $E_t X \in \mathcal{W}_t$, et les applications $E_t : \mathcal{W} \to \mathcal{W}_t$, forment une famille d'espérances conditionnelles qui vérifient la condition $E_t E_s = E_{s \wedge t}$.
Soit $x \in \mathcal{B}(H)$, on pose

$$A \cdot x = i[H, x] - \frac{1}{2}(JJ^* x + xJJ^* - 2JWxW^* J^*)$$

L'application $A \cdot$ est linéaire et bornée sur l'espace de Banach $\mathcal{B}(H)$.
On définit une famille de morphismes $j_t : \mathcal{B}(H) \to \mathcal{W}_t$, en posant

$$j_t(x) = U_t \circ (x \otimes Id) \circ U_t^*$$

Les U_t étant unitaires, il est clair que j_t est un morphisme à valeurs dans $\mathcal{W}$, qui préserve l'identité, et comme U est un processus adapté, il est à valeurs dans $\mathcal{W}_t$.
Nous pouvons maintenant énoncer le principal résultat de ce paragraphe.

THÉORÈME 9 – *Les applications $T_t = e^{tA}$ forment un semi-groupe d'applications complètement positives sur $\mathcal{B}(H)$, et $(\mathcal{W}, \omega, \mathcal{W}_t, E_t, j_t)_{t \in \mathbb{R}_+}$ est une dilatation de ce semi-groupe.*

Avant de démontrer ce théorème, nous allons établir deux résultats importants.
On définit des opérateurs bornés A^ε, $\varepsilon = -, 0, +$ sur $\mathcal{B}(H)$ par
$A^0 x = WxW^* - x$, $A^+ x = J^* x - WxW^* J^*$, $A^- x = xJ - JWxW^*$.

PROPOSITION 22 – *Pour tout $t \geq 0$ et tout $x \in \mathcal{B}(H)$, on a*

$$j_t(x) = x \otimes Id + \sum_\varepsilon \int_0^t j_t(A^\varepsilon x) da_t^\varepsilon$$

Démonstration
Il suffit de montrer que pour tous $a, b \in H$, tous $u, v \in S$ on a

$$< j_t(x)a \otimes \xi(u), b \otimes \xi(v) > = < (x \otimes Id + \sum_\varepsilon \int_0^t j_t(A^\varepsilon x) da_t^\varepsilon) a \otimes \xi(u), b \otimes \xi(v) >$$

or

$$< j_t(x)a \otimes \xi(u), b \otimes \xi(v) > = < (x \otimes Id) U_t^* a \otimes \xi(u), U_t^* b \otimes \xi(v) >$$

et un calcul long, mais sans difficulté utilisant la formule d'Itô et l'équation (108)
vérifiée par U^* permet de montrer la proposition.

PROPOSITION 23 – *Soit X_t un processus d'opérateurs bornés tel qu'il existe des
processus adaptés K_t^ε d'opérateurs bornés vérifiant $X_t = X_0 + \sum_\varepsilon \int_0^t K_s^\varepsilon da_s^\varepsilon$ pour
tout $t \leq 0$, alors $\forall s, t \in \mathbb{R}_+$ $E_t X_{t+s} = X_t + \int_t^{t+s} E_t K_r^\cdot dr$.*

Démonstration
Soient $a, b \in H$ et $u, v \in S$, par définition de E_t on a

$$< E_t X_{t+s} a \otimes \xi(u), b \otimes \xi(v) > = < X_{t+s} a \otimes \xi(u_{t]}), b \otimes \xi(v_{t]}) >< \xi(u_{[t}), \xi(v_{[t}) >$$

$$= \sum_{\varepsilon \neq \cdot} \int_0^t < K_r^\varepsilon a \otimes \xi(u_{t]}), b \otimes \xi(v_{t]}) >< \xi(u_{[t}), \xi(v_{[t}) > \kappa^\varepsilon(r) dr$$

$$+ \int_0^{t+s} < K_r^\cdot a \otimes \xi(u_{t]}), b \otimes \xi(v_{t]}) >< \xi(u_{[t}), \xi(v_{[t}) > dr$$

$$= \sum_{\varepsilon \neq \cdot} \int_0^t < K_r^\varepsilon a \otimes \xi(u), b \otimes \xi(v) > \kappa^\varepsilon(u) dr$$

$$+ \int_0^{t+s} < E_t K_r^\cdot a \otimes \xi(u), b \otimes \xi(v) > dr$$

$$= < X_t a \otimes \xi(u), b \otimes \xi(v) >$$

$$+ \int_t^{t+s} < E_t K_r^\cdot a \otimes \xi(u), b \otimes \xi(v) > dr$$

on en déduit la proposition.

Nous pouvons maintenant passer à la démonstration du théorème (9).
D'après les propositions (22) et (23) on a pour tous $s, t \in \mathbb{R}_+$ et $x \in \mathcal{B}(H)$

$$E_t j_{t+s}(x) = j_t(x) + \int_t^{t+s} E_t j_r(A^\cdot x) dr$$

On en déduit que $\frac{d}{ds}\big(E_t j_{t+s}(x)\big) = E_t j_{t+s}(A\cdot x)$. En résolvant cette équation différentielle, on voit que $E_t j_{t+s}(x) = j_t(e^{sA\cdot}x)$, d'où la propriété de Markov de j_t.

En particulier, $T_t(x) = \mathbb{E}_0 \circ j_t(x) \circ \mathbb{E}_0$. T_t est donc la composée de deux applications dont on voit facilement qu'elles sont complètement positives, d'une part $x \mapsto j_t(x)$ et d'autre part $X \mapsto \mathbb{E}_0 X \mathbb{E}_0$ de $\mathcal{W}$ dans $\mathcal{B}(H)$, donc T_t est un semi-groupe d'applications complètement positives sur $\mathcal{B}(H)$ et $(\mathcal{W}, \omega, \mathcal{W}_t, j_t, E_t)$ en est une dilatation.

6.5 Application aux processus de Markov

Les équations différentielles stochastiques ordinaires donnent un moyen de construire des diffusions, c'est à dire des processus de Markov à valeurs dans des variétés, dont le générateur est donné par un opérateur différentiel d'ordre 2 (voir [43]). Quitte à se placer dans une carte locale, une telle équation a la forme suivante

$$dX_t = \sigma(X_t)dB_t + b(X_t)dt \tag{109}$$

C'est une équation différentielle stochastique vectorielle dans laquelle le processus inconnu X est à valeurs dans un ouvert O de $\mathbb{R}^n$, σ est une fonction sur O à valeurs dans les matrices réelles $n \times d$, b une fonction sur O à valeurs dans $\mathbb{R}^n$, et B_t un mouvement brownien d-dimensionnel. En utilisant la formule d'Itô on peut écrire pour toute fonction f de classe C^2 sur la variété

$$df(X_t) = Lf(X_t)dB_t + Af(X_t)dt \tag{110}$$

Ici, $L = (L^1, \ldots, L^d)$ est une famille de d champs de vecteurs sur la variété, et A est un opérateur différentiel d'ordre 2, le générateur de la diffusion. Dans une carte locale, les champs de vecteurs L et l'opérateur A s'expriment au moyen de σ et b (cf [43]).

L'équation (110) peut être considérée comme une formulation intrinsèque de l'équation en coordonnées locales (109).

L'application qui à une fonction f sur la variété fait correspondre la variable aléatoire $f \circ X_t$, considérée comme opérateur de multiplication sur l'espace de Fock du mouvement brownien, est un morphisme d'algèbres. Désignons le par k_t. Cela permet d'écrire (109) sous la forme

$$dk_t(f) = k_t(A^+ f)da_t^+ + k_t(A^- f)da_t^- + k_t(A\cdot f)dt$$

avec $A^+ = A^- = L$ et $A\cdot = A$, réminiscente de la proposition 22 (mais ici on est sur un espace de Fock de multiplicité $d \geq 1$).

Maintenant que nous avons mis en évidence la similitude entre les calculs du paragraphe 6.4 et la construction des processus de diffusion, nous allons utiliser ces idées pour construire des processus de Markov discontinus en résolvant des équations différentielles du type (107). L'idée de base est la suivante, partant du processus de Markov X_t, on exhibe une martingale normale d-dimensionnelle, c'est à dire une famille de martingales M^j, $j = 1, \ldots, d$ telles que $< M^j, M^k >_t = \delta_{j,k} t$ qui permet de reconstruire le processus de Markov X_t par une équation du type (110) dans laquelle le mouvement brownien B est remplacé par la martingale normale M. L'inconvénient de la formulation de cette équation est que la martingale M n'est pas canonique. Nous verrons que si on interprète cette martingale comme une famille d'opérateurs de multiplication sur l'espace de Fock, ce qui est possible si elle possède la propriété de représentation chaotique, en utilisant le lemme 18 pour les exprimer comme des intégrales stochastiques non-commutatives d'opérateurs par rapport aux processus de base da^ε, on obtient une équation non-commutative du type de la proposition 24, que l'on résout grâce aux résultats du paragraphe précédent.

Nous allons traiter deux exemples simples. Commençons par le processus de naissance pur.

Il s'agit d'un processus de Markov sur $\mathbb{Z}$ de générateur

$$Af(x) = a_x\big(f(x+1) - f(x)\big)$$

Nous supposerons que les coefficients a_x vérifient une inégalité $0 < \frac{1}{c} < a_x < c$.
Le semi-groupe de générateur A est $T_t = e^{tA}$ sur $L^\infty(\mathbb{Z})$, et on peut décrire les trajectoires du processus correspondant de la façon suivante. Le processus part de X_0 où il reste pendant un temps de loi exponentielle de paramètre a_{X_0} puis il saute en $X_0 + 1$ où il reste un temps indépendant du premier temps de saut, de loi exponentielle de paramètre a_{X_0+1} avant de sauter en $X_0 + 2$, et ainsi de suite.
Supposons X_0 déterministe. Nous allons tout d'abord identifier l'espace L^2 du processus avec un espace de Fock en utilisant une martingale normale qui possède la propriété de représentation chaotique.
Cette martingale est

$$M_t = \sum_{s \leq t} 1_{\{X_{s-} \neq X_s\}} \frac{1}{\sqrt{a_{X_{s-}}}} - \int_0^t \sqrt{a_{X_{s-}}}\, ds$$

Elle est à variation finie, et le lemme facile suivant est laissé au lecteur.

LEMME – *M est une martingale de carré intégrable de crochets*

$$< M, M >_t = t$$

et

$$[M, M]_t = t + \int_0^t \frac{1}{\sqrt{a_{X_{s-}}}}\, dM_s$$

La martingale M est adaptée à la filtration du processus X, et réciproquement, le processus X peut être reconstruit à partir de M et de X_0 en résolvant l'équation

$$df(X_t) = \sqrt{a_{X_{t-}}}\big(f(X_{t-} + 1) - f(X_{t-})\big)dM_t + a_{X_{t-}}\big(f(X_{t-} + 1) - f(X_{t-})\big)dt$$

pour toute fonction f sur $\mathbb{Z}$.
Si on pose $Tf(x) = \sqrt{a_x}\big(f(x+1) - f(x)\big)$ on peut l'écrire

$$df(X_t) = Tf(X_{t-})dM_t + Af(X_{t-})dt$$

LEMME – *La martingale M possède la propriété de représentation chaotique.*

Ce lemme découle d'un résultat plus général dû à Emery ([27] Théorème 5).

Soit $\mathcal{X}$ la tribu du processus X, le lemme précédent entraine que $L^2(\mathcal{X}) \sim \Gamma(L^2_{\mathbb{C}}(\mathbb{R}_+))$. En appliquant la proposition 18 à M on obtient

PROPOSITION 24 – L'opérateur de multiplication par M_t en tant qu'opérateur sur l'espace de Fock s'écrit

$$M_t = \int_0^t \frac{1}{\sqrt{a_{X_{s-}}}} da_s^0 + a_t^+ + a_t^-$$

Soit $H = L^2(\mathbb{Z})$ ($\mathbb{Z}$ étant muni de la mesure de comptage), l'algèbre $L^\infty(\mathbb{Z})$ agissant par multiplication sur $L^2(\mathbb{Z})$ est une sous-algèbre de $\mathcal{B}(H)$. En utilisant l'expression précédente pour l'opérateur de multiplication par M, nous allons étudier l'équation différentielle suivante

$$dk_t(x) = k_t(A^0 x)da_t^0 + k_t(A^+ x)da_t^+ + k_t(A^- x)da_t^- + k_t(A^{\cdot}x)dt \qquad (111)$$

où k_t est un morphisme de $L^\infty(\mathbb{Z})$ dans $\mathcal{W}_t$ (avec les notations de 6.3), et les applications $A^\varepsilon : L^\infty(\mathbb{Z}) \to L^\infty(\mathbb{Z})$ sont données par $A^0(f)(n) = f(n+1) - f(n)$, $A^+ = T = A^-$, et $A^{\cdot} = A$.

Soit $W : L^2(\mathbb{Z}) \to L^2(\mathbb{Z})$ l'opérateur unitaire $Wf(n) = f(n+1)$, et J l'opérateur borné $Jf(n) = \sqrt{a_n}f(n)$, alors en tant qu'opérateurs sur $L^2(\mathbb{Z})$ on a pour tout $x \in L^\infty(\mathbb{Z})$

$A^0 x = WxW^* - x$, $A^+ x = J^* x - WxW^* J^*$, $A^- x = xJ - JWxW^*$

et $A^{\cdot} x = -\frac{1}{2}(JJ^* x + xJJ^* - 2JWxW^*J^*)$.

Les formules ci-dessus définissent en fait des applications sur $\mathcal{B}(H)$ tout entier. On peut maintenant utiliser les résultat du paragraphe précédent. Soit U_t la solution unitaire de l'équation

$$dU_t = U_t\big((W - I)da_t^0 - JWda_t^- + J^*da_t^+ - \frac{1}{2}JJ^* dt\big)$$

avec $U_0 = Id$.

Avec les notations du paragraphe 6.3, $(\mathcal{W}, \omega, \mathcal{W}_t, j_t, E_t)$ est une dilatation du semi-groupe complètement positif $T_t = e^{tA}$, T_t préserve l'algèbre $L^\infty(\mathbb{Z})$ et sa restriction à cette algèbre est le semi-groupe markovien de générateur A. Les restrictions k_t des morphismes j_t à $L^\infty(\mathbb{Z})$ sont solution de (111)

PROPOSITION 25 – Les algèbres $j_t(L^\infty(\mathbb{Z}))$, $t \in \mathbb{R}_+$ commutent.

Démonstration

Soient $x, y \in L^\infty(\mathbb{Z})$, pour tout $t \geq 0$, $j_t(x)$ et $j_t(y)$ commutent.

Soit $s \geq 0$, on a $j_{t+s}(x) - j_t(x) = \sum_\varepsilon \int_t^{t+s} j_u(A^\varepsilon x)da_u^\varepsilon$, et on vérifie en utilisant (77) que

$$(j_{t+s}(x) - j_t(x))j_t(y) = \sum_\varepsilon \int_t^{t+s} j_u(A^\varepsilon x)j_t(y)da_u^\varepsilon$$

et

$$j_t(y)(j_{t+s}(x) - j_t(x)) = \sum_\varepsilon \int_t^{t+s} j_t(y)j_u(A^\varepsilon x)da_u^\varepsilon$$

Posons

$$\alpha(r) = \sup_{\|x\|,\|y\|\leq 1 \ |a|\leq 1, \ s\leq r} |(j_t(y)j_{t+s}(x) - j_{t+s}(x)j_t(y))a \otimes \xi(u)|$$

D'après (86) il existe une constante K telle que $0 \leq \alpha(r) \leq K \int_0^r \alpha(s)ds$, d'où $\alpha \equiv 0$.
On en déduit que $j_t(L^\infty(\mathbb{Z}))$ et $j_{t+s}(L^\infty(\mathbb{Z}))$ commutent.

Considérons l'algèbre de von Neumann commutative $\mathcal{A}$ engendrée par les algèbres $j_t(L^\infty(\mathbb{Z}))$. Il existe un espace de probabilité $(\Omega, \mathcal{F}, P)$ et des variables aléatoires $X_t : \Omega \to \mathbb{Z}$ telles que $(\mathcal{A}, \omega) \sim (L^\infty(\Omega, \mathcal{F}, P), P)$, $j_t(f) = f \circ X_t$ pour $f \in L^\infty(\mathbb{Z})$.

PROPOSITION 26 – *Le processus X_t est un processus de Markov sur $\mathbb{Z}$ de semi-groupe T_t.*

Démonstration
Soient $t_1 < \ldots < t_n \in \mathbb{R}_+$ et $f_1, \ldots, f_n \in L^\infty(\mathbb{Z})$, on a

$$E[f_1(X_{t_1})\ldots f_n(X_{t_n})] = \omega(j_{t_1}(f_1)\ldots j_{t_n}(f_n))$$
$$= \omega\big(j_{t_1}(f_1)\ldots j_{t_{n-1}}(f_{n-1}T_{t_n-t_{n-1}}(f_n))\big)$$

car j_t est une dilatation de T_t.
On en déduit par récurrence sur n que l'on a bien

$$E[f_1(X_{t_1})\ldots f_n(X_{t_n})] = E[f_1 T_{t_2-t_1}(f_2 \ldots T_{t_n-t_{n-1}}(f_n)\ldots)(X_{t_1})]$$

ce qui permet de conclure facilement.

Passons maintenant à un second exemple plus intéressant d'un point de vue théorique, celui des chaînes de Markov à espace d'états fini. Je vais donner rapidement la méthode, les détails sont semblables à ceux du processus de naissance.

Soit X une chaîne de Markov en temps continu sur un espace d'états fini E. Cette chaîne a pour générateur l'opérateur $Af(x) = \sum_{y \in E} p(x, y)\big(f(y) - f(x)\big)$ où les coefficients $p(x, y)$ sont ≥ 0. Nous allons identifier E avec un groupe (pour fixer les idées, avec $\mathbb{Z}/d$ si E a d éléments), et supposer que $p(x, y) > 0$ pour $x \neq y$. On définit alors $d - 1$ martingales $(M^j)_{j \in E \setminus \{0\}}$ par

$$M_t^j = \sum_{s<t} \frac{1}{\sqrt{p(X_{s-}, X_s)}} 1_{\{X_s - X_{s-} = j\}} - \int_0^t \sqrt{p(X_s, X_s + j)}ds$$

Ces martingales sont normales, elles ont pour crochets

$$< M^j, M^k >_t = \delta_{jk} t$$

$$[M^j, M^k]_t = \delta_{jk}\big(t + \int_0^t \sqrt{p(X_{s-}, X_{s-} + j)}dM_s^j\big)$$

et de plus elles possèdent la propriété de représentation chaotique (cf [10]).
On en déduit que les opérateurs de multiplication sur l'espace de Fock correspondant sont donnés par les formules

$$M_t^j = \int_0^t \frac{1}{\sqrt{p(X_{s-}, X_{s-} + j)}}da_s^{0jj} + a_t^{+j} + a_t^{-j}$$

88

Le processus X peut être reconstruit à partir des martingales M^j en résolvant l'équation différentielle stochastique

$$df(X_t) = \sum_j \sqrt{p(X_{t-}, X_{t-} + j)}(f(X_{t-} + j) - f(X_{t-}))dM_t^j + Af(X_t)dt$$

pour toute fonction f sur E. En reportant l'expression des martingales M_t^j sur l'espace de Fock, on se ramène à résoudre l'équation

$$df(X_t) = Af(X_t)dt + \sum_{j \neq 0}(f(X_{t-} + j) - f(X_{t-}))da_t^{0jj}$$
$$+ \sqrt{p(X_{t-}, X_{t-} + j)}(f(X_{t-} + j) - f(X_{t-}))(da_t^{+j} + da_t^{-j})$$

Considérons $H = L^2(E)$ (avec la mesure de comptage sur E) et $L^\infty(E) \subset \mathcal{B}(H)$. On définit des opérateurs unitairess W^j, et des opérateurs bornés J^j, $j \in E \backslash \{0\}$ sur H par

$$W^j f(x) = f(x + j) \text{ et } J^j f(x) = \sqrt{p(x, x + j)}f(x)$$

L'équation

$$dU_t = U_t \sum_{j \neq 0}\left(W^d a_t^{0jj} + (J^j)^* da_t^{+j} - W^j J^j da_t^{-j} - \frac{1}{2} J^j(J^j)^* dt\right)$$

admet une solution composée d'opérateurs unitaire $(U_t)_{t \in \mathbb{R}_+}$.
On définit des morphismes $j_t : \mathcal{B}(H) \to \mathcal{W}$ par $j_t(x) = U_t \circ (x \otimes Id) \circ U_t^*$. On vérifie grâce à la formule d'Itô (multidimensionnelle cette fois-ci) que ces morphismes vérifient l'équation

$$dj_t(x) = j_t(A\dot{\,}x)dt + \sum_{j \neq 0}(j_t(A^{0jj}x)da_t^{0jj} + j_t(A^{+j}x)da_t^{+j} + j_t(A^jx)da_t^{-j})$$

avec $A^{0j}x = W^j x(W^j)^* - x$, $A^{+j}x = (J^j)^*x - W^j x(W^j)^*(J^j)^*$,
$A^{-j}x = xJ^j - J^j W^j x(W^j)^*$, et
$A\dot{\,}x = \sum_{j \neq 0} -\frac{1}{2}\left(J^j(J^j)^*x + xJ^j(J^j)^* - 2J^j W^j x(W^j)^*(J^j)^*\right)$.
On choisit un état S sur $\mathcal{B}(H)$ et on construit l'état ω sur $\mathcal{W}$ comme dans l'exemple précédent. On obtient alors que $(\mathcal{W}, \omega, \mathcal{W}, E_t, j_t)$ est une dilatation du semi-groupe $T_t = e^{tA}$. Il est facile de voir que ce semi-groupe laisse l'algèbre $L^\infty(E) \subset \mathcal{B}(H)$ invariante, et que sa restriction est le semi-groupe de générateur A. On montre encore que les $j_t(L^\infty(E))$ commutent et on en déduit qu'il existe un espace $(\Omega, \mathcal{F}, P)$, et des variables aléatoires $X_t : \Omega \to \mathbb{Z}$ tels que l'algèbre de von Neumann engendrée par les $j_t(L^\infty(E))$ soit isomorphe à $L^\infty(\Omega, \mathcal{F}, P)$, la restriction de ω à cette algèbre soit l'espérance pour P, et pour tout $f \in L^\infty(E)$, on ait $j_t(f) = f \circ X_t$. Les variables X_t forment un processus de Markov de générateur A et on a donc réussi à construire la chaîne de Markov de générateur A en résolvant une équation de la forme (99).

Commentaires et bibliographie

Pour écrire ce cours, j'ai utilisé abondamment le livre de K.R.Parthasarathy [60], ainsi que le gros article de P.A.Meyer [50], qui a été repris et considérablement modifié dans le livre [51]. J'espére que le présent cours pourra servir d'introduction à la lecture de ces sources qui couvrent de nombreux sujets que je n'ai pas pu traiter ici.

Les probabilités quantiques sont un vaste domaine, les commentaires et les références bibliographiques qui suivent en reflètent seulement ma connaissance partielle, et n'ont pas la prétention d'être exhaustifs.

Chapitre 1 − L'addition de variables de Bernoulli quantiques a été considérée par Meyer dans [50] comme donnant une approximation discrète de l'espace de Fock (le "bébé Fock"), d'après une idée de J.L.Journé.

La loi du processus de spin a été calculée dans [11], et [86], où on pourra trouver la solution de l'exercice.

Le théorème limite 2 est une version élémentaire d'un résultat de L.Accardi et A.Bach (non publié) exposé dans [52].

Chapitre 2 − Le contenu de ce chapitre est tout à fait classique, et peut se trouver dans tout bon ouvrage sur la théorie spectrale. La proposition 3 est inspirée de [20].

Chapitre 3 − Ce chapitre est inspiré du chapitre III de [50].

Les opérateurs de création et d'annihilation ont été introduits sous leur forme matricielle par W.Heisenberg. Le modèle des relations d'Heisenberg étudié dans ce chapitre n'est pas le plus couramment utilisé en physique, où l'on préfère utiliser les opérateurs x et $i\frac{d}{dx}$ sur $L^2(\mathbb{R}, dx)$ (dx étant la mesure de Lebesgue) à la place de Q et P. L'opérateur $i\frac{d}{dx}$ sert à calculer le moment d'une particule, et x sa position. Dans le paragraphe 3.8 pour tout élément de $SL(2, \mathbb{R})$ on a défini un opérateur unitaire U_τ (à une constante multiplicative près). Ces opérateurs vérifient $U_\tau U_\sigma = U_{\tau\sigma}\rho(\tau, \sigma)$ où ρ est un complexe de module 1. Il n'est pas possible de choisir les opérateurs U_τ de sorte que les nombres ρ soient tous égaux à 1, (on aurait alors une représentation de $SL(2, \mathbb{R})$), mais on peut en revanche construire une représentation du revêtement double de $SL(2, \mathbb{R})$ (le groupe métaplectique) appelée représentation de Segal-Shale-Weil (cf G.Lion, M.Vergne [48]).

La représentation de l'algèbre de Lie associée est composée d'opérateurs de la forme $\mathcal{P}(P, Q)$, où $\mathcal{P}$ est un polynôme homogène de degré 2 (voir aussi le chapitre I, paragraphes 15 et suivants, de V.Guillemin, S.Sternberg [37]).

Chapitre 4 − L'espace de Fock dont il est question ici est l'espace de Fock symétrique, ou bosonique dans la terminologie physique. Il existe aussi une notion d'espace de Fock antisymétrique (appelé aussi espace de Fock fermionique cf Meyer [50]), et d'espace de Fock complet (voir Speicher [82]), sur lesquels sont définis des opérateurs de création et d'annihilation qui satisfont de nouvelles relations de commutation. On peut également définir des déformations des relations de commutation (25) comme l'ont fait Bozejko et Speicher [15], qui permettent "d'interpoler" entre les relations des bosons et celles des fermions.

L'espace de Fock complet est très lié à la théorie de Voiculescu (cf [85]) des variables aléatoires libres qui a eu des applications remarquables dans la théorie des algèbres d'opérateurs.

La décompositions en chaos de l'espace L^2 du mouvement brownien est dûe à N.Wiener [88]. Ce résultat a été étendu aux processus de Lévy par K.Itô [44]. Pour les martingales normales, la théorie des décompositions chaotiques en est encore à ses débuts, les principaux résultats dans ce domaine sont dûs à M.Emery [26], [27], [28], (voir aussi le chapitre de [23] consacré à ces questions).

Chaque interprétation probabiliste de l'espace de Fock permet de considérer ses éléments commes des variables aléatoires, que l'on peut en particulier multiplier entre elles. Il existe des formules explicites pour le résultat d'une telle multiplication dans les cas des interprétations brownienne et poissonienne que l'on peut trouver dans [50]. Ces formules s'écrivent agréablement lorsqu'on utilise la notation de Guichardet pour l'espace de Fock ([38], [50]).

Chapitre 5 — Nous avons considéré l'espace de Fock construit sur $L^2(\mathbb{R}_+)$, qui possède une structure naturelle de produit tensoriel continu. On aurait pu, de façon plus intrinsèque se donner un espace de Hilbert H muni d'une famille croissante $(H_t)_{t\in\mathbb{R}_+}$ de sous-espaces. Ce point de vue est adopté dans le livre de Parthasarathy [60].

Une notion importante, que nous n'avons pas introduite ici est celle de martingale sur l'espace de Fock. Cette notion est discutée dans [50] et [60], citons aussi les références importantes [63] et [64] où des théorèmes de représentation des martingales analogues au théorème de représentation prévisible d'Itô sont démontrés. La situation est toutefois loin d'être aussi simple que dans le cas classique comme le montre un contre exemple de J.L.Journé [46].

La construction de l'intégrale stochastique non-commutative donnée dans le texte est inspirée d'un article de J.M.Lindsay [47]. Elle nécessite une certaine dose d'analyse fonctionnelle mais a l'avantage d'éviter le recours à des approximations de processus par des processus simples. Une autre construction, qui utilise la résolution d'une équation différentielle, a été donnée par Meyer dans [51].

Le lien entre intégrale non-commutative et intégrale de Skorokhod a été exploité par Belavkin [6], qui s'est inspiré des travaux de Maassen sur les opérateurs représentables par des noyaux ([49]).

L'inégalité sur les intégrales stochastiques non-commutatives est dûe à J.L.Journé (cf [50]). Nous l'avons établie ici en partant d'une inégalité semblable sur les intégrales de Skorokhod. Le problème de savoir si un processus de la forme $u_t v_t$, avec u_t mesurable par rapport au passé et v_t par rapport au futur, est intégrable au sens de Skorokhod a été étudié, entre autres, par M.Jolis et M.Sanz [45], qui ont donné des conditions suffisantes pour que ce soit le cas. Leurs calculs ont une forme assez semblable à celle employée dans le cours, mais dans un cadre plus général (la forme de v_t est y plus générale que dans le texte).

Le formalisme de Belavkin présenté à la fin du paragraphe 5.4 est développé dans [8] sous le nom d'algèbre d'Itô. Pour une application à la structure des fonctions de type positif, voir [7].

Les martingales d'Azéma ont été étudiées par Emery [26]. Ce sont les premières martingales qui ne soient pas des processus à accroissement indépendants pour lesquels on a démontré la propriété de représentation chaotique, néanmoins, un peu plus tard, M.Schürmann a montré que l'on pouvait considérer une martingale d'Azéma comme une composante d'un processus non-commutatif à accroissements indépendants (voir [77]).

Parallèlement au calcul stochastique bosonique présenté ici, il existe un calcul stochastique fermionique ([5]) et un calcul stochastique libre ([82]). Ces calculs stochastiques peuvent être unifiés comme l'on montré Parthasarathy et Sinha [65] (voir aussi [41]).

Chapitre 6 − Pour la notion d'espérance conditionnelle en probabilités quantiques, on peut lire D.Petz [67].

La construction de la dilatation présentée dans le paragraphe 6.2 se trouve dans [11] et [62]. Pour des exemples explicites de processus ainsi obtenus, voir [12], [13]. Le processus de Spin, considéré comme un processus à valeurs dans le dual de $SU(2)$ a été considéré par Eymard et Roynette [32].

Des réalisations en temps continu de ces processus, en lien avec le calcul stochastique non-commutatif ont été obtenues par Hudson et Parthasarathy [42].

Le problème de l'existence d'une solution unitaire à l'équation (99) lorsque les coefficients L^ε ne sont pas bornés et la multiplicité est infinie fait l'objet de recherches très actives. Récemment, des résultats importants ont été obtenus, notamment par Chebotarev, Fagnola, Mohari (voir l'exposé de Meyer au séminaire Bourbaki [53]). Ce problème est crucial pour la construction des processus de Markov au moyen des équations différentielles stochastiques non-commutatives. Le problème de l'existence et de l'unicité d'une solution unitaire est un analogue non-commutatif de la théorie des frontières pour les chaînes de Markov (cf [54]).

Le théorème 9 est le résultat principal de l'article [40], et son obtention était la motivation initiale du calcul stochastique non-commutatif. Le semi-groupe e^{tA} est complètement positif, et réciproquement, le théorème de Gorini, Kossakowski, Sudarshan, Lindblad (démontré dans [60]) énonce que tout semi-groupe complètemment positif sur $\mathcal{B}(H)$, uniformément continu a la forme e^{tB} où B est un opérateur qui s'exprime comme une somme d'opérateurs du type de $A^{\cdot}$.

En imitant la proposition 23 on peut se poser le problème de l'existence de morphismes $j_t : \mathcal{B}(H) \to \mathcal{W}$ satisfaisant l'équation

$$dj_t(x) = \sum_\varepsilon j_t(A^\varepsilon x) da_t^\varepsilon$$

où les A^ε sont des applications linéaires de $\mathcal{B}(H)$ dans $\mathcal{B}(H)$. Ces morphismes forment alors une *diffusion quantique*, et les A^ε sont des analogues des opérateurs différentiels qui apparaissent dans l'équation (110). La formule d'Itô, associée aux condition que les j_t soient multiplicatifs, préservent l'involution et l'identité, entraînent des conditions nécessaires algébriques sur les A^ε pour l'existence des j_t. Si ces conditions sont remplies, le problème de l'existence d'une solution a été résolu affirmativement si les A^ε sont des opérateurs bornés (cf [30], [31]). Dans le cas d'opérateurs A^ε non-bornés, on n'a que des résultats partiels ([33]).

Le contenu du paragraphe 6.5 est inspiré de [51], VI.3, et [66].

Citons encore d'autres travaux sur des thèmes voisins des diffusions quantiques, par Applebaum [3], [4], Davies et Lindsay [19], Sauvageot [75], [76].

Références

[1] L.Accardi, A.Frigerio, J.T.Lewis, Quantum Stochastic processes, *Publ. R.I.M.S. Kyoto*, 18, 1982, 94-133.

[2] L.Accardi, W.von Waldenfels Eds., Quantum Probability and applications, I-V Lect. Notes in Maths. Springer 1055, 1136, 1303, 1325, 1442, VI- World Scientific, Singapore.

[3] D.Applebaum, Unitary evolutions and horizontal lifts in quantum stochastic calculus, *Comm. Math. Phys.*, 140, 1991, 63-80.

[4] D.Applebaum, An operator theoretic approach to stochastic flows on manifolds, *Séminaire de Probabilités XXVI*, Lect. Notes in Maths. Springer 1526, 1992, 514-532.

[5] C.Barnett, R.F.Streater, I.F.Wilde, The Ito-Clifford integral I, *Jour. Funct. An.*, 48, 1982, 172-212.

[6] V.P.Belavkin, A quantum non-adapted Ito formula and stochastic analysis in Fock scale, *Jour. Funct. An.*, 102, 1991, 414-447.

[7] V.P.Belavkin, Chaotic states and stochastic integration in quantum systems, *Russian Math. Surv.* 14:1 1992, 53-116.

[8] V.P.Belavkin, The unified ito formula has the pseudo-Poisson structure, *preprint n° 98* Centro Vito Volterra, Universita degli studi di Roma II, 1992.

[9] B.V.R.Bhat, K.R.Parthasarathy, Markov dilations of non-conservative dynamical semi-groups and a quantum boundary theory, *preprint*, 1992.

[10] P.Biane, Chaotic representation for finite Markov chains, *Stochastics*, 30, 1990, 61-68.

[11] P.Biane, Marches de Bernoulli quantiques, *Séminaire de Probabilités XXIV*, Lect. Notes in Maths. Springer 1426, 1990, 329-344.

[12] P.Biane, Quantum random walks on the dual of SU(n), *Prob. Th. and Rel. Fields*, 89, 1991, 117-129.

[13] P.Biane, Minuscule weights and random walks on lattices, *Quantum probability and applications*, VII, World Scientific, Singapore, 1992, 51-65.

[14] N.Bourbaki, Groupes et algèbres de Lie, Chapitre 9, Hermann, Paris, 1969.

[15] M.Bozejko, R.Speicher, An example of generalized brownian motion, *Comm. Math. Phys.*, 137, 1991, 519-531.

[16] T.Chihara, An introduction to orthgonal polynomials, Gordon and Breach, New York, 1978.

[17] K.L.Chung, R.J.Williams, Introduction to stochastic integration, *Progress in Mathematics*, Birkhaüser, 1983.

[18] E.B.Davies, Quantum theory of open systems, Academic Press, 1976.

[19] E.B.Davies, J.M.Lindsay, Non-commutative Markov semi-groups, *preprint*, 1990.

[20] C.Dellacherie, P.A.Meyer, Probabilités et potentiel, Chapitre I à IV, Hermann, Paris, 1975.

[21] C.Dellacherie, P.A.Meyer, Probabilités et potentiel, Chapitre V à VIII, Hermann, Paris, 1980.

[22] C.Dellacherie, P.A.Meyer, Probabilités et potentiel, Chapitre IX à XI, Hermann, Paris, 1983.

[22] C.Dellacherie, P.A.Meyer, B.Maisonneuve, Probabilités et potentiel, Chapitre XVII à XXIV, Hermann, Paris, 1993.

94

[24] J.Dixmier, Les algèbres d'opérateurs dans l'espace hilbertien, (Algèbres de von Neumann), Gauthier-Villars, Paris, 1957.

[25] J.Dixmier, Les C*-algèbres et leurs représentations, Gauthier-Villars, Paris, 1964.

[26] M.Emery, On the Azéma martingales, *Séminaire de Probabilités XXIII*, Lect. Notes in Maths. Springer 1372, 1990, 66-87.

[27] M.Emery, Quelques cas de représentation chaotique, *Séminaire de Probabilités XXIV*, Lect. Notes in Maths. Springer 1426, 1991, 10-23.

[28] M.Emcry, On the chaotic representation property for martingales, *preprint* 1993.

[29] D.E.Evans, J.T.Lewis, Dilations of dynamical semi-groups, *Comm. Math. Phys.*, 50, 1976, 219-228.

[30] M.Evans, Existence of quantum diffusions, *Prob. Th. and Rel. Fields*, 81, 1989, 473-483.

[31] M.Evans, R.L.Hudson, Multidimensionnal quantum diffusions, *Quantum probability and applications*, III, Lect. Notes in Maths. Springer 1303, 1988, 69-88.

[32] P.Eymard, B.Roynette, Marches aléatoires sur le dual de SU(2), *Marches aléatoires sur les groupes*, Lect. Notes in Maths. Springer 624, 1977.

[33] F.Fagnola, K.B.Sinha, Quantum flows with unbounded structure maps and finite degrees of freedom, *à paraître dans Jour. Lond. Math. Soc.*

[34] W.Feller, An introduction to probability theory and its applications, Vol I, John Wiley & Sons, New York 1977.

[35] R.P.Feynman, R.B.Leighton, M.Sands, The Feynman Lectures on physics, Vol III, Addison Wesley, Reading Mass. 1965.

[36] B.Gaveau, P.Trauber, L'intégrale stochastique comme opérateur de divergence dans l'espace fonctionnel, *Jour. Funct. An.* 46, 1982, 230-238.

[37] V.Guillemin, S.Sternberg, Symplectic techniques in Physics, Cambridge University Press, 1984.

[38] A.Guichardet, Symmetric Hilbert space and related topics, *Lect. Notes in Maths. Springer* 261, 1970.

[39] R.Halmos, Introduction to Hilbert space, Chelsea, New York, 1951.

[40] R.L.Hudson, K.R.Parthasarathy, Quantum Ito's formula and stochastic evolutions, *Comm. Math. Phys.*, 93, 1984, 301-323.

[41] R.L.Hudson, K.R.Parthasarathy, Unification of Fermion and Boson stochastic calculus, *Comm. Math. Phys.*, 115, 1988, 47-53.

[42] R.L.Hudson, K.R.Parthasarathy, Casimir chaos in Boson Fock space, *preprint*, 1993.

[43] N.Ikeda, S.Watanabe, Stochastic differential equations and diffusion processes, North-Holland, Kodansha, 1981.

[44] K.Itô, Multiple Wiener integrals, *Jour. Math. Soc. Japan.* 3, 1951, 157-169.

[45] M.Jolis, M.Sanz, Integrator properties of the Skorokhod integral, *Stochastics,*, 41, 1992, 163-176.

[46] J.L.Journé, P.A.Meyer, Une martingale d'opérateurs bornés non représentable en intégrale stochastique, *Séminaire de Probabilités XX*, Lect. Notes in Maths. Springer 1204, 1986, 313-316.

[47] M.Lindsay, Quantum and non-causal stochastic calculus, *à paraître dans Prob. Theory and rel. Fields.*

[48] G.Lion, M.Vergne, The Weil representation, Maslov index, and theta series, Progress in Mathematics, Vol 6, Birkhaüser, 1980.

95

[49] H.Maassen, Quantum Markov processes on Fock space described by integral kernels, *Quantum probability and applications*, II, Lect. Notes in Maths. Springer 1136, 1985, 361-374.

[50] P.A.Meyer, Eléments de Probabilités quantiques, *Séminaire de Probabilités XX*, Lect. Notes in Maths. Springer 1204, 1986, 186-312.

[51] P.A.Meyer, Quantum Probability for Probabilists, Lect. Notes in Maths. Springer 1538, 1993.

[52] P.A.Meyer, Approximation de l'oscillateur harmonique (d'après L.Accardi et A.Bach), *Séminaire de Probabilités XXIII*, Lect. Notes in Maths. Springer 1372, 1990, 175-182.

[53] P.A.Meyer, Progrès récent en calcul stochastique quantique, *Séminaire Bourbaki*, exposé 761, 1992.

[54] A.Mohari, K.R.Parthasarathy, A quantum probabilistic analogue of Feller's condition for the existence of unitary markovian cocycles in Fock space, *I.S.I. preprint*, 1992.

[55] E.Nelson, Analytic vectors, *Ann. Math.* 70, 1959, 572-615.

[56] J.von Neumann, Mathematical Foundations of Quantum Mechanics, Princeton University Press, 1951.

[57] J.Neveu, Processus Aléatoires Gaussiens, Presses de l'Université de Montréal, 1968.

[58] J.Neveu, Processus Ponctuels, *Ecole d'élé de Probabilités de Saint-Flour VI*, Lect. Notes in Maths. Springer 598, 1976, 250-445.

[59] D.Nualart, Non-causal stochastic integrals and calculus, *Stochastic analysis and related topics*, Lect. Notes in Maths. Springer 1316, 1986, 80-129.

[60] K.R.Parthasarathy, An introduction to quantum stochastic calculus, Monographs in mathematics, Vol 85, Birkhäuser, 1992.

[61] K.R.Parthasarathy, Azéma martingales and quantum stochastic calculus, Proc. R.C.Bose Symposium, Wiley Eastern, 1990, 551-569.

[62] K.R.Parthasarathy, A generalized Biane's process, *Séminaire de Probabilités XXIV*, Lect. Notes in Maths. Springer 1426, 1990, 345-348.

[63] K.R.Parthasarathy, K.B.Sinha, Representation of bounded martingales in Fock space, *Jour. Funct. An.*, 67, 1986, 126-151

[64] K.R.Parthasarathy, K.B.Sinha, Representation of a class of bounded martingales, II, *Quantum probability and applications*, III, Lect. Notes in Maths. Springer 1303, 1988, 232-250.

[65] K.R.Parthasarathy, K.B.Sinha, Unification of quantum noise processes in Fock spaces, *Quantum probability and applications*, VI, World Scientific, Singapore, 1991.

[66] K.R.Parthasarathy, K.B.Sinha, Markov chains as Evans-Hudson diffusions in Fock space, *Séminaire de Probabilités XXIV*, Lect. Notes in Maths. Springer 1426, 1990, 362-369.

[67] D.Petz, Conditionnal expectations in quantum probability, *Quantum probability and applications*, III, Lect. Notes in Maths. Springer 1303, 1988, 251-260.

[68] J.Pitman, One dimensionnal brownian motion and the three dimensionnal Bessel process, *Adv. Appl. Prob.*, 7, 1975, 511-526.

[69] P.Protter, Stochastic integration and differential equations, a new approach, Springer, 1990.

96

[70] M.Reed, B.Simon, Methods of modern mathematical physics, II, Fourier analysis and self-adjointness, Academic press, 1970.

[71] S.Reynaud, Introduction à la réduction du bruit quantique, *Ann. Phys. Fr.* 15, 1990, 63-162.

[72] W.Rudin, Functionnal analysis, Mac Graw Hill, 1973.

[73] J.de Sam Lazaro, P.A.Meyer, Méthodes de martingales et théorie des flots, *Zeit. f. Wahr.* 18, 1971, 116-140.

[74] J.L.Sauvageot, Markov quantum semi-groups admit covariant Markov C*-dilations, *Comm. Math. Phys.* 106, 1986, 91-103.

[75] J.L.Sauvageot, Quantum Dirichlet forms, differential calculus and semi-groups, *Quantum probability and applications*, V, Lect. Notes in Maths. Springer 1442, 1990, 334-346.

[76] J.L.Sauvageot, Semi-groupe de la chaleur transverse sur la C*-algèbre d'un feuilletage riemannien, C.R.A.S. 310, Série I, 1990, 531-536.

[77] M.Schürmann, The Azéma martingales as components of quantum independent increment processes, *Séminaire de Probabilités XXV*, Lect. Notes in Maths. Springer 1485, 1991, 24-30.

[78] J.P.Serre, Algèbres de Lie semi-simples complexes, Benjamin, New York, 1966.

[79] D.Shale, Linear symmetries of free boson fields, *Trans. Am. Math. Soc.* 103, 1962, 149-167.

[80] A.V.Skorokhod, On a generalisation of a stochastic integral, *Teor. Ver.* 20, 1975, 219-233.

[81] R.Speicher, A new example of "independance" and "white noise", *Prob. Th. and Rel. Fields*, 84, 1990, 141-159.

[82] R.Speicher, Stochastic integration on the full Fock space with the help of a fernel calculus, *Publ. R.I.M.S. Kyoto*, 27, 1991, 149-184.

[83] N.J.Vilenkin, Special functions and the theory of group representations, *Translations of the A.M.S.*, 22, 1968.

[84] G.F.Vincent-Smith, Dilation of a dissipative quantum dynamical system into a quantum Markov process, *Proc. Lond. Math. Soc.* 49, 1984, 58-72.

[85] D.Voiculescu, Free non-commutative random variables, random matrices and the II_1 factors of free groups, *Quantum probability and applications*, VI, World Scientific, Singapore, 1991, 473-487.

[86] W.von Waldenfels, The Markov process of total spin, *Séminaire de Probabilités XXIV*, Lect. Notes in Maths. Springer 1426, 1990, 357-361.

[87] S.Watanabe, Stochastic differential equations and Malliavin calculus, Lectures on Mathematics and Physics, 73, Tat Institute of Fundamental Research, 1984.

[88] N.Wiener, The homogeneous chaos, *Amer. Jour. Math.* 55, 1938, 897-936.

CNRS, Laboratoire de Probabilités
Université Paris 6, Tour 56 3e étage
4 place Jussieu 75252 PARIS Cedex 05

LECTURES ON FREE PROBABILITY

THEORY

DAN VOICULESCU

Originally published in: *École d'Été de Probabilités de Saint-Flour XXVIII – 1998*,
Lecture Notes in Mathematics, Vol. **1738**, 279–349, DOI: 10.1007/ BFb0106719,
© Springer-Verlag Berlin Heidelberg 2000, Reprint by Springer-Verlag Berlin Heidelberg 2012

Contents

0   Introduction   283

**1 .   Noncommutative Probability and Operator
         Algebra Background**

   1.1   Noncommutative probability spaces   284
   1.2   C*-probability spaces   284
   1.3   W*-probability spaces   285
   1.4   Examples   285
   1.5   The distributions of noncommutative random variables   286
   1.6   Examples   287
   1.7   Usual independence   288
   1.8   Free independence   288
   1.9   Examples   290
   1.10   Further properties of free independence   291
   1.11   Free products   293

**2 .   Addition of Freely Independent Noncommutative
         Random Variables**

   2.1   Additive free convolution   294
   2.2   Canonical form   294
   2.3   Compactly suported probability measures on $\mathfrak{R}$ 295
   2.4   The R-transform   296
   2.5   The free central limit theorem   298
   2.6   Superconvergence in the free central limit theorem   299
   2.7   The free Poisson law   300
   2.8   The relation of free Poisson to be semicircle   301
   2.9   Free convolution of measures with unbounded support   302
   2.10   The Cauchy distribution   302
   2.11   Free infinite divisibility and the dree analogue of the   303
          Levy-Hincin theorem
   2.12   Free stable laws   304
   2.13   More free harmonic analysis on $\mathfrak{R}$ 305
   2.14   Spectra of convolution operators on free products of groups   307

**3   Multiplication of Freely Independent Noncommutative
        Random Variables**

   3.1   Multiplicative free convolution   308
   3.2   Probability measures on $\mathfrak{R} \geq 0$ and T   309
   3.3   The S-transform   310
   3.4   Examples   310
   3.5   Multiplicative free infinite divisibility on T   312
   3.6   Multiplicative free infinite divisibility on $\mathfrak{R} \geq 0$ 312

4   Generalized Canonical Form, Noncrossing Partitions

4.1	Combinatorial work	313
4.2	Generalized canonical form	313
4.3	Noncrossing partitions and the formula for θ	315

5   Free Independence with Amalgamation

5.1	Classical background on conditional independence	316
5.2	Conditional expectations in von Neumann algebras with trace state (background)	317
5.3	Noncommutative probability spaces over B and free independence over B	318
5.4	More on B-free probability theory	319

6   Some Basic Free Processes

6.1	The semicircular functor (free analogue of the Gaussian functor)	319
6.2	Free Poisson processes	321
6.3	Stationary processes with free increments	322
6.4	The Markov transitions property of processes with free increments	323
6.5	Free Markovianity	324
6.6	Further results	325

7   Random Matrices in the Large N Limit

7.1	Asymptotic free independence for Gaussian matrices	325
7.2	Asymptotic free independence for unitary matrices	328
7.3	Corollaries of the basic asymptotic free independence results	329
7.4	Further asymptotic free independence results	331

8   Free Entropy

8.1	Clarifications	332
8.2	Background on classical entropy via microstates	332
8.3	Free entropy via matricial microstates	333
8.4	Properties of $\chi(X_1, ..., X_n)$	334
8.5	Free entropy dimension	338
8.6	Classical Fisher information and the adjoint of the derivation	339
8.7	Free Fisher information of one variable	340
8.8	The underlying idea of the microstates-free-approach	341
8.9	Noncommutative Hilbert transforms	341
8.10	Free Fisher information and free entropy in the microstates-free approach	344

9   References   346

0. Introduction

What is free probability theory? It is not a euphemism for the advocacy of an unconstrained attitude in the practice of probability. It can rather be described by the exact formula

free probability theory = noncommutative probability theory + free independence

Around 1982, I realized that the right way to look at certain operator algebra problems, was by imitating some basic probability theory. More precisely: in noncommutative probability theory a new kind of independence can be defined by replacing tensor products with free products and this can help understand the von Neumann algebras of free groups. The subject has evolved into a kind of parallel to basic probability theory, which should be called free probability theory. On the way, links with random matrix theory, combinatorics, and some mathematical physics questions have appeared, along with the applications to operator algebras.

These lectures dwell on the noncommutative probability side, i.e., I do not discuss the applications to von Neumann algebras. Thus the operator algebra prerequisites are kept to a minimum and the emphasis is on the parallelism to classical probability and on random matrices. This includes presenting the free analogues of the central limit theorem, of Gaussian and Poisson processes, of the addition of independent variables, of infinite divisibility, and of some Markovianity questions. Large random matrices provide an asymptotic model of free probability theory and the application of free probability theory to the large N limit is explained. A substantial part at the end of these notes is about free entropy, my current research interest. Free entropy, as the name suggests is the free analogue of the entropy quantity in Shannon's information theory.

The reader who wishes to supplement this brief introduction with more detailed accounts should consult the book [67] which is the standard introduction to the subject including operator algebra applications up to 1991, the memoir [44] about the combinatorial approach via noncrossing partitions and the collection of papers [66]. Not discussed in these lectures is stochastic free integration for which we refer to [15],[22],[28]. The details on free entropy are in my original papers [57–62].

While working on these notes designed for probabilists (i.e., no operator algebra background presumed) I was helped by having had to talk about free probability in front of "true probabilists" audiences on various occasions, in particular the Minikurs I gave at ETH Zürich in 1997. The reader of these notes is invited to join the author in thanking Deborah Craig and Faye Yeager for the fine and speedy typing.

1. Noncommutative Probability and Operator Algebra Background

The basic data of a probability space can be encoded in the algebra of numerical random variables endowed with the expectation functional. Going noncommutative, we have the following definition.

1.1. Noncommutative probability spaces

Definition. A noncommutative probability space is a unital algebra $\mathcal{A}$ over $\mathbb{C}$ endowed with a linear functional $\varphi : \mathcal{A} \to \mathbb{C}$, $\varphi(1) = 1$. Elements $a \in \mathcal{A}$ are called random variables.

This definition is pure algebra. To add positivity to the picture we need to look at C^*-probability spaces.

1.2. C^*-probability spaces

Definition. A noncommutative probability space $(\mathcal{A}, \varphi)$ is a C^*-probability space if $\mathcal{A}$ is a C^*-algebra and φ is a state.

$\mathcal{A}$ is a C^*-algebra if it is a Banach algebra $(\mathcal{A}, \| \cdot \|)$ with an involution $a \to a^*$ which is isomorphic to an algebra of bounded operators on some Hilbert space with the usual operator norm and involution defined by taking the adjoint. This means we can identify $(\mathcal{A}, \| \cdot \|, *)$ with an algebra $I \in \mathcal{A} \subset B(\mathcal{H})$ ($B(\mathcal{H})$ the bounded operators) which is norm closed and such that $T \in \mathcal{A} \Rightarrow T^* \in \mathcal{A}$.

A *state* $\varphi : \mathcal{A} \to \mathbb{C}$ is a linear functional such that $\varphi(1) = 1$ and $\varphi(a) \geq 0$ if $a \geq 0$. Here $a \geq 0$ is in the sense of operator theory. Equivalent definitions for $a \geq 0$ are:

(1) $a = x^*x$ for some $x \in \mathcal{A}$, or

(2) $a = a^*$ and the spectrum $\sigma(a) \subset [0, \infty)$. or

(3) when $\mathcal{A} \subset B(\mathcal{H})$, $\langle ah, h \rangle \geq 0$ for all $h \in \mathcal{H}$.

By *the Gelfand-Naimark-Segal theorem* a C^*-probability space $(\mathcal{A}, \varphi)$ can always be realized in the form $\mathcal{A} \subset B(\mathcal{H})$ and $\varphi(a) = \langle a\xi, \xi \rangle$ for $a \in \mathcal{A}$, where $\xi \in \mathcal{H}$ is a unit vector $\|\xi\| = 1$.

Note that this is the situation encountered in *quantum mechanics*: A is an algebra of observables, ξ is the state-vector (wave-function) which gives the expectation of observables.

By a *theorem of Gelfand and Naimark* the unital commutative C^*-algebra is isomorphic to the algebra $C(X)$ of continuous complex-valued functions on some compact space X.

For probability theory we often want to go beyond continuous functions. This can be done using von Neumann algebras (W^*-algebras).

1.3. W^*-probability spaces

Definition. (A, φ) is a W^*-probability space if the pair is isomorphic to a W^*-algebra and some vector state $\langle \cdot \xi, \xi \rangle$.

A W^*-algebra or *von Neumann algebra* $I \in A \subset B(\mathcal{H})$ is a C^*-algebra of operators which is weakly closed, i.e., if $(T_i)_{i \in I} \subset A$ is a net such that $\langle T_i \eta_1, \eta_2 \rangle$ converges to $\langle T \eta_1, \eta_2 \rangle$ for all pairs $\eta_1, \eta_2 \in \mathcal{H}$, then $T \in A$.

1.4. Examples

1°. Let $(X, \Sigma, d\sigma)$ be a probability space. This data, up to sets of measure zero, can be encoded into $(L^\infty(X, \Sigma, d\mu), \varphi)$ where $\varphi(f) = \int f \, d\mu$. This is a W^*-probability space. Indeed, on $\mathcal{H} = L^2(X, \Sigma, d\mu)$ the multiplication operators $M(f)$ defined by L^∞-functions f, $M(f)g = fg$ form a von Neumann algebra and $\varphi(f) = \langle f1, 1 \rangle$, where 1 is the constant function taking value 1, viewed as an element of L^2.

2°. Let G be a discrete group. On $\ell^2(G)$ consider the left regular representation λ where $\lambda(g)e_h = e_{gh}$ (here $(e_g)_{g \in G}$ is the orthonormal basis of $\ell^2(G)$). Let $L(G)$ be the von Neumann algebra generated by $\lambda(G)$. Since $\lambda(G)$ is a group of unitary operators, $L(G)$ is the weak closure of the linear span of $\lambda(G)$. On $L(G)$ there is a canonical state: the von Neumann trace $\tau(T) = \langle Te_e, e_e \rangle$ ($T \in L(G)$, e_e the basis element for the neutral element $e \in G$.) This means $\tau(\lambda(g)) = \delta_{g,e}$ or for a linear combination $\tau(\sum_{g \in G} c_g \lambda(g)) = c_e$ (the constant term). τ is a trace-state, i.e., in addition to being a state, it is a trace, i.e., $\tau(T_1 T_2) = \tau(T_2 T_1)$. Indeed, $\tau(\lambda(g_1)\lambda(g_2)) = \tau(\lambda(g_2)\lambda(g_1))$ which is equivalent to $\tau(\lambda(g_1 g_2)) = \tau(\lambda(g_2 g_1))$, since $g_1 g_2 = e \iff g_2 g_1 = e$. The algebra $L(G)$ can be shown to consist of the bounded left convolution operators, i.e. formal sums $\sum_{g \in G} c_g \lambda(g)$ which define bounded operators on $\ell^2(G)$.

3°. For random matrices, the following is a convenient noncommutative probability context. Let $(X, \Sigma, d\sigma)$ be a probability space, let M_n denote the algebra of complex $n \times n$ matrices. Let further

$$\mathcal{A}_n = \bigcap_{1 \leq p < \infty} L^p(X, M_n)$$

be the algebra of $n \times n$ random matrices which are p-integrable for $1 \leq p < \infty$. Let $\varphi_n : \mathcal{A}_n \to \mathbb{C}$ be given by

$$\varphi_n(T) = \frac{1}{n} \int_X \mathrm{Tr}(T(x)) d\sigma(x) .$$

The algebra $\mathcal{A}_n$ is not a C^*-algebra (not a Banach algebra) but it has a natural involution $T(\cdot) \to T^*(\cdot)$.

So a random matrix $T \in \mathcal{A}_n$ gives rise to two kinds of random variables: a classical random variable $X \to M_n$ and a noncommutative random variable, by viewing T as an element of $(\mathcal{A}_n, \varphi_n)$.

1.5. The distribution of noncommutative random variables

The information about the distribution of a bounded random variable can be encoded in the moments. So in our algebraic context the distribution of $a \in \mathcal{A}$ will be described by the collection of moments $\varphi(a^n)$ ($n \geq 0$). Equivalently, let $\mu_a : \mathbb{C}[X] \to \mathbb{C}$ be the linear functional on the polynomials in the indeterminate X, given by

$$\mu_a(P) = \varphi(P(a)) .$$

We call μ_a the distribution of a. More generally, if $(a_i)_{i \in I}$ is a family of noncommutative random variables in $(\mathcal{A}, \varphi)$, let $\mathbb{C}\langle X_i \mid i \in I \rangle$ be the polynomials in the noncommuting indeterminates $(X_i)_{i \in I}$ and let

$$\mu_{(a_i)_{i \in I}} : \mathbb{C}\langle X_i \mid i \in I \rangle \to \mathbb{C}$$

be the linear map defined by

$$\mu_{(a_i)_{i \in I}}(P) = \varphi(P((a_i)_{i \in I})) .$$

We call $\mu_{(a_i)_{i \in I}}$ the distribution of the family $(a_i)_{i \in I}$. Clearly $\mu_{(a_i)_{i \in I}}$ is completely determined by the noncommutative moments

$$\varphi(a_{i_1} \ldots a_{i_p}) = \mu_{(a_i)_{i \in I}}(X_{i_1} \ldots X_{i_p}) .$$

If $(\mathcal{A}, \varphi)$ is a C^*-probability space and $a = a^* \in \mathcal{A}$ is a self-adjoint element than μ_a can be described by a compactly supported probability measure on $\mathbb{R}$. Indeed identifying $\mathcal{A}$ with an algebra of operators on a Hilbert space $\mathcal{H}$, by the spectral theorem there is a projection-valued compactly supported measure $E(\cdot\;; a)$ so that for every continuous function

$$f(a) = \int f(t) dE((-\infty, t]; a) .$$

Let $\xi \in \mathcal{H}$ be a unit vector such that $\varphi = \langle \cdot\,\xi, \xi \rangle$ and consider the scalar measure $\nu(\cdot) = \langle E(\cdot\;; a)\xi, \xi \rangle$. Then for any polynomial $P \in \mathbb{C}[X]$ we have

$$\mu_a(P) = \varphi(P(a)) = \langle (\int P(t) dE)\xi, \xi \rangle = \int P(t) d\langle E\xi, \xi \rangle = \int P(t) d\nu(t) .$$

We will often identify μ_a with ν.

1.6. Examples

1°. Let $(X, \Sigma, d\sigma)$ be a probability space and $f \in L^\infty(X, \Sigma, d\sigma)$ a bounded random variable. Selfadjointness $f = f^*$ means f is real-valued. The spectral projection $E(\omega; f)$ for a Borel set $\omega \subset \mathbb{R}$ is the indicator function $\chi_{f^{-1}(\omega)}$ of the set $f^{-1}(\omega)$. Since the expectation $\int \chi_{f^{-1}(\omega)} d\sigma = \sigma(f^{-1}(\omega))$, it is easily seen that μ_f corresponds precisely to the classical distribution of f.

2°. In the context of example 1.4.3, let $T = T^* \in \mathcal{A}_n$ be a self-adjoint $n \times n$ random matrix. Let $\lambda_1(x) \leq \cdots \leq \lambda_n(x)$ be the n eigenvalues of $T(x)$, $x \in X$. Then

$$\varphi_n(P(T)) = \frac{1}{n} \int \mathrm{Tr}(P(T(x))) d\sigma(x)$$

$$= \frac{1}{n} \int \sum_{1 \leq j \leq n} P(\lambda_j(x)) d\sigma(x)$$

$$= \int \left(\int_{\mathbb{R}} P(t) \left(d \sum_{1 \leq j \leq n} n^{-1} \delta_{\lambda_j(x)}\right)(t)\right) d\sigma(x)$$

$$= \int_{\mathbb{R}} P(t) d\nu(t)$$

where $\nu = \int \left(\sum_{1 \leq j \leq n} n^{-1} \delta_{\lambda_j(x)}\right) d\sigma(x)$. Thus the measure which is the expectation of counting measures on the eigenvalues of $T(x)$ gives the distribution μ_T.

288

Under reasonable conditions when the moment problem for the moments of ν has a unique solution, we identify ν and μ_T.

1.7. Usual independence

Definition. A family of subalgebras containing 1, $(\mathcal{A}_i)_{i \in I}$ in the noncommutative probability space $(\mathcal{A}, \varphi)$ is independent if the algebras $\mathcal{A}_i, \mathcal{A}_j$ for distinct indices $i, j \in I$ commute and $\varphi(a_1 \ldots a_n) = \varphi(a_1) \ldots \varphi(a_n)$ whenever $a_k \in \mathcal{A}_{i(k)}$, $1 \leq k \leq n$ and $k \neq \ell \Rightarrow i(k) \neq i(\ell)$.

This definition is modelled on tensor products. Indeed let $\mathcal{H}_1, \mathcal{H}_2$ be Hilbert spaces and endow $\mathcal{H}_1 \otimes \mathcal{H}_2$ with the scalar product such that

$$\langle h_1 \otimes h_2, k_1 \otimes k_2 \rangle = \langle h_1, k_1 \rangle \langle h_2, k_2 \rangle .$$

If $I \in \mathcal{A}_1 \subset \mathcal{B}(\mathcal{H}_1)$, $I \in \mathcal{A}_2 \subset \mathcal{B}(\mathcal{H}_2)$ consider $\mathcal{A}_1 \otimes I = \{T \otimes I : T \in \mathcal{A}_1\} \subset B(\mathcal{H}_1 \otimes \mathcal{H}_2)$ and $I \otimes \mathcal{A}_2 = \{I \otimes T : T \in \mathcal{A}_2\} \subset B(\mathcal{H}_1 \otimes \mathcal{H}_2)$. Consider on $B((\mathcal{H}_1 \otimes \mathcal{H}_2)$ a state $\varphi(\cdot) = \langle \cdot \xi_1 \otimes \xi_2, \xi_1 \otimes \xi_2 \rangle$ where $\xi_k \in \mathcal{H}_k$ are unit vectors. Then $\mathcal{A}_1 \otimes I$ and $I \otimes \mathcal{A}_2$ are independent in $(B(\mathcal{H}_1 \otimes \mathcal{H}_2), \varphi)$. In quantum mechanics this situation occurs in the description of noninteracting systems (bosonic case).

The independence studied in classical probability theory after passing to the L^∞-algebras (see 1.4, example 1) clearly fits into the framework of the above definition.

Note that since the algebras $(\mathcal{A}_i)_{i \in I}$ commute, the algebra they generate is spanned linearly by products $a_1 \ldots a_n$ where $a_k \in \mathcal{A}_{i(k)}$ and $i(1), \ldots, i(n)$ are distinct. Hence if the $(\mathcal{A}_i)_{i \in I}$ are independent, the expectation functional φ is completely determined on the algebra generated by the $(\mathcal{A}_i)_{i \in I}$ if the restrictions $(\varphi | \mathcal{A}_i)_{i \in I}$ are given.

1.8 Free independence

Definition. A family of subalgebras containing 1, $(\mathcal{A}_i)_{i \in I}$ in the noncommutative probability space $(\mathcal{A}, \varphi)$ is freely independent if $\varphi(a_1 \ldots a_n) = 0$ whenever $\varphi(a_k) = 0$, $1 \leq k \leq n$ and $a_k \in \mathcal{A}_{i(k)}$ where consecutive indices $i(k) \neq i(k+1)$ are distinct. A family of subsets in $(\mathcal{A}, \varphi)$ is freely independent if the subalgebras they generate with 1 are freely independent.

Like for usual independence, for free independence if the restrictions $(\varphi \mid \mathcal{A}_i)_{i \in I}$ are given then φ is completely determined on the algebra generated by the $\mathcal{A}_i$'s, $i \in I$. Indeed, the algebra is spanned by monomials $a_1 a_2 \ldots a_n$ where $a_k \in \mathcal{A}_{i_k}$ and $i(k) \neq i(k+1)$ for $1 \leq k < n$, since consecutive elements which are in the same algebra can be replaced by their product. We have by the freeness condition:

$$\varphi((a_1 - \varphi(a_1)1)(a_2 - \varphi(a_2)1) \ldots (a_n - \varphi(a_n)1)) = 0 \ .$$

Expanding the product we get a formula for $\varphi(a_1 \ldots a_n)$ in terms of expectations of products of $< n$ elements. Thus, by induction $\varphi(a_1 \ldots a_n)$ can be computed if we know the $\varphi | \mathcal{A}_i$.

Note also that freely independent variables in general do not commute. Thus classical numerical random variables, except for trivial cases, like that of a constant random variable, are not freely independent. To get an idea why this should be so, let $a, b \in \mathcal{A}$ be freely independent variables. If a, b would commute we would have $abab = a^2 b^2$ and by free independence:

$$\varphi(a^2 b^2) = \varphi((a^2 - \varphi(a^2)1)(b^2 - \varphi(b^2)1)) + \varphi(a^2)\varphi(b^2) = \varphi(a^2)\varphi(b^2) \ .$$

In general for a product of two elements free independence and usual independence yield the same expectation. On the other hand, computing the expectation of $abab$ and using the notation $a_0 = a - \varphi(a)1$, $b_0 = b - \varphi(b)1$ we have

$$\varphi(abab) = \varphi(ab_0 a_0 b) + \varphi(b)\varphi(a^2 b) + \varphi(a)\varphi(ab^2) - \varphi(a)\varphi(b)\varphi(ab)$$
$$= \varphi(ab_0 a_0 b) + \varphi(b)^2 \varphi(a^2) + \varphi(a)^2 \varphi(b^2) - \varphi(a)^2 \varphi(b)^2 \ .$$

Further,

$$\varphi(ab_0 a_0 b) = \varphi(a_0 b_0 a_0 b_0) + \varphi(a)\varphi(b_0 a_0 b_0)$$
$$+ \varphi(b)\varphi(a_0 b_0 a_0) - \varphi(a)\varphi(b)\varphi(b_0 a_0) = 0 \ .$$

Hence

$$\varphi(abab) = \varphi(b)^2 \varphi(a^2) + \varphi(a)^2 \varphi(b^2) - \varphi(a)^2 \varphi(b)^2 \ .$$

Thus

$$\varphi(a^2 b^2) - \varphi(abab) = (\varphi(a^2) - \varphi(a)^2)(\varphi(b^2) - \varphi(b)^2)$$

which is nonzero in general.

1.9 Examples. 1°. Let $G = \bigstar_{i \in I} G_i$ a group G which is the free product of its subgroups $(G_i)_{i \in I}$. This means the G_i's generate G and there is "no nontrivial relation among the G_i's". More precisely, we have $g_1 \ldots g_n \neq e$, whenever $g_j \in G_{i(j)} \backslash \{e\}$, $1 \leq j \leq n$ and $i(k) \neq i(k+1)$ for $1 \leq k < n$.

Let $(L(G), \tau)$ be the von Neumann algebra of the left regular representation λ of G and the von Neumann trace τ appearing in Example 1.4.2. Let further $\mathcal{A}_i$ be the von Neumann algebra generated by $\lambda(G_i)$. Then $\mathcal{A}_i$ can be shown to consist of those bounded convolution operators $\sum_{g \in G} c_g \lambda(g)$ such that $c_g = 0$ if $g \in G \backslash G_i$. If $a \in \mathcal{A}_i$ and $\tau(a) = 0$ then in addition to $c_g = 0$ for $g \notin G_i$ we also have $c_e = \tau(a) = 0$. Thus let $a_k \in \mathcal{A}_{i(k)}$, $\tau(a_k) = 0$, $1 \leq k \leq n$ and assume $i(k) \neq i(k+1)$ for $1 \leq k < n$. Then whatever the convergence problems in computing $a_1 \ldots a_n$ it is clear that the resulting convolution operator $\sum_{g \in G} c_g \lambda(g)$ will have $c_g \neq 0$ only if $g = g_1 \ldots g_n$ for some $g_j \in G_{i(j)} \backslash \{e\}$. In particular $c_e = 0$, since $g_1 \ldots g_n \neq e$, so that the family $(\mathcal{A}_i)_{i \in I}$ is freely independent.

The converse is even easier to prove, i.e., if the $(\mathcal{A}_i)_{i \in I}$ are freely independent then the subgroups $(G_i)_{i \in I}$ are free in the algebraic sense. Indeed, let $g_k \in G_{i(k)} \backslash \{e\}$, $1 \leq k \leq n$ and assume $i(k) \neq i(k+1)$ for $1 \leq k < n$. Then $\lambda(g_k) \in \mathcal{A}_{i(k)}$ and $\tau(\lambda(g_k)) = 0$. This implies $0 = \tau(\lambda(g_1) \ldots \lambda(g_n)) = \tau(\lambda(g_1 \ldots g_n))$, i.e. $g_1 \ldots g_n \neq e$.

2°. Let $\mathcal{H}$ be a complex Hilbert space and let

$$T\mathcal{H} = \bigoplus_{n \geq 0} \mathcal{H}^{\otimes n}, \quad \text{where} \quad \mathcal{H}^{\otimes 0} = \mathbb{C}1 \quad \text{and} \quad \mathcal{H}^{\otimes n} = \underbrace{\mathcal{H} \otimes \ldots \otimes \mathcal{H}}_{n\text{-times}},$$

be the Boltzmann-Fock space, i.e., the Fock-space without any symmetry. Let further $\ell(h)$ for $h \in \mathcal{H}$ be the left creation operator so that $\ell(h)\xi = h \otimes \xi$, and on the algebra $\mathcal{B}(T\mathcal{H})$ (=all bounded operators on $T\mathcal{H}$) consider the state given by the vacuum expectation, i.e.

$$\varphi(T) = \langle T1, 1 \rangle .$$

Let $(e_i)_{i \in I}$ be an orthonormal system in $\mathcal{H}$. We shall prove that the family of sets $(\{\ell(e_i), \ell(e_i)^*\})_{i \in I}$ is freely independent in $(\mathcal{B}(T\mathcal{H}), \varphi)$.

There is clearly no loss to enlarge the orthonormal system to a basis. To simplify notations put $\ell_i = \ell(e_i)$. Then $T(\mathcal{H})$ has an orthonormal basis given by

$$\{1\} \amalg \coprod_{n \geq 1} \{e_{i_1} \otimes \ldots \otimes e_{i_n} : (i_1, \ldots, i_n) \in I^n\} .$$

Then $\ell_i 1 = e_i$, $\ell_i e_{i_1} \otimes \ldots \otimes e_{i_n} = e_i \otimes e_{i_1} \otimes \ldots \otimes e_{i_n}$ and $\ell_i^* 1 = 0$, $\ell_i^* e_{i_1} \otimes \ldots \otimes e_{i_n} = \delta_{ii_1} e_{i_2} \otimes \ldots \otimes e_{i_n}$.

Since ℓ_i is an isometry and the ranges of different ℓ_i's are orthogonal (the range of ℓ_i has orthonormal basis given by the $e_i \otimes e_{i_1} \otimes \ldots \otimes e_{i_n}$) we have

$$\ell_i^* \ell_j = \delta_{ij} I \ .$$

The algebra generated by $\{I, \ell_i, \ell_i^*\}$ is spanned by the $\ell_i^q \ell_i^{*p}$, $p + q > 0$ and I. Note that $\varphi(\ell_i^q \ell_i^{*p}) = \langle \ell_i^{*p} 1, \ell_i^{*q} 1 \rangle = 0$ since at least one of the numbers p, q is > 0.

Free independence amounts to proving that

$$\varphi\left(\ell_{i_1}^{q_1} \ell_{i_1}^{*p_1} \ell_{i_2}^{q_2} \ell_{i_2}^{*p_2} \ldots \ell_{i_n}^{q_n} \ell_{i_n}^{*p_n}\right) = 0$$

if $p_k + q_k > 0$, $i_k \neq i_{k+1}$. If the above expectation is $\neq 0$ we must have $q_1 = 0$ since otherwise $\ell_{i_1}^{q_1} \ldots 1 \in e_{i_1} \otimes \mathcal{TH}$ which is orthogonal to 1. Then $p_1 > 0$ and we must have $q_2 = 0$ since otherwise $\ell_{i_1}^{*p_1} \ell_{i_2}^{q_2} = 0$. Hence $p_2 > 0$ and we continue getting in the end $q_1 = \cdots = q_n = 0$, $p_1 > 0, p_2 > 0, \ldots, p_n > 0$. Then however we have $\ell_{i_n}^{*p_n} 1 = 0$ so that the expectation is zero.

The two basic examples above are not unrelated. The basis we produced for $\mathcal{TH}$ shows that we deal with the free semigroup generated by I. Further connections will appear when we will discuss the free analogue of Gaussian processes. A different type of example of free independence will be provided by the asymptotic behavior of large random matrices we will examine later.

1.10 Further properties of free independence

The following three assertions are left as an exercise (proofs can be found in [67]).

(i) Let $(\mathcal{A}_i)_{i \in I}$, $1 \in \mathcal{A}_i$ be freely independent subalgebras in $(\mathcal{A}, \varphi)$. Let $I = \coprod_{j \in J} I_j$ be a partition and for each $j \in J$ let $\mathcal{B}_j$ be the subalgebra generated by $\bigcup_{i \in I_j} \mathcal{A}_i$. Then the family of subalgebras $(\mathcal{B}_j)_{j \in J}$ is freely independent in $(\mathcal{A}, \varphi)$.

(ii) If the family of subalgebras $(\mathcal{A}_i)_{i \in I}$ is freely independent in $(\mathcal{A}, \varphi)$ and if for each $i \in I$ there is a family of subalgebras $(\mathcal{C}_{i,k})_{k \in K(i)}$ which is freely independent in $\mathcal{A}_i$, then the family of subalgebras $(\mathcal{C}_{i,k})_{k \in K(i)}$ where $K = \coprod_{i \in I} \{i\} \times K(i)$ is freely independent in $(\mathcal{A}, \varphi)$.

(iii) Let $(\mathcal{A}_i)_{i \in I}$, $1 \in \mathcal{A}_i$ be freely independent subalgebras in $(\mathcal{A}, \varphi)$. Assume $\varphi | \mathcal{A}_i$ is a trace for all $i \in I$ (i.e. $\varphi(xy) = \varphi(yx)$ if $x, y \in \mathcal{A}_i$). Then φ is a trace on the subalgebra of A generated by $\bigcup_{i \in I} A_i$. In particular the conclusion holds if the A_i are commutative.

In the C^* and W^* contexts we also have the following.

(iv) *Let (A, φ) be a C^*- (resp. W^*-) probability space and let $(A_i)_{i \in I}$ be a freely independent family of subalgebras in (A, φ) so that $1 \in A_i$ and $T \in A_i \Rightarrow T^* \in A_i$ for each $i \in I$ (i.e., A_i are $*$-subalgebras). Then the C^*-subalgebras (resp. W^*-subalgebras) $C^*(A_i)$ (resp. $W^*(A_i)$) which are the norm closures (resp. weak closures) of the A_i, still form a freely independent family in (A, φ).*

The statement about C^*-algebras is an immediate consequence of the fact that a state $\varphi : A \to \mathbb{C}$ is a norm-continuous functional on a C^*-algebra. The W^*-algebra is easy assuming a basic technical fact on W^*-algebras (Kaplansky's density theorem):

If $M \subset \mathcal{B}(\mathcal{H})$ is a von Neumann algebra and $C \subset M$ is a weakly dense $*$-subalgebra, then the unit ball C_1 is $*$-strongly dense in the unit ball M_1, i.e., for every $T \in M$, $\|T\| \leq 1$ there is a net $(X_i)_{i \in I}$ of elements in C with $\|X_i\| \leq 1$ so that for all $h \in \mathcal{H}$ the nets $(X_i h)_{i \in I}$ and $(X_i^* h)_{i \in I}$ converge to T and respectively T^*.

The above C^*- and resp. W^*-closure result for free independence is frequently used in conjunction with the fact that:

If A is a C^*- (resp. W^*-) algebra in $\mathcal{B}(\mathcal{H})$ and $T \in A$ is a normal element, i.e., $T^*T = TT^*$, then for every continuous (resp. Borel) function $f : \sigma(T) \to \mathbb{C}$, $\sigma(T)$ the spectrum of T, then $f(T) \in A$. In particular the spectral projections of a normal element $T \in A$ are in A when A is a von Neumann algebra.

Note also that combining (i) and C^*-closure with example 1.9.2 we have:

Let $\mathcal{H}_j \subset \mathcal{H}$ ($j \in J$) be closed subspaces which are pairwise orthogonal. Let further $C^*(\ell(\mathcal{H}_j))$ be the C^*-subalgebra in $B(T\mathcal{H})$ generated by the creation operators $\ell(\mathcal{H}_j) = \{\ell(h) : h \in \mathcal{H}_j\}$. Then the $C^*(\ell(\mathcal{H}_j))$, $j \in J$ are freely independent in $(B(T\mathcal{H}), \langle \cdot 1, 1 \rangle)$.

1.11 Free products

The realization of independent random variables in classical probability is done via product spaces. In the noncommutative the corresponding construction for usual independence is via tensor products. For free independence there are corresponding free product constructions, depending on the chosen context (algebraic, C^* or W^*). We will outline here some of the constructions at the level of operators acting on Hilbert spaces with specified state vectors. A complete discussion can be found in [2],[49],[67].

Let $I \in \mathcal{A}_i \subset \mathcal{B}(\mathcal{H}_i)$, $i \in I$ be C^*-algebras acting on Hilbert spaces and let $\xi_i \in \mathcal{H}_i$ be unit vectors which define states $\varphi_i(\cdot) = \langle \cdot \xi_i, \xi_i \rangle$ on $\mathcal{A}_i$. Here is how to construct a C^*-probability space $(\mathcal{A}, \varphi)$ which contains the $\mathcal{A}_i$ as subalgebras which are freely independent and for which $\varphi | \mathcal{A}_i = \varphi_i$.

We first construct a free product of the pairs $(\mathcal{H}_i, \xi_i)$ $(i \in I)$. Let $\overset{\circ}{\mathcal{H}}_i = \mathcal{H} \ominus \mathbb{C}\xi_i$ and

$$\mathcal{H} = \mathbb{C}\xi \oplus \bigoplus_{n \geq 1} \bigoplus_{i_1 \neq \cdots \neq i_n} \overset{\circ}{\mathcal{H}}_{i_1} \otimes \ldots \otimes \overset{\circ}{\mathcal{H}}_{i_n}$$

where $\|\xi\| = 1$ and the direct sums are orthogonal. We denote this construction by

$$(\mathcal{H}, \xi) = \underset{i \in I}{\bigstar} (\mathcal{H}_i, \xi_i) .$$

Let further for each $i \in I$,

$$\mathcal{H} = \mathbb{C}\xi \oplus \bigoplus_{n \geq 1} \bigoplus_{\substack{i_1 \neq i_2 \neq \cdots \neq i_n \\ i_1 \neq i}} \overset{\circ}{\mathcal{H}}_{i_1} \otimes \ldots \otimes \overset{\circ}{\mathcal{H}}_{i_n} .$$

We define a unitary operator $V_i : \mathcal{H}_i \otimes \mathcal{H}(i) \to \mathcal{H}$ by identifying

$$\xi_i \otimes \xi \longrightarrow \xi ,$$
$$\overset{\circ}{\mathcal{H}}_i \otimes \xi \longrightarrow \overset{\circ}{\mathcal{H}}_i ,$$
$$\xi_i \otimes (\overset{\circ}{\mathcal{H}}_{i_1} \otimes \ldots \otimes \overset{\circ}{\mathcal{H}}_{i_n}) \to \overset{\circ}{\mathcal{H}}_{i_1} \otimes \ldots \otimes \overset{\circ}{\mathcal{H}}_{i_n}$$
$$\overset{\circ}{\mathcal{H}}_i \otimes (\overset{\circ}{\mathcal{H}}_{i_1} \otimes \ldots \otimes \overset{\circ}{\mathcal{H}}_{i_n}) \to \overset{\circ}{\mathcal{H}}_i \otimes \overset{\circ}{\mathcal{H}}_{i_1} \otimes \ldots \otimes \overset{\circ}{\mathcal{H}}_{i_n}$$

Then $\mathcal{A}_i$ acts on $\mathcal{H}_i \otimes \mathcal{H}(i)$ via $T \rightsquigarrow T \otimes I$, $T \in \mathcal{A}_i$. Using the V_i's we transport the $\mathcal{A}_i$'s on $\mathcal{H}$ by identifying $\mathcal{A}_i$ and $V_i(\mathcal{A}_i \otimes I)V_i^*$. Let $\mathcal{A}$ be the C^*-algebra generated by $\bigcup_{i \in I} V_i(\mathcal{A}_i \otimes I)V_i^*$ and φ the state on $\mathcal{A}$ defined by $\langle \cdot \xi, \xi \rangle$. If the $\mathcal{A}_i$ are W^*-algebras, to obtain a W^*-probability space $(\mathcal{A}, \varphi)$ one takes the W^*-algebra generated by

$$\bigcup_{i \in I} V_i(\mathcal{A}_i \otimes I)V_i^* .$$

Note that if the G_i's are groups and $(\mathcal{H}_i, \xi_i) = (\ell^2(G_i), e_e)$ then $\overset{\circ}{\mathcal{H}}_i = \ell^2(G_i \backslash \{e\})$ has basis $(e_g)_{g \in G_i \backslash \{e\}}$ and it is easy to identify the free product $\bigstar_{i \in I}(\mathcal{H}_i, \xi_i) = (\mathcal{H}, \xi)$ with $(\ell^2(G), e_e)$ where G is the free product of groups G_i.

2. Addition of Freely Independent Noncommutative Random Variables

2.1 Additive free convolution [49]

Classically, the distribution of the sum of two independent random variables is the convolution of their distributions. There is a parallel to this in free probability.

Let a, b be freely independent in (A, φ). Then by 1.8, the restriction of φ to the subalgebra generated by $\{1, a, b\}$ is determined by the restrictions of φ to the subalgebras generated by $\{1, a\}$ and respectively $\{1, b\}$. In particular the moments $\varphi((a + b)^n)$ $(n \geq 0)$ are completely determined by the $\varphi(a^p)$ and $\varphi(b^q)$ $(p, q \in \mathbb{N})$. Thus the distribution μ_{a+b} is completely determined by the distributions μ_a and μ_b. *Hence we may define a free convolution operation* $\boxplus$ *on the distributions of noncommutative random variables such that* $\mu_a \boxplus \mu_b = \mu_{a+b}$ *whenever a and b are freely independent in some noncommutative probability space.* Note that the space of distributions is the set of linear functionals $\mu :$ $\mathbb{C}[X] \to \mathbb{C}$ with $\mu(1) = 1$.

2.2 Canonical form

Using creation and annihilation operators on the Boltzmann Fock space there is a class of noncommutative random variables with a remarkable behavior under free addition. Such variables will be said to be in canonical form.

Lemma. [50] *Let e_1, e_2 be two orthogonal unit vectors in the Hilbert space $\mathcal{H}$ and let $\ell_1 = \ell(e_1)$ and $\ell_2 = \ell(e_2)$ the creation operators on $T\mathcal{H}$. Let further*

$$T_1 = \ell_1^* + \alpha_0 I + \alpha_1 \ell_1 + \alpha_2 \ell_1^2 + \dots$$
$$T_2 = \ell_2^* + \beta_0 I + \beta_1 \ell_2 + \beta_2 \ell_2^2 + \dots$$

Then $T_1 + T_2$ and

$$T_3 = \ell_1^* + (\alpha_0 + \beta_0)I + (\alpha_1 + \beta_1)\ell_1 + (\alpha_2 + \beta_2)\ell_1^2 + \dots$$

have the same distribution in $(\mathcal{B}(\mathcal{TH}), \langle \cdot 1, 1 \rangle)$.

SKETCH OF PROOF. Consider the expansions

$$T_3 = \ell_1^* + \sum_{j \geq 0} \alpha_j \ell_1^j + \sum_{j \geq 0} \beta_j \ell_1^j$$

$$T_1 + T_2 = (\ell_1 + \ell_2)^* + \sum_{j \geq 0} \alpha_j \ell_1^j + \sum_{j \geq 0} \beta_j \ell_2^j$$

Here $(\ell_1 + \ell_2)^*$ will be viewed as a "single element". There is an obvious bijection between the terms in the expansion of T_3 and $T_1 + T_2$. This bijection extends to a bijection between the terms in the expansions of T_3^m and $(T_1+T_2)^m$. The numerical coefficients of corresponding monomials are obviously equal so we are left with checking the equality of expectations of the creation and annihilation operators product: (For instance: $\alpha_1 \beta_2 \alpha_3 \ell_1 \ell_2^2 (\ell_1 + \ell_2)^* \ell_1^3$ corresponds to $\alpha_1 \beta_2 \alpha_3 \ell_1 \ell_1^2 \ell_1^* \ell_1^3$. We must check that $\langle \ell_1 \ell_2^2 (\ell_1 + \ell_2)^* \ell_1^3 1, 1 \rangle$ and $\langle \ell_1 \ell_1^2 \ell_1^* \ell_1^3 1, 1 \rangle$ are equal.)

Indeed, if in a product $\ell_1^{w_1} \ell_1^{w_2} \ldots \ell_1^{w_n} 1$ with $w_j \in \{1, *\}$ we replace every time ℓ_1 by ℓ_1 or ℓ_2 and we replace ℓ_1^* by $(\ell_1 + \ell_2)^*$ then the result equals 1 or a vector orthogonal to 1 at the same time (note that $(\ell_1 + \ell_2)^* \ell_2 = (\ell_1 + \ell_2)^* \ell_1 = 1$ like $\ell_1^* \ell_1 = I$, etc.). $\qquad \square$

Note that T_1, T_2 in the Lemma are freely independent.

We were somewhat imprecise about the sums defining the T_j's. To get bounded operators one may require that at most finitely many terms be nonzero. On the other hand, this doesn't really matter since there exists an algebraic variant of the Hilbert space construction, which accommodates infinite formal sums.

2.3 Compactly supported probability measures on $\mathbb{R}$

All this looks rather algebraic. Like for usual convolution there is an analysis side: free convolution gives an operation on compactly supported probability measures on $\mathbb{R}$.

Indeed, compactly supported probability measures on $\mathbb{R}$ are the distributions of self-adjoint elements in C^*-probability spaces. Consider for instance in $L^2(\mathbb{R}, d\nu)$ the multiplication operator by the identical function and the expectation given by the vector $1 \in L^2(\mathbb{R}, d\nu)$ (the C^*-algebra may be taken

$B(L^2(\mathbb{R}, d\nu))$. This yields a realization of ν as a distribution. Given two probability measures with compact support ν_1, ν_2 on $\mathbb{R}$, using 1.11 we get a pair a, b of freely independent random variables in a C^*-probability space (A, φ) so that $a = a^*$, $b = b^*$, $\mu_a = \nu_1$, $\mu_b = \nu_2$. Then μ_{a+b} also corresponds to a compactly supported probability measure on $\mathbb{R}$ since $a + b = (a + b)^*$.

2.4 The R-transform

We may compute free convolutions by computing moments $\varphi((a + b)^m)$. A better way, I found, is via a linearizing map. This is analogous to the use in classical probability theory of the logarithm of the Fourier transform, which is a linearizing map of usual convolution.

Theorem. [50] If $\mu : \mathbb{C}[X] \to \mathbb{C}$ is the distribution of a random variable let

$$G_\mu(z) = \sum_{n \geq 0} \mu(X^n) z^{-n-1}$$

and let K_μ be the formal inverse $G_\mu(K_\mu(z)) = z$ and let $R_\mu(z) = K_\mu(z) - z^{-1}$. Then

$$R_{\mu \boxplus \nu} = R_\mu + R_\nu \; .$$

Remarks. 1°. To compute $\mu \boxplus \nu$ using the theorem one proceeds as follows. One computes successively $G_\mu, G_\nu, K_\mu, K_\nu, R_\mu, R_\nu$ then $R_{\mu \boxplus \nu} = R_\mu + R_\nu$ and then one works backwards $K_{\mu \boxplus \nu}, G_{\mu \boxplus \nu}$.

2°. In case μ is a compactly supported probability measure on $\mathbb{R}$,

$$G_\mu(z) = \int \frac{d\mu(t)}{z - t}$$

is the Cauchy transform of μ, which is an analytic function in $\mathbb{C}\backslash\mathrm{supp}\ \mu$. The inversion of G_μ to get K is carried out in a neighborhood of ∞. The R-series is then an analytic function in a neighborhood of 0. To recover $\mu \boxplus \nu$ from $G_{\mu \boxplus \nu}$ amounts to solving a moment problem, or equivalently to taking boundary values, in the sense of distributions of $-\pi^{-1}\mathrm{Im}\ G(x + iy)$ as $y \downarrow 0$.

3°. If $R_\mu(z) = \sum_{n \geq 0} R_{n+1}(\mu) z^n$ then the formulae in the theorem imply the existence of universal polynomials so that $\mu(X^n) = P_n(R_1(\mu), \ldots, R_n(\mu))$ and $R_n(\mu) = Q_n(\mu(X), \ldots, \mu(X^n))$. Note that $R_n(\mu \boxplus \nu) = R_n(\mu) + R_n(\nu)$. which shows the $R_n(\mu)$ play the role of cumulants or semiinvariants for free convolution. We have in particular $R_1(\mu) = \mu(X)$, $R_2(\mu) = \mu(X^2) - (\mu(X))^2$ which are the same as the formulae for the classical case, but this is no longer true for higher n. We will return to this in the discussion of the combinatorial aspects.

THE IDEA OF THE PROOF OF THE THEOREM. When adding freely independent random variables with given distributions, we can choose among different realizations of the variables. Thus we could work in the context of creation and annihilation operators on the Boltzmann-Fock space with the vacuum expectation. If the freely independent variables are of the form considered in Lemma 2.2, i.e.,

$$T_1 = \ell_1^* + \sum_{j\geq 0} \alpha_j \ell_1^j, \qquad T_2 = \ell_2^* + \sum_{j\geq 0} \beta_j \ell_2^j$$

(don't worry about getting into formal sums; they can be handled) then by the lemma, $T_3 = \ell_1^* + \sum_{j\geq 0}(\alpha_j + \beta_j)\ell_1^j$ has the same distribution as $T_1 + T_2$. This means precisely for T_1 of the above form, the map

$$\mu_{T_1} \rightsquigarrow \sum_{j\geq 0} \alpha_j z^j$$

linearizes free convolution. Hence we will get a linearizing map for free convolution if for every distribution μ we find some

$$T = \ell_1^* + \sum_{j\geq 0} \alpha_j \ell_1^j$$

with distribution μ w.r.t. $\langle \cdot 1, 1 \rangle$. Since

$$\langle T^k 1, 1 \rangle = \alpha_{k-1} + \text{polynomial in } \alpha_0, \ldots, \alpha_{k-2}$$

for every $k \geq 1$ we infer such a T (formal sum) can be found.

To get the formulae in the theorem I used Toeplitz operators. This is based on the fact that the matrix of T in the orthonormal basis $1, e_1, e_1 \otimes e_1, e_1 \otimes e_1 \otimes e_1, \ldots$ is the Toeplitz matrix

$$
\begin{matrix}
\alpha_0 & 1 & 0 & 0 & 0 & \cdots \\
\alpha_1 & \alpha_0 & 1 & 0 & 0 & \cdots \\
\alpha_2 & \alpha_1 & \alpha_0 & 1 & 0 & \cdots \\
\alpha_3 & \alpha_2 & \alpha_1 & \alpha_0 & 1 & \cdots \\
\vdots & \vdots & \vdots & \vdots & \vdots &
\end{matrix}
$$

which has symbol $z^{-1} + \sum_{j\geq 0} \alpha_j z^j$.

116

2.5 The free central limit theorem

One of the things one can do using the R-transform is to prove a free central limit theorem (I actually proved such a theorem in [49] before having the formulae for the R-transform, but with the R-transform [50] this is just an exercise.)

Theorem. *Let $a_1, a_2, \cdots \in (A, \varphi)$ be a sequence of freely independent random variables. Assume $\varphi(a_j) = 0$ $(j \in \mathbb{N})$, $\lim\limits_{n \to \infty} \sum\limits_{1 \le j \le n} \varphi(a_j^2) = \alpha^2/4 > 0$ and $\sup\limits_{j \in \mathbb{N}} |\varphi(a_j^k)| = C_k < \infty$. Then*

$$\lim_{n \to \infty} \mu_{n^{-\frac{1}{2}}(a_1 + \cdots + a_n)}(P) = 2\pi^{-1}\alpha^{-2} \int_{-\alpha}^{\alpha} P(t)\sqrt{\alpha^2 - t^2}\, dt$$

SKETCH OF PROOF. 1. Let us first see how R behaves under dilations. We have

$$G_{\mu_{ra}}(z) = \varphi((z - ra)^{-1}) = \varphi(r^{-1}(r^{-1}z - a)^{-1}) = r^{-1}G_{\mu_a}(r^{-1}z) .$$

Hence $K_{\mu_{ra}}(z) = rK_{\mu_a}(rz)$ and

$$R_{\mu_{ra}}(z) = K_{\mu_{ra}}(z) - r(rz)^{-1} = rR_{\mu_a}(rz) .$$

$2°$. As pointed out in Remark $2.4.3°$, $R_n(\mu)$ is given by a universal polynomial $Q_n(\mu(X), \ldots, \mu(X^n))$. Hence in view of the boundedness assumption on the k-th order moments of the a_j's we also have bounds

$$|R_{n+1}(\mu_{a_j})| \le \rho_n \quad \text{for all } j \in \mathbb{N}$$

in the expansion

$$R_{\mu_{a_j}}(z) = \sum_{k \ge 0} R_{k+1}(\mu_{a_j})z^k .$$

$3°$. We have

$$R_{\mu_{n^{-\frac{1}{2}}(a_1 + \cdots + a_n)}} = \sum_{1 \le j \le n} n^{-\frac{1}{2}} R_{\mu_{a_j}}(n^{-\frac{1}{2}}z)$$

$$= \sum_{k \ge 0} n^{\frac{-k-1}{2}} \left(\sum_{1 \le j \le n} R_{k+1}(\mu_{a_j}) \right) z^k .$$

4°. Since $R_1(\mu_{a_j}) = \varphi(a_j) = 0$,
$R_2(\mu_{a_j}) = \varphi(a_j^2) - (\varphi(a_j))^2 = \varphi(a_j^2)$. Hence

$$R_{\mu_{n^{-\frac{1}{2}}(a_1+\cdots+a_n)}}(z) = n^{-1}\left(\sum_{1\leq j\leq n}\varphi(a_j^2)\right)z + \sum_{k\geq 2}O(n^{\frac{1-k}{2}})z^k \,.$$

Since the moments of a variable are given by universal polynomials in the coefficients of the R-transform, we infer $n^{-\frac{1}{2}}(a_1 + \cdots + a_n)$ has a limit distribution, the R-transform of which is $\frac{\alpha^2}{4}\,z$.

5°. To conclude the proof we have to find the distribution μ with R-transform $\frac{\alpha^2}{4}\,z$. We have

$$K_\mu = z^{-1} + \frac{\alpha^2}{4}\,z$$

$$G_\mu^{-1} + \frac{\alpha^2}{4}\,G_\mu = z\,,$$

$$\frac{\alpha^2}{4}\,G_\mu^2 - zG_\mu + 1 = 0$$

$$G_\mu = \frac{z + \sqrt{z^2 - \alpha^2}}{\alpha^2/2}$$

(The choice of branch of the square root is dictated by the requirement Im $z > 0 \Rightarrow$ Im $G_\mu < 0$.) The boundary values of $-\pi^{-1}G_\mu(x + iy)$ as $y \downarrow 0$ then gives that μ is the probability measure on $\mathbb{R}$ with density $\frac{2}{\pi\alpha^2}\sqrt{\alpha^2 - t^2}$ when $-\alpha \leq t \leq \alpha$ and 0 elsewhere. $\qquad\square$

Remarks. 1°. The distribution given by the density $\frac{2}{\pi\alpha^2}\sqrt{\alpha^2 - t^2}$, $-\alpha \leq t \leq \alpha$ is called a semicircle law (actually the graph is a semiellipse).

2°. Recall the fact that the R-transform of μ solves the problem of finding a noncommutative random variable in canonical form $\ell_1^* + \alpha_0 I + \alpha_1\ell_1 + \alpha_2\ell_1^2 + \cdots$, with distribution μ w.r.t. the vacuum (the underlying idea of the proof of Theorem 2.4). Hence the last part of the proof of the free central limit theorem shows that $\ell_1^* + \frac{\alpha^2}{4}\ell_1$ has distribution $\frac{2}{\pi\alpha^2}(\alpha^2 - t^2)^{\frac{1}{2}}dt$, $|t| \leq \alpha$. In particular, if $\alpha = 2$ we have a selfadjoint operator $\ell_1^* + \ell_1$.

2.6 Superconvergence in the free central limit theorem

The convergence to the semicircle law in Theorem 2.5 is in the sense of moments. This may leave the impression that due to the high noncommutativity of free independence, convergence in the free central limit theorem must be worse than in the classical theorem. Actually the opposite is true, as shown more recently: there is a superconvergence phenomenon for uniformly bounded self-adjoint variables. We state the result in the equivalent form of a theorem about free convolution.

Theorem. [6] *Let μ_j ($j \in \mathbb{N}$) be probability measures on R: Assume* supp $\mu_j \subset$ $[-C, C]$, $\int t d\mu_j(t) = 0$ *and* $\lim\limits_{n \to \infty} n^{-1} \sum\limits_{1 \leq j \leq n} \int t^2 d\mu_j(t) = \dfrac{\alpha^2}{4}$. *Let further D_λ denote the pushforward of a measure on $\mathbb{R}$ by the homotethy $t \to \lambda t$ and let σ_α denote the semicircle distribution supported in $[-\alpha, \alpha]$ given by the density $\frac{2}{\pi \alpha^2}(\alpha^2 - t^2)^{\frac{1}{2}}$. Then $D_{n^{-\frac{1}{2}}}(\mu_1 \boxplus \cdots \boxplus \mu_n) = \nu_n$ "superconverges" to σ_α in the sense that:*

(i) *There is $N \in \mathbb{N}$ such that if $n \geq N$ the measure ν_n is Lebesgue absolutely continuous and* supp $\nu_n = [a_n, b_n]$ *with* $\lim\limits_{n \to \infty} a_n = -\alpha$, $\lim\limits_{n \to \infty} b_n = \alpha$.

(ii) *The densities p_n of ν_n w.r.t. Lebesgue measure ($n \geq N$) converge uniformly to the density of σ_α.*

(iii) *For every $\varepsilon > 0$ there is $\delta > 0$ such that for some $N_1 \in \mathbb{N}$, if $n \geq N_1$ the densities p_n have analytic extensions to the rectangle $K_{\varepsilon,\delta} = \{z \in \mathbb{C} \mid |\mathrm{Re}\, z| \leq \alpha - \varepsilon, |\mathrm{Im} z| \leq \delta\}$ and these extensions converge uniformly on $K_{\varepsilon,\delta}$ to $\frac{2}{\pi\alpha^2}(\alpha^2 - z^2)^{\frac{1}{2}}$.*

The above superconvergence is in sharp contrast with the fact that for Bernoulli measures $\mu_j = \frac{1}{2}(\delta_1 + \delta_{-1})$ the classical convolutions $\mu_1 * \cdots * \mu_n$ are atomic measures for all $n \in \mathbb{N}$.

The proof of the above theorem is a complex analysis proof involving the R-transform (see [6]).

2.7 The free Poisson law

By analogy with classical Poisson laws, free Poisson laws are the limits of free convolutions

$$\left((1 - \frac{a}{n})\delta_0 + \frac{a}{n}\delta_b\right)^{\boxplus n}$$

as $n \to \infty$ (i.e., we replace $*$ by $\boxplus$).

The computation via R-transform goes as follows. Let $\mu_{1/n} = (1 - \frac{a}{n})\delta_0 + \frac{a}{n}\delta_b$, then

$$G = (1 - \frac{a}{n})z^{-1} + \frac{a}{n}(z - b)^{-1}$$

and

$$R = -\frac{1}{z} + \frac{bz + 1 + ((bz + 1)^2 - 4bz(1 - a/n))^{\frac{1}{2}}}{2z} = \frac{ab}{n(1 - bz)} + O(n^{-2}) .$$

Hence nR converges to $\frac{ab}{1-bz}$ in some neighborhood of 0. We infer the moments of $\mu_{1/n}^{\boxplus n}$ converge to the moments of a probability measure with R-transform $\frac{ab}{1-bz}$. Hence the K series of that measure is $z^{-1} + \frac{ab}{1-bz}$ and the Cauchy transform G is

$$\frac{z + b(1 - a) + ((z - b(1 + a))^2 - 4ab^2)^{\frac{1}{2}}}{2bz} .$$

We infer that the free Poisson distribution is given by

$$\mu = \begin{cases} (1 - a)\delta_0 + \nu & 0 \le a \le 1 \\ \nu & a > 1 \end{cases}$$

where ν is the measure with support $[b(1 - \sqrt{a})^2, b(1 + \sqrt{a})^2]$ with density

$$(2\pi bt)^{-1}(4ab^2 - (t - b(1 + a))^2)^{\frac{1}{2}} .$$

2.8 The relation of free Poisson to the semicircle

Except for an atom at zero, when $0 \le a < 1$, the free Poisson law is given by a Cauchy transform which is not too different from that of the semicircle law, has compact support and has a density which looks like a tilted semicircle. Actually, if $a = 1$ and $b > 0$ the free Poisson variable is the square of a semicircular variable. This is part of a more general result of [32], to which we will return in 3.4.2° (after introducing multiplicative free convolution). The case $a = 1$ (to keep formulas simple we also make the inessential assumption $b = 1$) is obtained by an easy inspection of Cauchy transforms.

Let X be a noncommutative random variable in (A, φ) with a $(0,1)$-semicircle distribution. Then we have the G-series

$$G_X(z) = \sum_{n \ge 0} z^{-n-1}\varphi(X^n) = \frac{z + (z^2 - 4)^{\frac{1}{2}}}{2} .$$

Since the distribution of X is symmetric, $\varphi(X^{2k+1}) = 0$ and hence

$$G_X(z) = \sum_{k \ge 0} z^{-2k-1}\varphi(X^{2k}) = zG_{X^2}(z^2)$$

so that

$$G_{X^2}(z) = \frac{G_X(z^{\frac{1}{2}})}{z^{\frac{1}{2}}} = \frac{z + ((z-2)^2 - 4)^{\frac{1}{2}}}{2z}$$

i.e., X^2 has a free Poisson distribution with parameters $a = b = 1$.

This is quite a departure from the classical situation where the square of a Gaussian variable has a χ^2 distribution.

2.9 Free convolution of measures with unbounded supports

The free convolution defined for measures with compact support on $\mathbb{R}$ [49] has an extension to arbitrary probability measures on $\mathbb{R}$ via an extension of the R-transform [5].

If μ is a probability measure on $\mathbb{R}$ the Cauchy transform

$$G_\mu(z) = \int \frac{d\mu(t)}{z - t}$$

is an analytic function in the upper half-plane $\{z \in \mathbb{C} \mid \text{Im } z > 0\}$. Then G_μ is univalent in some domain

$$\Gamma_{\alpha,\beta} = \{z = x + iy \mid y \geq \beta, \; |x| \leq \alpha y\} \; .$$

The inverse $K_\mu(z)$ is then defined in some angular domain of the form

$$\{z \in \mathbb{C} \mid 0 < |z| < r, \; \arg z \in \left(-\frac{\pi}{2} - \varepsilon, -\frac{\pi}{2} + \varepsilon \right)\} \; .$$

Then $R_\mu(z) = K_\mu(z) - z^{-1}$ is defined in the same angular domain.

To get the free convolution of μ_1, μ_2 one adds $R_{\mu_1} + R_{\mu_2}$ which is also defined in some angular domain and one then computes backward K and G for $R = R_{\mu_1} + R_{\mu_2}$. It turns out G is also defined in some $\Gamma_{\alpha,\beta}$ and has an analytic extension to the upper half-plane which is the Cauchy transform of a probability measure μ on $\mathbb{R}$. We then define this μ to be the free convolution $\mu_1 \boxplus \mu_2$.

2.10 The Cauchy distribution

Let γ be the probability measure on $\mathbb{R}$ with the Cauchy density $\frac{1}{\pi} \frac{a}{x^2 + a^2}$. Then

$$G_\gamma(z) = (z + ia)^{-1} \; , \quad K_\gamma(z) = z^{-1} - ia \; , \quad R_\gamma(z) = -ia \; .$$

Let μ be some other probability measure on $\mathbb{R}$ and $\nu = \mu \boxplus \gamma$. Then

$$K_\nu(z) = K_\mu(z) - ia$$

so that $G_\nu(z) = G_\mu(z+ia)$. Note that the classical convolution $\mu*\gamma$ has precisely this property $G_{\mu*\gamma}(z) = G_\mu(z+ia)$.

Thus the Cauchy distribution γ has the property that $\mu \boxplus \gamma = \mu * \gamma$ ([5]). More generally, this property also holds for uncentered Cauchy laws and Dirac probability measures on $\mathbb{R}$.

2.11 Free infinite divisibility and the free analogue of the Levy-Hinčin theorem

There is a parallel theory of infinite divisibility in the free context and an analogue of the Levy-Hinčin theorem. For compactly supported probability measures this was done in [50], the general case is in [5] (an intermediate case, measures with finite variance, was dealt with in [29]).

A probability measure μ is *freely infinitely divisible if for every $n \in \mathbb{N}$ there is a probability measure $\mu_{1/n}$ such that*

$$\underbrace{\mu_{1/n} \boxplus \cdots \boxplus \mu_{1/n}}_{n\text{-times}} = \mu$$

If μ is freely infinitely divisible then there exists a free convolution semigroup, i.e. $(\mu_t)_{t\geq 0}$, $\mu_{t+s} = \mu_t \boxplus \mu_s$, $t \to \mu_t$ weak*-continuous, so that $\mu_1 = \mu$.

From the formulae for the R-transform one derives that if $(\mu_t)_{t\geq 0}$ is a free convolution semigroup and ν an arbitrary probability measure on $\mathbb{R}$ then the Cauchy transform

$$G(z,t) = G_{\mu_t \boxplus \nu}(z)$$

satisfies the complex free convolution semigroup

$$\frac{\partial G}{\partial t} + R(G)\,\frac{\partial G}{\partial z} = 0 \quad \text{for} \quad \text{Im } z > 0, \ t \geq 0$$

with initial condition

$$G(z,0) = G_\nu(z) \,.$$

Here $R = R_{\mu_1}$.

Unlike real conservation laws, there are no singularities and the solutions exist for all $t \geq 0$ in the upper half-plane.

Classically the differential equation corresponding to convolution by the Gaussian semigroup is the heat equation. The free analogue of the heat equation is the complex conservation law for the Cauchy transforms of measures freely convoluted with the semicircular semigroup

$$\frac{\partial G}{\partial t} + G\,\frac{\partial G}{\partial z} = 0$$

which is the complex Burgers' equation.

The free analogue of the Levy-Hinčin theorem is in terms of R-transforms.

Theorem. (i) *A probability measure μ on $\mathbb{R}$ is freely infinitely divisible iff R_μ has an analytic extension to $\{z \in \mathbb{C} \mid \operatorname{Im} z < 0\}$ with values in $\{z \in \mathbb{C} \mid \operatorname{Im} z \le 0\}$.*

(ii) *An analytic function $R : \{z \in \mathbb{C} \mid \operatorname{Im} z < 0\} \to \{z \in \mathbb{C} \mid \operatorname{Im} z < 0\}$ is the R-transform of a probability measure on $\mathbb{R}$ iff for some $\varepsilon > 0$ we have*

$$\lim_{\substack{|z| \to 0 \\ |\arg z + \frac{\pi}{2}| < \varepsilon}} z R(z) = 0$$

(iii) *If R is the R-transform of some probability measure μ on $\mathbb{R}$, then there is $\beta \ge 0$ and a finite positive measure ν on $\mathbb{R}$ so that for $z \in \mathbb{C}$, $\operatorname{Im} z < 0$ we have*

$$R(z) = \beta + \int \frac{z+t}{1-tz}\,d\nu(t)$$

Note that the integrand for $t = 0$ is z, i.e., the R-transform of a semicircle while for $t \ne 0$,

$$\frac{z+t}{1-tz} = -t^{-1} + \frac{t+t^{-1}}{1-tz}$$

which is a shifted free Poisson law. Thus like in the classical case we get a combination of constant, free analogue of Gaussian and free analogue of Poisson.

2.12 Free stable laws

Adapting to the free context the definition of stable laws yields: a probability measure μ on $\mathbb{R}$ is freely stable if the free convolution of affine transforms of μ yields again affine transforms of μ.

Like for infinite divisibility there is a complete (unexplained) analogy between the classical and free classifications of stable laws.

Theorem. ([5]) *The following is a complete list of the R-transforms of freely stable laws on $\mathbb{R}$*

(i) $R(z) = a + ib$, *where $a \in \mathbb{R}$, $b \in \mathbb{R}$, $b \le 0$.*

(ii) $R(z) = a + bz^{\alpha-1}$, where $a \in \mathbb{R}$, $b \in \mathbb{C}$, $\alpha \in (1,2]$, $b \neq 0$ and $\arg b \in$ $[(\alpha - 2)\pi, 0]$.

(iii) $R(z) = a + bz^{\alpha-1}$, where $a \in \mathbb{R}$, $b \in \mathbb{C}$, $\alpha \in (0,1)$, $b \neq 0$ and $\arg b \in$ $[\pi, (1 + \alpha)\pi]$.

(iv) $R(z) = a + b \log z$ where $a \in \mathbb{C}$, $b \in \mathbb{R}$, and $\operatorname{Im} a \leq 0$, $b > 0$.

2.13 More free harmonic analysis on $\mathbb{R}$

Here is a brief enumeration of a few more results in the harmonic analysis which is developing around the operation $\boxplus$.

1°. **Supports.** If μ is a compactly supported probability measure on $\mathbb{R}$ let

$$N(\mu) = \sup_{t \in \operatorname{supp} \mu} |t| \quad \text{and} \quad R_2(\mu) = \int t^2 d\mu(t) - \left(\int t d\mu(t) \right)^2 .$$

In [50] for centered measures μ_j $(1 \leq j \leq n)$, $\int t d\mu_j(t) = 0$ it is shown that:

$$\left(\sum_{1 \leq j \leq n} R_2(\mu_j) \right)^{\frac{1}{2}} \leq N(\mu_1 \boxplus \cdots \boxplus \mu_n) \leq \max_{1 \leq j \leq n} N(\mu_j) + 2 \left(\sum_{1 \leq j \leq n} R_2(\mu_j) \right)^{\frac{1}{2}} .$$

2°. **Continuity.** In the process of extending free convolution to probability measures with unbounded support a continuity result for free convolution was obtained in [5].

If μ is a probability measure in $\mathbb{R}$ let $F_\mu(t) = \mu((-\infty, t))$ and let $d_\infty(\mu, \nu) = \sup_{t \in \mathbb{R}} |F_\mu(t) - F_\nu(t)|$. Then

$$d_\infty(\mu \boxplus \nu, \mu' \boxplus \nu') \leq d_\infty(\mu, \mu') + d_\infty(\nu, \nu') .$$

Applying this to $\mu' = \mu \boxplus \delta_\varepsilon$, $\nu' = \nu$ we get that if F_μ is Hölder continuous of order α $(0 < \alpha \leq 1)$ then $F_{\mu \boxplus \nu}$ is also a Hölder of order α.

3°. L^p-**norms.** If μ is absolutely continuous w.r.t. Lebesgue measure and $\frac{d\mu}{d\lambda} \in L^p$ where $1 < p \leq \infty$ then for any probability measure ν, the measure $\mu \boxplus \nu$ is Lebesgue absolutely continuous and

$$\left\| \frac{d(\mu \boxplus \nu)}{d\lambda} \right\|_p \leq \left\| \frac{d\mu}{d\nu} \right\|_p$$

This result and a similar one for Riesz energies were obtained in [57]. Subsequently these inequalities were shown in [10] to be the consequences of Markovian properties.

4°. **Atoms.** A free convolution of two probability measures μ and ν, has atoms only under very special circumstances, as shown by the following result from [7]:

$$(\mu \boxplus \nu)(\{t_0\}) = \alpha > 0 \text{ if and only if}$$
$$\text{there are } a, b \in \mathbb{R}, \text{ so that } a + b = t_0$$
$$\text{and } \mu(\{a\}) + \nu(\{b\}) = 1 + \alpha .$$

5°. **Failure of the free Cramer theorem.** The free analogue of Cramer's theorem does not hold, i.e. in [6] it is shown there are probability measures μ_1, μ_2 on $\mathbb{R}$, with compact supports, which are not semicircle laws, such that $\mu_1 \boxplus \mu_2$ is a semicircle law.

6°. $\mu^{\boxplus t}$ **when** $t \geq 1$. Though free infinite divisibility runs parallel to classical, the existence of non-integer convolution powers $\mu^{\boxplus t}$ when $t \geq 1$ is a quite different matter. As shown in [32], if R_μ is the R-transform of a probability measure μ on $\mathbb{R}$, then for all $t \geq 1$, tR_μ is the R-transform of a probability measure.

7°. **Regularization by free convolution with a semicircle law.** Let μ be a compactly supported probability measure on $\mathbb{R}$ and let $\sigma_{2\alpha}$ be the semicircle distribution with density $(2\pi\alpha^2)^{-1}(4\alpha^2 - t^2)^{\frac{1}{2}}$ on $[-2\alpha, 2\alpha]$. Then $\mu \boxplus \sigma_{2\alpha}$ is Lebesgue absolutely continuous with density u_α. In [63] it was shown that if D denotes the unbounded operator of derivation in $L^2(\mathbb{R})$, then u_α is in the Sobolev space of order $1/2$ and we have

$$\| |D|^{\frac{1}{2}} u_\alpha \|_2 \leq \alpha^{-\frac{1}{2}} .$$

On the other hand in [13] it is shown that

$$u_\alpha(t) \neq 0 \Rightarrow |Du_\alpha^3(t)| \leq 3(4\pi^3\alpha^2)^{-1}$$

8°. **Domains of attraction.** By analogy with the classical situation, domains of attraction w.r.t. $\boxplus$ were defined for the free stable laws and a surprising result was proved in [3], which can roughly be stated as follows:

> Under the natural correspondence between classical stable laws and
> free stable laws, the classical domain of attraction of a classical stable
> law equals the free domain of attraction of the corresponding free
> stable law.

2.14 Spectra of convolution operators on free products of groups

Free convolution of distributions can be used to compute the spectra of certain convolution operators on free products of groups. We need first some operator algebra preliminaries about separating vectors.

If $I \in M \subset B(\mathcal{H})$ is a von Neumann algebra, a vector $\xi \in \mathcal{H}$ is a separating vector for M, if the map

$$M \ni T \to T\xi \in \mathcal{H}$$

is an injection.

Let $M' = \{X \in B(\mathcal{H}) \mid XT = TX\}$ be the *commutant of M*. If $\xi \in \mathcal{H}$ is a *cyclic vector* for M', i.e., if $M'\xi$ is dense in $\mathcal{H}$, then ξ is separating for M. Indeed if $T \in M$ is such that $T\xi = 0$ then for $X \in M'$, $TX\xi = XT\xi = 0$ so that $T\eta = 0$ for all $\eta \in \overline{M'\xi} = \mathcal{H}$, i.e., $T = 0$. (Actually it is a basic fact in von Neumann algebra theory that the converse also holds, i.e. ξ is separating for M iff ξ is cyclic for M'.)

In particular, *if $T = T^* \in M$ and ξ is separating, let μ_T be the distribution of T w.r.t. to the state $\langle \cdot \xi, \xi \rangle$* (see 1.5) *then the spectrum $\sigma(T) = \text{supp } \mu_T$.* Indeed if $E(\cdot, T)$ is the spectral measure of T then $\sigma(T) = \text{supp } E(\cdot, T)$. On the other hand, supp $E(\cdot, T) = \text{supp } \mu_T$ since for a Borel set $\omega \subset \mathbb{R}$ we have

$$\mu_T(\omega) = \langle E(\omega, T)\xi, \xi \rangle = \|E(\omega, T)\xi\|^2$$

so that

$$\mu_T(\omega) = 0 \Rightarrow E(\omega, T)\xi = 0 \Rightarrow E(\omega, T) = 0$$

because ξ is separating.

Let G be a group and $L(G)$ the von Neumann algebra of the left regular representation λ on $\ell^2(G)$ and let $\tau(\cdot) = \langle \cdot e_e, e_e \rangle$ the trace (example 1.4.1°). Then the right regular representation ρ of G on $\ell^2(G)$, $\rho(g)e_h = e_{hg^{-1}}$ commutes with λ and hence $\rho(G)$ is in $(L(G))'$ (actually $\rho(G)$ generates $(L(G))'$). Then $\{\rho(g)e_e \mid g \in G\} = \{e_{g^{-1}} \mid\mid g \in G\}$ spans $\ell^2(G)$ so that e_e *is separating for* $L(G)$. In particular if $T = T^* \in L(G)$ then for the distribution μ_T w.r.t. τ we have $\sigma(T) = \operatorname{supp} \mu_T$.

In particular assume $G = \bigstar_{1 \le j \le n} G_j$ and let $T_j = T_j^*$ be in the von Neumann algebra of $\lambda(G_j)$ and $T = T_1 + \cdots + T_n$. Then $\mu_T = \mu_{T_1} \boxplus \cdots \boxplus \mu_{T_n}$ and $\sigma(T) = \operatorname{supp} \mu_T$.

Example. Let G be the free product of n copies of $\mathbb{Z}/2\mathbb{Z}$ and let g_j be the generator of the jth copy of $\mathbb{Z}/2\mathbb{Z}$. Let $T = \lambda(g_1) + \cdots + \lambda(g_n)$. Note that $g_j = g_j^{-1}$, $g_j^2 = 1$, $g_j \ne e$. Then if $T_j = \lambda(g_j)$ we have $\tau(T_j^{2k}) = 1$, $\tau(T_j^{2k+1}) = 0$ so that $\mu_{T_j} = 2^{-1}(\delta_{-1} + \delta_{+1})$ and $G_{T_j}(z) = \frac{z}{z^2+1}$. Then

$$R_{T_j} = \frac{-1 + (1 + 4z^2)^{\frac{1}{2}}}{2z} , \qquad R_T = n R_{T_1} ,$$

$$K_T = \frac{2-n}{2} z^{-1} + \frac{n}{2}(z^{-2} + 4)^{\frac{1}{2}}$$

$$G_T = \frac{(2-n)z + (z^2(2-n)^2 - 4(z^2 - n^2)(1-n))^{\frac{1}{2}}}{2(z^2 - n^2)}$$

By $2.13.4°$, μ_T has no atoms for $n \ge 2$, so that we do not need to worry about possible poles of G_T contributing Dirac measures to μ_T. Thus supp μ_T is the closure of the set where Im $G_T \ne 0$ on $\mathbb{R}$, i.e., $t^2(2-n)^2 - 4(n^2 - t^2)(n-1) < 0$. Hence

$$\sigma(T) = \operatorname{supp} \mu_T = [-2(n-1)^{\frac{1}{2}}, 2(n-1)^{\frac{1}{2}}] .$$

3. Multiplication of Freely Independent Noncommutative Random Variables

3.1 Multiplicative free convolution [49]

If a, b are freely independent in (A, φ) then μ_{ab} is completely determined by μ_a and μ_b. Hence *we define a multiplicative free convolution operation $\boxtimes$ on the distributions of noncommutative random variables, such that $\mu_a \boxtimes \mu_b = \mu_{ab}$ whenever a and b are freely independent in some noncommutative probability space.*

The additive free convolution is commutative because addition is commutative. Surprisingly, *multiplicative free convolution is also a commutative operation* $\mu \boxtimes \nu = \nu \boxtimes \mu$.

The property is a consequence of 1.10(iii). There are also various direct ways to see this. One is to compute $\varphi(abab\ldots ab)$. Working on this expression from the left or from the right (using the only rule given by free independence) inverts the roles of a and b, and we get symmetric expressions in the moments of a and b.

Another way to check $\mu \boxtimes \nu = \nu \boxtimes \mu$ is to notice that it suffices to do so for a sufficiently rich family of distributions. For instance, let $\mathcal{A} = L(F_2)$ where F_2 is the free group on generators g_1 and g_2. Let

$$a = \sum_{n \in \mathbb{Z}} \alpha_n \, \lambda(g_1^n) \, , \qquad b = \sum_{n \in \mathbb{Z}} \beta_n \, \lambda(g_2^n)$$

where only finitely many α_n, β_n are nonzero. Then $\tau((ab)^n) = \tau((ba)^n)$ because τ is a trace, which gives

$$\tau((ab)^n) = \tau(ab(ab)^{n-1}) = \tau(b(ab)^{n-1}a) = \tau((ba)^n) \, .$$

3.2 Probability measures on $R_{\geq 0}$ and $\mathbb{T}$

Multiplicative free convolution defines an operation on compactly supported probability measures on $[0, \infty)$.

Indeed, if μ_1, μ_2 are two such measures on $R_{\geq 0} = [0, \infty)$ there are Borel functions $f_j : \mathbb{T} \to \mathbb{R}_{\geq 0}$ such that μ_j is the pushforward by f_j of Haar measure on $\mathbb{T} = \{z \in \mathbb{C} \mid |z| = 1\}$, Consider again F_2 the free group on generators g_1, g_2. Then τ applied to the spectral measure of $\lambda(g_j)$ is Haar measure on $\mathbb{T}$ (indeed $\tau(\lambda(g_j)^k) = \tau(\lambda(g_j^k)) = \delta_{0,k}$). We infer that $a_j = f(\lambda(g_j))$, $j = 1, 2$ are two positive operators in $(L(F_2), \tau)$ with distributions μ_1, μ_2 and a_1, a_2 are freely independent. Then τ being a trace and the α_j's being positive, $a_j^{\frac{1}{2}}$ is defined and ab and $a^{\frac{1}{2}} b a^{\frac{1}{2}}$ have the same distribution. Indeed,

$$\tau((a^{\frac{1}{2}} b a^{\frac{1}{2}})^n) = \tau(a^{\frac{1}{2}} b(ab)^{n-1} a^{\frac{1}{2}})$$
$$= \tau(a^{\frac{1}{2}} a^{\frac{1}{2}} b(ab)^{n-1}) = \tau((ab)^n) \, .$$

Since $a^{\frac{1}{2}} b a^{\frac{1}{2}} = (a^{\frac{1}{2}} b^{\frac{1}{2}})(a^{\frac{1}{2}} b^{\frac{1}{2}})^* \geq 0$, we infer $\mu_{a^{\frac{1}{2}} b a^{\frac{1}{2}}}$ is a measure on $[0, \infty)$.

On the other hand, $\boxtimes$ also defines an operation on probability measures on $\mathbb{T}$.

Indeed, given such a measure μ there is a unitary U in a W^*-probability space $(\mathcal{A}, \varphi)$ such that (take $\mathcal{A} = L^\infty(\mathbb{T}, d\mu)$, U multiplication by z)

$$\varphi(U^k) = \int z^k \, d\mu(z) \, , \qquad k \in \mathbb{N}.$$

Conversely, given a unitary U in a W^*-probability space, there is a unique probability measure on $\mathbb{T}$, with the same moments as U (the expectation of the spectral measure of U).

The operation on probability measures on $\mathbb{T}$ is then a consequence of the fact that the product of two unitary operators is again a unitary operator.

3.3 The S-transform

For multiplicative free convolution there is a multiplicative map which plays the role of the Mellin transform.

Theorem [51]. *If* $\mu : \mathbb{C}[X] \to \mathbb{C}$ *is the distribution of a random variable with* $\mu(X) \neq 0$, *let*

$$\psi_\mu(z) = \sum_{k \geq 1} \mu(X^k) z^k$$

and let χ_μ *be the formal inverse* $\psi_\mu(\chi_\mu(z)) = z$. *Let further* $S_\mu(z) = \frac{1+z}{z} \chi_\mu(z)$. *Then we have*

$$S_{\mu \boxtimes \nu} = S_\mu S_\nu \ .$$

The original proof in [51] of the theorem is based on differential equations and on the results for additive free convolution. More recently in [25] another proof was found which uses some canonical operators like in the additive case. If μ is a distribution and $f(z) = \frac{1}{S_\mu(z)}$, the canonical operator is

$$(1 + \ell_1) f(\ell_1^*) \ .$$

3.4 Examples

1°. **Two projections.** Let $\mu = (1-a)\delta_0 + a\delta_1$ and $\nu = (1-b)\delta_0 + b\delta_1$. Computing $\mu \boxtimes \nu$ gives the distribution of PQP when P and Q are a pair of freely independent selfadjoint projections with $\varphi(P) = a$, $\varphi(Q) = b$ in a W^*-probability space $(\mathcal{A}, \varphi)$ where φ is a trace (this is actually inessential). Indeed

$$\varphi((PQP)^n) = \varphi((PQ)^n P) = \varphi(P(PQ)^n) = \varphi((PQ)^n) \ .$$

We have

$$\psi_\mu = \frac{az}{1-z} \ , \qquad \chi_\mu = \frac{z}{a+z} \ , \qquad S_\mu = \frac{1+z}{a+z} \ .$$

Similarly $S_\nu = \frac{1+z}{b+z}$ so that

$$S_{\mu \boxtimes \nu} = \frac{(1+z)^2}{(a+z)(b+z)} ,$$

$$\chi_{\mu \boxtimes \nu} = \frac{z(1+z)}{(a+z)(b+z)} ,$$

$$\psi_{\mu \boxtimes \nu} = \frac{1 - (a+b)z + ((a-b)^2 z^2 - (2a+2b-4ab)z + 1)^{\frac{1}{2}}}{2(z-1)}$$

Computing further $G_{\mu \boxtimes \nu}(z) = z^{-1}\psi_{\mu \boxtimes \nu}(z^{-1}) + z^{-1}$, one can show that $(\mu \boxtimes \nu)(\{1\}) = \max(a+b-1, 0)$, $(\mu \boxtimes \nu)(\{0\}) = \max(1-a-b, 0)$.

2°. **The realization of additive free Poisson variables.** In 2.8 we showed that X^2 has a free Poisson distribution with $a = 1$, $b = 1$ when X is a $(0,1)$ semicircle law. Here we show that *if X is a $(0,1)$ semicircle law and P is an idempotent with $\varphi(P) = a$ and X and P are free in (A, φ) then XPX has a free Poisson distribution with parameter a* ([32]).

We may assume (A, φ) is tracial, so that $\varphi((XPX)^n) = \varphi((PX^2)^n)$. Hence $\mu_{XPX} = \mu_P \boxtimes \mu_{X^2}$.

In 2.8 we found

$$G_{X^2}(z) = \frac{z + (z^2 - 4z)^{\frac{1}{2}}}{2z}$$

so that

$$\psi_{X^2}(z) = z^{-1} G_{X^2}(z^{-1}) - 1 = \frac{(1-2z) + (1-4z)^{\frac{1}{2}}}{2z} ,$$

$$\chi_{X^2} = \frac{z}{(z+1)^2} , \qquad S_{X^2} = \frac{1}{1+z} .$$

By Example 1, $S_P = \frac{1+z}{a+z}$ so that

$$S_{XPX} = \frac{1}{a+z} , \qquad \chi_{XPX} = \frac{z}{(1+z)(a+z)}$$

$$\psi_{XPX} = \frac{1 - (a+1)z + ((a-1)^2 z^2 - 2(a+1)z + 1)^{\frac{1}{2}}}{2z}$$

so that

$$G_{XPX}(z) = z^{-1}(1 + \psi(z^{-1})) = \frac{z - (a-1) + ((z-a-1)^2 - 4a)^{\frac{1}{2}}}{2z}$$

which is the free Poisson distribution with parameters a and 1.

3.5 Multiplicative free infinite divisibility on $\mathbb{T}$

With the obvious definitions of infinitely divisible measures and semigroups w.r.t. $\boxtimes$, there is a parallel to the classical theory. Let $\mathcal{P}_*(\mathbb{T})$ denote the probability measures on $\mathbb{T}$, with $\int z \, d\mu(z) \neq 0$. If $(\mu_t)_{t \geq 0}$ is a semigroup w.r.t. $\boxtimes$ in $\mathcal{P}_*(\mathbb{T})$, then $\psi(z,t) = \psi_{\mu_t}(z)$ satisfies the complex conservation law

$$\frac{\partial \psi}{\partial t} + u(\psi) z \, \frac{\partial \psi}{\partial z} = 0$$

where $S_{\mu_t}(z) = \exp(tu(z))$.

Theorem [4]. (i) $\mu \in \mathcal{P}_*(\mathbb{T})$ is infinitely divisible w.r.t. $\boxtimes$ iff there exists a function $u(z)$ analytic in $\{z \mid \mathrm{Re}\ z > -\frac{1}{2}\}$ such that $\mathrm{Re}\ u(z) \geq 0$ whenever $\mathrm{Re}\ z > -\frac{1}{2}$ and $S_\mu(z) = \exp u(z)$.

(ii) Every analytic function $u(z)$ on $\{z \mid \mathrm{Re}\ z > -\frac{1}{2}\}$ with $\mathrm{Re}\ u \geq 0$ is such that $\exp u(z)$ is the S_μ for some $\mu \in \mathcal{P}_*(\mathbb{T})$ infinitely divisible w.r.t. $\boxtimes$.

(iii) If $\mu \in \mathcal{P}_*(\mathbb{T})$ is infinitely divisible w.r.t. $\boxplus$ then there is $\alpha \in \mathbb{R}$ and a finite positive measure ν on $\mathbb{T}$ such that $S_\mu(z) = \exp u(z)$ where

$$u(z) = -i\alpha + \int_{\mathbb{T}} \frac{(1+z) + \zeta z}{(1+z) - \zeta z} \, d\nu(\zeta)$$

for $\mathrm{Re}\ z > -\frac{1}{2}$.

In particular, *the analogue of the normal distribution in this context is given by* $S_\mu(z) = \exp(t(z + \frac{1}{2}))$, $t > 0$ *and the analogue of the Poisson distribution is given by*

$$S_\mu(z) = \exp\left(\frac{t}{z + \frac{1}{2} + i\alpha}\right), \quad t > 0, \quad \alpha \in \mathbb{R},$$

(see [4]).

3.6 Multiplicative free infinite divisibility on $\mathbb{R}_{\geq 0}$

Like in the additive situation, in the multiplicative context on $\mathbb{R}_{\geq 0}$, there is an extension of free convolution $\boxtimes$ and of the S-transform machinery to measures with unbounded support ([5]). Moreover there is also a continuity inequality

$$d_\infty(\mu \boxtimes \nu, \mu' \boxtimes \nu') \leq d_\infty(\mu, \mu') + d_\infty(\nu, \nu')$$

analogous to 2.13.2° ([5]). In particular infinite divisibility results are for measures without the restriction of bounded supports. The differential equation for the ψ-function of a semigroup is of the same form as in the case of $\mathbb{T}$.

Theorem [5]. (i) *Let $\mu \neq \delta_0$ be a $\boxtimes$-infinitely divisible probability measure on $\mathbb{R}_{\geq 0}$. There exists an analytic function u on $\mathbb{C} \backslash (0, 1)$ such that $\mathrm{Im}\, u(z) \leq 0$ when $\mathrm{Im}\, z > 0$ and $S_\mu(z) = \exp(u(z))$.*

(ii) *Conversely, if u is an analytic function on $\mathbb{C} \backslash (0, 1)$ such that $\overline{u(z)} = u(\bar{z})$ and $\mathrm{Im}\, u(z) \leq 0$ when $\mathrm{Im}\, z > 0$, then there exists a $\boxtimes$-infinitely divisible probability measure μ on $\mathbb{R}_{\geq 0}$ such that $S_\mu(z) = \exp(u(z))$.*

(iii) *Let $\mu \neq \delta_0$ be a $\boxtimes$-infinitely divisible probability measure on $\mathbb{R}_{\geq 0}$. Then there is a finite positive measure ν on $[0, +\infty]$, $a \in \mathbb{R}$ and $b = \nu(\{+\infty\})$ so that $S_\mu(z) = \exp(u(z))$ where*

$$u\left(\frac{z}{1-z}\right) = a - bz + \int_0^\infty \frac{1+tz}{z-t}\, d\nu(t) \ .$$

In particular *the analogues of normal distributions are given by $S_\mu(z) = \exp(-t(z + \frac{1}{2}))$, $t > 0$ and the analogues of Poisson distributions are given by $S_\mu(z) = \exp\left(\frac{t}{z+\alpha}\right)$, $t > 0$, $\alpha \in \mathbb{R} \backslash [0, 1]$ (see [5]). Note that these measures have compact supports in $(0, \infty)$.*

4. Generalized Canonical Form, Noncrossing Partitions

4.1 Combinatorial work

There is a combinatorial approach to free independence due to R. Speicher ([44],[45],[46]) and further developed in ([31],[32],[33],[34]) in which the passage from classical to free amounts to replacing the lattice of all partitions of $\{1, \ldots, n\}$ by the lattice of noncrossing partitions. In particular the formulae giving the classical cumulants [41] of a distribution and the free ones, i.e., the coefficients of the R-transform, are the same modulo this replacement of all partitions by the noncrossing ones. Though we do not intend to give an introduction to the combinatorial approach in these notes, the generalization of the canonical form to n-variables and the derivation of combinatorial formulae for the general R-transform from this generalized canonical form found in [31] offers an occasion for a glimpse at the connection with noncrossing partitions.

4.2 Generalized canonical form [31]

Instead of writing the canonical form $\ell_1^* + \alpha_0 I + \alpha_1 \ell_1 + \ldots$ we will write $\ell_1^*(I + \alpha_0 \ell_1 + \alpha_1 \ell_1^2 + \ldots)$. Moreover we may consider the functional $\theta : \mathbb{C}[X] \to \mathbb{C}$ so that $\theta(X^n) = \alpha_{n-1}$ $(n \geq 1)$, $\theta(1) = 1$ and write

$$I + \alpha_0 \ell_1 + \alpha_1 \ell_1^2 + \cdots = \sum_{k \geq 0} \theta(X^k) \ell_1^k \ .$$

The generalized canonical form is given by a functional $\theta : \mathbb{C}\langle X_1, \ldots, X_n \rangle \to \mathbb{C}$, with $\theta(1) = 1$, which is used to construct the canonical n-tuple of variables $(\ell_1^* T, \ldots, \ell_n^* T)$ where

$$T = \sum_{m \geq 0} \sum_{1 \leq i_1, \ldots, i_m \leq n} \theta(X_{i_m} \ldots X_{i_1}) \ell_{i_1} \ldots \ell_{i_m} \, .$$

Here $\ell_1, \ldots, \ell_n$ are the n creation operators in $T\mathcal{H}_{\mathbb{C}}$ where $\mathcal{H}_{\mathbb{C}}$ is the complexification of $\mathbb{R}^n$ with orthonormal basis $e_1, \ldots, e_n$. The comments about formal sums in creation operators we made in the $n = 1$ case also apply here. Note that

$$\ell_j^* T = \sum_{m \geq 0} \sum_{1 \leq i_2, \ldots, i_m \leq n} \theta(X_{i_m} \ldots X_{i_2} X_j) \ell_{i_2} \ldots \ell_{i_m} + \ell_j^* \, .$$

Fact. *For every distribution* $\mu : \mathbb{C}\langle X_1, \ldots, X_n \rangle \to \mathbb{C}$ *there is exactly one* $\theta : \mathbb{C}\langle X_1, \ldots, X_n \rangle \to \mathbb{C}$ *such that the canonical n-tuple* $(\ell_1^* T, \ldots, \ell_n^* T)$ *determined by* θ *has distribution* μ *w.r.t. the vacuum expectation* φ.

Let $T_j = \ell_j^* T$. Then

$$\mu(X_{i_m} \ldots X_{i_1}) = \varphi(T_{i_m} \ldots T_{i_1}) = \theta(X_{i_m} \ldots X_{i_1}) + \text{polynomial in the } \theta(X_{j_k} \ldots X_{j_1})$$

with $k < m$. Clearly this assertion implies there is a bijection between θ and μ. Indeed when expanding $\varphi(T_{i_m} \ldots T_{i_1})$ we get

$$\varphi\big((\ell_{i_m}^* \ell_{j(1,m)} \ldots \ell_{j(k_m,m)}) \ldots (\ell_{i_1}^* \ell_{j(1,1)} \ldots \ell_{j(k_1,1)})\big)$$

with coefficient

$$\theta(X_{j(k_m,m)} \ldots X_{j(1,m)}) \ldots \theta(X_{j(k_1,1)} \ldots X_{j(1,1)}) \, .$$

Note that if $\varphi(\ldots)$ above is $\neq 0$ then $(k_1 - 1) + \cdots + (k_m - 1) = 0$. Thus $k_1 + \cdots + k_m = m$. If some $k_s = m$ then all other $k_p = 0$, but

$$\varphi(\ell_{i_m}^* \ldots \ell_{i_{s+1}}^* (\ell_{i_s}^* \ell_{j(1,s)} \ldots \ell_{j(k_s,s)}) \ell_{i_{s-1}}^* \ldots \ell_{i_1}^*)$$

equals zero unless $s = 1$, $k_1 = m$, $j(r,1) = i_r$ $(1 \leq r \leq m)$, in which case it $= 1$. This clearly proves the assertion.

Corollary. *Let μ and θ be as above and let μ', μ'' be the restrictions of μ to $\mathbb{C}\langle X_1, \ldots, X_k \rangle$ and $\mathbb{C}\langle X_{k+1}, \ldots, X_n \rangle$ and θ', θ'' the functional s corresponding to μ', μ''. Then $\{X_1, \ldots, X_k\}$, $\{X_{k+1}, \ldots, X_n\}$ are freely independent in $(\mathbb{C}\langle X_1, \ldots, X_n \rangle, \mu)$ iff*

$$\theta(X_{i_1} \ldots X_{i_m}) = \begin{cases} \theta'(X_{i_1} \ldots X_{i_m}) & \text{if all } i_j \leq k \\ \theta''(X_{i_1} \ldots X_{i_m}) & \text{if all } i_j > k \\ 0 & \text{otherwise.} \end{cases}$$

Indeed for θ of the above form $\{T_1, \ldots, T_k\}$ and $\{T_{k+1}, \ldots, T_n\}$ are free w.r.t. φ since the first involves only the first k creation and annihilation operators while the second only the last $n - k$.

Conversely, if $\{X_1, \ldots, X_k\}$ and $\{X_{k+1}, \ldots, X_n\}$ are freely independent then the canonical n-tuple $(T_1, \ldots, T_n)$ defined by the θ given by the formula will have $\{T_1, \ldots, T_k\}$ and $\{T_{k+1}, \ldots, T_n\}$ freely independent and with distributions μ' and μ''. It follows $(T_1, \ldots, T_n)$ has distribution μ.

4.3 Noncrossing partitions and the formula for θ

A partition $(P_1, \ldots, P_r)$ of $\{1, \ldots, n\}$ is noncrossing if there are no pairs $\{a, c\} \subset P_k$, $\{b, d\} \subset P_\ell$ with $k \neq \ell$, $a < b < c < d$.

We will evaluate

$$\varphi((\ell_{i_m}^* \ell_{j(1,m)} \ldots \ell_{j(k_m,m)}) \ldots (\ell_{i_1}^* \ell_{j(1,1)} \ldots \ell_{j(k_1,1)}))$$

which is the coefficient of

$$\theta(X_{j(k_m,m)} \ldots X_{j(1,m)}) \ldots \theta(X_{j(k_1,1)} \ldots X_{j(1,1)}) .$$

In the product of $\theta(\ldots)$'s note that the factors for which $k_s = 0$ make no contribution since $\theta(1) = 1$ and can be eliminated.

On the other hand, for the terms which make a nonzero contribution, if we compute $(\ell_{i_m}^* \ell_{j(1,m)} \ldots \ell_{j(k_m,m)}) \ldots (\ell_{i_1}^* \ell_{j(1,1)} \ldots \ell_{j(k_1,1)})$ starting from the right, we remark that in case $k_s > 0$ we must have $i_s = j(s, 1)$ and the remaining partial product $\ell_{j(2,s)} \ldots \ell_{j(k_s,s)}$ will be cancelled by $\ell_{i_{r(2,s)}}^*, \ell_{i_{r(3,s)}}^*, \ldots, \ell_{i_{r(k_s,s)}}^*$, where $i_{r(p,s)} = j(p, s)$ and $s < r(2, s) < r(3, s) < \ldots r(k_s, s)$ and $k_{r(p,s)} = 0$, $2 \leq p \leq k_s$. Thus to the factors $\theta(X_{j(k_s,s)} \ldots X_{j(1,s)})$ correspond subsets $\{s, r(2, s), \ldots, r(k_s, s)\} \subset \{1, \ldots, m\}$ which give a partition of $\{1, \ldots, m\}$ and $j(p, s) = i_{r(p,s)}$. So if for a subset $P_t \subset \{1, \ldots, m\}$, $P_t = \{r(1), \ldots r(k)\}$, $r(1) <$

$r(2) < \cdots < r(k)$ we denote $X(P_t \mid i_1 \ldots i_m) = X_{i_{r(1)}} \ldots X_{i_{r(k)}}$ then we get a formula

$$\varphi(T_{i_m} \ldots T_{i_1}) = \sum_{\substack{\text{certain} \\ \text{partitions} \\ (P_1 \ldots P_r)}} \prod_{1 \leq t \leq r} X(P_t \mid i_1 \ldots i_m) \ .$$

The partitions occurring can be shown to be exactly the noncrossing partitions. Roughly this can be seen building on the remark that when computing from the right

$$(\ell_{i_m}^* \ell_{j(1,m)} \cdots \ell_{j(k_m,m)}) \cdots (\ell_{i_1}^* \ell_{j(1,1)} \cdots \ell_{j(k_1,1)})$$

we draw a bond between the elements which cancel with $\ell_{j(p,r)}$ and $\ell_{j(p+1,r)}$ we see all the elements as well as complete groups between the two must have already cancelled so the bond can be drawn so that it does not intersect any of the preceding bonds. Also, conversely, the noncrossing of a partition guarantees that the cancellations don't get into one another's way.

What we just sketched shows that

Theorem. *If μ and θ are as above, then*

$$\mu(X_{i_m} \ldots X_{i_1}) = \sum_{\substack{\text{noncrossing} \\ \text{partitions} \\ (P_1 \ldots P_r) \\ \text{of } \{1, \ldots m\}}} \prod_{1 \leq t \leq r} \theta(X(P_t \mid i_1 \ldots i_m)) \ .$$

5. Free Independence with Amalgamation

5.1 Classical background on conditional independence

Conditional independence can be viewed as replacing the scalars by an algebra (of random variables) and using instead of the expectation functional the conditional expectation.

Indeed let $(X, \Sigma, d\mu)$ be a probability space and let $\Xi \subset \Sigma$ be a sub-σ-algebra. Let $B = L^\infty(X, \Xi, d\mu) \subset A = L^\infty(X, \Sigma, d\mu)$ and let $E_B : A \to B$ be the conditional expectation. Then E_B is a projection of A onto B and is a B-module map, i.e. $E_B|_B = \mathrm{id}_B$ and $E_B(ab) = bE_B(a)$.

Let $(A_i)_{i \in I}$ be a family of subalgebras of random variables in $L^\infty(X, \Sigma, d\mu)$ such that $A_i \supset B$, $i \in I$. Then the conditional independence of the $(A_i)_{i \in I}$ w.r.t. Ξ is equivalent to the requirement

$$E_B(a_1 \ldots a_n) = 0$$

whenever $a_j \in \mathcal{A}_{i(j)}$, $E_B(a_j) = 0$ and $i(1), \ldots, i(n)$ are distinct.

A family of sets of random variables $(\Omega_i)_{i \in I} \subset A$ is conditionally independent w.r.t. Ξ if the algebras $\mathcal{A}_i$ generated by $\Omega_i \cup B$ are conditionally independent as above.

This way of looking at conditional independence carries over to free probability theory.

5.2 Conditional expectations in von Neumann algebras with trace state

(background)

The class of von Neumann algebras with a faithful trace-state is an operator algebra context with good existence results for conditional expectations. We briefly sketch some basic facts to provide the essential examples for the discussion of free independence with amalgamation.

Thus, consistent with the point of view adopted in 1.3, let $A \subset B(\mathcal{H})$, $I \in A$ be a von Neumann algebra and $\xi \in \mathcal{H}$, $\|\xi\| = 1$ a unit vector which defines the state $\varphi(\cdot) = \langle \cdot \xi, \xi \rangle$ on A. We will assume φ is a trace, i.e. $\varphi(T_1 T_2) = \varphi(T_2 T_1)$ if $T_1, T_2 \in A$ and we will assume φ is faithful, a condition equivalent to the requirement that ξ is a separating vector for A (see 2.14).

Let $I \in B \subset A$ be a von Neumann subalgebra.

Fact. *Let P denote the orthogonal projection of $\overline{A\xi}$ onto the subspace $\overline{B\xi}$. Then, P maps $A\xi$ into $B\xi$ and $Pa\xi = b\xi$, $a \in A$, $b \in B$ implies $\|b\| \leq \|a\|$. Moreover if $b_1, b_2 \in B$ and $Pa\xi = b\xi$ then $Pb_1 ab_2\xi = b_1 bb_2\xi$.*

We will not give a proof of the above, but let us make some remarks about what the statement involves.

The assumption that φ is a trace implies $\varphi(a^*a) = \varphi(aa^*)$ which gives $\|a\xi\| = \|a^*\xi\|$. Similarly $\langle a_1\xi, a_2\xi \rangle = \varphi(a_2^*a_1) = \langle a_2^*\xi, a_1^*\xi \rangle$.

Since $\overline{B\xi}$ is an invariant subspace for b and b^* we have $(I - P)bP = (I - P)b^*P = 0$ which taking adjoints give $Pb(I - P) = 0$ and then $Pb = bP$. In particular $Pb_1 a\xi = b_1 Pa\xi$.

Note also that the fact that $a\xi \to a^*\xi$ is isometric and $\overline{B\xi}$ is invariant under this map implies that P commutes with this isometry. Thus, if $Pa\xi = b\xi$ then $Pa^*\xi = b^*\xi$ which implies $Pb_2^*a^*\xi = b_2^*b^*\xi$ and then $Pab_2\xi = P(b_2^*a^*)^*\xi = (b_2^*b^*)^*\xi = bb_2\xi$.

These remarks are what is used in the proof except for showing that $Pa\xi \in B\xi$ and $\|a\| \geq \|b\|$ which is more involved.

The map $E_B : A \to B$ such that $Pa\xi = (E_B a)\xi$ is called the canonical conditional expectation of A onto B.

Note that $\varphi(a_2^* a_1) = \langle a_1\xi, a_2\xi \rangle$ defines a pre-Hilbert space structure on A and the corresponding norm $(\varphi(a^*a))^{\frac{1}{2}} = \|a\xi\|$ is denoted $|a|_2$.

Among the properties of E_B we mention

(1) $\|E_B(a)\| \leq \|a\|$
(2) $|E_B(a)|_2 \leq |a|_2$
(3) $a \geq 0 \Rightarrow E_B(a) \geq 0$
(4) $E_B(b_1 a b_2) = b_1 E_B(a) b_2$
(5) $E_B b = b$
(6) $\varphi \circ E_B = \varphi$

where $a \in A$ and $b, b_1, b_2 \in B$.

Example 1°. In case $A = L(G)$ acting on $\ell^2(G)$ take B the von Neumann subalgebra generated by $\lambda(G_1)$ where G_1 is a subgroup of G. The von Neumann trace τ can be obtained using the vector $\xi = e_e$. It is immediate that for a sum with finitely many non-zero c_g we have

$$E_B \sum_{g \in G} c_g \lambda(g) = \sum_{g \in G_1} c_g \lambda(g) \ .$$

2°. The classical conditional expectation corresponds to $A = L^\infty(X, \Sigma, d\mu)$, $B = L^\infty(X, \Theta, d\mu)$, $\mathcal{H} = L^2(X, \Sigma, d\mu)$, ξ being the constant function 1 in $L^2(X, \Sigma, d\mu)$. Here A acts on $\mathcal{H}$ as multiplication operators.

5.3 Noncommutative probability spaces over B and free independence over B

Definition. If B is a unital algebra over $\mathbb{C}$, a noncommutative B-probability space is a pair $(\mathcal{A}, \Phi)$, where $\mathcal{A} \supset B$ is an algebra (same unit as B) and $\Phi : \mathcal{A} \to B$ is a B–B-bimodule map, i.e. linear and $\Phi(b_1 a b_2) = b_1 \Phi(a) b_2$ so that $\Phi(b) = b$ if $b \in B$. Elements in $\mathcal{A}$ are called B-random variables.

Example: If $(\mathcal{A}, \varphi)$ is a W^*-probability space with a faithful trace-state φ and $I \in B \subset \mathcal{A}$ is a von Neumann subalgebra, then $(\mathcal{A}, E_B)$ where E_B is the canonical conditional expectation, is a noncommutative B-probability space.

Definition. *A family of subalgebras* $(\mathcal{A}_i)_{i \in I}$*, such that* $B \subset \mathcal{A}_i$*, in a* B*-probability space* $(\mathcal{A}, \Phi)$ *is* B*-freely independent if*

$$\Phi(a_1 \ldots a_n) = 0$$

whenever $\Phi(a_j) = 0$*,* $1 \leq j \leq n$*,* $a_j \in \mathcal{A}_{i(j)}$*, and* $i(j) \neq i(j+1)$*,* $1 \leq j \leq n - 1$*.*

A family of subsets $(\Omega_i)_{i \in I} \subset \mathcal{A}$ is B-freely independent if the family of subalgebras $\mathcal{A}_i$ generated by $B \cup \Omega_i$ is B-freely independent.

Example. Let G be a group, $H \subset G$ a subgroup and $(G_i)_{i \in I}$ a family of subgroups in G such that $G_i \supset H$. The family $(G_i)_{i \in I}$ is free with amalgamation over H if $g_1 \ldots g_n \neq e$ whenever $g_j \in G_{i(j)}$, $i(j) \neq i(j+1)$, $1 \leq j \leq n - 1$ and $g_j \in G_{i(j)} \backslash H$. Then the $L(G_i)_{i \in I}$ viewed as subalgebras of $L(G)$ are $L(H)$-freely independent in $(L(G), E_{L(H)})$ if and only if the subgroups $(G_i)_{i \in I}$ are free with amalgamation over H.

5.4 More on B-free probability theory

Many of the basics of free probability theory have generalizations to the B-valued case. There are notions of distribution of variables and free product constructions ([49],[56]). The canonical form, free convolution, R-transform have been generalized to this context ([56]). There is an alternative approach to these questions based on the combinatorics of non-crossing partitions ([44]). (Section 3 in [67] also summarizes some of the results in this area.)

6. Some Basic Free Processes

6.1 The semicircular functor (free analogue of the Gaussian functor)

In the classical context Gaussian processes can all be realized using the Gaussian process over a Hilbert space, or what amounts to much the same, the bosonic second quantization functor. In free probability theory there is a perfect parallel to this situation: the semicircle replaces the Gaussian law and there is a corresponding semicircular functor.

Theorem [49]. *Let* $\mathcal{H}$ *be a real Hilbert space,* $\mathcal{H}_\mathbb{C}$ *its complexification and* $T\mathcal{H}_\mathbb{C}$ *the Boltzmann-Fock space over* $\mathcal{H}_\mathbb{C}$ *(1.9.2°) and let* $s(h) = 2^{-1}(\ell(h) + \ell(h)^*)$ *if*

$h \in \mathcal{H}$. If $\Phi(\mathcal{H})$ denotes the von Neumann algebra generated by $\{s(h) \mid h \in \mathcal{H}\}$ and $\tau_{\mathcal{H}}$ the state $\langle \cdot 1, 1 \rangle$ on $\Phi(\mathcal{H})$ given by the vacuum vector $1 \in T\mathcal{H}_{\mathbb{C}}$, then

(i) $(\Phi(\mathcal{H}), T\mathcal{H}_{\mathbb{C}}, \tau_{\mathcal{H}})$ is isomorphic to $(L(F_{\dim \mathcal{H}}), \ell^2(F_{\dim \mathcal{H}}), \tau)$ where $F_{\dim \mathcal{H}}$ is a free group on $\dim \mathcal{H}$ generators. In particular $\tau_{\mathcal{H}}$ is a faithful trace-state on $\Phi(\mathcal{H})$ and 1 is a cyclic and separating vector.

(ii) The $\mathbb{R}$-linear map $\mathcal{H} \ni h \to s(h) \in \Phi(\mathcal{H})$ has the property that for any orthogonal system $(h_i)_{i \in I}$ of vectors in $\mathcal{H}$, the family of random variables $(s(h_i))_{i \in I}$ is free in $(\Phi(\mathcal{H}), \tau_{\mathcal{H}})$. Moreover, for any $h \in \mathcal{H}$, $2s(h)$ has a $(0, \|h\|^2)$ semicircular distribution.

(iii) If $T : \mathcal{H}_1 \to \mathcal{H}_2$ is a linear contraction and $T_{\mathbb{C}}$ its complexification let $T(T_{\mathbb{C}}) : T\mathcal{H}_{1\mathbb{C}} \to T\mathcal{H}_{2\mathbb{C}}$ be the map naturally induced. There is a unique map
$$\Phi(T) : \Phi(\mathcal{H}_1) \to \Phi(\mathcal{H}_2) \text{ such that}$$
$$(\Phi(T)(X))1 = T(T_{\mathbb{C}})(X1), \quad \forall X \in \Phi(\mathcal{H}_1) .$$
Moreover $\|\Phi(T)\| = \|T\| \leq 1$, $\tau_{\mathcal{H}_2} \circ \Phi(T) = \tau_{\mathcal{H}_1}$, $X \geq 0 \Rightarrow \Phi(T)(X) \geq 0$, $\Phi(T)(I) = I$. If T is isometric then $\Phi(T)$ is an isometric injection and if T is a projection then $\Phi(T)$ is a canonical conditional expectation.

(iv) If $(\mathcal{H}_i)_{i \in I}$ is a family of orthogonal subspaces in $\mathcal{H}$, and $V_i : \mathcal{H}_i \to \mathcal{H}$ are the inclusions, then the family of subalgebras $(\Phi(V_i)\Phi(\mathcal{H}_i))_{i \in I}$ is free in $(\Phi(\mathcal{H}), \tau_{\mathcal{H}})$.

(v) Let $r(h)\xi = \xi \otimes h$, $h \in \mathcal{H}_{\mathbb{C}}$ be the right creation operator on $T\mathcal{H}_{\mathbb{C}}$ and $d(h) = 2^{-1}(r(h) + r(h)^*)$. Then the von Neumann algebra generated by $\{d(h) \mid h \in \mathcal{H}\}$ is the commutant of $\Phi(\mathcal{H})$.

We will not prove this theorem. Instead, here are some clarifying comments.

In the proof of the free central limit theorem (2.5) we saw that the $(0,1)$-semicircular variable has canonical form $\ell_1^* + \ell_1$. Moreover, the freeness assertion in (ii) follows from $1.9.2°$. These facts will give (ii) (and (iv)).

The isomorphism in (i) relies on the fact the the von Neumann algebras $W^*(\lambda(g))$, g a generator of a free group, and $W^*(s(h))$ are isomorphic: they are commutative and one is isomorphic to $L^\infty(\mathbb{T}, \text{Haar measure})$ and the other to $L^\infty([-\|h\|, \|h\|], \text{semicircular measure})$ and the isomorphism is a consequence of the isomorphism of the underlying measure spaces. Then each of the two algebras is generated by $\dim \mathcal{H}$ copies of such a commutative algebra, the copies being freely independent. Cyclicity of 1 is easy and 1 is separating because it is cyclic for the commutant described in (v).

321

(iii) is first proved for isometric T (obvious) and for T a projection (using the construction of the canonical conditional expectation). General T is dealt as a composition of inclusions and projections.

To understand (v) note the commutation relations

$$[\ell(h), r(k)] = [\ell(h)^*, r(k)^*] = 0$$
$$[\ell(k)^*, r(h)] = [r(k)^*, \ell(h)^*] = \langle h, k \rangle P$$

where P is the orthogonal projection onto $\mathbb{C}1$. In particular they imply the commutation of $s(h)$ and $d(k)$ if $h, k \in \mathcal{H}$.

If I is an index set and $C : I \times I \to \mathbb{C}$ is a non-negative kernel, the centered semicircular process with covariances given by C is obtained by considering a map $\gamma : I \to \mathcal{H}$ (some Hilbert space) so that $\langle \gamma(i), \gamma(j) \rangle = C(i,j)$ and taking $I \ni i \to 2s(\gamma(i)) \in \Phi(\mathcal{H})$.

In particular, Brownian motion corresponds classically to the map $[0, \infty) \ni t \to \chi_{[0,t]} \in L^2([0, \infty), d\lambda)$, where $\chi_{[0,t]}$ is the indicator function of $[0, t]$. The free analogue of Brownian motion is then given by $2s(\chi_{[0,t]})$ (this was used in [43]).

6.2 Free Poisson processes [32]

We explained in 2.8 and 3.3.2° how free Poisson variables arise from semicircular variables and compressions. This is part of a bigger picture giving a realization of general Poisson processes over a set.

Theorem. Let $(S_i)_{i \in I}$ be freely independent $(0, 1)$-semicircular variables and let $(\Omega_i)_{i \in I}$ be spaces of events with σ-algebras Σ_i and probability measures $d\omega_i$. Let further $(\Omega, \Sigma, d\omega)$ be the disjoint union of the $(\Omega_i, \Sigma_i, d\omega_i)$. Let (A_i, τ_i) be the W^*-probability space $L^\infty(\Omega_i, \Sigma_i, d\omega_i)$ with τ_i the expectation given by $d\omega_i$. Assume $(A_i)_{i \in I}$ and $(\{S_i\})_{i \in I}$ are all freely independent contained in (M, τ) a W^*-probability space with a faithful trace-state. Then if $\alpha_1, \ldots, \alpha_n \in \Sigma$ are disjoint with $\omega(\alpha_j) < \infty$, the variables

$$Y(\alpha_j) = \sum_{i \in I} S_i \chi_{\alpha_j \cap \Omega_i} S_i$$

($\chi_{\alpha_j \cap \Omega_i}$ the indicator function of $\alpha_j \cap \Omega_i$ as an element in A_i) are freely independent with free Poisson distributions with parameters $a = \omega(\alpha)$, $b = 1$.

6.3 Stationary processes with free increments

The additive and multiplicative free convolution semigroups (2.11, 3.5, 3.6) give rise via free product constructions (1.11) to stationary processes with free increments. Here are the basic types in the case of bounded operators, i.e., semigroups of measures with compact support.

$1°$. **Additive.** The process is a family $\{X_t \mid t \in [0, \infty)\}$ in a W^*-probability space such that $X_t = X_t^*$, the increments $X_{s_2} - X_{s_1}$ $(0 \leq s_1 < s_2)$ are freely independent w.r.t. $\{X_r \mid 0 \leq r \leq s_1\}$ and the distribution of $X_{s_2} - X_{s_1}$ depends only on $s_2 - s_1$. Such processes arise from an initial data X_0 and an additive free convolution semigroup $(\mu_t)_{t \geq 0}$, where μ_t is the distribution of $X_{s+t} - X_s$.

$2°$. **Unitary multiplicative.** The process is a family $\{U_t \mid t \in [0, \infty)\}$ in a W^*-probability space such that the U_t are unitary operators and the left multiplicative increments $U_{s_2} U_{s_1}^{-1}$ $(0 \leq s_1 < s_2)$ are $*$-freely independent w.r.t. $\{U_r \mid 0 \leq r \leq s_1\}$ and the distribution of $U_{s_2} U_{s_2}^{-1}$ depends only on $s_2 - s_1$. Such a process arises from an initial data U_0 and a multiplicative free convolution semigroup $(\mu_t)_{t \geq 0}$ on $\mathbb{T}$, where μ_t is the distribution of $U_{s+t} U_s^{-1}$. Instead of left increments one can also consider processes with right increments.

$3°$. **Positive multiplicative.** The process is a family $\{X_t \mid t \in [0, \infty)\}$ in a W^*-probability space such that $X_t \geq 0$ with bounded inverses X_t^{-1}, and the increments $X_{s_1}^{-\frac{1}{2}} X_{s_2} X_{s_1}^{-\frac{1}{2}}$ $(0 \leq s_1 < s_2)$ are freely independent w.r.t. $\{X_r \mid 0 \leq r \leq s_1\}$ and the distribution of $X_{s_1}^{-\frac{1}{2}} X_{s_2} X_{s_1}^{-\frac{1}{2}}$ depends only on $s_2 - s_1$. Such a process arises from an initial data X_0 and a multiplicative free convolution semigroup $(\mu_t)_{t \geq 0}$ on $\mathbb{R}_{>0}$, where μ_t is the distribution of $X_{s_1}^{-\frac{1}{2}} X_{s_2} X_{s_1}^{-\frac{1}{2}}$. (There are variants of this with "outer" increments which involve an auxiliary process in the description.)

With some technical modifications all this can also be done for the case of unbounded supports. The key to this extension is that all these processes give rise to von Neumann algebras with faithful trace states (see $1.10.3°$) and that in these algebras there is a good theory of affiliated unbounded operators. The affiliated unbounded operators can be described in various ways, one being as elements of the ring of fraction w.r.t. the multiplicative system of injective elements. The remarkable feature of these is that they can be added and multiplied without the usual headaches caused by domains of definition.

6.4 The Markov transitions property of processes with free increments [10]

One way to describe noncommutative processes, which extends the processes consisting of parametrized families of noncommutative random variables considered in the previous sections, is to deal with parametrized families of homomorphisms. This means to view a self-adjoint element $X_t \in M$ (M a von Neumann algebra) as giving a homomorphism $j_t : C_b(\mathbb{R}) \to M$ of the bounded continuous functions on $\mathbb{R}$ to M, where $j_t(f) = f(X_t)$ or a homomorphism $j_t : C(\mathbb{T}) \to M$ in the unitary case, so that $j_t(f) = f(X_t)$. The study of such more general processes, the algebras of continuous functions being replaced by general C^*-algebras, has received much attention in noncommutative probability theory (iterate the bibliography operator starting with [8],[10],[27],[36] on the present list of references).

In this framework, let $j_t : A \to M$, $t \geq 0$ be unital $*$-homomorphisms of the unital C^*-algebra A into the von Neumann algebra M which has a faithful trace-state.

One important Markovian property such processes may exhibit is the existence of a Markov transitions system, i.e., maps $\prod_{s,t} : A \to A$, $0 \leq s < t$, $\| \prod_{s,t} \| \leq 1$, $\prod_{s,t}(1) = 1$, $\prod_{s,t}$ completely positive (i.e., the induced maps $A \otimes M_n \to A \otimes M_n$, $\prod_{s,t} \otimes \mathrm{id}_{M_n}$ take positive elements to positive elements for all $n \geq 1$) so that $\prod_{s,t} \circ \prod_{t,r} = \prod_{s,r}$ if $s \leq t \leq r$. The required connection with the process $(j_t)_{t \geq 0}$ being that if $s \leq t$ then

$$E_s j_t(a) = j_s(\prod_{s,t}(a)) , \qquad a \in A$$

where E_s is the canonical conditional expectation of M onto $W^*(\bigcup_{0 \leq u \leq s} j_u(A))$.

Note that this is substantially more than the fact that $E_s j_t = E_{\{s\}} j_t$ where we denoted by $E_{\{s\}}$ the canonical conditional expectation onto $W^*(j_s(A))$.

The existence of Markov transitions for additive processes with free increments and multiplicative unitary processes with free increments (no stationarity) was proved in [10]. The key fact is a property of addition (respectively, multiplication) of freely independent random variables.

Theorem [10]. *Let $X = X^*$, $Y = Y^*$ be freely independent in (M, τ) where M is a von Neumann algebra with faithful trace-state. Let μ and ν be the distributions of X and Y and E the conditional expectation onto $W^*(\{X\})$. Then there is a Feller kernel $K = k(x, du)$ on $\mathbb{R} \times \mathbb{R}$ and an analytic function F on $\mathbb{C} \backslash \mathbb{R}$*

such that

(i) *For any bounded continuous function f on $\mathbb{R}$,*

$$Ef(X + Y) = (Kf)(X)$$

where $(Kf)(t) = \int f(u)k(t, du)$.

(ii) *$F(\bar{\zeta}) = \overline{F(\zeta)}$, $\operatorname{Im} \zeta > 0 \Rightarrow \operatorname{Im} F(\zeta) > 0$. $(iy)^{-1}F(iy) \to 1$ as $y \to +\infty$.*

(iii) *For all $\zeta \in \mathbb{C}\backslash\mathbb{R}$, $t \in \mathbb{R}$,*

$$\int_{\mathbb{R}} (\zeta - u)^{-1}k(t, du) = (F(\zeta) - t)^{-1}$$

(iv) *With $G_\mu, G_{\mu \boxplus \nu}$ denoting Cauchy transforms, we have*

$$G_\mu(F(\zeta)) = G_{\mu \boxplus \nu}(\zeta) \ .$$

Under genericity conditions, the analytic subordination property (iv) was proved earlier in [57], where it was used to prove inequalities on L^p-norms of densities for $\mu \boxplus \nu$ (2.13.3°).

6.5 Free Markovianity ([65])

The Markov transitions property we discussed in the previous section is a general Markovian feature not connected to any particular type of independence. In this section we look at the free correspondent of the weak Markovian feature that the future and past are conditionally independent over the present, which involves the type of independence considered.

Since conditional independence in free probability corresponds to free independence over a subalgebra it is clear what free Markovianity should be. Thus, in a W^*-probability space (M, τ) where τ is a faithful trace-state, a triple A, B, C of W^*-subalgebras containing I in M (past, present, future) is freely Markovian if A and C are B-free in (M, E_B).

A process $(j_t)_{t \geq 0}$ where $j_t : A \to M$ are unital $*$-homomorphisms is called freely Markovian if for every $t > 0$ the triple $W^*(\bigcup_{0 \leq s \leq t} j_s(A))$, $W^*(j_t(A))$, $W^*(\bigcup_{t \leq r < \infty} j_r(A))$ is freely Markovian.

The three types of processes with free increments in 6.3 (no stationarity required) are all freely Markovian.

6.6 Further results

1°. **Circular variables.** The free analogue of complex Gaussian variables are the circular variables $c = S_1 + iS_2$, where S_1, S_2 are freely independent and $(0, r)$-circular (see [53]). The realization of free Poisson processes in 6.2 also holds with the freely independent semicircular elements replaced by ∗-freely independent circular elements ([32]).

2°. **Deformation of the semicircular functor.** There is a deformation of the creation and annihilation operators and of the full Fock space so that the relations

$$\ell_i^* \ell_j - \mu \ell_j \ell_i^* = \delta_{ij} I$$

hold where $\mu \in [-1, 1]$. The free case is $\mu = 0$, $\mu = 1$ the classical (bosonic) case and $\mu = -1$ the anticommuting case (see [16],[17]).

3°. **Free dilation of Markov transitions.** A construction of an associated process based on free products for a system of Markov transitions is given in [36].

4°. **Another kind of realization of free Poisson variables** is given in [44].

7. Random Matrices in the Large N Limit

7.1 Asymptotic free independence for Gaussian matrices ([52])

The semicircle law, which plays the role of the normal distribution in free probability, had made a noted earlier appearance in Wigner's work [68],[69], as the limit distribution of eigenvalues of a large Gaussian random matrix. This is not a mere coincidence. The explanation I found is that free independence occurs asymptotically in large random matrices.

For random matrices we use the noncommutative probability framework we explained in Examples 1.4.3° and 1.6.2° i.e.

$$\mathcal{A}_n = \bigcap_{1 \le p < \infty} L^p(X, M_n, d\sigma)$$

$$\varphi_n(T) = \int n^{-1} \operatorname{Tr}(T(x)) d\sigma(x).$$

Moreover, if $T = T^*$ then $\mu_T = \int n^{-1} \left(\sum_{1 \le j \le n} \delta_{\lambda_j(x)} \right) d\sigma(x).$

Theorem. *Let*

$$G_{s,n} \in \mathcal{A}_n \quad s \in \mathbb{N}$$

be self-adjoint random matrices, which are independent as matrix-valued random variables and such that for the entries $a(i,j;n,s)$ of $G_{s,n}$ we have $(\operatorname{Re} a(i,j;n,s))_{1\leq i<j\leq n}$, $(\operatorname{Im} a(i,j;n,s))_{1\leq i<j\leq n}$, $(a(j,j;n,s))_{1\leq j\leq n}$ are Gaussian and independent, the variables in the first two groups being $(0,(2n)^{-1})$ and those in the third $(0,n^{-1})$. Let further $D_n \in \mathcal{A}_n$ be a diagonal matrix with constant entries $(d(j;n))_{1\leq j\leq n}$ so that $\sup \|D_n\| < \infty$ and μ_{D_n} has a weak $*$-limit. Then $D_n, G_{1,n}, G_{2,n}, \ldots$ are asymptotically freely independent as $n \to \infty$.

Before discussing the proof, a *definition of asymptotical free independence* is required. It means that *the joint distribution of $D_n, G_{1,n}, G_{2,n}, \ldots$ as a functional $\mu_n : \mathbb{C}\langle D, X_1, X_2, \ldots \rangle \to \mathbb{C}$ has a pointwise limit μ_∞ as $n \to \infty$ and $D_n, X_1, X_2, \ldots$ are freely independent in $(\mathbb{C}\langle D, X_1, X_2, \ldots \rangle, \mu_\infty)$* (equivalently μ_∞ is a distribution of freely independent variables).

The proof sketched below is the original proof in [52] which involves also a central limit process. (For a shorter recent proof with more combinatorics see [38].) To simplify matters I will ignore the diagonal matrix D (actually the theorem with a D can be derived from the theorem without one).

The rough idea of the proof is: by the assumptions $G_{1,n}, G_{2,n}, \ldots$ can be viewed trivially as the result of a central limit process. A central limit process gives rise to free random variables if some freeness up to second order holds. Then freeness up to second order (roughly) is checked directly.

Central Limit Lemma. *Let T_j $(j \in \mathbb{N})$ be noncommutative random variables in (A, φ), where φ is a trace. Assume:*

$1°$. $\sup_{(j_1, \ldots, j_k) \in \mathbb{N}^k} |\varphi(T_{j_1} \ldots T_k)| \leq C_k < \infty$.

$2°$. *Let $\alpha : \{1, \ldots, m\} \to \mathbb{N}$, then*

a) $|\alpha^{-1}(\alpha(1))| = 1$ *implies*

$$\varphi(T_{\alpha(1)} \ldots T_{\alpha(m)}) = 0.$$

b) $\alpha(1) = \alpha(2)$ *and* $|\alpha^{-1}(p)| \leq 2$ *for all p implies*

$$\varphi(T_{\alpha(1)} \ldots T_{\alpha(m)}) = \varphi(T_{\alpha(3)} \ldots T_{\alpha(m)}).$$

c) $\alpha(m) \neq \alpha(1)$, $\alpha(p) \neq \alpha(p+1)$ *and* $|\alpha^{-1}(p)| \leq 2$ *for all p, implies*

$$\varphi(T_{\alpha(1)} \ldots T_{\alpha(m)}) = 0.$$

Let $\beta : \mathbb{N} \times \mathbb{N} \to \mathbb{N}$ be a bijection and

$$X_{m,N} = N^{-1/2} \sum_{1 \le j \le n} T_{\beta(m,j)}.$$

Then:

(A) $(X_{k,N})_{k \in \mathbb{N}}$ has a limit distribution as $N \to \infty$, independent of the choice of the $T_{j,n}$.

(B) The assumptions of the lemma are satisfied by $(T_j)_{j \in \mathbb{N}}$ freely independent and $(0,1)$ semicircular. Hence:

(C) The limit distribution of $(X_{m,N})_{m \in \mathbb{N}}$ is that of a freely independent family of $(0,1)$-semicircular variables.

Part (A) in the lemma is proved by showing that a limit of a moment can be computed (without paying any attention to what the limit is) i.e. terms involving more than twice each T_j disappear as $N \to \infty$ etc.

Proof of the Theorem (Sketch). One shows $T_1 = G_{1,n}$, $T_2 = G_{2,n}, \dots$ satisfy the lemma asymptotically as $n \to \infty$. Then the central limit process yields freely independent semicircular variables. Note however that the central limit process applied to the $G_{j,n}$ yields a family of variables with the same distribution etc. As an example of what is involved in checking the assumptions of the lemma, here is how one deals with 2° c).

It must be shown

$$\varphi_n(G_{\alpha(1),n} \dots G_{\alpha(m),n}) \to 0$$

as $n \to \infty$ if $|\alpha^{-1}(p)| \le 2$ and $\alpha(p) \ne \alpha(p+1)$, $\alpha(m) \ne \alpha(1)$ for all p. For simplicity assume each $G_{\alpha(j),n}$ occurs exactly twice. So there is a permutation $\gamma : \{1, \dots, m\} \to \{1, \dots, m\}$, $\gamma^2 = $ id without fixed points, so that $\alpha(\gamma(j)) = \alpha(j)$ and $\gamma(j) - j \not\equiv \pm 1 \mod m$. Then

$$\varphi_n(G_{\alpha(1),n} \dots G_{\alpha(m),n}) =$$
$$= n^{-1} \sum a(i_1, i_2; n, \alpha(1)) a(i_2, i_3; n, \alpha(2)) \dots a(i_m, i_1; n, \alpha(m)).$$

Since $a(i_k, i_{k+1}; n, \alpha(k))$ and $a(i_l, i_{l+1}; n, \alpha(l))$ are independent unless $l = \gamma(k)$ and $i_k = i_{l+1}$, $i_l = i_{k+1}$, we infer $i_j = i_{\gamma(j)+1}$. Thus $\varphi_n(G_{\alpha(1),n}, \dots, G_{\alpha(m),n}) = n^{-1} \cdot n^{-m/2} \cdot n^{\#}$ independent indices. Consider the graph with vertices $1, \dots, m$ and edges $[1,2], [2,3], \dots, [m-1, m], [m,1]$. Identify $[j, j+1]$ with $[\gamma(j)+1, \gamma(j)]$ (orientation is reversed). Since adjacent sides are not identified, the quotient graph has at most $m/2$ vertices. The number of vertices of the quotient graph

is the number of independent indices. Hence
$$\varphi_n(G_{\alpha(1),n}, \ldots, G_{\alpha(m),n}) = O(n^{-1}) \text{ goes to zero as } n \to \infty. \qquad \square$$

7.2 Asymptotic free independence for unitary matrices ([52])

Having established asymptotic free independence for Gaussian matrices other results on random matrices can be obtained using functional calculus.

Theorem. *Let $(V(j,n)))_{j\in\mathbb{N}}$ be independent $n \times n$ unitary random matrices, uniformly distributed on $U(n)$ (according to Haar measure) and $W(n)$ a constant unitary $n \times n$ random matrix with limit distribution. Then $(\{V(j,n), V^*(j,n)\})_{j\in\mathbb{N}}$ and $\{W(n), W^*(n)\}$ are a family of pairs which is asymptotically freely independent. Moreover, the limit distribution of $V(j,n)$ is Haar measure on the unit circle.*

Idea of Proof. Like in the Gaussian case the diagonal matrix will be overlooked. Let $(G_{s,n})_{s\in\mathbb{N}}$ be Gaussian random matrices like in 7.1 and let

$$\Gamma_{s,n} = G_{2s-1,n} + \sqrt{-1}G_{2s,n},$$

which are complex independent random matrices with i.i.d. complex Gaussian entries for which the real and imaginary parts are independent and $(0, (2m)^{-1})$. The classical distribution of $\Gamma_{s,n}$ is the Gaussian probability measure on M_n corresponding to a Hilbert-space structure with scalar product proportional to the Hilbert–Schmidt norm. Such a measure is invariant under the action of $U(n)$ on M_n given by left multiplication. Hence if $\Gamma_{s,n}$ has polar decomposition $\Gamma_{s,n} = W_{s,n}(\Gamma_{s,n}^*\Gamma_{s,n})^{1/2}$ then $W_{s,n}$ is a.e. unitary and has distribution Haar measure on $U(n)$. Hence $(W_{j,n})_{j\in\mathbb{N}}$ is a family of random matrices with the same distribution as the $(V(j,n))_{j\in\mathbb{N}}$ and so it suffices to prove asymptotic free independence for the $(\{W_{j,n}, W_{j,n}^*\})_{j\in\mathbb{N}}$. Note that if the $W_{s,n}$ were polynomials in $\Gamma_{s,n}^*$ and $\Gamma_{s,n}$ this would follow from the Gaussian result. Since the polar decomposition is a limit of polynomials there are some technicalities to deal with this limit. $\qquad \square$

Remark. a) The Gaussian random matrix result implies that for every noncommutative polynomial P we have

$$\lim_{n\to\infty} \varphi_n(P(G_{1,n}, \ldots, G_{m,n})) = \langle P(2s(e_1), \ldots, 2s(e_m))1, 1\rangle$$

where $2s(e_j) = l(e_j) + l^*(e_j)$ on the Boltzmann-Fock-space.

b) Similarly, the unitary random matrix result implies that if

$$P(X_1, \ldots, X_m, X_1^*, \ldots, X_m^*)$$

is a noncommutative polynomial and $g_1, \ldots, g_m$ are the generators of a free group and τ is the von Neumann trace then for the left regular representation λ we have

$$\lim_{n \to \infty} \varphi_n(P(V(1,n), \ldots, V(m,n), V^*(1,n), \ldots, V^*(m,n)) =$$
$$= \tau(P(\lambda(g_1), \ldots, \lambda(g_m), \lambda(g_1^{-1}), \ldots, \lambda(g_m^{-1}))).$$

The asymptotic free independence result combined with the Gromov–Milman concentration results [24] give a stronger result.

Theorem. *Given $\varepsilon > 0$ and a non-trivial word*

$$g = g_{i_1}^{k_1} \cdots g_{i_m}^{k_m}$$

in the free group on k generators $F(k)$ ($m \geq 1, k_j \neq 0, i_s \neq i_{s+1}$) let $\Lambda_n(g) = \{(u_1, \ldots, u_k) \in (U(n))^k \| n^{-1} \operatorname{Tr}(u_{i_1}^{k_1} \ldots u_{i_m}^{k_m})| < \varepsilon\}$. Then $\lim_{n \to \infty} \mu_n(\Lambda_n(g)) = 1$ (μ_n Haar measure on $(U(n))^k$).

7.3 Corollaries of the basic asymptotic free independence results

For each n let $\lambda_1(n) \leq \lambda_2(n) \leq \cdots \leq \lambda_n(n)$ and $\mu_1(n) \leq \mu_2(n) \leq \cdots \leq \mu_n(n)$ in $[-C, C]$ be such that

$$n^{-1}(\delta_{\lambda_1(n)} + \cdots + \delta_{\lambda_n(n)}) \to \alpha$$
$$n^{-1}(\delta_{\mu_1(n)} + \cdots + \delta_{\mu_n(n)}) \to \beta$$

weak* as $n \to \infty$. Let $A(n), B(n)$ be independent random matrices uniformly distributed on the $n \times n$ hermitian matrices with eigenvalues $\lambda_1(n) \leq \cdots \leq \lambda_n(n)$ and respectively $\mu_1(n) \leq \cdots \leq \mu_n(n)$ (i.e. w.r.t. the $U(n)$-invariant measures on such matrices). The $A(n)$ and $B(n)$ are asymptotically freely independent.

Idea of Proof. $A(n) = V_1(n)D_1(n)V_1(n)^*$, $B(n) = V_2(n)D_2(n)V_2(n)^*$ where $V_1(n), V_2(n)$ are independent Haar distributed unitary random matrices and $D_1(n), D_2(n)$ are diagonal with the given eigenvalues. We can arrange to reduce to the case when $D_1(n) = f_1(D(n))$ and $D_2(n) = f_2(D(n))$ some suitable functions f_1, f_2. Then we can realize the limit distribution as the distribution of variables in $(L(F(3)), \tau)$, $F(3)$ a free group on generators g_1, g_2, g_3. Indeed $(V_1(n), V_2(n), D(n), A(n), B(n))$ converges in $*$-distribution to

$$(\lambda(g_1), \lambda(g_2), H(\lambda(g_3)), \lambda(g_1)f_1(H(\lambda(g_3)))\lambda(g_1)^{-1}, \lambda(g_2)f_2(H(\lambda(g_3)))\lambda(g_2)^{-1}),$$

where H is a suitable real-valued function on the unit circle. Note that the last two variables are $f_1(H(\lambda(g_1 g_3 g_1^{-1})))$, $f_2(H(\lambda(g_2 g_3 g_2^{-1})))$. The free independence of these variables follows from the algebraic freeness of the subgroups $((g_1 g_3 g_1^{-1})^n)_{n \in \mathbb{Z}}$ and $((g_2 g_3 g_2^{-1})^n)_{n \in \mathbb{Z}}$ in $F(3)$. $\qquad \Box$

Having these asymptotic free independence results, limit distributions of various random matrices can be computed using the free convolution machinery. This yields immediately as corollaries many such distributions which had been computed with ad hoc means.

Let $A(n), B(n), D_1(n), D_2(n)$ be as above with limit distributions α, β, then:

a) the limit distribution of $A(n) + B(n)$ is $\alpha \boxplus \beta$.

b) if the eigenvalues of $A(n), B(n)$ are nonnegative, then the limit distribution of
$A(n)^{1/2} B(n) A(n)^{1/2}$ is $\alpha \boxtimes \beta$.

item"c)" the limit distribution of $A(n) + D_2(n)$ is $\alpha \boxplus \beta$.

d) the limit distribution of $A(n)^{1/2} D_2(n) A(n)^{1/2}$ (when the eigenvalues are positive) is $\alpha \boxtimes \beta$.

Since Gaussian random matrices realize asymptotically freely independent semicircular variables, they can be used to give asymptotic realizations of semicircular processes. In general letting each of the entries not bound by self-adjointness requirements, carry out scaled independent Gaussian processes of the required type will asymptotically yield the corresponding semicircular process.

Similarly the results of [32] when combined with the asymptotic free independence results for random matrices yield asymptotic realizations of free Poisson processes.

Let for instance for each $n \in \mathbb{N}$, $a_{ij}(n)$, $1 \leq i \leq n$, $1 \leq j < \infty$ be independent complex Gaussian with independent real and imaginary parts which are $(0, \frac{1}{n})$. Let further $\Gamma_t(n)$ be the $n \times [tn]$ matrix with entries $a_{ij}(n)$, $1 \leq i \leq n$, $1 \leq j \leq [tn]$. Then $(\Gamma_t(n)\Gamma_t(n)^*)_{t \geq 0}$ as $n \to \infty$ converges in distribution to a free Poisson process.

7.4 Further asymptotic free independence results

1°. **Symmetric and orthogonal matrices.** There are variants of the basic results for symmetric instead of hermitian and orthogonal instead of unitary matrices [52].

2°. **Fermionic entries.** Instead of Gaussian entries fermionic entries can be considered with the same conclusion [52].

3°. **Non-Gaussian entries.** Asymptotic free independence has also been proved for matrices with i.i.d. entries with non-Gaussian distributions [21].

4°. **Permutation matrices.** Asymptotic freeness also holds for independent permutation matrices uniformly distributed on the symmetric group [30].

5°. **Asymptotic realizations of multiplicative processes.** By analogy with the Gaussian case, Brownian motions on the unitary group and on positive matrices provide asymptotic realizations of the corresponding multiplicative processes with free increments [11].

6°. **Asymptotics of group representations.** Asymptotic free independence also has connections with the asymptotics of representations of unitary and symmetric groups. In particular certain asymptotics of the decomposition into irreducibles of tensor products of irreducible representations and decompositions into irreducibles of restrictions are described by additive and respectively multiplicative free convolution [9],[14].

7°. **Large N Yang–Mills 2D QCD.** The Wilson loop variables in the 2-dimensional Yang–Mills large N quantum chromodynamics are asymptotically freely independent if the interiors of the loops are disjoint. In particular with parameter the area of the interior, they form a multiplicative process with free increments [42],[20],[23],[70].

8°. **Gaussian random band matrices.** Gaussian random band matrices in the large N limit can also be handled with free probability techniques, provided free independence with amalgamation over the diagonal matrices is used [40].

9°. **Asymptotic freeness with general constant matrices.** Strengthened asymptotic freeness results hold with the constant diagonal matrices replaced by general matrices [64].

8. Free Entropy

8.1 Clarifications

Free entropy is the free analogue of Shannon's entropy of a n-tuple of real random variables. To find a definition of free entropy I had to go back to the statistical mechanics roots of entropy, the Boltzmann formula. This together with the occurrence of freeness in the asymptotics of large random matrices led to the "matricial microstates" approach to free entropy [58]. From a physics point of view this seems a quite satisfactory definition. Many (but not all) of the expected properties of this quantity have been established and there have been striking applications to the solution of some old operator algebra problems. The technical difficulties to have a complete theory based on matricial microstates are however not to be overlooked, they involve in particular the solution to Alain Connes' well-known embedding of II_1 factors into the ultraproduct of the hyperfinite factor problem and many matricial problems.

In [61] I therefore began developing a second approach to free entropy which is "microstates-free". For this new approach I had to look at the statistical roots of entropy, Fisher's information. The free analogue of Fisher's information is related to a noncommutative generalization of the Hilbert transform. On this route there are other operator algebra problems which need to be solved.

At present free entropy theory encompasses two regions, a "matricial microstates" region and a "microstates-free" region with a number of bridges between the two. Ultimately my expectation is the solutions to the technical problems will be found and the two regions will be parts of a complete free entropy theory.

The exposition here will sketch the two approaches (mostly without proofs) accompanied by some classical background for a better perspective.

8.2 Background on classical entropy via microstates

The entropy of a n-tuple $(f_1, \ldots, f_n)$ of real-valued random variables is given by

$$H(f_1, \ldots, f_n) = -\int p(t_1, \ldots, t_n) \log p(t_1, \ldots, t_n) dt_1, \ldots, dt_n$$

in case their joint distribution on $\mathbb{R}^n$ is absolutely continuous w.r.t. Lebesgue measure, $p(t_1, \ldots, t_n)$ denoting the density and the entropy is $-\infty$ otherwise.

On the other hand, in statistical mechanics the entropy S of a state is given by the Boltzmann formula $S = k \log W$, where W is the Wahrscheinlichkeit of the state. Here the state is thought to be a "macro-state" the probability W of which is found by counting the "micro-states" that correspond to it.

The formula for $H(f_1, \ldots, f_n)$ can roughly be derived from the Boltzmann formula by assigning to each degree of approximation some set of approximating microstates and taking then a normalized limit of the logarithm of the volume of microstates as approximation improves.

For each $m \in \mathbb{N}$, $k \in \mathbb{N}$, $\varepsilon > 0$, $R > 0$ define $G_R(f_1, \ldots, f_n; \; m, k, \varepsilon)$ the set of approximating microstates to be the set of n-tuples $(a_j)_{1 \le j \le n}$ of functions $a_j : \{1, \ldots, k\} \to \mathbb{R}$, $\|a_j\|_\infty < R$, so that

$$|E_k(a_1^{m_1} a_2^{m_2} \ldots a_n^{m_n}) - E(f_1^{m_1} f_2^{m_2} \ldots f_n^{m_n})| < \varepsilon$$

for all $m_j \in \mathbb{N}$, $m_j \le m$ $(1 \le j \le n)$. Here E is the expectation on the space where $f_1, \ldots, f_n$ live, while E_k is the expectation for random variables on $\{1, \ldots, k\}$ for the measure which assigns probability k^{-1} to every atom. Moreover for simplicity assume $f_1, \ldots, f_n$ are bounded.

Then taking

$$\limsup_{k \to \infty}(k^{-1} \log \operatorname{vol} G_R(f_1, \ldots, f_n; \; m, k, \varepsilon) + c(n, k))$$

for some suitable normalization constants $c(n, k)$ (vol is the euclidean volume on $\mathbb{R}^k$ identified with the functions on $\{1, \ldots, k\}$) and then followed with

$$\sup_{R > 0} \inf_{m \in \mathbb{N}} \inf_{\varepsilon > 0}$$

one gets $H(f_1, \ldots, f_n)$. Note that the lim sup could actually be replaced with lim inf yielding the same result.

Remark also that instead of functions $a_j : \{1, \ldots, k\} \to \mathbb{R}$ we could have taken self-adjoint diagonal $k \times k$ matrices with the expectation E_k corresponding to the normalized trace $k^{-1} \operatorname{Tr}_k$ on the space Δ_k of such matrices.

8.3 Free entropy via matricial microstates [58]

The free entropy $\chi(X_1, \ldots, X_n)$ is defined for n-tuples of self-adjoint noncommutative random variables in a W^*-probability space (M, τ) where τ is a trace state. (The typical example being the W^*-algebras of discrete groups $(L(G), \tau)$.)

The set of approximating microstates $\Gamma_R(X_1, \ldots, X_n; \; m, k, \varepsilon)$ where $R > 0$, $m \in \mathbb{N}$, $k \in \mathbb{N}$, $\varepsilon > 0$ is defined to be the set of n-tuples $(A_1, \ldots, A_n) \in (M_k^{sa})^n$ so that

$$|\tau(X_{i_1} \ldots X_{i_p}) - k^{-1} \operatorname{Tr}(A_{i_1} \ldots A_{i_p})| < \varepsilon$$

for all $1 \le p \le m$, $(i_1, \ldots, i_p) \in \{1, \ldots, n\}^p$ and $\|A_j\| \le R$, $1 \le j \le n$. With vol denoting the volume on $(M_k^{sa})^n$ for the scalar product

$$\langle (A_j)_{1 \le j \le n}, (B_j)_{1 \le j \le n} \rangle = \sum_{1 \le j \le n} \operatorname{Tr} A_j B_j,$$

we take

$$\limsup_{k \to \infty} \left(k^{-2} \log \operatorname{vol} \Gamma_R(X_1, \ldots, X_n; \ m, k, \varepsilon) + \frac{n}{2} \log k \right)$$

and then define $\chi(X_1, \ldots, X_n)$ to be

$$\sup_{R>0} \ \inf_{m \in \mathbb{N}} \ \inf_{\varepsilon > 0}$$

of that quantity.

Remarks. 1°. The reason why matricial microstates define an entropy quantity with the right behavior w.r.t. free independence is related to the asymptotic freeness of large random matrices.

2°. Unfortunately our knowledge about the sets Γ_R, except for some important particular cases, is quite scarce and we don't know whether the lim sup in the definition of χ can be replaced with the lim inf without changing χ. A way around is to take a limit after an ultrafilter (see [64]).

3°. The cutoff given by R plays a minor role, as soon as $R \ge \max\{\|X_j\| : 1 \le j \le n\}$ the quantity we obtain is the same.

8.4 Properties of $\chi(X_1, \ldots, X_n)$

1°. $\chi(X_1, \ldots, X_n) \le \frac{n}{2} \log(2\pi e n^{-1} C^2)$ where $C'^2 = \tau(X_1^2 + \cdots + X_n^2)$ ([58]). This is the analogue of the Gaussian bound for H. It is obtained by binding vol Γ_R with the volume of the ball of radius C.

2°. **Semicontinuity** ([58]). If $(X_1^{(p)}, \ldots, X_n^{(p)})$ converges strongly to $(X_1, \ldots, X_n)$ then $\limsup_{p \to \infty} \chi(X_1^{(p)}, \ldots, X_n^{(p)}) \le \chi(X_1, \ldots, X_n)$.

3°. **Subadditivity** ([58]). $\chi(X_1, \ldots, X_{m+n}) \le \chi(X_1, \ldots, X_m) + \chi(X_{m+1}, \ldots, X_{m+n})$. This is just the inclusion

$$\Gamma_R(X_1, \ldots, X_{m+n}; \ldots) \subset \Gamma_R(X_1, \ldots, X_m; \ldots) \times \Gamma_R(X_{m+1}, \ldots, X_{m+n}; \ldots).$$

4°. **Additivity under freeness assumptions.** *If $X_1, \ldots, X_n$ are freely independent then* $\chi(X_1, \ldots, X_n) = \chi(X_1) + \cdots + \chi(X_n)$ ([58]). The proof uses the asymptotic freeness for random matrices. In this case the definition of $\chi(X_1, \ldots, X_n)$ with lim sup or lim inf gives the same result.

For groups of random variables we have (see [64]) using strengthened asymptotic freeness results for random matrices: *if* $\{X_1, \ldots, X_n\}$ *and* $\{Y_1, \ldots, Y_m\}$ *are freely independent then*

$$\chi_\omega(X_1, \ldots, X_n, Y_1, \ldots, Y_m) = \chi_\omega(X_1, \ldots, X_n) + \chi_\omega(Y_1, \ldots, Y_m).$$

Here χ_ω is the modified χ with lim sup replaced by the lim after the ultrafilter ω on $\mathbb{N}$.

5°. $n = 1$ ([58]). *If* X *has distribution* μ *then*

$$\chi(X) = \iint \log|s - t| d\mu(s) d\mu(t) + \frac{3}{4} + \frac{1}{2} \log 2\pi.$$

Up to constants, this is minus the logarithmic energy of μ.

Intuitively, such a formula can be expected for the following reason. Let $\mu_1 < \cdots < \mu_k$ be the eigenvalues of a matrix A such that $k^{-1} \sum_{1 \leq j \leq k} \delta_{\mu_j}$ is a "good approximant" of the distribution μ. Then the microstates of A are an ε-neighborhood of the unitary orbit $\{UAU^* \mid U \in \mathcal{U}(k)\}$ of A. The volume of the unitary orbit is, up to constants depending on k, given by

$$\prod_{1 \leq p < q \leq k} |\mu_p - \mu_q|^2.$$

Thus the normalized logarithm of the volume, up to additive constants, should be the limit of

$$k^{-2} \sum_{p \neq q} \log|\mu_p - \mu_q|$$

which is

$$\iint \log|s - t| d\mu(s) d\mu(t).$$

6°. **Change of variable formula** [58]. *Let* $F = (F_1, \ldots, F_n)$ *be a n-tuple of power series in the noncommuting indeterminates* $t_1, \ldots, t_n$

$$F_j(t_1, \ldots, t_n) = \sum_{k \geq 0} \sum_{1 \leq i_1, \ldots, i_k \leq n} c^{(j)}_{i_1, \ldots, i_k} t_{i_1} \ldots t_{i_k}.$$

Under suitable conditions on F *we have*

$$\chi(F_1(X_1), \ldots, F_n(X_n)) = \chi(X_1, \ldots, X_n) + \log|\det|(DF(X_1, \ldots, X_n)).$$

This statement requires several clarifications. The conditions on F are roughly:

(i) $F_j^* = F_j$ $(1 \leq j \leq n)$ i.e. (formally)

$$(F_j(t_1^*, \ldots, t_n^*))^* = F_j(t_1, \ldots, t_n).$$

(ii) Convergence radii conditions for the F_j's.

(iii) The transformation F has an inverse of the same kind (i.e. power series with radii of convergence conditions).

$DF(X_1, \ldots, X_n)$ is the differential of F viewed as an element in $\mathcal{M}_n \otimes M \otimes M^{\mathrm{op}}$ i.e. a $n \times n$ matrix with entries in $M \otimes M^{\mathrm{op}}$. To see why the partial differentials (i.e. of F_i w.r.t. X_j) which are the entries of the matrix are in $M \otimes M^{\mathrm{op}}$ consider a monomial $t_{i_1} \ldots t_{i_k}$ and the map it yields from n-tuples of operators. Its partial differential w.r.t. t_j at $X_1, \ldots, X_n$ is

$$\sum_{\{k : i_k = j, 1 \leq k \leq n\}} L_{X_{i_1} \ldots X_{i_{k-1}}} R_{X_{i_{k+1}} \ldots X_{i_n}}$$

where L and R denote left and respectively right multiplication operators. Identifying L_X with $X \in M$ and R_X with $X \in M^{\mathrm{op}}$ (the algebra with opposite multiplication) we get an element in $M \otimes M^{\mathrm{op}}$,

$$\sum_{i_k = j} X_{i_1} \ldots X_{i_{k-1}} \otimes (X_{i_{k+1}} \ldots X_{i_n})^{\mathrm{op}}.$$

The $|\det|$ is the Kadison–Fuglede positive determinant on $\mathcal{M}_n \otimes M \otimes M^{\mathrm{op}}$ endowed with $\mathrm{Tr} \otimes \tau \otimes \tau^{\mathrm{op}}$. This is if A is a W^*-algebra with a finite positive trace φ (we don't require $\varphi(1) = 1$) then for $a \in A$

$$|\det|(a) = \exp(\varphi(\tfrac{1}{2} \log(a^* a))).$$

The formula is (very roughly) obtained by checking that under the assumptions F yields a map of the microstates of $(X_1, \ldots, X_n)$ to those of $F(X_1, \ldots, X_n)$, which has an inverse and one uses the change of variable formula for the integral giving the volume of microstates.

Note that *in particular for a linear F, given by a real matrix $C = (C_i^{(j)})_{1 \leq i, j \leq n}$ the positive Jacobian $|\det|(DF)$ is just $|\det C|$.*

7°. **Semicircular maximum** ([60]). *If $\tau(X_1^2) = \cdots = \tau(X_n^2) = 1$ then $\chi(X_1, \ldots, X_n)$ is maximal if and only if $X_1, \ldots, X_n$ are $(0, 1)$ semicircular and freely independent.*

That the free semicircular n-tuple maximizes χ follows from 1°, 4°, 5°.

The converse is obtained using an infinitesimal version of the change of variable formula. If $P_1, \ldots, P_n$ are self-adjoint noncommutative polynomials in $X_1, \ldots, X_n$ applying the change of variable formula to $F_j = X_j + \varepsilon P_j$ for small ε one derives

$$\frac{d}{d\varepsilon}\chi(X_1 + \varepsilon P_1, \ldots, X_n + \varepsilon P_n) = \sum_j (\tau \otimes \tau)(\partial_j P_j(X_1, \ldots, X_n))$$

where

$$\partial_j X_{i_1} \ldots X_{i_n} = \sum_{i_k = j} X_{i_1} \ldots X_{i_{k-1}} \otimes X_{i_{k+1}} \ldots X_{i_n}.$$

8°. **Additivity implies freeness** [60]. *If $\chi(X_1, \ldots, X_n) = \chi(X_1) + \cdots + \chi(X_n)$ and $\chi(X_j) > -\infty$ $(1 \leq j \leq n)$, then $X_1, \ldots, X_n$ are freely independent.*

Very roughly after some approximations, $X_1, \ldots, X_n$ are transformed into $(f_1(X_1), \ldots, f_n(X_n))$ which are semicircular elements. Then the additivity together with suitable extensions of the change of variable formula imply $f_1(X_1), \ldots, f_n(X_n)$ maximize χ and hence by 7° are freely independent.

9°. **The free analogue of Shannon's entropy power inequality.** *If $X = X^*$, $Y = Y^*$ are freely independent in (M, τ) then*

$$\exp(2\chi(X)) + \exp(2\chi(Y)) \leq \exp(2\chi(X + Y)).$$

This fact is proved in [48] via a geometric inequality applied to the sets of microstates. Roughly, the inequality we want to prove amounts to majorizing

$$(\operatorname{vol}\Gamma(X; m, k, \varepsilon))^{2/k} + (\operatorname{vol}\Gamma(X; m, k, \varepsilon))^{2/k}$$

by some $(\operatorname{vol}\Gamma(X + Y; m_1, k, \varepsilon_1))^{2/k}$. If $\Gamma(X; \ldots) + \Gamma(Y; \ldots) \subset \Gamma(X + Y; \ldots)$ we could use the Minkowski inequality and get the better inequality with exponent $1/k$. However by asymptotic freeness results, all we have is that there is a set $\Theta \subset \Gamma(X; \ldots) \times \Gamma(Y; \ldots)$ of addable pairs, i.e. for which the sum is in $\Gamma(X + Y; \ldots)$ and that

$$\frac{\operatorname{vol}\Theta}{\operatorname{vol}(\Gamma(X; \ldots) \times \Gamma(Y; \ldots))} \to 1.$$

The inequality is precisely such a Minkowski type inequality with exponent $2/k$ for a restricted sum $A +_\Theta B$ where the addable pairs $\Theta \subset A \times B$ have

$\frac{\text{vol}\,\Theta}{\text{vol}(A\times B)} \to 1$. The proof of the geometric inequality relies on a "rearrangement inequality" of Brascamp–Lieb–Lüttinger, whose origins are in classical entropy theory.

Combining the same geometric inequality with the strengthened asymptotic freeness results, the result is generalized to freely independent $\{X_1,\ldots,X_n\}$ and $\{Y_1,\ldots,Y_n\}$ in the form

$$\exp\left(\frac{2}{n}\chi_\omega(X_1,\ldots,X_n)\right)$$
$$+\exp\left(\frac{2}{n}\chi_\omega(Y_1,\ldots,Y_n)\right) \le \exp\left(\frac{2}{n}\chi_\omega(X_1+Y_1,\ldots,X_n+Y_n)\right)$$

where χ_ω is the modified free entropy following an ultrafilter and $\chi_\omega(X_1,\ldots,X_n)$, $\chi_\omega(Y_1,\ldots,Y_n)$ are assumed $> -\infty$ ([64]).

8.5 Free entropy dimension

Since χ is a normalized logarithm of volume of microstates, it can be used to define a kind of normalized dimension, of Minkowski type, for the microstates.

Definition [58]. The free entropy dimension of $X_1,\ldots,X_n$ is

$$\delta(X_1,\ldots,X_n) = n + \limsup_{\varepsilon\downarrow 0} \frac{\chi(X_1+\varepsilon S_1,\ldots,X_n+\varepsilon S_n)}{|\log\varepsilon|}$$

where $S_1,\ldots,S_n$ are $(0,1)$-semicircular and $\{S_1\},\ldots,\{S_n\}$, $\{X_1,\ldots,X_n\}$ are freely independent.

Properties of δ ([58])

1°. a) $\delta(X_1,\ldots,X_n) \le n$.

b) $\delta(X_1,\ldots,X_n) \ge 0$ if for every $m\in\mathbb{N}$, $\varepsilon>0$, $R>\|X_j\|$ for sufficiently large k $\Gamma_R(X_1,\ldots,X_n; m,k,\varepsilon) \ne \emptyset$. (That this property holds for all $(X_1,\ldots,X_n)$ is equivalent to Connes' problem on embedding into the ultraproduct of the hyperfinite II_1 factor.)

2°. $\delta(X_1,\ldots,X_{p+1}) \le \delta(X_1,\ldots,X_p) + \delta(X_{p+1},\ldots,X_{p+q})$.

3°. If $X_1,\ldots,X_n$ are freely independent then $\delta(X_1,\ldots,X_n) = \delta(X_1) + \cdots + \delta(X_n)$.

4°. $\delta(X) = 1 - \sum_{t\in\mathbb{R}}(\mu(\{t\}))^2$ where μ is the distribution of X.

Variants of δ with limits following ultrafilters have been studied in [64]. *For certain variants of δ if $(X_1, \ldots, X_n)$ and $(Y_1, \ldots, Y_m)$ generate the same algebra (algebraically) then the free entropy dimensions of $(X_1, \ldots, X_n)$ and $(Y_1, \ldots, Y_m)$ are equal ([64]).*

8.6 Classical Fisher information and the adjoint of the derivation

If f is a bounded real random variable the Fisher information $\mathcal{J}(f)$ is defined by

$$\mathcal{J}(f) = \lim_{\varepsilon \downarrow 0} \varepsilon^{-1}(H(f + \varepsilon^{1/2}g) - H(f))$$

where f and g are independent and g is $(0, 1)$-Gaussian. If the distribution of f is Lebesgue absolutely continuous, with density p then

$$\mathcal{J}(f) = \int \frac{(p'(t))^2}{p(t)} \, dt.$$

Another way to arrive at the formula for $\mathcal{J}(f)$ ([61]) is to consider the derivation $\frac{d}{dt}$ as a densely defined operator on $L^2(\mathbb{R}, p d\lambda)$ ($d\lambda$ Lebesgue measure) with domain of definition the polynomial functions and values in $L^2(\mathbb{R}, p d\lambda)$. If 1 is in the domain of the adjoint of $\frac{d}{dt}$ then

$$\left(\frac{d}{dt}\right)^* 1 = -\frac{\frac{dp}{dt}}{p}$$

provided the right-hand side is in $L^2(\mathbb{R}, p d\lambda)$. Note that we then have

$$\mathcal{J}(f) = \left\| \left(\frac{d}{dt}\right)^* 1 \right\|^2_{L^2(\mathbb{R}, p d\lambda)}.$$

Let us also recall that to recover H from $\mathcal{J}$ one considers a Brownian motion starting at f. Equivalently one has $f + t^{1/2}g$ $(t \geq 0)$ and

$$\lim_{\varepsilon \downarrow 0} \varepsilon^{-1}(H(f + (t + \varepsilon)^{1/2}g) - H(f + t^{1/2}g)) = \mathcal{J}(f + t^{1/2}g).$$

Together with

$$H((1 + t)^{-1/2}(f + t^{1/2}g)) = H(f + t^{1/2}g) - \frac{1}{2}\log(1 + t)$$

this gives

$$H(g) - H(f) = \int_0^\infty \left(\mathcal{J}(f + t^{1/2}g) - \tfrac{1}{2}(1 + t)^{-1}\right) dt.$$

8.7 Free Fisher information of one variable [57]

By analogy with the classical case, the free Fisher information of one self-adjoint variable X with distribution μ is defined by

$$\Phi(X) = \lim_{\varepsilon \downarrow 0} \varepsilon^{-1}(\chi(X + \varepsilon^{1/2}S) - \chi(X))$$

where X and S are freely independent and S is $(0,1)$ semicircular.

As in the classical case this gives a formula of $\Phi(X)$ in terms of the density of μ. Here is roughly how this is done (we overlook the smoothing that is done by replacing $\mathbb{R}$ by $\mathbb{R} + i\varepsilon$, see [57]).

Le $\mu(t)$ be the distribution of $X + t^{1/2}S$ and $G(z,t)$ its Cauchy transform. Then on $\mathbb{R}$, $G(x,t) = u(x,t) + iv(x,t)$, where $v(\cdot, t)$ is the density of $-\pi\mu(t)$. We have (see 2.11)

$$\frac{\partial G}{\partial t} + G\frac{\partial G}{\partial z} = 0.$$

This gives on $\mathbb{R}$ the system

$$\begin{cases} u_t + (uu_x - vv_x) = 0 \\ v_t + (uv_x + u_x v) = 0 \\ u = -Hv \end{cases}$$

where H denotes the Hilbert transform

$$(Hv)(x) = \lim_{\varepsilon \downarrow 0} \frac{1}{\pi} \int \frac{(x-s)v(s)}{(x-s)^2 + \varepsilon^2} ds.$$

Up to constants (see 8.4.5°) $\chi(X + t^{1/2}S)$ is minus the logarithmic energy of $\mu(t)$. So we have

$$\begin{aligned} \pi^2 \Phi(X) &= \left(\frac{d}{d\varepsilon} \iint v(x,\varepsilon)v(y,\varepsilon) \log|x-y| dx dy \right)\Big|_{\varepsilon=0} \\ &= -2 \iint (uv)_x(x,\varepsilon)v(y,\varepsilon) \log|x-y| dx dy \Big|_{\varepsilon=0} \\ &= 2 \int (uv)(x,\varepsilon) \left(\int \log|x-y|v(y,\varepsilon)dy \right)_x \Big|_{\varepsilon=0} dx \\ &= 2\pi \int (uv)(x,\varepsilon)u(x,\varepsilon)dx \Big|_{\varepsilon=0} \\ &= 2\pi \int (u^2 v)(x,0)dx. \end{aligned}$$

On the other hand for Cauchy transforms one has $0 = \int G^3(x)dx$ because of the zero at ∞ which gives (under L^3-assumptions)

$$3 \int u^2 v dx = \int v^3 dx.$$

Thus if μ has density p w.r.t. Lebesgue measure then

$$\Phi(X) = \frac{2\pi^2}{3} \int p^3(s)ds = 2\pi^2 \int (Hp)^2 p \ ds.$$

8.8 The underlying idea of the microstates-free approach [61]

The formula for $\Phi(X)$ as an integral of p^3 is not a good starting point for generalizations. On the other hand the integral of $(Hp)^2 p$ will do the job. This integral is the square of the L^2-norm of Hp viewed as an element of $L^2(\mathbb{R}, pdt) = L^2(\mathbb{R}, d\mu)$. Thus, what we need, is a way to view Hp as an element of $L^2(\mathbb{R}, pdt)$ which can be generalized. Roughly, Hp is given by

$$\int \frac{d\mu(t)}{s-t}.$$

This in turn can be expressed using the difference quotient derivation which is defined on the polynomials viewed as a dense subset in $L^2(\mathbb{R}, d\mu)$, by

$$\partial f = \frac{f(s) - f(t)}{s-t},$$

the values being in $L^2(\mathbb{R}, d\mu) \otimes L^2(\mathbb{R}, d\mu)$. Up to constants Hp is given by

$$\partial^*(1 \otimes 1)$$

where ∂^* is the adjoint of ∂.

Note that $\|\partial^*(1 \otimes 1)\|_{L^2(\mathbb{R}, d\mu)}$ is quite similar to the formula $\left\| \left(\frac{d}{dt} \right)^* 1 \right\|^2_{L^2(\mathbb{R}, d\mu)}$ for classical Fisher information in 8.6. Another clue that we are on the right track is the occurrence of difference quotient type derivations in the infinitesimal change of variable formula for χ (see 8.4.7°).

8.9 Noncommutative Hilbert transforms [61]

In (M, τ) let $1 \in B \subset M$ be a *-subalgebra and $X = X^* \in M$ such that X and B are algebraically free (i.e. no nontrivial algebraic relation between X and B). The noncommutative analogue of the difference quotient derivation with "constants" B is the linear map

$$\partial_X : B[X] \to B[X] \otimes B[X]$$

such that

$$\partial_X(b_0 X b_1 X \ldots b_n) = \sum_{1 \leq j \leq n} b_0 X \ldots b_{j-1} \otimes b_j X \ldots b_n$$

($B[X]$ is the algebra generated by B and X.)

Thus ∂_X is a densely defined unbounded operator from $L^2(B[X], \tau)$ to $L^2(B[X], \tau) \otimes L^2(B[X], \tau)$ ($L^2(\cdot, \tau)$ is the completion w.r.t. the scalar product $\langle T_1, T_2 \rangle = \tau(T_2^* T_1)$).

We define the *conjugate of* X *w.r.t.* B to be $J(X : B) = \partial_X^*(1 \otimes 1)$ *if it exists as an element in* $L^2(B[X], \tau)$.

Equivalently $J(X : B) = \xi \in L^2(B[X], \tau)$ if

$$\tau(\xi b_0 X b_1 X \ldots b_n) = \sum_{1 \leq j \leq n} \tau(b_0 X \ldots b_{j-1}) \tau(b_j X \ldots b_n)$$

(remark that the right-hand side is $(\tau \otimes \tau)(\partial_X(b_0 X b_1 X \ldots b_n))$, further the formula is with ξ instead of ξ^*, since it turns out that we must have $\xi = \xi^*$).

Remark. To avoid operator algebra technicalities we restricted the definition of $J(X : B)$ to the L^2 case.

Here are some basic properties of $J(X : B)$.

1°. $J(\lambda X : B) = \lambda^{-1} J(X : B)$ *if* $\lambda \in \mathbb{R}$, $\lambda \neq 0$.

2°. $J(X : \mathbb{C}) = g(X)$ *where* $g = 2\pi H p$ *when the distribution of* X *has density* $p \in L^3$ *w.r.t. Lebesgue measure* (Hp *the Hilbert transform of* p).

3°. *If* S *is* $(0, 1)$-*semicircular then* $J(S : \mathbb{C}) = S$. This is analogous to the fact in the classical context that the Gaussian distribution $p(t) = c \exp(-t^2/2)$ has the property

$$\left(\left. \frac{d}{dt} \right|_{L^2(\mathbb{R}, p \, dt)} \right)^* 1 = -\frac{p'}{p} = t.$$

4°. *Let* $1 \in C \subset M$ *be a* *-subalgebra such that* $B[X]$ *and* C *are freely independent. Then*

$$J(X : B) = J(X : W^*(B \cup C)).$$

The conclusion also holds under the weaker assumption that $\{X\}$ and C are freely independent over B (see [39] for this strengthening).

5°. *If* $B[X]$ *and* $C[Y]$ *are freely independent, then*

$$J(X + Y : B \vee C) = E_{(B \vee C)[X+Y]} J(X : B).$$

$6°$. *If S is $(0,1)$ semicircular and $B[X]$ and S are freely independent and $\varepsilon \neq 0$ then*

$$J(X + \varepsilon S : B) = \varepsilon^{-1} E_{B[X+\varepsilon S]} S.$$

In particular

$$\|J(X + \varepsilon S : B)\| \leq 2\varepsilon^{-1}.$$

This is obtained by combining $5°$, $3°$, $1°$ and $\|S\| = 2$.

Thus small semicircular perturbations regularize the noncommutative Hilbert transform. *Moreover $J(X : B)$ exists iff*

$$\sup_{\varepsilon > 0} |J(X + \varepsilon S : B)|_2 < \infty$$

and if the supremum is finite it equals $|J(X : B)|_2$.

$7°$. *Assume S is $(0,1)$ semicircular, $B[X]$ and S are freely independent, $\|J(X : B)\| < \infty$ and $\varepsilon > 0$. Then*

$$\tau\left(b_0\left(X + \frac{\varepsilon}{2}J(X : B)\right) b_1\left(X + \frac{\varepsilon}{2}J(X : B)\right) \ldots b_n\right)$$
$$= \tau(b_0(X + \varepsilon^{1/2}S)b_1(X + \varepsilon^{1/2}S)\ldots b_n) + O(\varepsilon^2).$$

Note that $X + t^{1/2}S$ is the same from the distribution point of view as a free Brownian motion starting at X, at time t. *This suggests as a possible alternative name for $J(X : B)$, to call it the free Brownian gradient of X w.r.t. B.*

$8°$. *Let $X_j = X_j^* \in M$, $1 \leq j \leq n$ be such that $\chi(X_1, \ldots, X_n) > -\infty$ and assume $J(X_k : \mathbb{C}[X_1, \ldots, X_{k-1}, X_{k+1}, \ldots, X_n])$ $1 \leq k \leq n$ exist. Let further $P_j = P_j^* \in \mathbb{C}[X_1, \ldots, X_n]$. Then*

$$\frac{d}{d\varepsilon}\chi(X_1 + \varepsilon P_1, \ldots, X_n + \varepsilon P_n)|_{\varepsilon=0} = \sum_{1 \leq j \leq n} \tau(P_j J(X_j : \mathbb{C}[X_1, \ldots \widehat{X_j}, \ldots X_n])).$$

This is a consequence of the infinitesimal change of variable formula for the free entropy χ (see [60] and 8.4.7°). It is a confirmation that the noncommutative Hilbert transforms are the right ingredients for a microstates-free approach.

8.10 Free Fisher information and free entropy in the microstates-free approach [61]

By analogy with the classical case *the relative free Fisher information* $\Phi^*(X_1, \ldots, X_n : B)$ *of a n-tuple of self-adjoint variables* $X_1, \ldots, X_n$ *w.r.t. the subalgebra B is defined by*

$$\Phi^*(X_1, \ldots, X_n : B) = \sum_{1 \leq j \leq n} |\mathcal{J}(X_j : B[X_1, \ldots, \widehat{X_j}, \ldots, X_n])|_2^2$$

if the right-hand side is defined and ∞ *otherwise.*

The asterisk in $\Phi^*(\ldots)$ is to distinguish quantities in the microstates-free approach from their counterparts in the matricial microstates approach, denoted Φ in this case.

Here are some of the properties of Φ^*, based essentially on those of the non-commutative Hilbert transform.

$\Phi^*1°$. **Scaling.** $\lambda \in \mathbb{R}$, $\lambda \neq 0$ *then*

$$\Phi^*(\lambda X : B) = \lambda^{-2} \Phi(X : B).$$

$\Phi^*2°$. **Superadditivity.**

$$\Phi^*(X_1, \ldots, X_n, Y_1, \ldots, Y_m : B) \geq \Phi^*(X_1, \ldots, X_n : B) + \Phi^*(Y_1, \ldots, Y_m : B).$$

$\Phi^*3°$. **Increasing in** B. *If* $B_1 \subset B_2$ *then*

$$\Phi^*(X_1, \ldots, X_n : B_1) \leq \Phi^*(X_1, \ldots, X_n : B_2).$$

$\Phi^*4°$. **Free additivity.** *If* $B[X_1, \ldots, X_n]$ *and* $C[Y_1, \ldots, Y_m]$ *are freely independent, then*

$$\Phi^*(X_1, \ldots, X_n, Y_1, \ldots, Y_m : W^*(B \cup C)) = \Phi^*(X_1, \ldots, X_n : B) + \Phi^*(Y_1, \ldots, Y_m : C).$$

$\Phi^*5°$. **Free analogue of the Cramer–Rao inequality.**

$$\Phi^*(X_1, \ldots, X_n : B)\tau(X_1^2 + \cdots + X_n^2) \geq n^2.$$

Equality holds iff the X_j's *are centered semicircular and* $B, \{X_1\}, \ldots, \{X_n\}$ *are freely independent.*

$\Phi^*6°$. **Free analogue of the Stam inequality.** *If $B[X_1,\ldots,X_n]$ and $C[Y_1,\ldots,Y_n]$ are freely independent then*

$$(\Phi^*(X_1+Y_1,\ldots,X_n+Y_n : B\vee C))^{-1} \geq (\Phi^*(X_1,\ldots,X_n : B))^{-1}+(\Phi^*(Y_1,\ldots,Y_n : C))^{-1}.$$

$\Phi^*7°$. **Semicontinuity.** *If $X_j^{(k)} = X_j^{(k)*} \in M$ and $s-\lim_{k\to\infty} X_j^{(k)} = X_j$, then*

$$\liminf_{k\to\infty} \Phi^*(X_1^{(k)},\ldots,X_n^{(k)} : B) \geq \Phi^*(X_1,\ldots,X_n : B).$$

$\Phi^*8°$. *If $\Phi^*(X_1,\ldots,X_n : B) = \Phi^*(X_1,\ldots,X_n : \mathbb{C}) < \infty$ then $\{X_1,\ldots,X_n\}$ and B are freely independent. If $\Phi^*(X_1,\ldots,X_n,Y_1,\ldots,Y_m : \mathbb{C}) = \Phi^*(X_1,\ldots,X_n : \mathbb{C})+ \Phi^*(Y_1,\ldots,Y_m : \mathbb{C}) < \infty$ then $\{X_1,\ldots,X_n\}$, $\{Y_1,\ldots,Y_m\}$ are freely independent [62].*

The *free entropy* of $X_1,\ldots,X_n$ w.r.t. B in the microstates-free approach is defined by

$$\chi^*(X_1,\ldots,X_n : B)$$
$$= \tfrac{1}{2}\int_0^\infty \left(\frac{n}{1+t} - \Phi^*(X_1 + t^{1/2}S_1,\ldots,X_n + t^{1/2}S_n : B)\right) dt + \frac{n}{2}\log 2\pi e$$

where the S_j's are $(0,1)$-semicircular and $B[X_1,\ldots,X_n]$, $\{S_1\},\ldots,\{S_n\}$ are freely independent.

(Again the asterisk in χ^* distinguishes the microstates-free and the matricial microstates approaches.)

Here are the main properties of χ^* which have been proved at this time.

$\chi^*1°$. $\chi^*(X : \mathbb{C}) = \chi(X)$.

$\chi^*2°$. $\chi^*(X_1,\ldots,X_n : B) \leq \frac{n}{2}\log(2\pi e n^{-1}C^2)$.

$\chi^*3°$. *If $B[X_1,\ldots,X_n]$ and C are freely independent, then*

$$\chi^*(X_1,\ldots,X_n : B) = \chi^*(X_1,\ldots,X_n : W^*(B\cup C)).$$

$\chi^*4°$. $\chi^*(X_1,\ldots,X_n,Y_1,\ldots,X_m : B\vee C) \leq \chi^*(X_1,\ldots,X_n : B)+\chi^*(Y_1,\ldots,Y_m : C)$.

$\chi^*5°$. *If* $B[X_1, \ldots, X_n]$ *and* $C[Y_1, \ldots, Y_m]$ *are freely independent, then*

$$\chi^*(X_1, \ldots, X_n, Y_1, \ldots, Y_m : B \vee C) = \chi^*(X_1, \ldots, X_n : B) + \chi^*(Y_1, \ldots, Y_m : C).$$

$\chi^*6°$. *If* $X_j^{(k)} = X_j^{(k)*}$ *and* $s - \lim_{k\to\infty} X_j^{(k)} = X_j$ *then*

$$\limsup_{k\to\infty} \chi^*(X_1^{(k)}, \ldots, X_n^{(k)} : B) \leq \chi^*(X_1, \ldots, X_n : B).$$

$\chi^*7°$. *If* $\Phi^*(X_1, \ldots, X_n : B) < \infty$ *then*

$$\chi^*(X_1, \ldots, X_n : B) \geq \frac{n}{2} \log \left(\frac{2\pi n e}{\Phi^*(X_1, \ldots, X_n : B)} \right).$$

In particular $\chi^*(X_1, \ldots, X_n : B) > -\infty$.

(The inequality $\chi^*7°$. is the free analogue of the isoperimetric inequality for classical entropy.)

REFERENCES

1. Anshelevich, M., *The linearization of the central limit operator in free probability theory*, preprint.
2. Avitzour, D., *Free products of C^*-algebras*, Trans. Amer. Math. Soc. **271** (1982), 423–465.
3. Bercovici, H. and Pata, V. (with an appendix by P. Biane), *Stable laws and domains of attraction in free probability theory*, Annals of Math., to appear.
4. Bercovici, H. and Voiculescu, D., *Levy–Hinčin type theorems for multiplicative and additive free convolution*, Pacific J. Math. **153** (1992), no. 2, 217–248.
5. Bercovici, H. and Voiculescu, D., *Free convolution of measures with unbounded support*, Indiana Univ. Math. J. **42** (1993), no. 3, 733–773.
6. Bercovici, H. and Voiculescu, D., *Superconvergence to the central limit and failure of the Cramer Theorem for free random variables*, Probab. Th. and Rel. Fields **102** (1995), 215–222.
7. Bercovici, H. and Voiculescu, D., *Regularity questions for free convolution*, Preprint, Berkeley (1996).
8. Bhat, B.V.R. and Parthasarathy, K.R., *Markov dilations of nonconservative dynamical semigroups and a quantum boundary theory*, Ann. Inst. H. Poincaré **31** (1995), no. 4, 601–651.
9. Biane, P., *Representations of unitary groups and free convolution*, Publ. RIMS Kyoto Univ. **31** (1995), 63–79.
10. Biane, P., *Processes with free increments*, Math. Z. (1998), no. 1, 143–174.
11. Biane, P., *Free Brownian motion, free stochastic calculus and random matrices*, in [66], 1–19.
12. Biane, P., *Free hypercontractivity*, Comm. Math. Phys. **184** (1997), 457–474.
13. Biane, P., *On the free convolution with a semicircular distribution*, Indiana Univ. Math. J. **46** (1997), no. 3, 705–718.
14. Biane, P., *Representations of symmetric groups and free probability*, Preprint (1998).
15. Biane, P. and Speicher, R., *Stochastic calculus with respect to free Brownian motion and analysis on Wigner space*, Preprint ENS (1997).

16. Bozejko, M. and Speicher, R., *An example of generalized Brownian motion*, Commun. Math. Phys. **137** (1991), 519–531.

17. Bozejko, M. and Speicher, R., *An example of generalized Brownian motion II*, Quantum Probability and Related Topics (L. Accardi, ed.), vol. VI, World Scientific, Singapore, 1991, pp. 219–236.

18. Dixmier, J., *Les C*-Algebres et leur Représentations*, Gauthier–Villars, Paris, 1964.

19. Dixmier, J., *Les Algebres d'Opérateurs dans l'Espace Hilbertien*, Gauthier–Villars, Paris, 1969.

20. Douglas, M.R., *Stochastic master fields*, Phys. Lett. **B344**, 117–126.

21. Dykema, K.J., *On certain free product factors via an extended matrix model*, J. Funct. Anal. **112**, 31–60.

22. Fagnola, F., *On quantum stochastic integration with respect to "free" noises*, Quantum Probability and Related Topics (L. Accardi, ed.), vol. VI, World Scientific, Singapore, 1991, pp. 285–304.

23. Gopakumar, R. and Gross, D.J., *Mastering the master field*, Nucl. Phys. **B451**, 379–415.

24. Gromov, M. and Milman, V.D., *A topological application of the isoperimetric inequality*, Amer. J. Math. **105** (1983), 843–854.

25. Haagerup, U., *On Voiculescu's R- and S-transforms for free noncommuting random variables*, in [66], 127–148.

26. Kadison, R. and Ringrose, J., *Fundamentals of the Theory of Operator Algebras*, (3 Volumes) Birkhäuser, Boston.

27. Kummerer, B., *Markov dilations on W*-algebras*, J. Funct. Anal. **63** (1985), 139–177.

28. Kummerer, B. and Speicher, R., *Stochastic integration on the Cuntz algebra*, J. Funct. Anal. **103** (1992), 372–408.

29. Maassen, H., *Addition of freely independent random variables*, J. Funct. Anal. **106** (1992), 409–438.

30. Nica, A., *Asymptotically free families of random unitaries in symmetric groups*, Pacific J. Math. **157** (1993), no. 2, 295–310.

31. Nica, A., *R-transforms of free joint distributions and non-crossing partitions*, J. Funct. Anal. **135** (1996), 271–296.

32. Nica, A. and Speicher, R. (with an appendix by D. Voiculescu), *On the multiplication of free N-tuples of noncommutative random variables*, Amer. J. Math. **118** (1996), 799–837.

33. Nica, A. and Speicher, R., *A "Fourier transform" for multiplicative functions on non-crossing partitions*, J. of Algebraic Combinatorics **6** (1997), 141–160.

34. Nica, A. and Speicher, R., *Commutators of free random variables*, Duke Math. J., to appear.

35. Nica, A.; Shlyakhtenko, D. and Speicher, R., *Some minimization problems for the free analogue of the Fisher information*, Preprint (1998).

36. Sauvageot, J.L., *Markov quantum semigroups admit covariant Markov C*-dilations*, Commun. Math. Phys. **106** (1986), 91–103.

37. Shannon, C.E. and Weaver, W., *The Mathematical Theory of Communications*, University of Illinois Press, 1963.

38. Shlyakhtenko, D., *Limit distributions of matrices with bosonic and fermionic entries*, in [66], 241–252.

39. Shlyakhtenko, D., *Free entropy with respect to a completely positive map*, Preprint (1998).

40. Shlyakhtenko, D., *Random Gaussian band matrices and freeness with amalgamation*, International Math. Res. Notices (1996), no. 20, 1013–1025.

41. Shiryayev, A.N., *Probability*, Springer, 1984.

42. Singer, I.M., *On the master field in two dimensions*, Functional Analysis on the Eve of the 21st Century in Honor of the 80th Birthday of I.M. Gelfand, Progress in Mathematics, vol. 131, pp. 263–283.

43. Speicher, R., *A new example of "independence" and "white noise"*, Prob. Th. Rel. Fields **84** (1990), 141–159.

166

44. Speicher, R., *Combinatorial theory of the free product with amalgamation and operator-valued free probability theory*, Memoirs of the AMS **627** (1998).
45. Speicher, R., *Multiplicative functions on the lattice of non-crossing partitions and free convolution*, Math. Ann. **298** (1994), 611–628.
46. Speicher, R., *Free probability theory and non-crossing partitions*, Seminaire Lotharingien de Combinatoire **B39c** (1997).
47. Stratila, S. and Zsido, L., *Lectures on von Neumann Algebras*, Editura Academia and Abacus Press, 1979.
48. Szarek, S.V. and Voiculescu, D., *Volumes of restricted Minkowski sums and the free analogue of the entropy power inequality*, Comun. Math. Phys. **178** (1996), 563–570.
49. Voiculescu, D., *Symmetries of some reduced free product C^*-algebras*, Operator Algebras and their Connections with Topology and Ergodic Theory, Lecture Notes in Math. **1132** (1985), Springer Verlag, 556–588.
50. Voiculescu, D., *Addition of certain non-commuting random variables*, J. Funct. Anal. **66** (1986), 323–346.
51. Voiculescu, D., *Multiplication of certain non-commuting random variables*, J. Operator Theory **18** (1987), 223–235.
52. Voiculescu, D., *Limit laws for random matrices and free products*, Invent. Math. **104** (1991), 201–220.
53. Voiculescu, D., *Circular and semicircular systems and free product factors*, Progr. Math. **92** (1990), Birkhäuser, Boston, 45–60.
54. Voiculescu, D., *Free non-commutative random variables, random matrices and the II_1-factors of free groups*, Quantum Probability and Related Topics (L. Accardi, ed.), vol. VI, World Scientific, Boston, 1991, pp. 473–487.
55. Voiculescu, D., *Free probability theory: random matrices and von Neumann algebras*, Proceedings of the International Congress of Mathematicians, Zürich 1994, Birkhäuser, Boston (1995), 227–241.
56. Voiculescu, D., *Operations on certain non-commuting operator-valued random variables*, Astérisque (1995), no. 232, 243–275.
57. Voiculescu, D., *The analogues of entropy and of Fisher's information measure in free probability theory I*, Commun. Math. Phys. **155** (1993), 71–92.
58. Voiculescu, D., *The analogues of entropy and of Fisher's information measure in free probability theory II*, Invent. Math. **118** (1994), 411–440.
59. Voiculescu, D., *The analogues of entropy and of Fisher's information measure in free probability theory III: the absence of Cartan subalgebras*, Geometric and Funct. Anal. **6** (1996), no. 1, 172–199.
60. Voiculescu, D., *The analogues of entropy and of Fisher's information measure in free probability theory IV: Maximum entropy and freeness*, in [66], 293–302.
61. Voiculescu, D., *The analogues of entropy and of Fisher's information measure in free probability theory V: Noncommutative Hilbert transforms*, Invent. Math. **132** (1998), 182–227.
62. Voiculescu, D., *The analogues of entropy and of Fisher's information measure in free probability theory VI: liberation and mutual free information,*, preprint 1998.
63. Voiculescu, D., *The derivative of order 1/2 of a free convolution by a semicircle distribution*, Indiana Univ. Math. J. **46** (1997), no. 3, 697–703.
64. Voiculescu, D., *A strengthened asymptotic freeness result for random matrices with applications to free entropy*, International Math. Res. Notices (1998), no. 1, 41–63.
65. Voiculescu, D., *A note on free Markovianity*, (in preparation).
66. Voiculescu, D. (editor), *Free Probability Theory*, Fields Institute Communications, Vol. 12, American Math. Soc., 1997.
67. Voiculescu, D.; Dykema, K.J. and Nica, A., *Free Random Variables*, CRM Monograph Series, Vol. 1, American Math. Soc., 1992.
68. Wigner, E., *Characteristic vectors of bordered matrices with infinite dimensions*, Ann. Math. **62** (1955), 548–564.

69. Wigner, E., *On the distribution of the roots of certain symmetric matrices*, Ann. Math. **67** (1958), 325–327.
70. Xu, F., *A random matrix model from two-dimensional Yang–Mills theory*, Commun. Math. Phys. **190** (1997), 287–307.

Alice Guionnet

Large Random Matrices: Lectures on Macroscopic Asymptotics

École d'Été de Probabilités
de Saint-Flour XXXVI – 2006

Originally published in: *École d'Été de Probabilités de Saint-Flour XXXVI – 2006*,
Lecture Notes in Mathematics, Vol. **1957**, III–XII, 1–294, DOI: 10.1007/978-3-540-69897-5,
© Springer-Verlag Berlin Heidelberg 2009, Reprint by Springer-Verlag Berlin Heidelberg 2012

Preface

These notes include the material from a series of nine lectures given at the Saint-Flour probability summer school in 2006. The two other lecturers that year were Maury Bramson and Steffen Lauritzen.

The topic of these lectures was large random matrices, and more precisely the asymptotics of their macroscopic observables such as the empirical measure of their eigenvalues. The interest in such questions goes back to Wishart and Wigner, in the twenties and fifties respectively. Large random matrices have been since then intensively studied in theoretical physics, in connection with various fields such as QCD, quantum chaos, string theory or quantum gravity.

Since the nineties, several key mathematical results have been obtained and the theory of large random matrices expanded in various directions, in connection with combinatorics, operator algebra theory, number theory, algebraic geometry, integrable systems etc. I felt that the time was right to summarize some of them, namely those which connect with the asymptotics of macroscopic observables, with a particular emphasis on their relation with combinatorics and operator algebra theory.

I wish to thank Jean Picard for organizing the Saint-Flour school and helping me through the preparation of these notes, and the other participants of the school, in particular for their useful comments to improve these notes. I am very grateful to several collaborators with whom I consulted on various points, in particular Greg Anderson, Edouard Maurel Segala, Dima Shlyakhtenko and Ofer Zeitouni.

Lyon, France   *Alice Guionnet*
July 2008

Contents

Introduction 1

Part I   Wigner Matrices and Moments Estimates   5

1   Wigner's Theorem 7
   1.1   Catalan Numbers, Non-crossing Partitions and Dick Paths.... 7
   1.2   Wigner's Theorem 16
   1.3   Weak Convergence of the Spectral Measure 20
   1.4   Relaxation of the Hypotheses over the Entries–Universality ... 22

2   Wigner's Matrices; More Moments Estimates 29
   2.1   Central Limit Theorem 29
   2.2   Estimates of the Largest Eigenvalue of Wigner Matrices 33

3   Words in Several Independent Wigner Matrices 41
   3.1   Partitions of Colored Elements and Stars 41
   3.2   Voiculescu's Theorem 42

Part II   Wigner Matrices and Concentration Inequalities   47

4   Concentration Inequalities and Logarithmic Sobolev Inequalities 49
   4.1   Concentration Inequalities for Laws Satisfying Logarithmic Sobolev Inequalities 49
   4.2   A Few Laws Satisfying a Log-Sobolev Inequality 52

174

5   Generalizations 59
    5.1   Concentration Inequalities for Laws Satisfying Weaker
        Coercive Inequalities 59
    5.2   Concentration Inequalities by Talagrand's Method 60
    5.3   Concentration Inequalities on Compact Riemannian Manifold
        with Positive Ricci Curvature 61
    5.4   Local Concentration Inequalities.......................... 62

6   Concentration Inequalities for Random Matrices 65
    6.1   Smoothness and Convexity of the Eigenvalues
        of a Matrix 65
    6.2   Concentration Inequalities for the Eigenvalues
        of Random Matrices 70
    6.3   Concentration Inequalities for Traces of Several Random
        Matrices 72
    6.4   Concentration Inequalities for the Haar Measure
        on $O(N)$ 74
    6.5   Brascamp–Lieb Inequalities; Applications
        to Random Matrices 77

Part III   Matrix Models   **89**

7   Maps and Gaussian Calculus 93
    7.1   Combinatorics of Maps and Non-commutative
        Polynomials 93
    7.2   Non-commutative Polynomials 93
    7.3   Maps and Polynomials 97
    7.4   Formal Expansion of Matrix Integrals 99

8   First-order Expansion 109
    8.1   Finite-dimensional Schwinger–Dyson Equations 109
    8.2   Tightness and Limiting Schwinger–Dyson Equations 110
    8.3   Convergence of the Empirical Distribution 113
    8.4   Combinatorial Interpretation of the Limit 114
    8.5   Convergence of the Free Energy 118

9   Second-order Expansion for the Free Energy 121
    9.1   Rough Estimates on the Size of the Correction $\tilde{\delta}_t^N$ 122
    9.2   Central Limit Theorem 124
    9.3   Comments on the Results 137
    9.4   Second-order Correction to the Free Energy 140

Part IV   Eigenvalues of Gaussian Wigner Matrices
and Large Deviations   147

10 Large Deviations for the Law of the Spectral Measure
    of Gaussian Wigner's Matrices................................149

11 Large Deviations of the Maximum Eigenvalue 159

Part V   Stochastic Calculus   165

12 Stochastic Analysis for Random Matrices 167
    12.1 Dyson's Brownian Motion 167
    12.2 Itô's Calculus 175
    12.3 A Dynamical Proof of Wigner's Theorem 1.13...............176

13 Large Deviation Principle for the Law of the Spectral
    Measure of Shifted Wigner Matrices.......................183
    13.1 Large Deviations from the Hydrodynamical Limit
         for a System of Independent Brownian Particles 186
    13.2 Large Deviations for the Law of the Spectral Measure
         of a Non-centered Large Dimensional Matrix-valued
         Brownian Motion..192

14 Asymptotics of Harish–Chandra–Itzykson–Zuber
    Integrals and of Schur Polynomials 211

15 Asymptotics of Some Matrix Integrals 217
    15.1 Enumeration of Maps from Matrix Models..................220
    15.2 Enumeration of Colored Maps from Matrix Models 222

Part VI   Free Probability   225

16 Free Probability Setting 227
    16.1 A Few Notions about Algebras and Tracial States 227
    16.2 Space of Laws of m Non-commutative Self-adjoint Variables ..228

17 Freeness 231
    17.1 Definition of Freeness 231
    17.2 Asymptotic Freeness 232
    17.3 The Combinatorics of Freeness 236

18 Free Entropy 245

Part VII   Appendix   **261**

19 Basics of Matrices 263
    19.1 Weyl's and Lidskii's Inequalities 263
    19.2 Non-commutative Hölder Inequality 264

20 Basics of Probability Theory..............................267
    20.1 Basic Notions of Large Deviations 267
    20.2 Basics of Stochastic Calculus.............................270
    20.3 Proof of (2.3) 274

References...275

Index..287

List of Participants of the Summer School 289

List of Short Lectures Given at the Summer School............293

Notation

- $\mathcal{C}_b(\mathbb{R})$ (resp. $\mathcal{C}_b^1(\mathbb{R}^N, \mathbb{R})$) denotes the space of bounded continuous functions on $\mathbb{R}$ (resp. k times continuously differentiable functions from $\mathbb{R}^N$ into $\mathbb{R}$). If f is a real-valued function on a metric space (X, d),

$$\|f\|_\infty = \sup_{x \in X} |f(x)|$$

denotes its supremum norm, whereas we set the Lipschitz norms to be

$$\|f\|_{\mathcal{L}} = \sup_{x \neq y} \frac{|f(x) - f(y)|}{d(x, y)} + \sup_x |f(x)|, \quad |f|_{\mathcal{L}} = \sup_{x \neq y} \frac{|f(x) - f(y)|}{d(x, y)}$$

For $x \in \mathbb{R}^N$, and $f \in \mathcal{C}_b^1(\mathbb{R}^N, \mathbb{R})$, we let

$$\|x\|_2 = \left(\sum_{i=1}^N (x_i)^2\right)^{\frac{1}{2}}, \quad \|\nabla f\|_2 = \left(\sum_{i=1}^N (\partial_{x_i} f(x))^2\right)^{\frac{1}{2}}.$$

- $\mathcal{P}(X)$ denotes the set of probability measures on the metric space (X, d). $\mu(f)$ is a shorthand for $\int f(x) d\mu(x)$. We shall call the weak topology on $\mathcal{P}(X)$ the topology so that $\mu \to \mu(f)$ is continuous if f is bounded continuous on (X, d). The moments topology refers to the continuity of $\mu \to \mu(x^k)$ for all $k \in \mathbb{N}$. Even though both topologies coincide if X is compact subset of $\mathbb{R}$, they can be different in general.
- If (X, d) is a metric space, Dudley's distance d_D on $\mathcal{P}(X)$ (which is compatible with the weak topology on $\mathcal{P}(X)$) is given by

$$d_D(\mu, \nu) := \sup_{\|f\|_{\mathcal{L}} \le 1} \left| \int f(x) d\mu(x) - \int f(x) d\nu(x) \right| \tag{0.1}$$

- $\mathcal{M}_N(\mathbb{C})$ (resp. $\mathcal{H}_N^{(1)}$, resp. $\mathcal{H}_N^{(2)}$) denotes the set of $N \times N$ (resp. symmetric, resp. Hermitian) matrices with complex (resp. real, resp. complex) coefficients. $\mathcal{M}_N(\mathbb{C})$ is equipped with the trace Tr:

XII   Notation

$$\text{Tr}(A) = \sum_{i=1}^{N} A_{ii}.$$

- If A is an $N \times N$ Hermitian matrix, we denote by $(\lambda_k(A))_{1 \leq k \leq N}$ its eigenvalues.
- For A an $N \times N$ matrix, we define

$$\|A\|_2 = \left(\sum_{i,j=1}^{N} |A_{ij}|^2 \right)^{\frac{1}{2}} \quad \text{and} \quad \|A\|_\infty = \lim_{n \to \infty} (\text{Tr}((AA^*)^n))^{\frac{1}{2n}}.$$

The latter norm also coincides with the spectral radius of A which we denote by $\lambda_{\max}(A)$. 1 or I will denote the identity in $\mathcal{M}_N(\mathbb{C})$ and when no confusion is possible, for any constant c, c denotes $c1$.

- $\mathbb{C}\langle X_1, \ldots, X_m \rangle$ denotes the set of polynomials in m non-commutative indeterminates $(X_1, \ldots, X_m)$, $\mathbb{C}\langle X_1, \ldots, X_m \rangle_{sa}$ the subset of polynomials such that $P = P^*$ for some involution $*$ defined on $\mathbb{C}\langle X_1, \ldots, X_m \rangle$.

- Often, bold symbols will indicate vectors, e.g., $\mathbf{X} = (X_1, \ldots, X_m)$ or matrices e.g., $\mathbf{A} = (A_{ij})_{1 \leq i,j \leq N}$. The letters $(\mathbf{A}, \mathbf{B})$ in general refer to random matrices, whereas (X, Y, Z), to generic (eventually non-commutative) indeterminates.

Introduction

Random matrix theory was introduced in statistics by Wishart [206] in the thirties, and then in theoretical physics by Wigner [205]. Since then, it has developed separately in a wide variety of mathematical fields, such as number theory, probability, operator algebras, convex analysis etc.

Therefore, lecture notes on random matrices can only focus on special aspects of the theory; for instance, the well-known book by Mehta [153] displays a detailed analysis of the classical matrix ensembles, and in particular of their eigenvalues and eigenvectors, the recent book by Bai and Silverstein [10] emphasizes the results related to sample covariance matrices, whereas the book by Hiai and Petz [117] concentrates on the applications of random matrices to free probability and operator algebras. The book in progress [6] in collaboration with Anderson and Zeitouni will try to take a broader and more elementary point of view, but still without relations to number theory or Riemann Hilbert approach for instance. The first of these topics is reviewed briefly in [126] and the second is described in [73].

The goal of these notes is to present several aspects of the asymptotics of random matrix "macroscopic" quantities (RMMQ) such as

$$L_N(X_{i_1} \cdots X_{i_p}) := \frac{1}{N} \text{Tr}(\mathbf{A}_{i_1}^N \cdots \mathbf{A}_{i_p}^N)$$

when $(i_k \in \{1, \ldots, m\}, 1 \leq k \leq p)$ and $(\mathbf{A}_p^N)_{1 \leq p \leq m}$ are some $N \times N$ random matrices whose size N goes to infinity. We will study their convergence, their fluctuations, their concentration towards their mean and, as much as possible in view of the states of the art, their large deviations and the asymptotics of their Laplace transforms. We will in particular stress the relation of the latest to enumeration questions. We shall focus on the case where $(\mathbf{A}_p^N)_{1 \leq p \leq m}$ are Wigner matrices, that is Hermitian matrices with independent entries, although several results of these notes can be extended to other classical random matrices such as Wishart matrices.

When $m = 1$, $L_N(X_1^p)$ is the normalized sum of the pth power of the eigenvalues of X_1, that is the pth moment of the spectral measure of X_1.

2   Introduction

In his seminal article [205], Wigner proved that $E[L_N(X_1^p)]$ converges for any integer number p provided the entries of $\sqrt{N}\mathbf{A}_1$ have all their moments finite, are centered and have constant variance equal to one. We shall investigate this convergence in Chapter 1. We will show that it holds almost surely and that the hypothesis on the entries can be weakened. This result extends to several matrices, as shown by Voiculescu [197], see Section 3.2.

One of the interesting aspects of this convergence is the relation between the limits of the RMMQ and the enumeration of interesting graphs. Indeed, a key observation is that the empirical moment $L_N(X_1^{2p})$ converges towards the Catalan number C_p, the number of rooted trees with p edges, or equivalently the number of non-crossing pair partitions of $2p$ ordered points. As shown by Voiculescu [197], words in several matrices lead to the enumeration of *colored* trees. Considering the central limit theorem for such macroscopic quantities, we shall see also that their limiting covariances can be expressed in terms of numbers of certain planar graphs. It turns out that if the matrices have complex Gaussian entries, this relation extends far beyond the first two moments. Harer and Zagier [113] showed that the expansion of $E[L_N(X_1^p)]$ in terms of the dimension N can be seen as a topological expansion (i.e., as a generating function with parameter N^{-2} and coefficients which count graphs sorted by their genus). We shall see in these notes that also Laplace transforms of RMMQ's can be interpreted as generating functions (with parameters the dimension and the parameters of the Laplace transform) of interesting numbers.

This idea goes back to Brézin, Itzykson, Parisi and Zuber [50] (see also 't Hooft [187]) who considered matrix integrals given by

$$Z_N(P) = E[e^{N\mathrm{Tr}(P(\mathbf{A}_1^N,...,\mathbf{A}_m^N))}]$$

with a polynomial function P and independent copies $\mathbf{A}_i^N$ of an $N \times N$ matrix $\mathbf{A}^N$ with complex Gaussian entries. Then, they showed that if $P = \sum t_i q_i$ with some (complex) parameters t_i and some monomials q_i, $\log Z_N(P)$ expands formally (as a function of the parameters t_i and the dimension N of the matrices). The weight $N^{-2g} \prod_i (t_i)^{k_i}/k_i!$ will have a coefficient which counts the number of graphs with k_i vertices depending on the monomial q_i, for $i \geq 0$, that can be embedded properly in a surface of genus g. This relation is based on Feynman diagrams (see a review by A. Zvonkin [211]) and we shall describe it more precisely in Section 7.4. Matrix integrals were used widely in physics to solve problems in connection with the enumeration of maps [42, 50, 76, 77, 125, 209]. Part of these notes (mostly Part III) will show that, under appropriate assumptions, such formal equalities can be proved to hold as well asymptotically. In particular, we will see that $N^{-2} \log Z_N(\lambda P)$ converges for sufficiently small λ and the limit is a generating function for graphs embedded into the sphere.

In the second part of these notes, we show how to estimate Laplace transforms of traces of matrices in non-perturbative situation (that is estimate

$N^{-2} \log Z_N(\lambda P)$ for large λ's). In this case, it is no longer clear whether matrix integrals are related to the enumeration of graphs (except when P satisfies some convexity property, in which case it was shown in [106] that the free energy is the analytic extension of the enumeration of planar maps found for $Z_N(\lambda P)$ and λ small). Thus, different tools have to be introduced to estimate $Z_N(P)$ in general. First we consider one-matrix integrals and derive the large deviation principle for the spectral measure of Gaussian Wigner matrices. We then introduce a dynamical point of view to extend the previous result to *shifted* Gaussian Wigner matrices. The latter is applied to estimate some two-matrix integrals (such as the Ising model on random graphs) and Schur functions. We show in the last part of these notes how dynamics and large deviations techniques can be used to study the more general problem of estimating free entropy (see Chapter 18). The question of computing the free entropy remains open.

The outline of this book is as follows.

In the first part of these notes, we study the convergence of the RMMQ's and more precisely the convergence of the spectral measure of a Wigner matrix. We follow Wigner's original approach to study this question and estimate moments. This moments method can be refined to prove a central limit theorem (Section 2.1) or study the largest eigenvalue of random matrices, as proposed initially by Sinaï and Soshnikov (Section 2.2). Finally, we show that Wigner's theorem can be generalized to several matrices.

In the second part of these notes, we study concentration inequalities. These inequalities have provided a very powerful tool to control the probability of deviations of diverse random variables from their mean or their median (see some applications in [188]). After introducing some basic notions and results of concentration of measure theory, we specialize them to random matrices. In particular, we deduce concentration of the spectral measure or of the largest eigenvalue of Wigner matrices with nice entries. We also apply Brascamp–Lieb inequalities to random matrices.

In the third part, we study Gaussian matrix integrals in a perturbative regime. We give sufficient conditions so that they converge as the size of the matrices goes to infinity, study the first order correction to this convergence and relate the limits to the enumeration of graphs. The inequalities developed in Part II are important tools for this analysis.

In the fourth part of these notes, we concentrate on the eigenvalues of Gaussian random matrices (mainly the so-called Gaussian unitary or orthogonal ensembles). We remind the reader that their joint law is given as the law of Gaussian random variables interacting via a Coulomb gas potential. This joint law is key to many detailed analysis of the spectrum of the Gaussian ensembles, such has the study of the spacing fluctuations in the bulk or at the edge [153, 191], the interpretation of the limit has a determinantal process [39, 119, 182] etc. In these notes, we will only focus again on the RMMQ and deduce large deviation principles for the spectral measure and the largest eigenvalue.

4   Introduction

In the fifth part, we start addressing the question of proving large deviation principles for the laws of RMMQ's in a multi-matrix setting. We obtain a large deviations principle for the law of the Hermitian Brownian motion, from which we deduce estimates on Schur functions and Harish–Chandra–Itzykson–Zuber integrals. We apply these results to the related enumeration questions of the Ising model on random graphs for instance.

In the last part, we discuss the natural generalization of these questions to a general multi-matrix setting, namely analyzing free entropy. We introduce a free probability set-up and the notion of freeness. We then obtain bounds on free entropy.

As a conclusion, the goal of these notes is to present an overview of the study of macroscopic quantities of random matrices (law of large numbers, central limit theorems etc.) with a special emphasis on large deviations questions. I tried to give proofs as elementary and complete as possible, based on "standard tools" of probability (concentration, large deviations, etc.) which we shall, however, recall in some detail to help non-probabilists to understand proofs. Some proofs are new, some are improved versions of the proofs taken from articles and others are inspired from a book in progress with G. Anderson and O. Zeitouni [6]. In comparison with that book, these notes focus on matrix models and large deviations questions, whereas [6] attempts to give a more complete picture of random matrix theory, including local properties of the spectrum.

Part I

Wigner Matrices and Moments Estimates

In this part, we follow the strategy introduced by Wigner [205] to study the spectrum of random matrices: we estimate moments of traces of polynomials in these random matrices. We prove in this way several key results. First, we obtain the convergence (in expectation and almost surely) of the spectral measure (for the moments or the weak topology) of Wigner matrices. We also study its fluctuations around the limit. We generalize the convergence to a multi-matrix setting by showing that the trace of words in several matrices converges in the limit where the dimension goes to infinity. Finally, we generalize the estimation of moments to the case where their degree blows up with the dimension N of the matrices, but more slowly than $\sqrt{N}$. This is enough to bound the distance between the largest eigenvalue and its limit.

1

Wigner's Theorem

We consider in this section an $N \times N$ matrix $\mathbf{A}^N = \left(A_{ij}^N\right)_{1\leq i,j\leq N}$ with real or complex entries such that $\left(A_{ij}^N\right)_{1\leq i\leq j\leq N}$ are independent and $\mathbf{A}^N$ is self-adjoint; $A_{ij}^N = \bar{A}_{ji}^N$. We assume further that

$$\mathbb{E}[A_{ij}^N] = 0, \quad \lim_{N\to\infty} \frac{1}{N^2} \sum_{1\leq i,j\leq N} |N\mathbb{E}[|A_{ij}^N|^2] - 1| = 0.$$

We shall show in this chapter that the eigenvalues $(\lambda_1, \ldots, \lambda_N)$ of $\mathbf{A}^N$ satisfy the almost sure convergence

$$\lim_{N\to\infty} \frac{1}{N} \sum_{i=1}^{N} f(\lambda_i) = \int f(x) d\sigma(x)$$

where f is a bounded continuous function or a polynomial function, when the entries have finite moments. σ is the semicircle law

$$\sigma(dx) = \frac{1}{2\pi} \sqrt{4 - x^2} 1_{|x|\leq 2} dx.$$

We shall first prove this convergence for polynomial functions and rely on the fact that for all $k \in \mathbb{N}$, $\int x^k d\sigma(x)$ is null when k is odd and given by the Catalan number $C_{k/2}$ when k is even. We thus start this chapter by discussing the properties and characterizations of Catalan numbers.

1.1 Catalan Numbers, Non-crossing Partitions and Dick Paths

We will encounter first the Catalan numbers as the number of (oriented) rooted trees. We shall define more precisely this object in the next paragraph. Actually, Catalan numbers count many other combinatorial objects. In a first

A. Guionnet, *Large Random Matrices: Lectures on Macroscopic Asymptotics*, Lecture Notes in Mathematics 1957, DOI: 10.1007/978-3-540-69897-5_1, © 2009 Springer-Verlag Berlin Heidelberg, Reprint by Springer-Verlag Berlin Heidelberg 2012

part, we shall see that they also enumerate non-crossing partitions as well as Dick paths, a fact which we shall use later. As a warm-up to matrix models, we will also state the bijection with planar maps with one star. Then, we will study the Catalan numbers, their generating function, and relate them to the moments of the semicircle law.

1.1.1 Catalan Numbers Enumerate Oriented Rooted Trees

Let us recall that a graph is given by a set of vertices (or nodes) $V = \{i_1, \ldots, i_k\}$ and a set E of edges $(e_i)_{i \in I}$. An edge is a couple $e = (i_{j_1}, i_{j_2})$ for some $j_1, j_2 \in \{1, \ldots, k\}^2$. An edge $e = (i_p, i_\ell)$ is directed if (i_p, i_ℓ) and (i_ℓ, i_p) are distinct when $i_p \neq i_\ell$, which amounts to writing edges as directed arrows. It is undirected otherwise. A cycle (or loop) is a collection of distinct undirected edges $e_i = (v_i, v_{i+1})$, $1 \leq i \leq p$ such that $v_1 = v_{p+1}$ for some $p \geq 1$. A graph is connected if any two vertices (v_1, v_2) of the graph are connected by a path (that is that there exists a collection of edges $e_i = (a_i, b_i)$, $1 \leq i \leq n$ such that $v_1 = a_1$, $b_i = a_{i+1}$, $b_n = v_2$).

A *tree* is a connected graph with no loops (or cycles).

We will say that a tree is oriented if it is drawn (or embedded) into the plane; it then inherits the orientation of the plane. A tree is rooted if we specify one oriented edge, called the root. Note that if each edge of an oriented tree is seen as a double (or fat) edge, the connected path drawn from these double edges surrounding the tree inherits the orientation of the plane (see Figure 1.1). A root on this oriented tree then specifies a starting point in this path. This path will be intimately connected with the Dick path that we consider next.

Let us give the following well-known characterization of trees among connected graphs.

Lemma 1.1. *Let $G = (V, E)$ be a connected graph with E a set of undirected edges, and denote by $|A|$ the number of distinct elements of a finite discrete set A. Then,*

$$|V| \leq |E| + 1. \tag{1.1}$$

Moreover, $|V| = |E| + 1$ iff G is a tree.

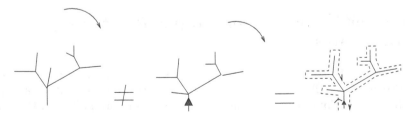

Fig. 1.1. Embedding rooted trees into the plane

Proof. (1.1) is straightforward when $|V| = 1$ and can be proven by induction as follows. Assume $|V| = n$ and consider one vertex v of V. This vertex is contained in $l \geq 1$ edges of E that we denote $(e_1, \ldots, e_l)$. The graph G then decomposes into $(v, e_1, \ldots, e_l)$ and $r \leq l$ connected graphs $(G_1, \ldots, G_r)$. We denote $G_j = (V_j, E_j)$ for $j \in \{1, \ldots, r\}$. We have

$$|V| - 1 = \sum_{j=1}^{r} |V_j|, \quad |E| - l = \sum_{j=1}^{r} |E_j|.$$

Applying the induction hypothesis to the connected graphs $(G_j)_{1 \leq j \leq r}$ gives

$$|V| - 1 \leq \sum_{i=1}^{r} (|E_j| + 1) = |E| + r - l \leq |E|, \tag{1.2}$$

which proves (1.1). In the case where $|V| = |E| + 1$, we claim that G is a tree, namely does not have any loop. In fact, for the equality to hold, we need to have equalities when performing the previous decomposition of the graph, a decomposition which can be reproduced until all vertices have been considered. If the graph contains a loop, the first time that the decomposition considers a vertex v of this loop, v must be the end point of at least two different edges, with end points belonging to the same connected graph (because they belong to the loop). Hence, we must have $r < l$ so that a strict inequality occurs in the right-hand side of (1.2). $\qquad\square$

Definition 1.2. *We denote by C_k the number of rooted oriented trees with k edges.*

Equivalently, we shall see in the following two paragraphs that C_k is the number of Dick paths of length $2k$, or the number of non-crossing pair partitions of $2k$ elements, or the number of planar maps with one star of type x^{2k}.

Exercise 1.3. *Show that $C_2 = 2$ and $C_3 = 5$ by drawing the corresponding graphs.*

1.1.2 Bijection with Dick Paths

Definition 1.4. *A Dick path of length $2n$ is a path starting and ending at the origin, with increments $+1$ or -1, and that stays above the non-negative real axis.*

We shall prove:

Property 1.5. *There exists a bijection between the set of rooted oriented trees and the set of Dick paths.*

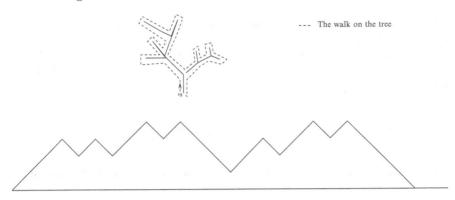

Fig. 1.2. Bijection between trees and Dick paths

Proof. To construct a Dick path from a rooted oriented tree, let us define a walk on the tree (or a closed path around the tree) as follows. We regard the oriented tree as a *fat tree*, which amounts to replacing each edge by a double edge (the double edge is made of two parallel edges surrounding the original edge, see Figure 1.1) while keeping the same set of vertices. The union of these double edges defines a path that surrounds the tree. The walk on the tree is defined by putting the orientation of the plane on this curve and starting from the root as the first step of the Dick path (see Figure 1.2). To define the Dick path, one follows the walk and counts a unit of time each time one meets a vertex; then adds +1 to the Dick path when one meets an (non-oriented) edge that has not yet been visited and −1 otherwise. Since to add a −1, one must have added a +1 corresponding to the first visit of the edge, the Dick path is non-negative and since at the end all edges are visited exactly twice, the path constructed will come back at 0 at time $2n$. This defines a bijection (see Figure 1.2) since, given a Dick path, we can recover the rooted tree by first gluing the couples of steps where one step up is followed by one step down and representing each couple of glued steps by one edge; we then obtain a path decorated with edges. Continuing this procedure until all steps have been glued two by two provides a rooted tree.

1.1.3 Bijection with Non-crossing Pair Partitions

Let us recall the following definition:

Definition 1.6. • *A partition of the set* $S := \{1, \ldots, n\}$ *is a decomposition*

$$\pi = \{V_1, \ldots, V_r\}$$

of S into disjoint and non-empty sets V_i.
• *The set of all partitions of S is denoted by* $\mathcal{P}(S)$, *and for short by* $\mathcal{P}(n)$ *if* $S := \{1, \ldots, n\}$.

- *The $V_i, 1 \le i \le r$ are called the blocks of the partition and we say that $p \sim_\pi q$ if p, q belong to the same block of the partition π.*
- *A partition π of $\{1, \ldots, n\}$ is said to be crossing if there exist $1 \le p_1 < q_1 < p_2 < q_2 \le n$ with*

$$p_1 \sim_\pi p_2 \not\sim_\pi q_1 \sim_\pi q_2.$$

It is non-crossing otherwise.
- *A partition is a pair partition if all blocks have cardinality two.*

The bijection between oriented rooted trees with n edges and non-crossing pair-partitions of $2n$ elements goes as follows. On each edge of the tree we draw an arc going from one side of the edge to the other side and that does not cross the tree. We start from the root and draw one arc in such a way that the part of the tree visited by the walk before arriving for the second time at the first edge is contained in the ball with boundary given by the arc. We then continue this procedure, drawing the arcs in such a way that they do not cross, till no edge is left. Finally, we think of the tree as being drawn by the folding of a cord with both ends at the root; in other words, we replace the tree by the fat tree designed from the trajectory of the walk as shown in Figure 1.2. Unfolding the cord while keeping the arcs gives a pair-partition. A less colorful way to say the same thing is to label each side of the edges starting from the root and following the orientation and to write down the pair-partition with pairings given by the labels of the two sides of the edges. For instance, the drawing below represents the pair-partition of $\{1, \ldots, 24\}$ given by $(1, 24)$, $(2, 13)$, $(3, 4)$, $(5, 6)$, $(7, 12)$, $(8, 9)$, $(10, 11)$, $(14, 23)$, $(15, 16)$, $(17, 22)$, $(18, 19)$, $(20, 21)$.

We claim that the resulting partition is non-crossing. Indeed, if we take two edges of a tree, say $e_1 = (a_1, b_1)$ and $e_2 = (a_2, b_2)$, let T_1 be the subtree visited, when one follows the orientation on the tree, between the time it visits

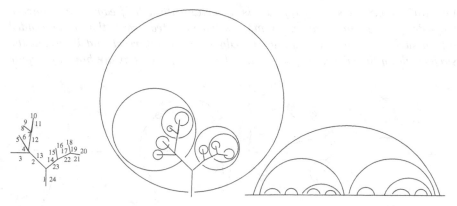

Fig. 1.3. Drawing the partitions on the tree and unfolding the tree

the two sides of the edge e_1. Then, either $e_2 \in T_1$, and then $a_1 < a_2 < b_2 < b_1$, or $e_2 \notin T_1$, corresponding either to $a_2 < b_2 < a_1 < b_1$ or $a_1 < b_1 < a_2 < b_2$. We have thus shown:

Property 1.7. *To each oriented rooted tree with n edges we can associate bijectively a non-crossing pair partition of 2n elements.*

Remark 1. Observe that in the bijection, the elements of the partition are the edges of the tree seen as double (or fat) edges, as for the definition of the walk on the tree (see Figure 1.2).

Let us finally remark that there is an alternative way to draw non-crossing partitions that we shall use later. Instead of drawing the points of the partition on the real line, we can draw them on the circle, provided we mark, say, the place where we put the first element and provide the circle with an orientation corresponding to the orientation on the real line. With this mark and orientation, we have again a bijection. The drawing of the partition then becomes a series of arcs which can be drawn either outside of the annulus or inside (see Figure 1.4). As a matter of fact, the circle is irrelevant here, the only thing that matters are the points, the marked point and the orientation. So, we can also see one such point as the end point of a half-edge, all the half-edges intersecting in one vertex in the center of the previous circle. Thus, we can draw our set of the k points on the real line as a vertex with k half-edges, one marked half-edge and an orientation. We shall later call the set of these edges, marked half-edge and orientation a star. In this picture, the pair partition corresponds to the gluing of these half-edges two by two and the fact that the partition is non-crossing exactly means that the edges (obtained by the gluing of two half-edges) do not cross.

The last drawing in Figure 1.4 is a planar map; that is, a connected graph that is embedded into the sphere.

Definition 1.8. *A star of type x^k is a vertex with k half-edges, one marked half-edge and an orientation. A map is a connected graph that is embedded into a surface in such a way that the edges do not intersect and if we cut the surface along the edges, we get a disjoint union of sets that are homeomorphic*

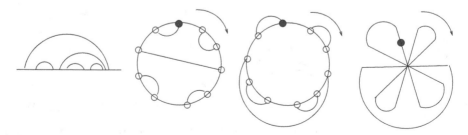

Fig. 1.4. Non-crossing partitions and stars

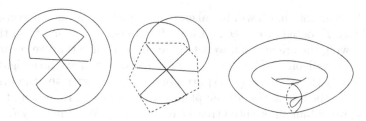

Fig. 1.5. Partitions and maps

to an open disk (these sets are called the faces of the map). A map with stars $x^{q_1}, \ldots, x^{q_p}$ *is a graph where the half-edges of the stars* $x^{q_1}, \ldots, x^{q_p}$ *have been glued pair-wise, the orientation of each pair of edge agreeing, hence providing to the full graph one orientation.*

The genus g of the map is the genus of such a surface; it satisfies

$$2 - 2g = \sharp vertices + \sharp faces - \sharp edges.$$

A planar map is a map with genus zero.

For more details on maps, we refer to the review [211]. Note that once a graph is embedded into a surface, the natural orientation of the surface induces an orientation around each vertex of the graph (more precisely a cyclic order on the end points of the half-edges of its vertices). This fact has its counterpart since (cf. [211, Proposition 4.7]) an orientation around each vertex of a graph uniquely determines its embedding into a surface. This shows that, modulo the notion of marked points, the notion of a star is intimately related to the idea of embedding the corresponding graph into a surface. Prescribing a marked half-edge will be useful later to describe how we will count these graphs (the orientation and the marked point of the stars being equivalent to a labeling of its half-edges).

To find out the genus of a map with only one vertex of degree k, one can also recall that the end points of the half-edges of the star represent the middle of the edges of the fat tree. Drawing these edges and gluing them pairwise according to the map allows one to visualize the surface on which one can embed the map (in the figure below, the lines on the surface are now the boundary of the polygon).

1.1.4 Induction Relation

We next show that the Catalan numbers satisfy the following induction relation.

Property 1.9. $C_0 = 1$ *and for all* $k \geq 1$

$$C_k = \sum_{l=0}^{k-1} C_{k-l-1} C_l. \tag{1.3}$$

Proof. By convention, C_0 will be taken to be equal to one and we consider an oriented tree T rooted at $r = (i_1, i_2)$ with $k \geq 1$ edges. Starting from the root r and following the orientation, we let t_1 be the first time that we return to i_1 following the walk on T. The subgraph T_1 of the tree we have investigated is a tree, with only the edge $r = (i_1, i_2)$ attached to i_1. We let r_1 be the first edge (according to the orientation of the plane) attached to i_2. Removing the edge r from T_1, we obtain an oriented tree T_1' rooted at r_1. We denote by $l_1 \leq k - 1$ the number of its edges. $T_2 = T \backslash T_1$ is an oriented rooted tree (at the first edge attached to i_1) with $k - 1 - l_1$ edges. Therefore, any oriented rooted tree with k edges can be decomposed into an edge and two oriented rooted trees with respectively l_1 and $k - l_1 - 1$ vertices for some $l_1 \in \{0, \ldots, k-1\}$. This proves (using $C_0 = 1$) that (1.3) holds with $l = l_1$. □

Property (1.9) defines uniquely the Catalan numbers by induction. We can also give the more explicit formula:

Property 1.10. *For all $k \geq 0$, $C_k \leq 2^{2k}$ and*

$$C_k = \frac{\binom{2k}{k}}{k+1}.$$

Proof. Note that since C_k is also the number of Dick paths with length $2k$, it is smaller than the number of walks (that is, the number of connected paths with steps equal to $+1$ or -1) starting at the origin with length $2k$, that is, 2^{2k}. In particular, if we define

$$S(z) := \sum_{k=0}^{\infty} C_k z^k,$$

$S(z)$ is absolutely convergent in $|z| < 4^{-1}$. We can therefore multiply both sides of equality (1.3) by z^k and sum the resulting equalities for $k \in \mathbb{N} \backslash \{0\}$. We arrive at

$$S(z) - 1 = z S(z)^2.$$

As a consequence,

$$S(z) = \frac{1 \pm \sqrt{1 - 4z}}{2z}.$$

Since $S(0) = 1$, we conclude that

$$S(z) = \frac{1 - \sqrt{1 - 4z}}{2z}. \tag{1.4}$$

We can now expand $\sqrt{1 - 4z}$ in a Taylor series around the origin to obtain

$$\sqrt{1 - 4z} = 1 - 2z - \sum_{k=1}^{n} \frac{(2^{-1})^{k+1}(2k-1)(2k-3)\cdots(1)}{(k+1)!}(4z)^{k+1} + o(z^{n+1})$$

yielding

$$S(z) = 1 + 2 \sum_{k=1}^{n} \frac{(2^{-1})^{k+1}(2k-1)(2k-3)\cdots(1)}{(k+1)!}(4z)^k + o(z^n).$$

Therefore, by identifying each term of the series we find

$$C_k = 2\frac{4^k(2^{-1})^{k+1}(2k-1)(2k-3)\cdots(1)}{(k+1)!} = \frac{2k!}{(k+1)!k!} = \frac{\binom{2k}{k}}{k+1}.$$

$\square$

1.1.5 The Semicircle Law and Catalan Numbers

The standard semicircle law is given by

$$\sigma(dx) = \frac{1}{2\pi}\sqrt{4-x^2}\,1_{|x|\leq 2}dx.$$

Property 1.11. *Let $m_k = \int x^k d\sigma(x)$. Then for all $k \geq 0$,*

$$m_{2k} = C_k.$$

Proof. By the change of variables $x = 2\sin(\theta)$

$$m_{2k} = \int_{-2}^{2} x^{2k}\sigma(x)dx = \frac{2\cdot 2^{2k}}{\pi}\int_{-\pi/2}^{\pi/2} \sin^{2k}(\theta)\cos^2(\theta)d\theta$$

$$= \frac{2\cdot 2^{2k}}{\pi}\int_{-\pi/2}^{\pi/2} \sin^{2k}(\theta)d\theta - (2k+1)m_{2k}.$$

Hence,

$$(2k+2)m_{2k} = \frac{2\cdot 2^{2k}}{\pi}\int_{-\pi/2}^{\pi/2} \sin^{2k}(\theta)d\theta = 4(2k-1)m_{2k-2},$$

from which, together with $m_0 = 1$, one concludes that

$$m_{2k} = \frac{4(2k-1)}{(2k+2)}m_{2k-2}, \tag{1.5}$$

leading to the claimed assertion that $m_{2k} = C_k$ by Property 1.10. $\square$

Corollary 1.12. *For $z \in \mathbb{C}\backslash\mathbb{R}$, let*

$$G_\sigma(z) := \int \frac{1}{z-x}d\sigma(x)$$

be the Stieltjes transform of the semicircle law. Then, for $z \in \mathbb{C}\backslash[-2,2]$

$$G_\sigma(z) = \frac{1}{2}\left(z - \sqrt{z^2 - 4}\right).$$

Proof. When $|z| > 2$, we can write

$$G_\sigma(z) = \frac{1}{z} \int \frac{1}{1 - z^{-1}x} d\sigma(x) = \frac{1}{z} \sum_{k \geq 0} z^{-k} \int x^k d\sigma(x)$$

$$= \frac{1}{z} \sum_{k \geq 0} z^{-2k} C_k = \frac{1}{z} S(z^{-2})$$

$$= \frac{1}{z} \left(\frac{1 - \sqrt{1 - 4z^{-2}}}{2z^{-2}} \right) = \frac{1}{2} \left(z - \sqrt{z^2 - 4} \right)$$

where we finally used (1.4). This equality extends to the whole domain of analyticity of G_σ, i.e., $\mathbb{C} \backslash [-2, 2]$. $\square$

1.2 Wigner's Theorem

We consider an $N \times N$ matrix $\mathbf{A}^N$ with real or complex entries such that $(A_{ij}^N)_{1 \leq i \leq j \leq N}$ are independent and $\mathbf{A}^N$ is self-adjoint; $A_{ij}^N = \bar{A}_{ji}^N$. We assume that

$$\mathbb{E}[A_{ij}^N] = 0, \ 1 \leq i, j \leq N, \ \lim_{N \to \infty} \frac{1}{N^2} \sum_{1 \leq i,j \leq N} |N\mathbb{E}[|A_{ij}^N|^2] - 1| = 0. \quad (1.6)$$

In this section, we use the same notation for complex and for real entries since both cases will be treated at once and yield the same result. The aim of this section is to prove the convergence of the quantities $N^{-1} \mathrm{Tr} \left((\mathbf{A}^N)^k \right)$ as N goes to infinity and k is any positive integer number. Since $\mathrm{Tr} \left((\mathbf{A}^N)^k \right) = \sum_{i=1}^N \lambda_i^k$ if $(\lambda_1, \ldots, l_N)$ are the eigenvalues of $\mathbf{A}_N$, this prove the convergence in moments of the spectral measure of $\mathbf{A}_N$.

Theorem 1.13 (Wigner's theorem). *[205] Assume that (1.6) holds and for all $k \in \mathbb{N}$,*

$$B_k := \sup_{N \in \mathbb{N}} \sup_{ij \in \{1,\ldots,N\}^2} \mathbb{E}[|\sqrt{N} A_{ij}^N|^k] < \infty. \quad (1.7)$$

Then,

$$\lim_{N \to \infty} \frac{1}{N} \mathrm{Tr} \left((\mathbf{A}^N)^k \right) = \begin{cases} 0 & \text{if } k \text{ is odd,} \\ C_{\frac{k}{2}} & \text{otherwise,} \end{cases}$$

where the convergence holds in expectation and almost surely. $(C_k)_{k \geq 0}$ are the Catalan numbers.

Proof. We start the proof by showing the convergence in expectation. The strategy is simply to expand the expectation of the trace of the matrix in terms of the expectation of its entries. We then use some (easy) combinatorics on trees to find out the main contributing term in this expansion. The almost sure convergence is obtained by estimating the variance of the considered random variables.

1. *Expanding the expectation.*
   Setting $\mathbf{B}^N = \sqrt{N}\mathbf{A}^N = (B_{ij})_{1\leq i,j\leq N}$, we have

$$\mathbb{E}\left[\frac{1}{N}\mathrm{Tr}\left((\mathbf{A}^N)^k\right)\right] = \sum_{i_1,\ldots,i_k=1}^{N} N^{-\frac{k}{2}-1}\mathbb{E}[B_{i_1 i_2}B_{i_2 i_3}\cdots B_{i_k i_1}] \quad (1.8)$$

where $(B_{ij})_{1\leq i,j\leq N}$ denote the entries of $\mathbf{B}^N$ (which may eventually depend on N). We denote $\mathbf{i} = (i_1,\ldots,i_k)$ and set

$$P(\mathbf{i}) := \mathbb{E}[B_{i_1 i_2}B_{i_2 i_3}\cdots B_{i_k i_1}].$$

By (1.7) and Hölder's inequality, $P(\mathbf{i})$ is bounded uniformly by B_k. Since the random variables $(B_{ij})_{1\leq i\leq j\leq N}$ are independent and centered, $P(\mathbf{i})$ vanishes unless for any edge (i_p, i_{p+1}), $p \in \{1,\ldots,k\}$, there exists $l \neq p$ such that $(i_p, i_{p+1}) = (i_l, i_{l+1})$ or (i_{l+1}, i_l). Here, we used the convention $i_{k+1} := i_1$. We next show that the set of indices that contributes to the first order in the right-hand side of (1.8) is described by trees.

2. *Connected graphs and trees.*
   $V(\mathbf{i}) = \{i_1,\ldots,i_k\}$ will be called the vertices. An edge is a pair (i,j) with $i,j \in \{1,\ldots,N\}^2$. At this point, edges are directed in the sense that we distinguish (i,j) from (j,i) when $j \neq i$. We denote by $E(\mathbf{i})$ the collection of the k edges $(e_p)_{p=1}^k = (i_p, i_{p+1})_{p=1}^k$ with $i_{k+1} = i_1$.
   We consider the graph $G(\mathbf{i}) = (V(\mathbf{i}), E(\mathbf{i}))$. $G(\mathbf{i})$ is connected since there exists an edge between any two vertices i_ℓ and $i_{\ell+1}$, $\ell \in \{1,\ldots,k-1\}$. Note that $G(\mathbf{i})$ may contain loops (e.g., cycles, for instance edges of type (i,i)) and multiple undirected edges.
   The skeleton $\tilde{G}(\mathbf{i})$ of $G(\mathbf{i})$ is the graph $\tilde{G}(\mathbf{i}) = \left(\tilde{V}(\mathbf{i}), \tilde{E}(\mathbf{i})\right)$ where $\tilde{V}(\mathbf{i})$ is the set of different vertices of $V(\mathbf{i})$ (without multiplicities) and $\tilde{E}(\mathbf{i})$ is the set of undirected edges of $E(\mathbf{i})$, also taken without multiplicities.

3. *Convergence in expectation.*
   Since we noticed that $P(\mathbf{i})$ equals zero unless each edge in $E(\mathbf{i})$ is repeated at least twice, we have that

$$|\tilde{E}(\mathbf{i})| \leq \frac{k}{2} \Rightarrow |\tilde{E}(\mathbf{i})| \leq \left[\frac{k}{2}\right],$$

and so by (1.1) applied to the skeleton $\tilde{G}(\mathbf{i})$ we find

$$|\tilde{V}(\mathbf{i})| \leq \left[\frac{k}{2}\right] + 1$$

where $[x]$ is the integer part of x. Thus, since the indices are chosen in $\{1, \ldots, N\}$, there are at most $N^{[\frac{k}{2}]+1}$ indices that contribute to the sum (1.8) and so we have

$$\left| \mathbb{E}\left[\frac{1}{N} \operatorname{Tr}\left((\mathbf{A}^N)^k\right) \right] \right| \leq B_k N^{[\frac{k}{2}]-\frac{k}{2}}.$$

where we used (1.7). In particular, if k is odd,

$$\lim_{N \to \infty} \mathbb{E}\left[\frac{1}{N} \operatorname{Tr}\left((\mathbf{A}^N)^k\right) \right] = 0.$$

If k is even, the only indices that will contribute to the first order asymptotics in the sum are those such that

$$|\tilde{V}(\mathbf{i})| = \frac{k}{2} + 1,$$

which, by Lemma 1.1, implies that:

a) $\tilde{G}(\mathbf{i})$ is a tree.

b) $|\tilde{E}(\mathbf{i})| = 2^{-1}|E(\mathbf{i})| = \frac{k}{2}$ and so each edge in $E(\mathbf{i})$ appears exactly twice. Thus, $G(\mathbf{i})$ appears as a fat tree where each edge of $\tilde{G}(\mathbf{i})$ is repeated exactly twice.

$G(\mathbf{i})$ is rooted (a root is given by the directed edge (i_1, i_2)). These edges are directed by the natural order on the indices. Because $G(\mathbf{i})$ is a tree, we see that each pair of directed edges corresponding to the same undirected edge in $\tilde{E}(\mathbf{i})$ is of the form $\{(i_p, i_{p+1}), (i_{p+1}, i_p)\}$. Moreover, the order on the indices induces a cyclic order on the fat tree that uniquely prescribes the way this fat tree can be embedded into the plane, the orientation of the plane agreeing with the orientation on the fat tree (see Figure 1.1). Therefore, for these indices, $P(\mathbf{i}) = \prod_{e \in \tilde{E}(\mathbf{i})} E[|\sqrt{N} A_e^N|^2]$. We write $G(\mathbf{i}) \simeq G(\mathbf{j})$ if $G(\mathbf{i})$ and $G(\mathbf{j})$ corresponds to the same rooted tree (but with eventually different values of the indices). By (1.6), for any fixed rooted tree G,

$$\frac{1}{N^{\frac{k}{2}+1}} \sum_{\mathbf{i}:G(\mathbf{i}) \simeq G} | \prod_{e \in \tilde{E}(\mathbf{i})} E[|\sqrt{N} A_e^N|^2] - 1| \leq \frac{k B_2^{\frac{k}{2}-1}}{N^2} \sum_{i,j=1}^{N} |E[|B_{ij}|^2] - 1|$$

goes to zero as N goes to infinity. Hence, we deduce that

$$\lim_{N \to \infty} \mathbb{E}\left[\frac{1}{N} \operatorname{Tr}\left((\mathbf{A}^N)^k\right) \right] = \sharp\{\text{rooted oriented trees with } k/2 \text{ edges}\}.$$

4. *Almost sure convergence.* To prove the almost sure convergence, we estimate the variance and then use the Borel–Cantelli lemma. The variance is given by

$$\mathrm{Var}((\mathbf{A}^N)^k) := \mathbb{E}\left[\frac{1}{N^2}\left(\mathrm{Tr}\,((\mathbf{A}^N)^k)\right)^2\right] - \mathbb{E}\left[\frac{1}{N}\mathrm{Tr}\,((\mathbf{A}^N)^k)\right]^2$$

$$= \frac{1}{N^{2+k}}\sum_{\substack{i_1,\ldots,i_k=1 \\ i'_1,\ldots,i'_k=1}}^{N}[P(\mathbf{i},\mathbf{i}') - P(\mathbf{i})P(\mathbf{i}')]$$

with

$$P(\mathbf{i},\mathbf{i}') := \mathbb{E}[B_{i_1 i_2}B_{i_2 i_3}\cdots B_{i_k i_1}B_{i'_1 i'_2}\cdots B_{i'_k i'_1}].$$

We denote by $G(\mathbf{i},\mathbf{i}')$ the graph with vertices $V(\mathbf{i},\mathbf{i}') = \{i_1,\ldots,i_k,i'_1,\ldots,$ $i'_k\}$ and edges $E(\mathbf{i},\mathbf{i}') = \{(i_p,i_{p+1})_{1\le p\le k},(i'_p,i'_{p+1})_{1\le p\le k}\}$. For the indices $(\mathbf{i},\mathbf{i}')$ to contribute to the leading order of the sum, $G(\mathbf{i},\mathbf{i}')$ must be connected. Indeed, if $E(\mathbf{i})\cap E(\mathbf{i}') = \emptyset$, $P(\mathbf{i},\mathbf{i}') = P(\mathbf{i})P(\mathbf{i}')$. Moreover, as before, each edge must appear at least twice to give a non-zero contribution so that $|\tilde{E}(\mathbf{i},\mathbf{i}')| \le k$. Therefore, we are in the same situation as before, and if $\tilde{G}(\mathbf{i},\mathbf{i}') = (\tilde{V}(\mathbf{i},\mathbf{i}'),\tilde{E}(\mathbf{i},\mathbf{i}'))$ denotes the skeleton of $G(\mathbf{i},\mathbf{i}')$, we have the relation

$$|\tilde{V}(\mathbf{i},\mathbf{i}')| \le |\tilde{E}(\mathbf{i},\mathbf{i}')| + 1 \le k + 1. \tag{1.9}$$

This already shows that the variance is at most of order N^{-1} (since $P(\mathbf{i},\mathbf{i}') - P(\mathbf{i})P(\mathbf{i}')$ is bounded by $2B_{2k}$ uniformly), but we need a slightly better bound to prove the almost sure convergence. To improve our bound let us show that the case where $|\tilde{V}(\mathbf{i},\mathbf{i}')| = |\tilde{E}(\mathbf{i},\mathbf{i}')| + 1 = k + 1$ cannot occur. In this case, we have seen that $\tilde{G}(\mathbf{i},\mathbf{i}')$ must be a tree since then equality holds in (1.9). Also, $|\tilde{E}(\mathbf{i},\mathbf{i}')| = k$ implies that each edge appears with multiplicity exactly equals to 2. For any contributing set of indices $\mathbf{i},\mathbf{i}'$, $\tilde{G}(\mathbf{i},\mathbf{i}')\cap G(\mathbf{i})$ and $\tilde{G}(\mathbf{i},\mathbf{i}')\cap G(\mathbf{i}')$ must share at least one edge (i.e., one edge must appear with multiplicity one in each of this subgraph) since otherwise $P(\mathbf{i},\mathbf{i}') = P(\mathbf{i})P(\mathbf{i}')$. This is a contradiction. Indeed, if we equip $\tilde{G}(\mathbf{i},\mathbf{i}')$ with the orientation of the indices from $\mathbf{i}$ and the root (i_1,i_2), we may define the walk on $\tilde{G}(\mathbf{i},\mathbf{i}')\cap G(\mathbf{i})$ as in Figure 1.2 (it is simply the path $i_1 \to i_2 \cdots \to i_k \to i_1$). Since this walk comes back to i_1, either it visits each edge twice, which is impossible if $\tilde{G}(\mathbf{i},\mathbf{i}')\cap G(\mathbf{i})$ and $\tilde{G}(\mathbf{i},\mathbf{i}')\cap G(\mathbf{i}')$ share one edge (and all edges have multiplicity two), or it has a loop, which is also impossible since $\tilde{G}(\mathbf{i},\mathbf{i}')$ is a tree. Therefore, we conclude that for all contributing indices,

$$|\tilde{V}(\mathbf{i},\mathbf{i}')| \le k$$

which implies

$$\mathrm{Var}((\mathbf{A}^N)^k) \le 2B_k N^{-2}.$$

Applying Chebychev's inequality gives for any $\delta > 0$

$$\mathbb{P}\left(\left|\frac{1}{N}\mathrm{Tr}\left((\mathbf{A}^N)^k\right) - \mathbb{E}\left[\frac{1}{N}\mathrm{Tr}\left((\mathbf{A}^N)^k\right)\right]\right| > \delta\right) \leq \frac{2B_k}{\delta^2 N^2},$$

and so the Borel–Cantelli lemma implies

$$\lim_{N\to\infty}\left|\frac{1}{N}\mathrm{Tr}\left((\mathbf{A}^N)^k\right) - \mathbb{E}\left[\frac{1}{N}\mathrm{Tr}\left((\mathbf{A}^N)^k\right)\right]\right| = 0 \quad a.s.$$

The proof of the theorem is complete.

$\square$

Exercise 1.14. *Take for $L \in \mathbb{N}$, $\mathbf{A}^{N,L}$ the $N \times N$ self-adjoint matrix such that $\mathbf{A}^{N,L}_{ij} = (2L)^{-\frac{1}{2}}1_{|i-j|\leq L}A_{ij}$ with $(A_{ij}, 1 \leq i \leq j \leq N)$ independent centered random variables having all moments finite and $E[A^2_{ij}] = 1$. The purpose of this exercise is to show that for all $k \in \mathbb{N}$,*

$$\lim_{L\to\infty}\lim_{N\to\infty}\mathbb{E}\left[\frac{1}{N}\mathrm{Tr}((\mathbf{A}^{N,L})^k)\right] = C_{k/2}$$

with C_x null if x is not integer. Hint: Show that for $k \geq 2$

$$\mathbb{E}\left[\frac{1}{N}\mathrm{Tr}((\mathbf{A}^{N,L})^k)\right] = (2L)^{-k/2}\sum_{\substack{|i_2 - L|\leq L, \\ |i_{p+1}-i_p|\leq L, p\geq 2}} E[A_{Li_2}\cdots A_{i_k L}] + o(1).$$

Then prove that the contributing indices to the above sum correspond to the case where $G(L, i_2, \cdot, i_k)$ is a tree with $k/2$ vertices and show that being given a tree there are approximately $(2L)^{\frac{k}{2}}$ possible choices of indices $i_2, \ldots, i_k$.

1.3 Weak Convergence of the Spectral Measure

Let $(\lambda_i)_{1\leq i\leq N}$ be the N (real) eigenvalues of $\mathbf{A}^N$ and define

$$L_{\mathbf{A}^N} := \frac{1}{N}\sum_{i=1}^N \delta_{\lambda_i}$$

to be the *spectral measure* of $\mathbf{A}^N$. $L_{\mathbf{A}^N}$ belongs to the set $\mathcal{P}(\mathbb{R})$ of probability measures on $\mathbb{R}$. We claim the following:

Theorem 1.15. *Assume that (1.7) holds for all $k \in \mathbb{N}$. Then, for any bounded continuous function f,*

$$\lim_{N\to\infty}\int f(x)dL_{\mathbf{A}^N}(x) = \int f(x)d\sigma(x) \quad a.s.$$

Proof. Let $B > 2$ and $\delta > 0$ be fixed. By Weierstrass' theorem, we can find a polynomial P_δ such that

$$\sup_{|x| \leq B} |f(x) - P_\delta(x)| \leq \delta.$$

Then

$$\left| \int f(x) d(L_{\mathbf{A}^N}(x) - \sigma(x)) \right| \leq \left| \int P_\delta(x) d(L_{\mathbf{A}^N}(x) - \sigma(x)) \right|$$

$$+ \delta + \left| \int_{|x| \geq B} (f - P_\delta)(x) dL_{\mathbf{A}^N}(x) \right| \quad (1.10)$$

where we used that $1_{|x| \geq B} d\sigma(x) = 0$ since $B > 2$. By Theorem 1.13,

$$\lim_{N \to \infty} \left| \int P_\delta(x) d(L_{\mathbf{A}^N}(x) - \sigma(x)) \right| = 0 \quad \text{a.s.} \quad (1.11)$$

Moreover, using that f is bounded, if p denotes the degree of P_δ, we can find a finite constant $C = C(\delta, B)$ so that

$$\left| \int_{|x| \geq B} (f - P_\delta)(x) dL_{\mathbf{A}^N}(x) \right| \leq C \int_{|x| \geq B} (1 + |x|^p) dL_{\mathbf{A}^N}(x)$$

$$\leq 2CB^{-p-2q} \int x^{2(p+q)} dL_{\mathbf{A}^N}(x)$$

where we finally used Chebychev's inequality with some $q \geq 0$. Using again Theorem 1.13, we find that

$$\limsup_{N \to \infty} \left| \int_{|x| \geq B} (f - P_\delta)(x) dL_{\mathbf{A}^N}(x) \right| \leq 2CB^{-p-2q} \int x^{2(p+q)} d\sigma(x)$$

$$\leq CB^{-p-2q} 2^{2(p+q+1)} \quad \text{a.s.}$$

We let q go to infinity to conclude, since $B > 2$, that

$$\limsup_{N \to \infty} \left| \int_{|x| \geq B} (f - P_\delta)(x) dL_{\mathbf{A}^N}(x) \right| = 0 \quad \text{a.s.}$$

Finally, let δ go to zero to conclude from (1.10) and (1.11) that

$$\limsup_{N \to \infty} \left| \int f(x) d(L_{\mathbf{A}^N}(x) - \sigma(x)) \right| = 0 \quad \text{a.s.}$$

$\square$

1.4 Relaxation of the Hypotheses over the Entries–Universality

In this section, we relax the assumptions on the moments of the entries while keeping the hypothesis that $(A_{ij}^N)_{1 \leq i \leq j \leq N}$ are independent. Generalizations of Wigner's theorem to possibly mildly dependent entries can be found for instance in [45].

1.4.1 Relaxation over the Number of Finite Moments

A nice, simple, but finally optimal way to relax the assumption that the entries of $\sqrt{N}\mathbf{A}^N$ possess all their moments, relies on the following observation.

Lemma 1.16. *Let A, B be $N \times N$ Hermitian matrices, with eigenvalues $\lambda_1(A) \geq \lambda_2(A) \geq \cdots \geq \lambda_N(A)$ and $\lambda_1(B) \geq \lambda_2(B) \geq \cdots \geq \lambda_N(B)$. Then,*

$$\sum_{i=1}^{N} |\lambda_i(A) - \lambda_i(B)|^2 \leq \mathrm{Tr}(A - B)^2 \,.$$

Proof. Since $\mathrm{Tr}A^2 = \sum_i (\lambda_i(A))^2$ and $\mathrm{Tr}B^2 = \sum_i (\lambda_i(B))^2$, the lemma amounts to showing that

$$\mathrm{Tr}(AB) \leq \sum_{i=1}^{N} \lambda_i(A)\lambda_i(B)$$

for all A, B as above, or equivalently, since if $A = U\mathrm{diag}(\lambda_1(A), \ldots, \lambda_N(A))U^*$ with a unitary matrix U,

$$\mathrm{Tr}(AB) = \sum_{i,j=1}^{N} \lambda_k(A)\lambda_j(B)|U_{ij}|^2,$$

that

$$\sum_{i=1}^{N} \lambda_i(A)\lambda_i(B) = \sup_{v_{ij} \geq 0 : \sum_j v_{ij}=1, \sum_i v_{ij}=1} \sum_{i,j} \lambda_i(A)\lambda_j(B)v_{ij} \,. \tag{1.12}$$

An elementary proof can be given (see [6]) by showing by induction over N that the optimizing matrix v above has to be the identity matrix. Indeed, this is true for $N = 1$, and one proceeds by induction: if $v_{11} = 1$ then the problem is reduced to $N - 1$, while if $v_{11} < 1$, there exists a j and a k with $v_{1j} > 0$ and $v_{k1} > 0$. Set $v = \min(v_{1j}, v_{k1}) > 0$ and define $\bar{v}_{11} = v_{11} + v$, $\bar{v}_{kj} = v_{kj} + v$ and $\bar{v}_{1j} = v_{1j} - v$, $\bar{v}_{k1} = v_{k1} - v$, and $\bar{v}_{ab} = v_{ab}$ for all other pairs ab. Then,

$$\sum_{i,j} \lambda_i(A)\lambda_j(B)(\bar{v}_{ij} - v_{ij})$$

$$= v(\lambda_1(A)\lambda_1(B) + \lambda_k(A)\lambda_j(B) - \lambda_k(A)\lambda_1(B) - \lambda_1(A)\lambda_j(B))$$

$$= v(\lambda_1(A) - \lambda_k(A))(\lambda_1(B) - \lambda_j(B)) \geq 0.$$

Thus, $\bar{V} = \{\bar{v}_{ij}\}$ satisfies the constraints, is also a maximum, and the number of zero elements in the first row and column of $\bar{V}$ is larger by 1 at least from the corresponding one for V. If $\bar{v}_{11} = 1$, the conclusion follows by the induction hypothesis, while if $\bar{v}_{11} < 1$, one repeats this (at most $2N - 2$ times since the operation sends one entry to zero in the first column or the first line) to conclude. □

Corollary 1.17. *Assume that the entries $\{\sqrt{N}A_{ij}^N, i \leq j\}$ are independent and are either equidistributed with finite variance or such that*

$$\sup_{N\in\mathbb{N}} \sup_{1\leq i,j\leq N} \mathbb{E}[|\sqrt{N}A_{ij}^N|^4] < \infty. \tag{1.13}$$

Assume also that $\{\sqrt{N}A_{ij}^N, i \leq j\}$ are centered and for all

$$\lim_{N\to\infty} \max_{1\leq i\leq j\leq N} |\mathbb{E}[(\sqrt{N}A_{ij}^N)^2] - 1| = 0.$$

Then, for any bounded continuous function f

$$\lim_{N\to\infty} \int f(x)dL_{\mathbf{A}^N}(x) = \int f(x)d\sigma(x) \quad a.s.$$

Remark. When the entries are not equidistributed, the convergence in probability can be proved when $(\sqrt{N}A_{ij}^N)_{1\leq i\leq j\leq N}$ are uniformly integrable. We strengthen here the hypotheses to have the almost sure convergence of the law of large numbers theorem.

Proof. Fix a constant C and consider the matrix $\hat{A}_N$ whose elements satisfy, for $i \leq j$ and $i = 1, \ldots, N$,

$$\hat{A}_{ij}^N = \frac{1}{\sigma_{ij}^N(C)} \left(A_{ij}^N \mathbf{1}_{\sqrt{N}|A_{ij}^N|\leq C} - E(A_{ij}^N \mathbf{1}_{\sqrt{N}|A_{ij}^N|\leq C}) \right)$$

with

$$\sigma_{ij}^N(C)^2 := \mathbb{E}\left[\left(A_{ij}^N \mathbf{1}_{\sqrt{N}|A_{ij}^N|\leq C} - E(A_{ij}^N \mathbf{1}_{\sqrt{N}|A_{ij}^N|\leq C}) \right)^2 \right].$$

$\hat{\mathbf{A}}^N$ satisfies the hypothesis of Theorem 1.15 for any $C \in \mathbb{R}^+$, so that

$$\lim_{N\to\infty} \int f(x)dL_{\hat{\mathbf{A}}^N}(x) = \int f(x)d\sigma(x) \text{ a.s.} \tag{1.14}$$

Assume now that f is bounded Lipschitz, with Lipschitz constant

$$\|f\|_{\mathcal{L}} = \sup_{x \neq y} \frac{|f(x) - f(y)|}{|x - y|} + \sup_x |f(x)|.$$

Then,

$$\left| \int f(x) dL_{\hat{\mathbf{A}}^N}(x) - \int f(x) dL_{\mathbf{A}^N}(x) \right|$$

$$\leq \frac{\|f\|_{\mathcal{L}}}{N} \sum_{i=1}^N |\lambda_i(\mathbf{A}^N) - \lambda_i(\hat{\mathbf{A}}^N)|$$

$$\leq \|f\|_{\mathcal{L}} \left(\frac{1}{N} \sum_{i=1}^N |\lambda_i(\mathbf{A}^N) - \lambda_i(\hat{\mathbf{A}}^N)|^2 \right)^{\frac{1}{2}}$$

regardless of the order on the eigenvalues. We conclude that

$$\left| \int f(x) dL_{\hat{\mathbf{A}}^N}(x) - \int f(x) dL_{\mathbf{A}^N}(x) \right| \leq \|f\|_{\mathcal{L}} \left(\frac{1}{N} \mathrm{Tr}(\mathbf{A}^N - \hat{\mathbf{A}}^N)^2 \right)^{\frac{1}{2}}$$

with

$$(\mathbf{A}^N - \hat{\mathbf{A}}^N)_{ij} = \frac{1}{\sigma_{ij}^N(C)} \left(A_{ij}^N 1_{\sqrt{N} A_{ij}^N \geq C} - \mathbb{E}[A_{ij}^N 1_{\sqrt{N} A_{ij}^N \geq C}] \right) + (1 - \sigma_{ij}^N(C)) \hat{A}_{ij}^N \tag{1.15}$$

where we used that $\mathbb{E}[A_{ij}^N] = 0$ for all i, j. Under the assumption (1.13) or when $\{\sqrt{N} A_{ij}^N, i \leq j\}$ are independent and equidistributed and with finite variance, we can use the strong law of large numbers to get that

$$\limsup_{N \to \infty} \frac{1}{N} \sum_{i,j=1}^N |(A^N - \hat{A}^N)_{ij}|^2 \leq \limsup_{N \to \infty} \max_{1 \leq i \leq j \leq N} \mathbb{E}[|\sqrt{N}(A^N - \hat{A}^N)_{ij}|^2] \quad \text{a.s.} \tag{1.16}$$

Thus, by Lemma 1.16,

$$\limsup_{N \to \infty} \left| \int f(x) dL_{\hat{\mathbf{A}}^N}(x) - \int f(x) dL_{\mathbf{A}^N}(x) \right|$$

$$\leq \|f\|_{\mathcal{L}} \limsup_{N \to \infty} \max_{1 \leq i \leq j \leq N} \mathbb{E}[((A^N - \hat{A}^N)_{ij})^2] \quad \text{a.s.}$$

Letting C go to infinity shows that the above right-hand side goes to zero (by (1.15) and since $(\sqrt{N} A_{ij}^N)_{i \leq j}$ is uniformly integrable under our assumptions) and therefore

$$\limsup_{C \to \infty} \limsup_{N \to \infty} \left| \int f(x) dL_{\hat{\mathbf{A}}^N}(x) - \int f(x) dL_{\mathbf{A}^N}(x) \right| = 0 \quad \text{a.s.}$$

We conclude with (1.14) that for all Lipschitz functions f,

$$\lim_{N\to\infty} \int f(x)dL_{\mathbf{A}^N}(x) = \int f(x)d\sigma(x) \quad \text{a.s.}$$

Now, taking any non negative Lipschitz function that vanishes on $[-2,2]$ and equals one on $[-3,3]^c$, we deduce that

$$\lim_{N\to\infty} L_{\mathbf{A}^N}([-3,3]^c) = 0 \quad \text{a.s.}$$

Since by the Weierstrass theorem, Lipschitz functions are dense in the set of continuous functions on the compact set $[-3,3]$, we can approximate any bounded continuous function f on $[-3,3]$ by a sequence of Lipschitz functions f_δ up to an error δ (for the supremum norm on $[-3,3]$). We choose f_δ with uniform norm bounded by that of f on the whole real line. We now conclude that for any bounded continuous function f,

$$\limsup_{N\to\infty}\left|\int f(x)dL_{\mathbf{A}^N}(x) - \int f(x)d\sigma(x)\right|$$
$$\leq 2\|f\|_\infty \limsup_{N\to\infty}(L_{\mathbf{A}^N}([-3,3]^c) + \sigma([-3,3]^c))$$
$$+ \limsup_{N\to\infty}\left|\int f_\delta(x)dL_{\mathbf{A}^N} - \int f_\delta(x)d\sigma(x)\right| + \delta$$
$$= \delta.$$

Letting δ go to zero finishes the proof. □

Remark. Let us remark that if $\sqrt{N}A^N(ij)$ has no moments of order 2, the theorem is no longer valid (see the heuristics of Cizeau and Bouchaud [64] and rigorous studies in [208] and [26]). Even though under appropriate assumptions the spectral measure of the matrix A^N, once properly normalized, converges, its limit is not the semicircle law but a heavy-tailed law with unbounded support.

1.4.2 Relaxation of the Hypothesis on the Centering of the Entries

The last generalization concerns the hypothesis on the mean of the variables $\sqrt{N}A^N_{ij}$ which, as we shall see, is irrelevant in the statement of Corollary 1.17. More precisely, we shall prove the following lemma (taken from [109]).

Lemma 1.18. *Let* $\mathbf{A}^N, \mathbf{B}^N$ *be* $N \times N$ *Hermitian matrices for* $N \in \mathbb{N}$ *such that* $\mathbf{B}^N$ *has rank* $r(N)$. *Assume that* $N^{-1}r(N)$ *converges to zero as* N *goes to infinity. Then, for any bounded continuous function* f *with compact support,*

$$\limsup_{N\to\infty}\left|\int f(x)dL_{\mathbf{A}^N+\mathbf{B}^N}(x) - \int f(x)dL_{\mathbf{A}^N}(x)\right| = 0.$$

If moreover $(L_{\mathbf{A}^N}, N \in \mathbb{N})$ *is tight in* $\mathcal{P}(\mathbb{R})$, *equipped with its weak topology, the above holds for any bounded continuous function.*

Proof. We first prove the statement for bounded increasing functions. To this end, we shall first prove that for any Hermitian matrix $\mathbf{Z}^N$, any $e \in \mathbb{C}^N$, $\lambda \in \mathbb{R}$, and for any bounded measurable increasing function f,

$$\left| \int f(x) dL_{\mathbf{Z}^N}(x) - \int f(x) dL_{\mathbf{Z}^N + \lambda ee^*}(x) \right| \leq \frac{2}{N} \|f\|_\infty. \tag{1.17}$$

We denote by $\lambda_1^N \leq \lambda_2^N \cdots \leq \lambda_N^N$ (resp. $\eta_1^N \leq \eta_2^N \cdots \leq \eta_N^N$) the eigenvalues of $\mathbf{Z}^N$ (resp. $\mathbf{Z}^N + \lambda ee^*$). By Lidskii's theorem 19.3, the eigenvalues λ_i and η_i are interlaced;

$$\lambda_1^N \leq \eta_2^N \leq \lambda_3^N \cdots \leq \lambda_{2[\frac{N-1}{2}]+1}^N \leq \eta_{2[\frac{N}{2}]}^N,$$

$$\eta_1^N \leq \lambda_2^N \leq \eta_3^N \cdots \leq \eta_{2[\frac{N-1}{2}]+1}^N \leq \lambda_{2[\frac{N}{2}]}^N.$$

Therefore, if f is an increasing function,

$$\sum_{i=1}^N f(\lambda_i^N) \leq \sum_{i=2}^N f(\eta_i^N) + \frac{1}{N}\|f\|_\infty \leq \sum_{i=1}^N f(\eta_i^N) + \frac{2}{N}\|f\|_\infty$$

but also

$$\sum_{i=1}^N f(\lambda_i^N) = f(\lambda_1^N) + \sum_{i=2}^N f(\lambda_i^N) \geq f(\lambda_1^N) + \sum_{i=2}^N f(\eta_{i-1}^N)$$

$$= f(\lambda_1^N) - f(\eta_N^N) + \sum_{i=1}^N f(\eta_i^N).$$

These two bounds prove (1.17).

Now, let us denote by $(e_1^N, \ldots, e_{r(N)}^N)$ an orthonormal basis of the vector space of eigenvectors of $\mathbf{B}^N$ with non-zero eigenvalues so that

$$\mathbf{B}^N = \sum_{i=1}^{r(N)} \eta_i^N e_i^N (e_i^N)^*$$

with some real numbers $(\eta_i^N)_{1 \leq i \leq r(N)}$. Iterating (1.17) shows that for any bounded increasing function f,

$$\left| \int f(x) dL_{\mathbf{A}^N}(x) - \int f(x) dL_{\mathbf{A}^N + \mathbf{B}^N}(x) \right| \leq \frac{2r(N)}{N}\|f\|_\infty. \tag{1.18}$$

Therefore, for any increasing bounded continuous function, when $N^{-1}r(N)$ goes to zero,

$$\limsup_{N \to \infty} \left| \int f(x) dL_{\mathbf{A}^N + \mathbf{B}^N}(x) - \int f(x) dL_{\mathbf{A}^N}(x) \right| = 0. \tag{1.19}$$

Of course, the result immediately extends to decreasing functions by $f \to -f$. Now, note that any Lipschitz function f that vanishes outside of a compact set $K = [-k, k]$ can be written as the difference of two bounded increasing continuous functions (this is in fact true as soon as f has bounded variations) since it is almost surely (with respect to Lebesgue measure) differentiable with derivative bounded by $|f|_{\mathcal{L}}$ and

$$f(x) - f(0) = \int_0^x f'(x) 1_{f'(x) \geq 0} dx - \int_0^x (-f'(x)) 1_{f'(x) < 0} dx.$$

Hence, (1.19) extends to the case of compactly supported Lipschitz functions, and then to any bounded compactly supported continuous functions (by density for the supremum norm).

To remove the assumption that f is compactly supported we assume $(L_{\mathbf{A}^N})_{N \in \mathbb{N}}$ tight so that $\sup_N L_{\mathbf{A}^N}([-k, k]^c)$ goes to zero as k goes to infinity. Now, taking $f(x) = (x - k) \wedge 1 \vee 0$ for some finite k, we deduce that

$$\limsup_{N \to \infty} L_{\mathbf{A}^N + \mathbf{B}^N}([k+1, \infty[) \leq \limsup_{N \to \infty} \int (x - k) \wedge 1 \vee 0 \, dL_{\mathbf{A}^N + \mathbf{B}^N}(x)$$

$$= \limsup_{N \to \infty} \int (x - k) \wedge 1 \vee 0 \, dL_{\mathbf{A}^N}(x)$$

$$\leq \limsup_{N \to \infty} L_{\mathbf{A}^N}([k, \infty[) \leq \varepsilon_k$$

where ε_k is a sequence going to zero with k, which exists by the assumption that $(L_{\mathbf{A}^N}, N \in \mathbb{N})$ is tight. We apply the same argument for $L_{\mathbf{A}^N + \mathbf{B}^N}(] - \infty, -k - 1])$ with the decreasing function $f(x) = (-k - x) \wedge 1 \vee 0$ and deduce that

$$\limsup_{k \to \infty} \limsup_{N \to \infty} L_{\mathbf{A}^N + \mathbf{B}^N}([-k, k]^c) = 0.$$

This allows us to finish the proof of the lemma for any bounded continuous function f since we also have $\limsup_{k \to \infty} \limsup_{N \to \infty} L_{\mathbf{A}^N}([-k, k]^c) = 0$. □

By Corollary 1.17 and Lemma 1.18, we find the following:

Corollary 1.19. *Assume that the matrix* $(\mathbb{E}[A_{ij}^N])_{1 \leq i,j \leq N}$ *has rank* $r(N)$ *so that* $N^{-1} r(N)$ *goes to zero as* N *goes to infinity, and that the variables* $\sqrt{N}(A_{ij}^N - \mathbb{E}[A_{ij}^N])$ *satisfy (1.13) and have variance 1. Then, for any bounded continuous function* f,

$$\lim_{N \to \infty} \int f(x) dL_{\mathbf{A}^N}(x) = \int f(x) d\sigma(x) \quad a.s.$$

This result holds in particular if $\mathbb{E}[A_{ij}^N] = x^N$ *is independent of* $i, j \in \{1, \ldots, N\}^2$, *in that case* $r(N) = 1$.

Bibliographical Notes. Since the convergence of the spectral measure was proved by Wigner [205] when the entries possess moments of all orders, many papers have improved this result. The optimal hypothesis for the convergence of the spectral measure of Wigner matrices to the semicircle law is that the entries have a finite second moment, since if they do not, the asymptotics of the spectral measure described in [26] show that the renormalization of the eigenvalues must depend on the tail of the entries and the limit is a heavy tailed law rather than the semicircle law. More precise results have been derived; for instance, when the entries have only a finite fourth moment, Bai [11] proved the convergence of the spectral measure and showed that some distance to the limit is at most of order $N^{-\frac{1}{4}}$ (this result was improved to N^{-1} under stronger hypotheses in [96]). Bai used a method directly based on estimations of the Cauchy–Stieljes transform of the spectral measure, rather than on moments. The convergence of the spectral measure of diverse classical ensembles of matrices were shown; for instance for Wishart matrices [145], for Wigner matrices with correlated entries [45], Toeplitz matrices [53, 112], or for non symmetric matrices (with a complex spectrum) such as Ginibre ensemble [12, 95]. We refer the reader to [13] for more examples.

2

Wigner's Matrices; More Moments Estimates

In this chapter, we elaborate upon the previous computation of moments in two directions. First we give a better estimate of the error to the previous limit and prove a central limit theorem. Second, we consider the case where moments are taken at powers that blow up with the dimension of the matrices; we basically show that if this power is small compared to the square root of the dimension, the first-order contribution is still given, in the moment expansion, by graphs that are trees.

2.1 Central Limit Theorem

In the previous section, we proved Wigner's theorem by evaluating $\int x^p dL_{\mathbf{A}^N}(x)$ for $p \in \mathbb{N}$. We shall push this computation one step further here and prove a central limit theorem. Namely, setting

$$\int x^k d\bar{L}_{\mathbf{A}^N}(x) := \mathbb{E}\left[\int x^k dL_{\mathbf{A}^N}(x)\right],$$

we shall prove that

$$M_k^N := N\left(\int x^k dL_{\mathbf{A}^N}(x) - \int x^k d\bar{L}_{\mathbf{A}^N}(x)\right) = \sum_{i=1}^N \left(\lambda_i^k - \mathbb{E}[\lambda_i^k]\right)$$

converges in law to a centered Gaussian variable. Since in Part III we shall give a complete and detailed proof of the central limit theorem in the case of Gaussian entries with a weak interaction, we will be rather sketchy here. We refer to [7] for a complete and clear treatment and [6] for a simplified exposition of the full proof of the theorem we state below. To simplify, we assume here that $\mathbf{A}^N$ is a Wigner matrix with

$$A_{ij}^N = \frac{B_{ij}}{\sqrt{N}},$$

A. Guionnet, *Large Random Matrices: Lectures on Macroscopic Asymptotics*, 29
Lecture Notes in Mathematics 1957, DOI: 10.1007/978-3-540-69897-5_2,
© 2009 Springer-Verlag Berlin Heidelberg, Reprint by Springer-Verlag Berlin Heidelberg 2012

where $(B_{ij}, 1 \leq i \leq j \leq N)$ are independent real equidistributed random variables. Their marginal distribution μ has all moments finite (in particular (1.7) is satisfied) and satisfies

$$\int x d\mu(x) = 0 \quad \text{and} \quad \int x^2 d\mu(x) = 1.$$

We shall show why the following statement holds.

Theorem 2.1. *Let*

$$\sigma_k^2 = k^2 [C_{\frac{k-1}{2}}]^2 + \frac{k^2}{2} [C_{\frac{k}{2}}]^2 \left[\int x^4 d\mu(x) - 1 \right] + \sum_{r=3}^{\infty} \frac{2k^2}{r} \left(\sum_{\substack{k_i \geq 0 \\ 2\sum_{i=1}^r k_i = k-r}} \prod_{i=1}^r C_{k_i} \right)^2,$$

In this formula, C_x equals zero if x is not an integer and otherwise is equal to the Catalan number.

Then, M_k^N converges in moments to the centered Gaussian variable with variance σ_k^2, i.e., for all $l \in \mathbb{N}$,

$$\lim_{N \to \infty} \mathbb{E}\left[(M_k^N)^l\right] = \frac{1}{\sqrt{2\pi}\sigma_k} \int x^l e^{-\frac{x^2}{2\sigma_k^2}} dx.$$

Remark. Unlike the standard central limit theorem for independent variables, the variance here depends on $\mu(x^4)$.

Outline of the proof.

- *We first prove that the statement is true when $l = 2$.* (It is clearly true for $k = 1$ since A_k^N is centered.) We thus want to show

$$\sigma_k^2 = \lim_{N \to \infty} \mathbb{E}\left[(M_k^N)^2\right]. \tag{2.1}$$

Below (1.9), we proved that $\mathbb{E}\left[(A_k^N)^2\right]$ is bounded, uniformly in N. Furthermore, we can write

$$\mathbb{E}\left[(M_k^N)^2\right] = \frac{1}{N^k} \sum_{\mathbf{i}, \mathbf{i}'} [P(\mathbf{i}, \mathbf{i}') - P(\mathbf{i})P(\mathbf{i}')]$$

where the sum over $\mathbf{i}, \mathbf{i}'$ will hold on graphs $\tilde{G}(\mathbf{i}, \mathbf{i}') = (\tilde{V}(\mathbf{i}, \mathbf{i}'), \tilde{E}(\mathbf{i}, \mathbf{i}'))$ so that

$$|\tilde{V}(\mathbf{i}, \mathbf{i}')| \leq k, \quad |\tilde{E}(\mathbf{i}, \mathbf{i}')| \leq k.$$

Since $[P(\mathbf{i}, \mathbf{i}') - P(\mathbf{i})P(\mathbf{i}')]$ is uniformly bounded, the only contributing graphs to the leading order will be those such that $|\tilde{V}(\mathbf{i}, \mathbf{i}')| = k$. Then, since we always have $|\tilde{V}(\mathbf{i}, \mathbf{i}')| \leq |\tilde{E}(\mathbf{i}, \mathbf{i}')| + 1$, we have two cases:

- $|\tilde{E}(\mathbf{i}, \mathbf{i}')| = k - 1$ in that case the skeleton $\tilde{G}(\mathbf{i}, \mathbf{i}')$ will again be a tree but with one edge less than the total number possible; this means that one edge appears with multiplicity four and belongs to $\tilde{E}(\mathbf{i}) \cap \tilde{E}(\mathbf{i}')$, the other edges appearing with multiplicity 2. Hence, the graphs of $\tilde{E}(\mathbf{i})$ and $\tilde{E}(\mathbf{i}')$ are both trees (so that k must be even); there are $C_{\frac{k}{2}}^2$ such trees, and they are glued by a common edge, to choose among $\frac{k}{2}$ edges in each of the tree. Finally, there are two possible choices to glue the two trees according to the orientation. Thus, there are

$$2 \left(\frac{k}{2} \right)^2 C_{\frac{k}{2}}^2 = \left(\frac{k^2}{2} \right) C_{\frac{k}{2}}^2$$

such graphs and then

$$P(\mathbf{i}, \mathbf{i}') - P(\mathbf{i}) P(\mathbf{i}') = \int x^4 d\mu(x) - 1.$$

We hence obtain the contribution $(\frac{k^2}{2}) C_{\frac{k}{2}}^2 (\int x^4 d\mu(x) - 1)$ to the variance.

• $|\tilde{E}(\mathbf{i}, \mathbf{i}')| = k$. In this case, the graph is no longer a tree and because $|\tilde{E}(\mathbf{i}, \mathbf{i}')| - |\tilde{V}(\mathbf{i}, \mathbf{i}')| = 1$, it contains exactly one cycle. This can be seen either by closer inspection of the arguments given after (1.1) or by using the formula that relates the genus of a graph and its number of vertices, faces and edges:

$$\sharp\text{vertices} + \sharp\text{faces} - \sharp\text{edges} = 2 - 2g \leq 2.$$

The faces are defined by following the boundary of the graph; each of these boundaries are exactly one cycle of the graph except one (since a graph has always one boundary) and therefore

$$\sharp\text{faces} = 1 + \sharp\text{cycles}.$$

So we get, for a connected graph with skeleton $(\tilde{V}, \tilde{E})$,

$$|\tilde{V}| \leq |\tilde{E}| + 1 - \sharp\text{cycles}. \tag{2.2}$$

In our case, $\sharp\text{vertices} = \sharp\text{edges} = k$ and $\sharp\text{cycles} \geq 1$ (since the graph is not a tree), so that the number of cycles must be exactly one. Counting the number of such graphs completes the proof of the convergence of $\mathbb{E}\left[(M_k^N)^2\right]$ to σ_k^2 (see [7] for more details).

• *Convergence to the Gaussian law.*

We next show that M_k^N is asymptotically Gaussian. This amounts to proving that $\lim_{N \to \infty} \mathbb{E}[(M_k^N)^{2l+1}] = 0$ whereas,

$$\lim_{N \to \infty} \mathbb{E}[(M_k^N)^{2l}] = \sharp\{\text{number of pair partitions of } 2l \text{ elements}\} \times \sigma_k^{2l}.$$

Again, we shall expand the expectation in terms of graphs and write for $l \in \mathbb{N}$,

$$\mathbb{E}[(M_k^N)^l] = \frac{1}{N^{\frac{kl}{2}}} \sum_{\mathbf{i}_1,\dots,\mathbf{i}_l} P(\mathbf{i}^1,\dots,\mathbf{i}^l)$$

with $P(\mathbf{i}^1,\dots,\mathbf{i}^l)$ given by

$$\mathbb{E}\left[\left(B_{i_1^1 i_2^1} \cdots B_{i_k^1 i_1^1} - \mathbb{E}[B_{i_1^1 i_2^1} \cdots B_{i_k^1 i_1^1}]\right)\right.$$
$$\left.\cdots \left(B_{i_1^l i_2^l} \cdots B_{i_k^l i_1^l} - \mathbb{E}[B_{i_1^l i_2^l} \cdots B_{i_k^l i_1^l}]\right)\right].$$

We denote by $G(\mathbf{i}^1,\dots,\mathbf{i}^l) = (V(\mathbf{i}^1,\dots,\mathbf{i}^l), E(\mathbf{i}^1,\dots,\mathbf{i}^l))$ the corresponding graph; $V(\mathbf{i}^1,\dots,\mathbf{i}^l) = \{i_n^j, 1 \le j \le l, 1 \le n \le k\}$ and $E(\mathbf{i}^1,\dots,\mathbf{i}^l) = \{(i_n^j, i_{n+1}^j), 1 \le j \le l, 1 \le n \le k\}$ with the convention $i_{l+1}^j = i_1^j$. As before, $P(\mathbf{i}^1,\dots,\mathbf{i}^l)$ equals zero unless each edge appears with multiplicity 2 at least. Also, because of the centering, it vanishes if there exists a $j \in \{1,\dots,l\}$ so that $E(\mathbf{i}^1,\dots,\mathbf{i}^l) \cap E(\mathbf{i}^j)$ does not intersect $E(\mathbf{i}^1,\dots,\mathbf{i}^{j-1},\mathbf{i}^{j+1},\dots,\mathbf{i}^l)$. Let us decompose $G(\mathbf{i}^1,\dots,\mathbf{i}^l)$ into its connected components $(G_1,\dots,G_c)$. We claim that

$$|V(\mathbf{i}^1,\dots,\mathbf{i}^l)| \le c - l + \left[\frac{l(k+1)}{2}\right]. \tag{2.3}$$

This type of bound is rather intuitive; if a connected component G_i contains $G(\mathbf{i}^{j_1}),\dots,G(\mathbf{i}^{j_p})$, each gluing of the $G(\mathbf{i}^{j_i})$ should create either a cycle or an edge with multiplicity 4, the total number of vertices decreasing at least by one in each gluing. Hence, $|V(\mathbf{i}^1,\dots,\mathbf{i}^l)|$ should grow linearly with the number of connected components. The proof is given in Appendix 20.3 for completeness (see [6] or [7]). With (2.3), we conclude that the only indices that will contribute are such that

$$c - l + \left[\frac{l(k+1)}{2}\right] \ge \frac{kl}{2}$$

with $c \le [\frac{l}{2}]$. This implies that

$$\frac{kl}{2} \le \left[\frac{l}{2}\right] - l + \left[\frac{l(k+1)}{2}\right] \le \frac{l}{2} - l + \frac{l(k+1)}{2} = \frac{kl}{2}$$

resulting in all inequalities being equalities. Thus, to get a first-order contribution we must have l even and $c = \frac{l}{2}$. In that case, we write $(s_j, r_j)_{1 \le j \le l}$ the pairing so that $(G(\mathbf{i}_{s_j}), G(\mathbf{i}_{r_j}))_{1 \le j \le l}$ are connected for all $1 \le j \le l$ (with the convention $s_j < r_j$). By independence of the entries, we have

$$P(\mathbf{i}_1,\dots,\mathbf{i}_{2l}) = \prod_{j=1}^{l} P(\mathbf{i}_{s_j}, \mathbf{i}_{r_j})$$

and so we have proved that

$$N^{-kl} \sum_{\mathbf{i}_1,\ldots,\mathbf{i}_{2l}} P(\mathbf{i}_1,\ldots,\mathbf{i}_{2l}) = \sum_{\substack{s_1 < \cdots < s_l \\ r_j > s_j}} \left(N^{-k} \sum_{\mathbf{i}_1,\mathbf{i}_2} P(\mathbf{i}_1,\mathbf{i}_2) \right)^l + o(1)$$

$$= \sigma_k^{2l} \sum_{\substack{s_1 < \cdots < s_l \\ r_j > s_j}} 1 + o(1)$$

which proves the claim since

$$\frac{1}{\sqrt{2\pi}} \int x^{2l} e^{-\frac{x^2}{2}} dx = \sum_{\substack{s_1 < \cdots < s_l \\ r_j > s_j}} 1 = (2l-1)(2l-3)(2l-5)\cdots 1.$$

This completes the proof of the moments convergence.

□

Exercise 2.2. *Show that Theorem 2.1 implies that M_k^N converges weakly to the centered Gaussian variable with variance σ_k^2. Hint: control tails to approximate bounded continuous functions by polynomials.*

Bibliographical Notes. Johansson [120] proved a rather general central limit theorem for the spectral measure of Gaussian random matrices (and more generally for particles interacting via a Coulomb gas potential). It was generalized to β-ensembles and Laguerre ensembles in [82] by using tri-diagonal representation of the classical ensembles [81]. The strategy of moments developed here follows an article of Anderson and Zeitouni [7] (see a generalization in [177]). Central limit theorems were also obtained in the case of Ginibre ensembles (with spectral measure converging to the so-called circular law) in [169].

We shall see in Part III that this kind of theorem generalizes to the multi-matrix setting that we shall introduce in the next chapter.

2.2 Estimates of the Largest Eigenvalue of Wigner Matrices

In this section, we derive estimates on the largest eigenvalue of a Wigner matrix with real entries $A_{ij}^N = N^{-\frac{1}{2}} B_{ij}$ with $(B_{ij}, 1 \le i \le j \le N)$ independent equidistributed centered random variables with marginal distribution P. The idea is to improve the moments estimates of the previous chapter.

We shall assume that P is a symmetric law (see the recent article [166] for a relaxation of this hypothesis):

$$P(-x \in .) = P(x \in .).$$

We take the normalization $E[x^2] = 1$. Further, we assume that P has sub-Gaussian tail, i.e., that there exists a finite constant c such that for all $k \in \mathbb{N}$,

$$E[x^{2k}] \leq (ck)^k.$$

We follow the article of S. Sinaï and A. Soshnikov [179] to prove the following result:

Theorem 2.3 (S. Sinaï–A. Soshnikov [179]). *For all $\epsilon > 0$, all $N \in \mathbb{N}$, there exists a finite function $o(s, N)$ such that $\lim_{N \to \infty} \sup_{N^\epsilon \leq s \leq N^{\frac{1}{2} - \epsilon}} o(s, N) = 0$ and*

$$\mathbb{E}[\mathrm{Tr}((A^N)^{2s})] = \frac{N 2^{2s}}{\sqrt{\pi s^3}}(1 + o(s, N)). \tag{2.4}$$

As a consequence, for all $\epsilon > 0$, if we let $\lambda_{max}(A^N)$ denote the spectral radius of A^N,

$$\lim_{N \to \infty} P(|\lambda_{\max}(A^N) - 2| \geq \epsilon) = 0.$$

A previous result of the same nature (but under weaker hypothesis (the symmetry hypothesis of the distribution of the entries being removed) under which the moments estimate (2.4) holds for a smaller range of s) was proved by Komlós and Füredi [93]. A later result of Soshnikov [180] improves the range of s under which (2.4) holds to s of order less than $n^{\frac{2}{3}}$, a result that captures the fluctuations of $\lambda_{\max}(A^N)$. We emphasize here that the proof below heavily depends on the assumption that the distribution of the entries is symmetric.

Proof. Let us first derive the convergence in probability from the moment estimates. First, note that

$$P(\lambda_{\max}(A^N) \leq 2 - \epsilon) \leq P\left(\int f(x)dL_{A^N} = 0\right)$$

for all functions f supported on $]2 - \epsilon, \infty[$. Taking f bounded continuous, null on $]-\infty, 2-\epsilon]$ and strictly positive in $[2-\frac{\epsilon}{2}, 2]$, we see that $P(\int f(x)dL_{A^N} = 0)$ goes to zero by Theorem 1.15. For the upper bound on $\lambda_{\max}(A^N)$, we shall use Chebychev's inequality and the moment estimates (2.4) as follows:

$$P(\lambda_{\max}(A^N) \geq 2 + \epsilon) \leq \frac{1}{(2 + \epsilon)^{2s}} \mathbb{E}[\lambda_{\max}(A^N)^{2s}] \leq \frac{1}{(2 + \epsilon)^{2s}} \mathbb{E}[\mathrm{Tr}((A^N)^{2s})]$$

$$\leq \frac{N 2^{2s}}{(2 + \epsilon)^{2s}\sqrt{\pi s^3}}(1 + o(s, N))$$

where the right-hand side goes to zero with N when $s = N^\epsilon$ for some $\epsilon > 0$.

To prove the moment estimates we shall again expand the moments and count contributing paths, in particular estimate more precisely contributions from paths that are not trees. Yet, the central point of the proof is to show that these paths give a negligible contribution. We follow the presentation of [179].

1. *Moments expansion.* As usual, we write

$$\mathbb{E}[\mathrm{Tr}\,((\mathbf{A}^N)^{2s})] = \frac{1}{N^s} \sum_{i_0,\ldots,i_{2s-1}=1}^{N} \mathbb{E}[B_{i_0 i_1} \cdots B_{i_{2s-1},i_0}]. \tag{2.5}$$

We let E denote the set of edges of the graph, i.e., the undirected collection of couples $\{(i_p, i_{p+1}), p = 0, \ldots, 2s-1\}$. Because we assumed the law of the B_{ij}'s symmetric, only indices such that each edge in E appears an even number of times will contribute. We call a *closed path* the sequence $P : i_0 \to i_1 \to \cdots \to i_{2s-1} \to i_0$. An *even path* is a closed path where each edge appears with even multiplicity; they are the only contributing paths.

2. *Descriptions of paths.* We will say that the ℓth step $i_{\ell-1} \to i_\ell$ of a path P is *marked* if during the first ℓ steps of P, the edge $\{i_{\ell-1}, i_\ell\}$ appears an odd number of times (note here that the ℓth step is counted, and so a step is marked iff the edge $\{i_{\ell-1}, i_\ell\}$ appears an even number of times in the previous steps, in particular if it does not appear). The step is *unmarked* otherwise. For even paths, the number of marked and unmarked edges is equal to s. The complete set of vertices V is the collection $\{1, \ldots, N\}$ of all possible values of the points $(i_k, 0 \le k \le 2s-1)$. We say that a vertex $i \in V$ belongs to the subset $\mathcal{N}_k = \mathcal{N}_k(P)$ if the number of times we arrive at i via marked edges equals k. Note that no vertex of the path except i_0 can belong to $\mathcal{N}_0$. Moreover, $\mathcal{N}_p = 0$ for $p > s$ (since there are at most s edges). Note that if we let $n_k = \sharp \mathcal{N}_k$, since $(\mathcal{N}_0, \ldots, \mathcal{N}_s)$ is a partition of V, $\sum_{k=0}^{s} n_k = N$. Moreover, $(\mathcal{N}_0, \ldots, \mathcal{N}_s)$ also induces a partition of the edges and hence

$$\sum_{k=0}^{s} k n_k = s.$$

We say that P is of type $(n_0, n_1, \ldots, n_s)$ if $n_k = \sharp \mathcal{N}_k = \sharp \mathcal{N}_k(P)$ for all $k \in \{0, \ldots, s\}$. We finally say that a path is a *simple even path* if $i_0 \in \mathcal{N}_0$ and P is of type $(N-s, s, 0, \ldots, 0)$. Observe that in a simple even path, each edge appears only twice (since there are at most s different edges in P and here exactly s since there are s different vertices in $\mathcal{N}_1$). Also, we see that the graph corresponding to P has exactly s vertices in $\mathcal{N}_1$ plus $i_0 \in \mathcal{N}_0$ and so exactly $s+1$ vertices. Hence, the skeleton $(V, \tilde{E})$ of the graph drawn by P satisfies the relation $|V| = |\tilde{E}| + 1$ and hence is a tree. The strategy of the proof will be to show that simple even paths dominate the expectation when $s = o(\sqrt{N})$.

3. *Contribution of simple even paths.* Considering (2.5), we see that for simple even paths, $\mathbb{E}[B_{i_0 i_1} \cdots B_{i_{2s-1} i_0}] = 1$. Moreover, given a simple even path, we have N possible choices for i_0, $N-1$ for the first new vertex encountered when following P, $N-2$ for the second new vertex encountered,

etc. Since we have $C_s = (2s)!/s!(s+1)!$ simple even paths (see Property 1.10), we get the contribution

$$C_1^N = \frac{1}{N^s} N(N-1) \cdots (N-s) \frac{(2s)!}{s!(s+1)!} = \frac{2^{2s} N}{\sqrt{\pi s^3}} (1 + o_1(s, N))$$

where we have used Stirling's formula and found

$$o_1(s, N) = -\frac{1}{N} \sum_{k=1}^{s} k + \frac{1}{s} \approx \frac{s^2}{2N} + \frac{1}{s}.$$

In the case where $i_0 \notin \mathcal{N}_0$ but $n_1 = s, n_2 = 0 \cdots, n_s = 0$, we must have $i_0 \in \mathcal{N}_1$. This means that we have one cycle and one different vertex less in the graph of an even path. Note that if we split the vertex i_0 into two vertices as in Figure 2.1, the new vertex being attached to the marked edge, then the old i_0 belongs to $\mathcal{N}_0$ and the new vertex to $\mathcal{N}_1$ and we are back to the case where $i_0 \in \mathcal{N}_0$.

There are s possibilities for the position of the marked edge incoming in i_0, but we are losing $N - s$ possibilities to choose a different vertex. Hence, the contribution to this term is bounded by

$$C_2^N \leq \frac{s}{N-s} E[x^4] C_1^N$$

where the last term comes from the possibility that one edge attached to i_0 now has multiplicity 4.

4. *Contribution of paths that are not simple.* If a path is not as in the previous paragraph, there must be an $n_k \geq 1$ for $k \geq 2$. Let us count the number of these paths.

Given $n_0, n_1, \ldots, n_s$, we have $\frac{N!}{n_0! n_1! \cdots n_s!}$ ways to choose the values of the vertices. Then, among the n_0 vertices in $\mathcal{N}_0$, we have at most n_0 ways to choose the vertex corresponding to i_0 (if $i_0 \in \mathcal{N}_0$).

Being given the values of the vertices, a path is uniquely described if we know the order of appearance of the vertices at the marked steps, the times when the marked steps occur and the choice of end points of the unmarked steps. The moments of time when marked steps occur can be coded by

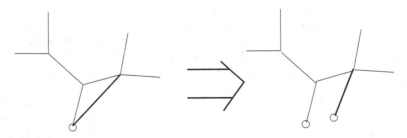

Fig. 2.1. Splitting of the graph

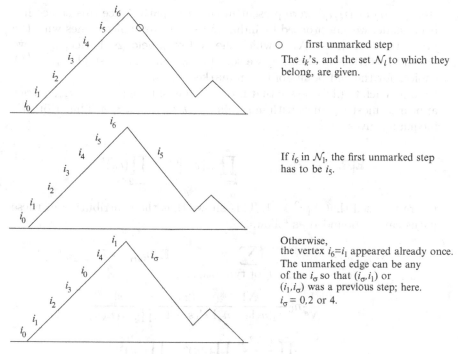

○   first unmarked step
The i_k's, and the set $\mathcal{N}_l$ to which they belong, are given.

If i_6 in $\mathcal{N}_1$, the first unmarked step has to be i_5.

Otherwise,
the vertex $i_6 = i_1$ appeared already once. The unmarked edge can be any of the i_σ so that (i_σ, i_1) or (i_1, i_σ) was a previous step; here. $i_\sigma = 0, 2$ or 4.

Fig. 2.2. Counting unmarked steps

a Dick path by adding $+1$ when the step is marked and -1 otherwise. Hence, there are $C_s = (2s)!/s!(s+1)!$ choices for the times of marked steps. Once we are given this path, we have s marked steps. The marked steps are partitioned into s sets corresponding to the $\mathcal{N}_k$, $1 \leq k \leq s$, with cardinality $n_k k$ each. Hence, we have $\frac{s!}{\prod_{k=1}^s (n_k k)!}$ possibilities to assign the sets into which the end points of the marked steps are. Finally, we have $(n_k k)!/(k!)^{n_k}$ ways to partition the set $\mathcal{N}_k$ into k copies of the same point of $\mathcal{N}_k$. So far, we have prescribed uniquely the marked steps and the set to which they belong.

To prescribe the unmarked steps, we still have an indeterminate. In fact, let us follow the Dick path of the marked steps till the first decreasing part corresponding to unmarked steps. Let i_ℓ be the vertex assigned to the last step. Then, if i_ℓ appeared only once in the past path (in the edge $(i_{\ell-1}, i_\ell)$), we have no choice and the next vertex in the path has to be $i_{\ell-1}$. This is the case in particular if $i_\ell \in \mathcal{N}_1$. If now $i_\ell \in \mathcal{N}_k$ for $k \geq 2$, the undirected step (i_p, i_ℓ) for some i_p may have occurred already at most $2k$ times (since it could occur either as a step (i_p, i_ℓ) or a step (i_ℓ, i_p), the later happening also less than k times since it requires that a marked step arrived at i_ℓ before). We have thus at most $2k$ choices now for the next vertex; one of the i_p among the at most $2k$ vertices such that the step (i_p, i_ℓ) or (i_ℓ, i_p) were present in the past path. Once this choice has been made, we can proceed by induction since this choice comes with the prescription of the set $\mathcal{N}_l$ in which the vertex i_p belongs. Hence, since we have kn_k vertices in each set, we see that we have at most $\prod_{k=2}^{s}(2k)^{kn_k}$ choices for the end points of the unmarked steps.

Coming back to (2.5) we see that if the path is of type $(n_0, \ldots, n_s)$, entries appear at most n_k times with multiplicity $2k$ for $1 \leq k \leq s$. Thus Hölder's inequality gives

$$\mathbb{E}[B_{i_0 i_1} \cdots B_{i_{2s-1} i_0}] \leq \prod_{k=1}^{s} \mathbb{E}[x^{2k}]^{n_k} \leq \prod_{k=2}^{s}(ck)^{kn_k}$$

where we used that $\mathbb{E}[x^2] = 1$. This shows that the contribution of these paths can be bounded as follows.

$$E_{n_0,\ldots,n_s} = \sum_{i_0,\cdots i_{2s-1}: P \text{ of type}(n_0,\ldots,n_s)} \mathbb{E}[B_{i_0 i_1} \cdots B_{i_{2s-1} i_0}]$$

$$\leq \frac{1}{N^s} n_0 \frac{N!}{n_0! n_1! \cdots n_s!} \frac{(2s)!}{s!(s+1)!} \frac{s!}{\prod_{k=1}^{s}(n_k k)!}$$

$$\prod_{k=1}^{s} \frac{(n_k k)!}{(k!)^{n_k}} \prod_{k=2}^{s}(2k)^{kn_k} \prod_{k=2}^{s}(ck)^{kn_k}$$

$$\leq n_0 \frac{N(N-1)\cdots(n_0+1)}{N^s} \frac{(2s)!}{s!(s+1)!} \frac{1}{n_1! \cdots n_s!}$$

$$\frac{s!}{\prod_{k=1}^{s}(ke^{-1})^{n_k k}} \prod_{k=1}^{s}(2ck^2)^{kn_k}$$

$$\leq NN^{N-n_0-s} \frac{(2s)!}{s!(s+1)!} \frac{s!}{n_1! \cdots n_s!} \prod_{k=2}^{s}(2cek)^{kn_k}$$

where we have used that $(k!)^{n_k} \geq (ke^{-1})^{kn_k}$. Since $s = \sum_{k=1}^{s} kn_k$ and $N = \sum_k n_k$, we have $N - n_0 - s = \sum_{k=2}^{s}(1-k)n_k$. Using $s! \leq (s)^s$, we obtain the bound

$$E_{n_0,\ldots,n_s} \leq N \frac{(2s)!}{s!(s+1)!} \prod_{k=2}^{s} \frac{1}{n_k!} (N^{1-k}(2ceks)^k)^{n_k}.$$

We next sum over all $n_i \geq 0$ so that at least one $n_i \geq 1$ for $i \in \{2, \ldots, s\}$. This gives, with $\gamma_k := N^{1-k}(2ceks)^k$,

$$\sum_{n_0,\ldots,n_s:\max_{j\geq 2} n_j\geq 1} E_{n_0,\ldots,n_s} \leq N\frac{(2s)!}{s!(s+1)!}\sum_{k=2}^{s}(e^{\gamma_k}-1)\prod_{\ell\neq k}e^{\gamma_\ell}$$

$$\leq N\frac{(2s)!}{s!(s+1)!}e^{\sum_{\ell\geq 2}\gamma_\ell}\left(\sum_{\ell\geq 2}\gamma_\ell\right)$$

where we used that $e^x - 1 \leq xe^x$ for all $x \geq 0$. Note that in the range of s where $s^2 \leq N^{1-\epsilon}$, if we choose K big enough so that $K\epsilon \geq 1$,

$$\sum_\ell \gamma_\ell = \sum_{2\leq\ell\leq s}N^{1-\ell}(2ce\ell s)^\ell$$

$$\leq NK(2cesKN^{-1})^2 + N\sum_{K+1\leq\ell\leq s}(2ces^2N^{-1})^\ell$$

$$\leq \text{constant}(N^{-1}K^2s^2 + N(2ceN^{-\epsilon})^{K+1}) \leq \text{constant } N^{-\epsilon}$$

goes to zero as N goes to infinity. Thus, we conclude that

$$\sum_{n_0,\ldots,n_s} E_{n_0,\ldots,n_s} \leq CC_1^N N^{-\epsilon}.$$

Hence, in the regime s^2/N going to zero, the contribution of the indices $\{i_0,\ldots,i_{2s-1}\}$ associated with a path of type $(n_0,\ldots,n_s)$ with some $n_k \geq 1$ for some $k \geq 2$ is negligible compared to the contribution of simple even paths.

$\square$

Exercise 2.4. *The extension of Theorem 2.3 to Hermitian Wigner matrices satisfying the same type of hypotheses is left to the reader as an exercise.*

Bibliographical Notes. Soshnikov [181] elaborated on his combinatorial estimation of moments to prove that the largest eigenvalue fluctuations follow the Tracy–Widom law, by estimating moments of order $N^{\frac{2}{3}}$ when the entries are symmetrically distributed and have sub-Gaussian tails. By approximation, Ruzmaikina [174] could weaken the later hypothesis to the case where the entries have only the eighteenth (thirty-sixth according to [9]) moment finite. The case where the entries are not symmetrically distributed is still mysterious, despite recent progress by Péché and Soshnikov [166] who prove the universality of moments of order much larger than $\sqrt{N}$ (but still much smaller than $N^{\frac{2}{3}}$). A rather different result was proved by Johansson [121]; he showed the universality of the fluctuations of the eigenvalues in the bulk for matrices whose entries are the convolution of a Gaussian law with a law with finite six moments. Similar results are expected to hold for the largest eigenvalues. It is well known [15] that the largest eigenvalue of a Wigner matrix converges to 2 if and only if the entries have fourth moments. It is expected that the fluctuations follow the Tracy–Widom law when the fourth moment is finite. What happens when the entries have less finite moments is described in [9, 184]. Also, the case where one adds a finite rank perturbation to the matrix was studied in [16]; if the perturbation is sufficiently small the fluctuations still follows the Tracy–Widom law, whereas if it is large, they will be Gaussian.

Other classical ensembles were studied; for instance Wishart matrices [14, 27, 183, 190].

In the next chapter, we shall consider polynomials in several random matrices; it was shown in [111] that the spectral radius of polynomials in several independent matrices following the GUE converge to the expected limit (that is the edge of the support of the limiting spectral measure of this polynomial). This was generalized to the case of matrices interacting via a convex potential in [106].

3

Words in Several Independent Wigner Matrices

In this chapter, we consider m independent Wigner $N \times N$ matrices $\{\mathbf{A}^{N,\ell}, 1 \leq \ell \leq m\}$ with real or complex entries. That is, the $\mathbf{A}^{N,\ell}$ are self-adjoint random matrices with independent entries $\left(A_{ij}^{N,\ell}, 1 \leq i \leq j \leq N\right)$ above the diagonal that are centered and with variance one. Moreover, the $\left(A_{ij}^{N,\ell}, 1 \leq i \leq j \leq N\right)_{1 \leq \ell \leq m}$ are independent. We shall generalize Theorem 3.3 to the case where one considers words in several matrices, that is show that $N^{-1}\mathrm{Tr}\left(\mathbf{A}^{N,\ell_1}\mathbf{A}^{N,\ell_2}\cdots\mathbf{A}^{N,\ell_k}\right)$ converges for all choices of $\ell_i \in \{1, \ldots, m\}$ and give a combinatorial interpretation of the limit. In Part VI, we describe the non-commutative framework proposed by D. Voiculescu to see the limit in the more natural framework of free probability. Here, we simply generalize Theorem 1.13 as a first step towards Part III. Let us first describe the combinatorial objects that we shall need.

3.1 Partitions of Colored Elements and Stars

Because we now have m different matrices, the partitions that will naturally show up are partitions of elements with m different colors. In the following, each $\ell \in \{1, \ldots, m\}$ will be assigned to a different color. Also, because matrices do not commute, the order of the elements is important. This leads us to the following definition.

Definition 3.1. *Let* $q(X_1, \ldots, X_m) = X_{\ell_1}X_{\ell_2}\cdots X_{\ell_k}$ *be a monomial in* m *non-commutative indeterminates.*

We define the set $S(q)$ *associated with the monomial* q *as the set of* k *colored points on the real line so that the first point has color* ℓ_1, *the second one has color* ℓ_2, *till the last one that has color* ℓ_k.

$NP(q)$ *is the set of non-crossing pair partitions of* $S(q)$ *such that two points of* $S(q)$ *cannot be in the same block if they have different colors.*

Note that S defines a bijection between non-commutative monomials and the set of colored points on the real line.

A. Guionnet, *Large Random Matrices: Lectures on Macroscopic Asymptotics*, Lecture Notes in Mathematics 1957, DOI: 10.1007/978-3-540-69897-5_3, © 2009 Springer-Verlag Berlin Heidelberg, Reprint by Springer-Verlag Berlin Heidelberg 2012

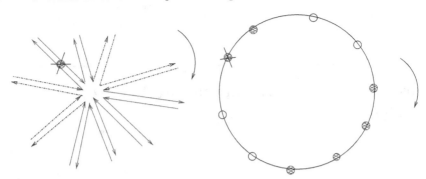

Fig. 3.1. The star of type $q(X) = X_1^2 X_2^2 X_1^4 X_2^2$

Even though the language of non-crossing partitions is very much adapted to generalization in free probability (see the last part of these notes) where partitions can eventually be not pair partitions, it seems to us that it is more natural to consider the bijective point of view of stars when considering matrix models in Part III. The definition we give below is equivalent to the above definition according to Figure 1.4 (with colors).

Definition 3.2. *Let* $q(X_1, \ldots, X_m) = X_{\ell_1} X_{\ell_2} \cdots X_{\ell_k}$ *be a monomial in* m *non-commutative indeterminates.*

We define a star of type q *as a vertex equipped with* k *colored half-edges, one marked half-edge and an orientation such that the marked half-edge is of color* ℓ_1, *the second (following the orientation) is of color* ℓ_2, *etc., until the last half-edge that is of color* ℓ_k.

$PM(q)$ is the set of planar maps (see Definition 1.8) with one star of type q such that the half-edges can be glued only if they have the same color.

Equivalently, a star can be represented by an annulus with an orientation, colored dots and a marked dot (see Figure 3.1; color 1 is blue and color 2 is dashed).

Remark 2. Planar maps with one colored star are also in bijection with trees with colored edges. However, when we deal with planar maps with several stars (see, e.g., Part III), the language of trees will become less transparent and we will no longer use it.

3.2 Voiculescu's Theorem

The aim of this chapter is to prove the following:

Theorem 3.3 (Voiculescu [197]). *Assume that for all* $k \in \mathbb{N}$,

$$B_k := \sup_{1 \leq \ell \leq m} \sup_{N \in \mathbb{N}} \sup_{ij \in \{1, \ldots, N\}^2} \mathbb{E}[|\sqrt{N} A_{ij}^{N, \ell}|^k] < \infty \tag{3.1}$$

and

$$\max_{1\le i,j\le N} |\mathbb{E}[A_{ij}^{N,\ell}]| = 0, \quad \lim_{N\to\infty} \max_{\ell} \frac{1}{N^2} \sum_{1\le i,j\le N} |N\mathbb{E}[|A_{ij}^{N,\ell}|^2] - 1| = 0.$$

Then, for any $\ell_j \in \{1,\dots,m\}, 1 \le j \le k$,

$$\lim_{N\to\infty} \frac{1}{N}\mathrm{Tr}\left(\mathbf{A}^{N,\ell_1}\mathbf{A}^{N,\ell_2}\cdots\mathbf{A}^{N,\ell_k}\right) = \sigma^m(X_{\ell_1}\cdots X_{\ell_k})$$

where the convergence holds in expectation and almost surely. $\sigma^m(X_{\ell_1}\cdots X_{\ell_k})$ is the number $|NP(X_{\ell_1}\cdots X_{\ell_k})| = |PM(X_{\ell_1}\cdots X_{\ell_k})|$ of planar maps with one star of type $X_{\ell_1}\cdots X_{\ell_k}$.

Remark 3. • Because a star has a marked edge and an orientation, each edge can equivalently be labeled. The counting is therefore performed for these labeled objects, regardless of possible symmetries.

• σ^m, once extended by linearity to all polynomials, is called the law of m free semi-circular variables since they satisfy the freeness property (17.1) and the moments of each variable are given by the moments of the semicircle law.

Proof. The proof is very close to that of Theorem 1.13.

1. *Expanding the expectation.*
   Setting $\mathbf{B}^N = \sqrt{N}\mathbf{A}^N$, we have

$$\mathbb{E}\left[\frac{1}{N}\mathrm{Tr}\left(\mathbf{A}^{N,\ell_1}\mathbf{A}^{N,\ell_2}\cdots\mathbf{A}^{N,\ell_k}\right)\right]$$

$$= \frac{1}{N^{\frac{k}{2}+1}} \sum_{i_1,\dots,i_k=1}^{N} \mathbb{E}[B_{i_1 i_2}^{\ell_1} B_{i_2 i_3}^{\ell_2}\cdots B_{i_k i_1}^{\ell_k}] \tag{3.2}$$

where $B_{ij}^\ell, 1 \le i,j \le N$ denotes the entries of $\mathbf{B}^{N,\ell}$ (which may possibly depend on N). We denote by $\mathbf{i} = (i_1,\dots,i_k)$ and set

$$P(\mathbf{i},\ell) = \mathbb{E}[B_{i_1 i_2}^{\ell_1} B_{i_2 i_3}^{\ell_2}\cdots B_{i_k i_1}^{\ell_k}].$$

By hypothesis, $P(\mathbf{i},\ell)$ is uniformly bounded by B_k. We let, as in the proof of Theorem 1.13, $V(\mathbf{i}) = \{i_1,\dots,i_k\}$ be the set of vertices, $E(\mathbf{i})$ the collection of the k half-edges $(e_p)_{p=1}^k = (i_p, i_{p+1})_{p=1}^k$ and consider the graph $G(\mathbf{i}) = (V(\mathbf{i}), E(\mathbf{i}))$. $G(\mathbf{i})$ is, as before, a rooted, oriented and connected graph.

$P(\mathbf{i},\ell)$ equals zero unless any edge has at least multiplicity two in $G(\mathbf{i})$. Therefore, by the same considerations as in the proof of Theorem 1.13, the indices that contribute to the first order in (3.2) are such that $G(\mathbf{i})$ is a rooted oriented tree. In particular, the limit equals zero if k is odd. This is equivalent to saying (see the bijection between trees and non-crossing partitions, Figure 1.3) that if we draw the points $i_1, \cdots i_k, i_1$ on the real line, we can draw a non-crossing pair-partition between the edges of $E(\mathbf{i})$. We write $E(\mathbf{i}) = \{(i_{s_l}, i_{s_{l+1}}); (i_{r_l}, i_{r_{l+1}})\}_{1 \leq l \leq \frac{k}{2}}$ for the corresponding partition. Since $G(\mathbf{i})$ is a tree we have again that $(i_{s_l}, i_{s_{l+1}}) = (i_{r_{l+1}}, i_{r_l})$ for $l \in \{1, \ldots, \frac{k}{2}\}$. Thus,

$$P(\mathbf{i}, \boldsymbol{\ell}) = \prod_{l=1}^{\frac{k}{2}} \mathbb{E}[B_{i_{s_l} i_{s_{l+1}}}^{N, \ell_{s_l}} B_{i_{r_l} i_{r_{l+1}}}^{N, \ell_{r_l}}].$$

By our hypothesis, we can replace $\mathbb{E}[B_{i_{s_l} i_{s_{l+1}}}^{N, \ell_{s_l}} B_{i_{r_l} i_{r_{l+1}}}^{N, \ell_{r_l}}]$ by $1_{\ell_{s_l} = \ell_{r_l}}$ up to a small error in the sum of the $P(\mathbf{i}, \boldsymbol{\ell})$'s. Therefore, if the edges of $E(\mathbf{i})$ are colored according to which matrix they came from, the only contributing indices will come from a non-crossing pair-partition where only edges of the same color can belong to the same block. This proves that

$$\mathbb{E}\left[\frac{1}{N} \text{Tr}\left(\mathbf{A}^{N, \ell_1} \mathbf{A}^{N, \ell_2} \cdots \mathbf{A}^{N, \ell_k}\right)\right] = |\sigma^m(X_{\ell_1} \cdots X_{\ell_k})| + o(1).$$

2. *Almost sure convergence.* To prove the almost sure convergence, we estimate the variance and then use the Borel–Cantelli lemma. The variance is given by

$$\text{Var}(\mathbf{A}^{N, \ell_1} \mathbf{A}^{N, \ell_2} \cdots \mathbf{A}^{N, \ell_k}) := \mathbb{E}\left[\frac{1}{N^2} \text{Tr}\left(\mathbf{A}^{N, \ell_1} \mathbf{A}^{N, \ell_2} \cdots \mathbf{A}^{N, \ell_k}\right)^2\right]$$

$$- \mathbb{E}\left[\frac{1}{N} \text{Tr}\left(\mathbf{A}^{N, \ell_1} \mathbf{A}^{N, \ell_2} \cdots \mathbf{A}^{N, \ell_k}\right)\right]^2$$

$$= \frac{1}{N^{2+k}} \sum_{\substack{i_1, \ldots, i_k = 1 \\ i'_1, \ldots, i'_k = 1}}^{N} [P(\mathbf{i}, \mathbf{i}') - P(\mathbf{i})P(\mathbf{i}')]$$

with $P(\mathbf{i})$ as before and

$$P(\mathbf{i}, \mathbf{i}') := \mathbb{E}[B_{i_1 i_2}^{\ell_1} B_{i_2 i_3}^{\ell_2} \cdots B_{i_k i_1}^{\ell_k} B_{i'_1 i'_2}^{\ell_1} \cdots B_{i'_k i'_1}^{\ell_k}].$$

We denote by $G(\mathbf{i}, \mathbf{i}')$ the graph with vertices $V(\mathbf{i}, \mathbf{i}')$ given by $\{i_1, \ldots, i_k, i'_1, \ldots, i'_k\}$ and edges $E(\mathbf{i}, \mathbf{i}')$ equal to $\{(i_p, i_{p+1})_{1 \leq p \leq k}, (i'_p, i'_{p+1})_{1 \leq p \leq k}\}$. For $\mathbf{i}, \mathbf{i}'$ to contribute to the sum, $G(\mathbf{i}, \mathbf{i}')$ must be connected (otherwise $P(\mathbf{i}, \mathbf{i}') = P(\mathbf{i})P(\mathbf{i}')$), and so it is an oriented rooted connected graph. Again, each edge must appear twice and the walk on $G(\mathbf{i})$ begins and finishes at the root i_1. Therefore, exactly the same arguments that we used in the proof of Theorem 1.13 show that

$$|V(\mathbf{i}, \mathbf{i}')| \leq k.$$

By boundedness of $P(\mathbf{i}, \mathbf{i}') - P(\mathbf{i})P(\mathbf{i}')$ we conclude that

$$\mathrm{Var}(\mathbf{A}^{N,\ell_1}\mathbf{A}^{N,\ell_2}\cdots\mathbf{A}^{N,\ell_k}) \leq D_k N^{-2}$$

for some finite constant D_k. The proof is thus complete by a further use of the Borel–Cantelli lemma.

$\square$

Exercise 3.4. *The next exercise concerns a special case of what is called "Asymptotic freeness" and was proved in greater generality by D. Voiculescu (see Theorem 17.5).*

Let $(A_{ij}^N, 1 \leq i \leq j \leq N)$ be independent real variables and consider $\mathbf{A}^N$ the self-adjoint matrix with these entries. Assume

$$\mathbb{E}[A_{ij}^N] = 0 \qquad \mathbb{E}[(\sqrt{N}A_{ij}^N)^2] = 1 \quad \forall i \leq j.$$

Assume that for all $k \in \mathbb{N}$,

$$B_k = \sup_{N \in \mathbb{N}} \sup_{ij \in \{1,\dots,N\}^2} \mathbb{E}[|\sqrt{N}A_{ij}^N|^k] < \infty.$$

Let D^N be a deterministic diagonal matrix such that

$$\sup_{N \in \mathbb{N}} \max_{i \leq j} |D_{ii}^N| < \infty \qquad \lim_{N \to \infty} \frac{1}{N}\mathrm{Tr}((D^N)^k) = m_k \text{ for all } k \in \mathbb{N}.$$

Show that:

1. *for any $k \in \mathbb{N}$,*

$$\lim_{N \to \infty} \mathbb{E}\left[\frac{1}{N}\mathrm{Tr}(D^N(\mathbf{A}^N)^k)\right] = C_{k/2}m_1,$$

2. *for any $k_1, k_2 \in \mathbb{N}$,*

$$\lim_{N \to \infty} \mathbb{E}\left[\frac{1}{N}\mathrm{Tr}((D^N)^{l_1}(\mathbf{A}^N)^{k_1}(D^N)^{l_2}(\mathbf{A}^N)^{k_2})\right]$$
$$= C_{k_1/2}C_{k_2/2}m_{l_1+l_2} + C_{(k_1+k_2)/2}m_{l_1}m_{l_2},$$

3. *for any $l_1, k_1, \cdots, l_p, k_p \in \mathbb{N}$,*

$$\lim_{N \to \infty} \mathbb{E}\left[\left(\frac{1}{N}\mathrm{Tr}((D^N)^{l_1}) - \frac{1}{N}\mathrm{Tr}(D^N)^{l_1}\right)\left((\mathbf{A}^N)^{k_1} - \mathbb{E}\left[\frac{1}{N}\mathrm{Tr}(\mathbf{A}^N)^{k_1}\right]\right)\right.$$
$$\left.\cdots\left((D^N)^{l_p} - \frac{1}{N}\mathrm{Tr}(D^N)^{l_p}\right)\left((\mathbf{A}^N)^{k_p} - \mathbb{E}\left[\frac{1}{N}\mathrm{Tr}(\mathbf{A}^N)^{k_p}\right]\right)\right]$$

goes to zero as N goes to infinity for any integer numbers $l_1, \dots, l_p$, $k_1, \dots, k_p$.

Hint: Expand the trace in terms of a weighted sum over the indices and show that the main contribution comes from indices whose associated graph is a tree. Fixing the tree, average out the quantities in the D^N and conclude (be careful that the D^N's can come with the same indices but show then that the main contribution comes from independent entries of the $(A^N)_{ii}^k$'s because of the tree structure).

In [6, 84], the previous exercise is generalized to prove the convergence of any words in $\{\mathbf{A}_1^N, \ldots, \mathbf{A}_m^N\}$ and $\{D_1^N, \ldots, D_m^N\}$ when the trace of words in the deterministic matrices $\{D_1^N, \ldots, D_m^N\}$ are assumed to converge and $\{\mathbf{A}_1^N, \ldots, \mathbf{A}_m^N\}$ are independent Wigner matrices. This can also be deduced from Theorem 17.5 in the case of complex Gaussian matrices (by using their invariance under multiplication by unitary matrices).

Bibliographical Notes. After the seminal article [197] of Voiculescu, Theorem 3.3 was generalized to non-Gaussian entries by Dykema [84]. Generalizations of Exercise 3.4 are given in [6].

Part II

Wigner Matrices and Concentration Inequalities

In the last twenty years, concentration inequalities have developed into a very powerful tool in probability theory. They provide a general framework to control the probability of deviations of smooth functions of random variables from their mean or their median. We begin this section by providing some general framework where concentration inequalities are known to hold. We first consider the case where the underlying measure satisfies a log-Sobolev inequality; we show how to prove this inequality in a simple situation and then how it implies concentration inequalities. We then review a few other situations where concentration inequalities hold. To apply these techniques to random matrices, we show that certain functions of the eigenvalues of matrices, such as $\int f(x)dL_{\mathbf{A}^N}(x)$ with f Lipschitz, are smooth functions of the entries of the matrix $\mathbf{A}^N$ so that concentration inequalities hold as soon as the joint law of the entries satisfies one of the conditions seen in the first two chapters of this part. Another useful *a priori* control is provided by Brascamp–Lieb inequalities; we shall apply them to the setting of random matrices at the end of this part.

To motivate the reader, let us state the type of result we want to obtain in this part.

To this end, we introduce some extra notations. Let us recall that if X is a symmetric (resp. Hermitian) matrix and f is a bounded measurable function, $f(X)$ is defined as the matrix with the same eigenvectors than X but with eigenvalues that are the image by f of those of X; namely, if e is an eigenvector of X with eigenvalue λ, $Xe = \lambda e$, $f(X)e := f(\lambda)e$. In terms of the spectral decomposition $X = UDU^*$ with U orthogonal (resp. unitary) and D diagonal real, one has $f(X) = Uf(D)U^*$ with $f(D)_{ii} = f(D_{ii})$. For $M \in \mathbb{N}$, we denote by $\langle \cdot, \cdot \rangle$ the Euclidean scalar product on $\mathbb{R}^M$ (resp. $\mathbb{C}^M$), $\langle x, y \rangle = \sum_{i=1}^M x_i y_i$ ($\langle x, y \rangle := \sum_{i=1}^M x_i y_i^*$), and by $|| \cdot ||_2$ the associated norm $||x||_2^2 := \langle x, x \rangle$.

Throughout this section, we denote the Lipschitz constant of a function $G : \mathbb{R}^M \to \mathbb{R}$ by

$$|G|_{\mathcal{L}} := \sup_{x \neq y \in \mathbb{R}^M} \frac{|G(x) - G(y)|}{\|x - y\|_2},$$

and call G a *Lipschitz function* if $|G|_{\mathcal{L}} < \infty$.

Lemma II.1. *Let $g : \mathbb{R}^N \to \mathbb{R}$ be Lipschitz with Lipschitz constant $|g|_{\mathcal{L}}$. Then, with $\mathbf{A}^N$ denoting the Hermitian (or symmetric) matrix with entries $(A_{ij}^N)_{1 \leq i,j \leq N}$, the map $\{A_{ij}^N\}_{1 \leq i \leq j \leq N} \mapsto \mathrm{Tr}(g(\mathbf{A}^N))$ is a Lipschitz function with constant $\sqrt{N}|g|_{\mathcal{L}}$. Therefore, if the joint law of $(A_{ij}^N)_{1 \leq i \leq j \leq N}$ is "good", there exists $\alpha > 0$, constants $c > 0$ and $C < \infty$ so that for all $N \in \mathbb{N}$*

$$\mathbb{P}\left(\left|\mathrm{Tr}(g(\mathbf{A}^N)) - \mathbb{E}[\mathrm{Tr}(g(\mathbf{A}^N))]\right| > \delta|g|_{\mathcal{L}}\right) \leq C e^{-c|\delta|^{\alpha}}.$$

"Good" here means for instance that the law satisfies a log-Sobolev inequality; an example is when the $\{A_{ij}^N\}_{1 \leq i \leq j \leq N}$ are independent Gaussian variables with uniformly bounded covariance (see Theorem 6.6).

The interest of results such as Lemma II.1 is that they provide bounds on deviations that do not depend on the dimension. They can be used to show laws of large numbers (reducing the proof of the almost sure convergence to the prove of the convergence in expectation) or to ease the proof of central limit theorems (indeed, when $\alpha = 2$ in Lemma II.1, $\mathrm{Tr}(g(\mathbf{A}^N)) - \mathbb{E}[\mathrm{Tr}(g(\mathbf{A}^N))]$ has a sub-Gaussian tail, providing tightness arguments for free).

We shall recall below the elements of the theory of concentration we shall need. In fact, we will mostly use concentration inequalities related to log-Sobolev inequalities; we shall therefore provide details on this point and give full proofs. We will then review other classical settings where concentration inequalities are known to apply. Finally, we will apply this theory to random matrices and provide for instance sufficient hypotheses so that Lemma II.1 holds.

4

Concentration Inequalities and Logarithmic Sobolev Inequalities

We first derive concentration inequalities based on the logarithmic Sobolev inequality and then give some generic and classical examples of laws that satisfy this inequality. Since we shall use it in these notes for Wigner's matrices, we focus first on concentration for laws in $\mathbb{R}^N$. We then briefly generalize the results to compact Riemannian manifolds in order to state concentration inequalities for probability measures on the orthogonal or unitary group.

4.1 Concentration Inequalities for Laws Satisfying Logarithmic Sobolev Inequalities

Throughout this section an integer number N will be fixed.

Definition 4.1. *A probability measure P on $\mathbb{R}^N$ is said to satisfy the logarithmic Sobolev inequality (LSI) with constant c if, for any differentiable function $f : \mathbb{R}^N \to \mathbb{R}$,*

$$\int f^2 \log \frac{f^2}{\int f^2 dP} dP \leq 2c \int \|\nabla f\|_2^2 dP. \tag{4.1}$$

Here, $\|\nabla f\|_2^2 = \sum_{i=1}^N (\partial_{x_i} f)^2$.

The interest in the logarithmic Sobolev inequality, in the context of concentration inequalities, lies in the following argument, that among other things, shows that LSI implies sub-Gaussian tails. This fact and a general study of logarithmic Sobolev inequalities may be found in [107], [171] or [138]. The Gaussian law, and any probability measure ν absolutely continuous with respect to the Lebesgue measure satisfying the Bobkov and Götze [38] condition (including $\nu(dx) = Z^{-1} e^{-|x|^\alpha} dx$ for $\alpha \geq 2$, where $Z = \int e^{-|x|^\alpha} dx$), as well as any distribution absolutely continuous with respect to such laws possessing a bounded above and below density, satisfies the LSI [138], [107, Property 4.6].

A. Guionnet, *Large Random Matrices: Lectures on Macroscopic Asymptotics*,   49
Lecture Notes in Mathematics 1957, DOI: 10.1007/978-3-540-69897-5_4,
© 2009 Springer-Verlag Berlin Heidelberg, Reprint by Springer-Verlag Berlin Heidelberg 2012

Lemma 4.2 (Herbst). *Assume that P satisfies the LSI on $\mathbb{R}^N$ with constant c. Let G be a Lipschitz function on $\mathbb{R}^N$, with Lipschitz constant $|G|_{\mathcal{L}}$. Then, for all $\lambda \in \mathbb{R}$, we have*

$$\int e^{\lambda(G - E_P(G))} dP \le e^{c\lambda^2 |G|_{\mathcal{L}}^2/2}, \tag{4.2}$$

and so for all $\delta > 0$

$$P\left(|G - E_P(G)| \ge \delta\right) \le 2e^{-\delta^2/2c|G|_{\mathcal{L}}^2}. \tag{4.3}$$

Note that Lemma 4.2 also implies that $E_P G$ is finite.

Proof of Lemma 4.2. We denote by E_P the expectation $E_P[f] = \int f dP$. Note first that (4.3) follows from (4.2). Indeed, by Chebychev's inequality, for any $\lambda > 0$,

$$\begin{aligned}
P\left(|G - E_P G| \ge \delta\right) &\le e^{-\lambda\delta} E_P[e^{\lambda|G - E_P G|}] \\
&\le e^{-\lambda\delta}(E_P[e^{\lambda(G - E_P G)}] + E_P[e^{-\lambda(G - E_P G)}]) \\
&\le 2e^{-\lambda\delta} e^{c|G|_{\mathcal{L}}^2 \lambda^2/2}.
\end{aligned}$$

Optimizing with respect to λ (by taking $\lambda = \delta/c|G|_{\mathcal{L}}^2$) yields the bound (4.3).

Turning to the proof of (4.2), let us first assume that G is a bounded differentiable function such that

$$|| \, ||\nabla G||_2^2||_\infty := \sup_{x \in \mathbb{R}^N} \sum_{i=1}^{N} (\partial_{x_i} G(x))^2 < \infty.$$

Define

$$X_\lambda = \log E_P e^{2\lambda(G - E_P G)}.$$

Then, taking $f = e^{\lambda(G - E_P G)}$ in (4.1), some algebra reveals that for $\lambda > 0$,

$$\frac{d}{d\lambda}\left(\frac{X_\lambda}{\lambda}\right) \le 2c|| \, ||\nabla G||_2^2||_\infty.$$

Now, because $G - E_P(G)$ is centered,

$$\lim_{\lambda \to 0^+} \frac{X_\lambda}{\lambda} = 0$$

and hence integrating with respect to λ yields

$$X_\lambda \le 2c|| \, ||\nabla G||_2^2||_\infty \lambda^2,$$

first for $\lambda \ge 0$ and then for any $\lambda \in \mathbb{R}$ by considering the function $-G$ instead of G. This completes the proof of (4.2) in the case where G is bounded and differentiable.

Let us now assume only that G is Lipschitz with $|G|_{\mathcal{L}} < \infty$. For $\epsilon > 0$, define $\bar{G}_\epsilon = G \wedge (-1/\epsilon) \vee (1/\epsilon)$, and note that $|\bar{G}_\epsilon|_{\mathcal{L}} \leq |G|_{\mathcal{L}} < \infty$. Consider the regularization $G_\epsilon(x) = p_\epsilon * \bar{G}_\epsilon(x) = \int \bar{G}_\epsilon(y) p_\epsilon(x-y) dy$ with the Gaussian density $p_\epsilon(x) = e^{-|x|^2/2\epsilon} dx / \sqrt{(2\pi\epsilon)^N}$ such that $p_\epsilon(x) dx$ converges weakly to the atomic measure δ_0 as ϵ converges to 0. Since for any $x \in \mathbb{R}^N$,

$$|G_\epsilon(x) - \bar{G}_\epsilon(x)| \leq |G|_{\mathcal{L}} \int \|y\|_2 p_\epsilon(y) dy = |G|_{\mathcal{L}} \sqrt{\epsilon N},$$

G_ϵ converges pointwise to G. G_ϵ is also continuously differentiable and

$$
\begin{aligned}
\||\nabla G_\epsilon\|_2^2\|_\infty &= \sup_{x \in \mathbb{R}^M} \sup_{u \in \mathbb{R}^M} \{2\langle \nabla G_\epsilon(x), u \rangle - \|u\|_2^2\} \\
&\leq \sup_{u,x \in \mathbb{R}^M} \sup_{\delta > 0} \{2\delta^{-1}(G_\epsilon(x + \delta u) - G_\epsilon(x)) - \|u\|_2^2\} \\
&\leq \sup_{u \in \mathbb{R}^M} \{2|G|_{\mathcal{L}}\|u\|_2 - \|u\|_2^2\} = |G|_{\mathcal{L}}^2.
\end{aligned}
\tag{4.4}
$$

Thus, we can apply the previous result to find that for any $\epsilon > 0$ and all $\lambda \in \mathbb{R}$

$$E_P[e^{\lambda G_\epsilon}] \leq e^{\lambda E_P G_\epsilon} e^{c\lambda^2 |G|_{\mathcal{L}}^2 / 2}. \tag{4.5}$$

Therefore, by Fatou's lemma,

$$E_P[e^{\lambda G}] \leq e^{\liminf_{\epsilon \to 0} \lambda E_P G_\epsilon} e^{c\lambda^2 |G|_{\mathcal{L}}^2 / 2}. \tag{4.6}$$

We next show that $\lim_{\epsilon \to 0} E_P G_\epsilon = E_P G$, which, in conjunction with (4.4), will conclude the proof. Indeed, (4.5) implies that

$$P\left(|G_\epsilon - E_P G_\epsilon| > \delta\right) \leq 2e^{-\delta^2/2c|G|_{\mathcal{L}}^2}. \tag{4.7}$$

Consequently,

$$
\begin{aligned}
E[(G_\epsilon - E_P G_\epsilon)^2] &= 2 \int_0^\infty x P\left(|G_\epsilon - E_P G_\epsilon| > x\right) dx \leq 4 \int_0^\infty x e^{-\frac{x^2}{2c|G|_{\mathcal{L}}^2}} dx \\
&= 4c|G|_{\mathcal{L}}^2
\end{aligned}
\tag{4.8}
$$

so that the sequence $(G_\epsilon - E_P G_\epsilon)_{\epsilon \geq 0}$ is uniformly integrable. Now, G_ϵ converges pointwise to G and therefore there exists a constant K, independent of ϵ, such that for $\epsilon < \epsilon_0$, $P(|G_\epsilon| \leq K) \geq \frac{3}{4}$. On the other hand, (4.7) implies that $P(|G_\epsilon - E_P G_\epsilon| \leq r) \geq \frac{3}{4}$ for some r independent of ϵ. Thus,

$$\{|G_\epsilon - E_P G_\epsilon| \leq r\} \cap \{|G_\epsilon| \leq K\} \subset \{|E_P G_\epsilon| \leq K + r\}$$

is not empty, providing a uniform bound on $(E_P G_\epsilon)_{\epsilon < \epsilon_0}$. We thus deduce from (4.8) that $\sup_{\epsilon < \epsilon_0} E_P G_\epsilon^2$ is finite, and hence $(G_\epsilon)_{\epsilon < \epsilon_0}$ is uniformly integrable. In particular,

$$\lim_{\epsilon \to 0} E_P G_\epsilon = E_P G < \infty,$$

which finishes the proof. $\square$

4.2 A Few Laws Satisfying a Log-Sobolev Inequality

In the sequel, we shall be interested in laws of variables that are either independent or in interaction via a potential. We shall give sufficient conditions to ensure that a log-Sobolev inequality is satisfied.

- *Laws of independent variables.*
  One of the most important properties of the log-Sobolev inequality is the product property:

Lemma 4.3. *Let $(\mu_i)_{i=1,2}$ be two probability measures on $\mathbb{R}^N$ and $\mathbb{R}^M$, respectively, satisfying the logarithmic Sobolev inequalities with coefficients $(c_i)_{i=1,2}$. Then, the product probability measure $\mu_1 \otimes \mu_2$ on $\mathbb{R}^{M+N}$ satisfies the logarithmic Sobolev inequality with coefficient $\max(c_1, c_2)$.*
Consequently, if μ is a probability measure on $\mathbb{R}^M$ satisfying a logarithmic Sobolev inequality with a coefficient $c < \infty$, then the product probability measure $\mu^{\otimes n}$ satisfies the logarithmic Sobolev inequality with the same coefficient c for any integer n.

Proof. Let f be a continuously differentiable function on $\mathbb{R}^N \times \mathbb{R}^M$. Then, using the logarithmic Sobolev inequality under the probability measure μ_1 applied to $f(., x_2)$ and under μ_2 applied to $\mu_1(f^2)(.) = \int f^2(x_1, .) d\mu(x_1)$, we obtain

$$
\mu_1 \otimes \mu_2 \left(f^2 \log \frac{f^2}{\mu_1 \otimes \mu_2(f^2)} \right)
$$

$$
= \mu_2 \left(\mu_1(f^2 \log \frac{f^2}{\mu_1(f^2)}) + \mu_1(f^2) \log \frac{\mu_1(f^2)}{\mu_1 \otimes \mu_2(f^2)} \right)
$$

$$
\leq \mu_2 \left(2c_1 \mu_1 [\|\nabla_{x_1} f\|_2^2] \right) + 2c_2 \mu_2 \left(\|\nabla_{x_2} \sqrt{\mu_1(f^2)}\|_2^2 \right)
$$

$$
\leq \mu_2 \otimes \mu_1 \left(2c_1 \|\nabla_{x_1} f\|_2^2 + 2c_2 \|\nabla_{x_2} f\|_2^2 \right) \leq 2\max(c_1, c_2) \mu_2 \otimes \mu_1 \left(\|\nabla f\|_2^2 \right)
$$

where we have used in the last line that $\|\nabla f\|_2^2 = \|\nabla_{x_1} f\|_2^2 + \|\nabla_{x_2} f\|_2^2$ and

$$
\|\nabla_{x_2} \sqrt{\mu_1(f^2)}\|_2^2 = \sum_{i=1}^{M} (\partial_{x_2^i} (\int f(x_1, x_2)^2 d\mu_1(x_1))^{\frac{1}{2}})^2
$$

$$
= \sum_{i=1}^{M} \left(\frac{\int f(x_1, x_2) \partial_{x_2^i} f(x_1, x_2) d\mu_1(x_1)}{(\int f(x_1, x_2)^2 d\mu_1(x_1))^{\frac{1}{2}}} \right)^2
$$

$$
\leq \sum_{i=1}^{M} \int (\partial_{x_2^i} f(x_1, x_2))^2 d\mu_1(x_1)
$$

by the Cauchy–Schwarz inequality. $\qquad\square$

- *Log-Sobolev inequalities for variables in strictly convex interaction.*
Below, we follow below [8, chapter 5] and [107, chapter 4], which we recommend for more details. We show that a log-Sobolev inequality holds if the so-called Bakry–Emery condition is satisfied. We then give sufficient conditions for the latter to be true. Let dx denote the Lebesgue measure on $\mathbb{R}$ and Φ be a smooth function (at least twice continuously differentiable) from $\mathbb{R}^N$ into $\mathbb{R}$ going to infinity fast enough so that the probability measure

$$\mu_\Phi(dx) := \frac{1}{Z} e^{-\Phi(x_1,\dots,x_N)} dx_1 \cdots dx_N$$

is well defined. We consider the operator on the set $C_b^2(\mathbb{R}^N)$ of twice continuously differentiable functions defined by

$$\mathcal{L}_\Phi = \Delta - \nabla\Phi.\nabla = \sum_{i=1}^{N}(\partial_i^2 - \partial_i\Phi\partial_i).$$

Here and below, we shall write for short $\partial_i = \partial_{x_i}$ for $i \in \{1,\dots,N\}$. By integration by parts, one sees that $\mathcal{L}_\Phi$ is symmetric in $L^2(\mu_\Phi)$, i.e., for any functions $f, g \in C_b^2(\mathbb{R}^N)$,

$$\mu_\Phi\left(f\mathcal{L}_\Phi g\right) = \mu_\Phi\left(g\mathcal{L}_\Phi f\right).$$

By the Hille–Yoshida theorem (see, e.g., [107, Chapter 1]), we can associate to the operator $\mathcal{L}_\Phi$ a Markov contractive semi-group $(P_t)_{t\geq 0}$, i.e. a family of linear operators on $C_b^0(\mathbb{R}^N)$ such that $P_t : C_b^0(\mathbb{R}^N) \to C_b^0(\mathbb{R}^N)$ satisfies:
(1) $P_0 f = f$ for all $f \in C_b^0(\mathbb{R}^N)$.
(2) The map $t \to P_t$ is continuous in the sense that for all $f \in C_b^0(\mathbb{R}^N)$, $t \to P_t f$ is a continuous map from $\mathbb{R}^+$ into $C_b^0(\mathbb{R}^N)$.
(3) For any $f \in C_b^0(\mathbb{R}^N)$ and $(t, s) \in (\mathbb{R}^+)^2$,

$$P_{t+s}f = P_t P_s f.$$

(4) $P_t 1 = 1$ for all $t \geq 0$.
(5) P_t preserves positivity, i.e., for any $f \geq 0$, $P_t f \geq 0$. In particular, by (4), for all $t \geq 0$,

$$\|P_t f\|_\infty \leq \|f\|_\infty$$

(6)

$$\mathcal{L}_\Phi(f) = \lim_{t\downarrow 0} t^{-1}(P_t f - f)$$

for any function f for which this limit exists.

Exercise 4.4. *Let Φ be a twice continuously differentiable function, with uniformly bounded second derivatives. Then, by Theorem 20.16, there exists a unique solution to the stochastic differential equation*

$$dx_t^i = dB_t^i - \partial_i\Phi(x_t)dt$$

such that $x_0^i = z^i$ for $1 \leq i \leq N$. Denote by $x^{i,z}$ this solution.

1. *Show that the law P_t^z of x_t^z obeys*

$$\partial_t \mathbb{E}[f(x_t^z)] = \mathbb{E}[\mathcal{L}_\Phi f(x_t^z)].$$

   Hint: Use Itô's calculus.
2. *Let $P_t f(z) := \mathbb{E}[f(x_t^z)]$. Show that P_t satisfies conditions (1)–(6) above.*
3. *Assume Hess $\Phi(x) \geq (1/c)I$ for all x with some $c > 0$ and show that $P_t f(x) - \mu_\Phi(f)$ goes to zero exponentially fast for any C^1 function f. Hint: Write $d(x_t^z - x_t^y) = -(\nabla\Phi(x_t^z) - \nabla\Phi(x_t^y))dt$ and use that*

$$< \nabla\Phi(x_t^z) - \nabla\Phi(x_t^y), x_t^z - x_t^y > \geq \frac{1}{c}\|x_t^z - x_t^y\|_2^2$$

We define the operator "carré du champ" Γ_1 by

$$\Gamma_1(f, f) = \frac{1}{2}\left(\mathcal{L}_\Phi f^2 - 2f\mathcal{L}_\Phi f\right).$$

Simple algebra shows that $\Gamma_1(f, f) = \sum_{i=1}^N (\partial_i f)^2 = \|\nabla f\|_2^2$. We define $\Gamma_1(f, g)$ by bilinearity:

$$\Gamma_1(f, g) = \Gamma_1(g, f) = \frac{1}{2}\left(\Gamma_1(f + g, f + g) - \Gamma_1(f, f) - \Gamma_1(g, g)\right).$$

Note that because $\mathcal{L}_\Phi$ is symmetric in $L^2(\mu_\Phi)$, P_t is reversible in $L^2(\mu_\Phi)$, i.e.,

$$\mu_\Phi(fP_t g) = \mu_\Phi(gP_t f)$$

for any smooth functions f, g. In particular, since $P_t 1 = 1$, $\mu_\Phi P_t = \mu_\Phi$ and so μ_Φ is invariant under P_t. We expect that P_t is ergodic in the sense that for all $f \in C_b^0(\mathbb{R}^N)$,

$$\lim_{t\to\infty} \mu_\Phi(P_t f - \mu_\Phi f)^2 = 0. \tag{4.9}$$

We shall not prove this point in the most general context here but only when the Bakry–Emery condition holds, see (4.12).
Finally, let us introduce the 'carré du champ itéré'

$$\Gamma_2(f, f) = \frac{1}{2}\frac{d}{dt}\left(P_t(\Gamma_1(f, f)) - \Gamma_1(P_t f, P_t f)\right)|_{t=0}$$

$$= \frac{1}{2}\{\mathcal{L}_\Phi \Gamma_1(f, f) - 2\Gamma_1(f, \mathcal{L}_\Phi f)\}.$$

We define the Bakry–Emery condition as follows.

Definition 4.5. *We say that the Bakry–Emery condition (denoted (BE)) is satisfied if there exists a positive constant $c > 0$ such that*

$$\Gamma_2(f, f) \geq \frac{1}{c}\Gamma_1(f, f) \tag{4.10}$$

for any function f for which $\Gamma_1(f, f)$ and $\Gamma_2(f, f)$ are well defined.

233

In our case,

$$\Gamma_2(f,f) = \sum_{i,j=1}^{m} (\partial_i \partial_j f)^2 + \sum_{i,j=1}^{m} \partial_i f \mathrm{Hess}(\Phi)_{ij} \partial_j f$$

with $\mathrm{Hess}(\Phi)$ the Hessian of Φ; $\mathrm{Hess}(\Phi)_{ij} = \partial_i \partial_j \Phi$. Thus, (BE) is equivalent to $\mathrm{Hess}(\Phi)(x) \geq c^{-1} I$ (observe that the choice $f = \sum v_i x_i$ shows that (BE) implies the latter).

Theorem 4.6 (Bakry–Emery theorem). *Bakry–Emery condition implies that μ_Φ satisfies the logarithmic Sobolev inequality with constant c.*

Before going into the proof of this theorem, let us observe the following:

Corollary 4.7. *If for all $x \in \mathbb{R}^N$,*

$$\mathrm{Hess}(\Phi)(x) \geq \frac{1}{c} I$$

in the sense of the partial order on self-adjoint operators, then (BE) holds and μ_Φ satisfies the logarithmic Sobolev inequality with constant c.
In particular, if μ is the law of N independent Gaussian variables with variance bounded above by c, then μ satisfies the logarithmic Sobolev inequality with constant c.

Proof of Theorem 4.6. Let us first prove (4.9) when **(BE)** is satisfied. Let f be a continuously differentiable function such that $\|\nabla f\|_2$ is uniformly bounded. Fix $t > 0$ and consider, for $s \in [0, t]$, $\psi(s) = P_s \Gamma_1(P_{t-s} f, P_{t-s} f)$. We shall assume hereafter that $P_t f$ is sufficiently smooth so that $\Gamma_1(P_{t-s} f, P_{t-s} f)$ and $P_{t-s} f$ are in the domain of the generator. We refer to [6] or [171] for details about this assumption. Then, we find

$$\partial_s \psi(s) = 2 P_s \Gamma_2(P_{t-s} f, P_{t-s} f)$$
$$\geq \frac{2}{c} P_s \Gamma_1(P_{t-s} f, P_{t-s} f) = \frac{2}{c} \psi(s)$$

where we finally used **(BE)**. Thus, for all $t \geq 0$,

$$\Gamma_1(P_t f, P_t f) \leq e^{-\frac{2}{c} t} P_t \Gamma_1(f, f). \tag{4.11}$$

Since $\Gamma_1(f, f) = \|\nabla f\|_2^2$ is uniformly bounded, we deduce that $\Gamma_1(P_t f, P_t f) = \|\nabla P_t f\|_2^2$ goes to zero as t goes to infinity, ensuring that $P_t f$ converges almost surely to a constant. Indeed, for all x, y in $\mathbb{R}^N$, (4.11) implies that

$$|P_t f(x) - P_t f(y)| = \left| \int_0^1 \langle \nabla P_t f(\alpha x + (1-\alpha)y), (x-y) \rangle d\alpha \right|$$

$$\leq \max_{z \in \mathbb{R}^N} \|\nabla P_t f\|_2(z) \|x-y\|_2$$

$$\leq e^{-\frac{2}{c}t} \max_{z \in \mathbb{R}^N} P_t \|\nabla f\|_2(z) \|x-y\|_2$$

$$\leq e^{-\frac{2}{c}t} \|\|\nabla f\|_2\|_\infty \|x-y\|_2$$

where we used the fifth property of the Markov semi-group. Thus, for $f \in C_b^1(\mathbb{R}^N)$,

$$\lim_{t \to \infty} P_t f = \lim_{t \to \infty} \mu_\Phi(P_t f) = \mu_\Phi(f) \quad \text{a.s.} \tag{4.12}$$

The convergence also holds in $L^2(\mu_\Phi)$ since $P_t f$ is uniformly bounded by property (5) of Markov processes, yielding (4.9).

Let f be a positive bounded continuous function so that $\mu_\Phi f = 1$. We set $f_t = P_t f$ and let

$$S_f(t) = \mu_\Phi(f_t \log f_t).$$

Since f_t converges to $\mu_\Phi f$ and $f_t \log f_t$ is uniformly bounded, we have

$$\lim_{t \to \infty} S_f(t) = \mu_\Phi(f) \log \mu_\Phi(f) = 0.$$

Hence,

$$S_f(0) = -\int_0^\infty dt \frac{d}{dt} S_f(t) = \int_0^\infty dt \mu_\Phi \Gamma_1(f_t, \log f_t). \tag{4.13}$$

Next using the fact that P_t is symmetric together with the Cauchy–Schwarz inequality, we get

$$\mu_\Phi [\Gamma_1(f_t, \log f_t)] = \mu_\Phi [\Gamma_1(f, P_t(\log f_t))] \tag{4.14}$$

$$\leq \left(\mu_\Phi \frac{\Gamma_1(f, f)}{f} \right)^{\frac{1}{2}} (\mu_\Phi [f \Gamma_1(P_t \log f_t, P_t \log f_t)])^{\frac{1}{2}}.$$

Applying (4.11) to the function $\log f_t$, we obtain

$$(\mu_\Phi(f \Gamma_1(P_t \log f_t, P_t \log f_t)))^{\frac{1}{2}} \leq \left(\mu_\Phi(f e^{-\frac{2}{c}t} P_t \Gamma_1(\log f_t, \log f_t)) \right)^{\frac{1}{2}} \tag{4.15}$$

$$= e^{-\frac{1}{c}t} (\mu_\Phi(f_t \Gamma_1(\log f_t, \log f_t)))^{\frac{1}{2}}$$

$$= e^{-\frac{1}{c}t} (\mu_\Phi(\Gamma_1(f_t, \log f_t)))^{\frac{1}{2}}$$

where in the last stage we have used symmetry of the semigroup and the fact that $\Gamma_1(f, \log f) = f \Gamma_1(\log f, \log f)$. The inequalities (4.14) and (4.15) imply the following bound:

$$\mu_\Phi \Gamma_1(f_t, \log f_t) \le e^{-\frac{2}{c}t} \mu_\Phi \frac{\Gamma_1(f,f)}{f} = 4e^{-\frac{2}{c}t} \mu_\Phi \Gamma_1(f^{\frac{1}{2}}, f^{\frac{1}{2}}). \qquad (4.16)$$

Plugging this into (4.13), one arrives at

$$S_f(0) \le \int_0^\infty 4e^{-\frac{2t}{c}} dt \mu_\Phi \Gamma_1(f^{\frac{1}{2}}, f^{\frac{1}{2}})) = 2c\mu_\Phi \Gamma_1(f^{\frac{1}{2}}, f^{\frac{1}{2}})$$

which completes the proof. $\qquad\qquad\qquad\qquad\qquad\qquad\qquad\qquad\qquad$ $\square$

Bibliographical Notes. The reader interested in the theory of concentration inequalities and log-Sobolev inequalities can find more material for instance in the articles [8, 107, 136, 138, 140]. The Bakry–Emery condition was introduced in [18, 19].

5

Generalizations

5.1 Concentration Inequalities for Laws Satisfying Weaker Coercive Inequalities

Concentration inequalities under log-Sobolev inequalities are optimal in the sense that they provide a Gaussian tail for statistics that are expected to satisfy a central limit theorem. However, for that very same reason, laws satisfying a log-Sobolev inequality must have a sub-Gaussian tail. One way to weaken this hypothesis is to weaken both requirements and hypotheses, for instance to assume a weaker coercivity inequality such as a Poincaré inequality. In this section, we keep the notations of the previous section. Let us recall that a probability measure μ on $\mathbb{R}^M$ satisfies Poincaré's inequality with coefficient $m > 0$ iff for any test function $f \in \mathcal{C}_b^2(\mathbb{R}^M)$

$$\mu_\Phi(\Gamma_1(f, f)) \geq m\mu_\Phi[(f - \mu(f))^2].$$

Exercise 5.1. *Show that Poincaré's inequality satisfies a product property similar to the product property of LSI that we saw in Lemma 4.3.*

We have:

Lemma 5.2. *[1, Theorem 2.5] Assume that μ_Φ satisfies Poincaré's inequality with constant m. Then, for any Lipschitz function f*

$$\mu_\Phi\left(\exp\{\sqrt{2m}\frac{f - \mu_\Phi(f)}{|f|_\mathcal{L}}\}\right) \leq K$$

with $K = 2\prod_1^\infty(1 - 4^{-m})^{-2^m}$. As a consequence, for all $\delta > 0$,

$$\mu_\Phi(|f - \mu_\Phi(f)| \geq \delta) \leq 2Ke^{-\sqrt{\frac{m}{2}}\frac{\delta}{|f|_\mathcal{L}}}.$$

Note that the lemma shows that measures satisfying Poincaré's inequality must have a sub-exponential tail.

A. Guionnet, *Large Random Matrices: Lectures on Macroscopic Asymptotics*, 59
Lecture Notes in Mathematics 1957, DOI: 10.1007/978-3-540-69897-5_5,
© 2009 Springer-Verlag Berlin Heidelberg, Reprint by Springer-Verlag Berlin Heidelberg 2012

Exercise 5.3. *Prove the above lemma by showing that for any* $\lambda > 0$, *any continuously differentiable function* f,

$$E[e^{\lambda f}] \leq E[e^{\frac{\lambda}{2}f}]^2 + \frac{|f|_{\mathcal{L}}^2 \lambda^2}{4m} E[e^{\lambda f}].$$

Exercise 5.4. *Show that a log-Sobolev inequality with coefficient* c *implies a spectral gap inequality with coefficient bounded below by* c^{-1}. *Hint: Put* $f = 1 + \epsilon g$ *in (LSI) and let* ϵ *go to zero.*

5.2 Concentration Inequalities by Talagrand's Method

Talagrand's concentration inequality does not require that the underlying measure satisfies a coercive inequality. It holds for the law of independent equidistributed uniformly bounded variables. The price to pay is that one needs to assume that the test function is convex and also to consider concentration with respect to the median rather than the mean.

Let us recall that the median M_Y of a random variable Y is defined as the largest real number such that $P(Y \leq x) \leq 2^{-1}$. Then, let us state the following easy consequence of a theorem due to Talagrand [189, Theorem 6.6].

Theorem 5.5 (Talagrand). *Let* K *be a connected compact subset of* $\mathbb{R}$ *with diameter* $|K| = \sup_{x,y \in K} |x - y|$. *Consider a convex real-valued function* f *defined on* K^N. *Assume that* f *is Lipschitz on* K^N, *with constant* $|f|_{\mathcal{L}}$. *Let* P *be a probability measure on* K *and* $X_1, \dots, X_N$ *be* N *independent copies with law* P. *Then, if* M_f *is the median of* $f(X_1, \dots, X_N)$, *for all* $\epsilon > 0$,

$$P\left(|f(X_1, \dots, X_N) - M_f| \geq \epsilon\right) \leq 4 e^{-\frac{\epsilon^2}{16|K|^2|f|_{\mathcal{L}}^2}}.$$

Theorem 6.6 of [189] deals with the case where $K \subset [-1, 1]$ (which easily generalizes in the above statement by rescaling) and functions f that can be Lipschitz only in a subset of K^N (in which case the above statement has to be corrected by the probability that $(X_1, \dots, X_N)$ belongs to this subset).

Under the hypotheses of the above theorem,

$$E[|f(X_1, \dots, X_N) - M_f|] = \int_0^\infty P\left(|f(X_1, \dots, X_N) - M_f| \geq t\right) dt$$

$$\leq 4 \int_0^\infty e^{-\frac{t^2}{16|K|^2|f|_{\mathcal{L}}^2}} dt = 16|K||f|_{\mathcal{L}}.$$

Hence, we obtain as an immediate corollary to Theorem 5.5:

Corollary 5.6. *Under the hypotheses of Theorem 5.5, for all* $t \in \mathbb{R}^+$,

$$P\left(|f(X_1, \dots, X_N) - E[f(X_1, \dots, X_N)]| \geq (t + 16)|K||f|_{\mathcal{L}}\right) \leq 4 e^{-\frac{t^2}{16}}.$$

5.3 Concentration Inequalities on Compact Riemannian Manifold with Positive Ricci Curvature

Let M be a compact connected manifold of dimension N equipped with a Riemannian metric g. g is a differentiable map on M such that for each $x \in M$, g_x is a scalar product on the tangent space of M at x and therefore can be identified with a positive $N \times N$ matrix $((g_x)_{ij})_{1 \leq i,j \leq N}$. We shall denote by μ the Lebesgue measure on (M, g), that is, the normalized volume measure; it is seen that locally

$$d\mu(x) = \sqrt{\det(g_x)} dx.$$

On (M, g), one can define the Laplace–Baltrami operator Δ (which generalizes the usual Laplace operator on $\mathbb{R}^N$) and a gradient ∇ such that for any smooth real-valued function f all $x \in M$, all y in the tangent space $T_x M$ at x,

$$df_x(y) = g_x(\nabla f, y).$$

We let Φ be a smooth function on M and define

$$\mu_\Phi(dx) = \frac{1}{Z} e^{-\Phi(x)} d\mu(x)$$

as well as the operator $\mathcal{L}_\Phi$ such that for all smooth functions (h, f)

$$\mu_\Phi(f\mathcal{L}_\Phi h) = \mu_\Phi(h\mathcal{L}_\Phi f) = \mu_\Phi(g_x(\nabla f, \nabla h)).$$

By integration by parts, $\mathcal{L}_\Phi$ can be written in local coordinates:

$$\mathcal{L}_\Phi = \sum_{i,j=1}^{N} g_x^{ij} \partial_i \partial_j + \sum_{i=1}^{N} b_i^\Phi(x) \partial_i$$

with some b_i that can be explicitly computed in terms of Φ and g_x. We can define the "opérateurs carré du champ" as before. Simple algebra shows that

$$\Gamma_1(f, f)(x) := \left(\mathcal{L}_\Phi(f^2) - 2f\mathcal{L}_\Phi(f) \right)(x)$$

$$= \sum_{i,j=1}^{N} (g_x)_{ij} \partial_i f(x) \partial_j f(x) = g_x^{-1}(\nabla f(x), \nabla f(x))$$

and,

$$\Gamma_2(f, f)(x) := \left(\mathcal{L}_\Phi(\Gamma_1(f, f)) - 2\Gamma_1(\mathcal{L}_\Phi f, f) \right)(x)$$

$$= (\text{Hess}_x f, \text{Hess}_x f)_{g_x} + (\text{Ric}_x + \text{Hess}_x \Phi)(\nabla f(x), \nabla f(x))$$

Here, in local coordinates, the Hessian $(\text{Hess} f)_{ij}$ of f at x is equal to $(\partial_{ij} - \Gamma_{ij}^k \partial_k) f$ where Γ_{ij}^k are the Christofell symbols,

$$(\text{Hess} f, \text{Hess} f)_g = \sum g_{ij} g_{jl} (\text{Hess} f)_{ij} (\text{Hess} f)_{jl},$$

$(\text{Hess}\varPhi)(\nabla f, \nabla f)$ is obtained by differentiating twice $\varPhi$ in the direction ∇f and Ric denotes the Ricci tensor. An analytic definition of Ric is actually given above as the term due to the non-commutativity of derivatives on the manifold.

The arguments of Theorem 4.6 extend to the setting of a compact Riemannian manifold. Indeed, they were mainly based on the facts that g_x is positive definite and that ∇ obeys the Leibniz property

$$\nabla(h(f)) = \nabla f(\nabla h)(f)$$

for any differentiable functions $f, h : M \to M$. Since these properties still hold, the proof of Theorem 4.6 can be generalized to this setting yielding:

Corollary 5.7. *If for all $x \in M$ and $v \in T_x M$,*

$$(\text{Ric}_x + \text{Hess}\varPhi_x)(v, v) \geq c^{-1} g_x^{-1}(\nabla f, \nabla f),$$

$\mu_\varPhi$ *satisfies a log-Sobolev inequality with constant c, i.e., for any function $f : M \to \mathbb{R}$ we have*

$$\mu_\varPhi \left(f^2 \log \frac{f^2}{\mu_\varPhi(f^2)} \right) \leq 2c\mu_\varPhi(\Gamma_1(f, f)).$$

A straightforward generalization of the proof of Lemma 4.2 shows:

Corollary 5.8. *Assume that for all $x \in M$ and $v \in T_x M$,*

$$(\text{Ric}_x + \text{Hess}\varPhi_x)(v, v) \geq c^{-1} g_x^{-1}(v, v).$$

Then, for any differentiable function f on M, if we set

$$\||\nabla f|\|_2 := \sup_{x \in M} \Gamma_1(f, f)^{\frac{1}{2}}(x),$$

for all $\delta > 0$

$$\mu_\varPhi \left(|f - \mu_\varPhi(f)| \geq \delta \right) \leq 2e^{-\frac{\delta^2}{2c\||\nabla f|\|_2^2}}.$$

Exercise 1. Prove the corollary. *Hint: Prove and use Leibniz rule*

$$\Gamma_1(e^f, e^f) = e^{2f} \Gamma_1(f, f)$$

and follow the proof of Lemma 4.2.

5.4 Local Concentration Inequalities

In many instances, one may need to obtain concentration inequalities for functions that are only locally Lipschitz. To this end we state (and prove) the following lemma. Let (X, d) be a metric space and set for $f : X \to \mathbb{R}$

$$|f|_{\mathcal{L}} := \sup_{x,y \in X} \frac{|f(x) - f(y)|}{d(x,y)}.$$

Define, for a subset B of X, $d(x, B) = \inf_{y \in B} d(x, y)$. Then :

Lemma 5.9. *Assume that a probability measure μ on (X, d) satisfies a concentration inequality; for all $\delta > 0$, for all $f : X \to \mathbb{R}$,*

$$\mu(|f - \mu(f)| \geq \delta) \leq e^{-g(\frac{\delta}{|f|_{\mathcal{L}}})}$$

for some increasing function g on $\mathbb{R}^+$. Let B be a subset of X and let $f : B \to \mathbb{R}$ such that

$$|f|_{\mathcal{L}}^B := \sup_{x,y \in B} \frac{|f(x) - f(y)|}{d(x,y)}$$

is finite. Then, with $\delta(f) := \mu\left(1_{B^c}(\sup_{x \in B} |f(x)| + |f|_{\mathcal{L}}^B d(x, B))\right)$, we have

$$\mu(\{|f - \mu(f 1_B)| \geq \delta + \delta(f)\} \cap B) \leq e^{-g(\frac{\delta}{|f|_{\mathcal{L}}^B})}.$$

Proof. It is enough to define a Lipschitz function $\tilde{f}$ on X, whose Lipschitz constant $|\tilde{f}|_{\mathcal{L}}$ is bounded above by $|f|_{\mathcal{L}}^B$ and so that $\tilde{f} = f$ on B. We set

$$\tilde{f}(x) = \sup_{y \in B}\{f(y) - |f|_{\mathcal{L}}^B d(x, y)\}.$$

Note that, if $x \in B$, since $f(y) - f(x) - |f|_{\mathcal{L}}^B d(x, y) \leq 0$, the above supremum is taken at $y = x$ and $\tilde{f}(x) = f(x)$. Moreover, using the triangle inequality, we get that for any $x, z \in X$,

$$\tilde{f}(x) \geq \sup_{y \in B}\{f(y) - |f|_{\mathcal{L}}^B(d(x, z) + d(z, y))\}$$
$$= -|f|_{\mathcal{L}}^B d(x, z) + \tilde{f}(z) \tag{5.1}$$

and hence $\tilde{f}$ is Lipschitz, with constant $|f|_{\mathcal{L}}^B$. Therefore, we find that

$$\mu(\{|f - \mu(f 1_B)| \geq \delta\} \cap B) \leq \mu(|\tilde{f} - \mu(\tilde{f})| \geq \delta + \mu(|1_B f - \tilde{f}|))$$

Note that $\mu(|1_B f - \tilde{f}|) = \mu(1_{B^c}|\tilde{f}|)$. (5.1) with $z \in B$ shows that

$$|\tilde{f}(x)| \leq |f(z)| + |f|_{\mathcal{L}}^B d(z, x)$$

and so optimizing over $z \in B$ gives

$$|\tilde{f}(x)| \leq \max_{z \in B} |f(z)| + |f|_{\mathcal{L}}^B d(B, x).$$

Hence,

$$\mu(|1_B f - \tilde{f}|) \leq \mu\left(1_{B^c}(\sup_{x \in B} |f(x)| + |f|_{\mathcal{L}}^B d(., B))\right) =: \delta(f)$$

gives the desired estimate. $\qquad\square$

64   5 Generalizations

Bibliographical notes. Since the generalization of concentration inequalities to laws satisfying Poincaré's inequalities by Aida and Stroock [1], many recent results have considered the case where the decay at infinity is intermediate [21, 94] or even is very slow with heavy tails [22]. The generalization to Riemannian manifolds of the Bakry–Emery condition was already introduced in [18]. The case of discrete-valued random variables was considered by Talagrand [188].

6

Concentration Inequalities for Random Matrices

In this chapter, we shall apply the previous general results on concentration inequalities to random matrix theory, in particular to the eigenvalues of random matrices. To this end, we shall first study the regularity of the eigenvalues of matrices as a function of their entries (since the idea will be to apply concentration inequalities to the entries of the random matrices and then see the eigenvalues as nice functions of these entries).

6.1 Smoothness and Convexity of the Eigenvalues of a Matrix

We shall not follow [108] where smoothness and convexity were mainly proved by hand for smooth functions of the empirical measure and for the largest eigenvalue. We will rather, as in [6], rely on Weyl and Lidskii inequalities (see Theorems 19.1 and 19.4). We recall that we denote, for $\mathbf{B} \in \mathcal{M}_N(\mathbb{C})$, $\|\mathbf{B}\|_2$ its Euclidean norm:

$$\|\mathbf{B}\|_2 := \left(\sum_{i,j=1}^{N} |B_{ij}|^2 \right)^{\frac{1}{2}}.$$

From Weyl and Löwner Theorem 19.4, we will deduce that each eigenvalue of the matrix is a Lipschitz function of the entries of the matrix. We define $\mathcal{E}_N^{(1)} = \mathbb{R}^{N(N+1)/2}$ (resp. $\mathcal{E}_N^{(2)} = \mathbb{C}^{N(N-1)/2} \times \mathbb{R}^N$) and denote by $\mathbf{A}$ the symmetric (resp. Hermitian) $N \times N$ Wigner matrix such that $\mathbf{A} = \mathbf{A}^*; (\mathbf{A})_{ij} = A_{ij}, 1 \leq i \leq j \leq N$ for $(A_{ij})_{1 \leq i \leq j \leq N} \in \mathcal{E}_N^{(\beta)}$, $\beta = 1$ (resp. $\beta = 2$).

Lemma 6.1. *We denote by $\lambda_1(\mathbf{A}) \leq \lambda_2(\mathbf{A}) \leq \cdots \leq \lambda_N(\mathbf{A})$ the eigenvalues of $\mathbf{A} \in \mathcal{H}_N^{(2)}$. Then for all $k \in \{1, \ldots, N\}$, all $\mathbf{A}, \mathbf{B} \in \mathcal{H}_N^{(2)}$,*

$$|\lambda_k(\mathbf{A} + \mathbf{B}) - \lambda_k(\mathbf{A})| \leq \|\mathbf{B}\|_2.$$

A. Guionnet, *Large Random Matrices: Lectures on Macroscopic Asymptotics*, 65
Lecture Notes in Mathematics 1957, DOI: 10.1007/978-3-540-69897-5_6,
© 2009 Springer-Verlag Berlin Heidelberg, Reprint by Springer-Verlag Berlin Heidelberg 2012

In other words, for all $k \in \{1, \ldots, N\}$,

$$(A_{ij})_{1 \leq i \leq j \leq N} \in \mathcal{E}_N^{(2)} \to \lambda_k(\mathbf{A})$$

is Lipschitz with constant one.

For all Lipschitz functions f with Lipschitz constant $|f|_{\mathcal{L}}$, the function

$$(A_{ij})_{1 \leq i \leq j \leq N} \in \mathcal{E}_N^{(2)} \to \sum_{k=1}^{N} f(\lambda_k(\mathbf{A}))$$

is Lipschitz with respect to the Euclidean norm with a constant bounded above by $\sqrt{N}|f|_{\mathcal{L}}$. When f is continuously differentiable we have

$$\lim_{\epsilon \to 0} \epsilon^{-1} \left(\sum_{k=1}^{N} f(\lambda_k(\mathbf{A} + \epsilon\mathbf{B})) - \sum_{k=1}^{N} f(\lambda_k(\mathbf{A})) \right) = \mathrm{Tr}(f'(\mathbf{A})\mathbf{B}).$$

Proof. The first inequality is a direct consequence of Theorem 19.4 and entails the same control on $\lambda_{\max}(\mathbf{A})$. For the second we only need to use the Cauchy–Schwarz inequality:

$$\left| \sum_{i=1}^{N} f(\lambda_i(\mathbf{A})) - \sum_{i=1}^{N} f(\lambda_i(\mathbf{A} + \mathbf{B})) \right| \leq |f|_{\mathcal{L}} \sum_{i=1}^{N} |\lambda_i(\mathbf{A}) - \lambda_i(\mathbf{A} + \mathbf{B})|$$

$$\leq \sqrt{N}|f|_{\mathcal{L}} \left(\sum_{i=1}^{N} |\lambda_i(\mathbf{A}) - \lambda_i(\mathbf{A} + \mathbf{B})|^2 \right)^{\frac{1}{2}}$$

$$\leq \sqrt{N}|f|_{\mathcal{L}}\|\mathbf{B}\|_2$$

where we used Theorem 19.4 in the last line. For the last point, we check it for $f(x) = x^k$ where the result is clear since

$$\mathrm{Tr}((\mathbf{A} + \epsilon\mathbf{B})^k) = \mathrm{Tr}(\mathbf{A}^k) + \epsilon k \mathrm{Tr}(\mathbf{A}^{k-1}\mathbf{B}) + O(\epsilon^2) \tag{6.1}$$

and complete the argument by density of the polynomials. □

We can think of $\sum_{i=1}^{N} f(\lambda_i(\mathbf{A}))$ as $\mathrm{Tr}(f(\mathbf{A}))$. Then, the second part of the previous lemma can be extended to several matrices as follows.

Lemma 6.2. *Let P be a polynomial in m non-commutative indeterminates. For $1 \leq i \leq m$, we denote by D_i the cyclic derivative with respect to the ith variable given, if P is a monomial, by*

$$D_i P(X_1, \ldots, X_m) = \sum_{P = P_1 X_i P_2} P_2(X_1, \ldots, X_m) P_1(X_1, \ldots, X_m)$$

where the sum runs over all decompositions of P into $P_1 X_i P_2$ for some monomials P_1 and P_2. D_i is extended linearly to polynomials. Then, for all $(\mathbf{A}_1, \cdots, \mathbf{A}_m)$ and $(\mathbf{B}_1, \ldots, \mathbf{B}_m) \in \mathcal{H}_N^{(2)}$,

$$\lim_{\epsilon \to 0} \epsilon^{-1} \left(\mathrm{Tr}(P(\mathbf{A}_1 + \epsilon \mathbf{B}_1, \cdots, \mathbf{A}_m + \epsilon \mathbf{B}_m)) - \mathrm{Tr}(P(\mathbf{A}_1, \ldots, \mathbf{A}_m)) \right)$$

$$= \sum_{i=1}^m \mathrm{Tr}(D_i P(\mathbf{A}_1, \ldots, \mathbf{A}_m) \mathbf{B}_i).$$

In particular, if $(\mathbf{A}_1, \cdots, \mathbf{A}_m)$ belong to the subset Λ_M^N of elements of $\mathcal{H}_N^{(2)}$ with spectral radius bounded by $M < \infty$,

$$((A_k)_{ij})_{\substack{1 \le i \le j \le N \\ 1 \le k \le m}} \in \mathbb{C}^{N(N+1)m/2}, \mathbf{A}_k \in \mathcal{H}_N^{(2)} \cap \Lambda_M^N \to \mathrm{Tr}(P(\mathbf{A}_1, \ldots, \mathbf{A}_m))$$

is Lipschitz with a Lipschitz norm bounded by $\sqrt{N} C(P, M)$ for a constant $C(P, M)$ that depends only on M and P. If P is a monomial of degree d, one can take $C(P, M) = d M^{d-1}$.

Proof. We can assume without loss of generality that P is a monomial. The first equality is due to the simple expansion

$$\mathrm{Tr}(P(\mathbf{A}_1 + \epsilon \mathbf{B}_1, \cdots, \mathbf{A}_m + \epsilon \mathbf{B}_m)) - \mathrm{Tr}(P(\mathbf{A}_1, \ldots, \mathbf{A}_m))$$

$$= \epsilon \sum_{i=1}^m \sum_{P = P_1 X_i P_2} \mathrm{Tr}(P_1(\mathbf{A}_1, \ldots, \mathbf{A}_m) \mathbf{B}_i P_2(\mathbf{A}_1, \ldots, \mathbf{A}_m)) + O(\epsilon^2)$$

together with the trace property $\mathrm{Tr}(\mathbf{AB}) = \mathrm{Tr}(\mathbf{BA})$.

For the estimate on the Lipschitz norm, observe that if P is a monomial containing d_i times X_i, $\sum_{i=1}^m d_i = d$ and $D_i P$ is the sum of exactly d_i monomials of degree $d - 1$. Hence, $D_i P(\mathbf{A}_1, \ldots, \mathbf{A}_m)$ has spectral radius bounded by $d_i M^{d-1}$ when $(\mathbf{A}_1, \ldots, \mathbf{A}_m)$ are Hermitian matrices in Λ_M^N. Hence, by the Cauchy–Schwarz inequality, we obtain

$$\left| \sum_{i=1}^m \mathrm{Tr}(D_i P(\mathbf{A}_1, \ldots, \mathbf{A}_m) \mathbf{B}_i) \right|$$

$$\le \left(\sum_{i=1}^m \mathrm{Tr}(|D_i P(\mathbf{A}_1, \ldots, \mathbf{A}_m)|^2) \right)^{\frac{1}{2}} \left(\sum_{i=1}^m \mathrm{Tr}(\mathbf{B}_i^2) \right)^{\frac{1}{2}}$$

$$\le \left(N \sum_{i=1}^m d_i^2 M^{2(d-1)} \right)^{\frac{1}{2}} \left(\sum_{i=1}^m \|\mathbf{B}_i\|_2^2 \right)^{\frac{1}{2}}$$

$$\le \sqrt{N} d M^{d-1} \left(\sum_{i=1}^m \|\mathbf{B}_i\|_2^2 \right)^{\frac{1}{2}}.$$

□

Exercise 6.3. *Prove that when* $m = 1$, $D_1 P(x) = P'(x)$.

We now prove the following result originally due to Klein.

Lemma 6.4 (Klein's lemma). *Let* $f : \mathbb{R} \to \mathbb{R}$ *be a convex function. Then, if* $\mathbf{A}$ *is the* $N \times N$ *Hermitian matrix with entries* $(A_{ij})_{1 \leq i \leq j \leq N}$ *on and above the diagonal,*

$$\psi_f : (A_{ij})_{1 \leq i \leq j \leq N} \in \mathbb{C}^N \to \sum_{i=1}^{N} f(\lambda_i(\mathbf{A}))$$

is convex. Moreover, if f *is twice continuously differentiable with* $f''(x) \geq c$ *for all* x, ψ_f *is twice continuously differentiable with Hessian bounded below by* cI.

Proof. Let $X, Y \in \mathcal{H}_N^{(2)}$. We shall show that if f is a convex continuously differentiable function

$$\operatorname{Tr} \left(f(X) - f(Y) \right) \geq \operatorname{Tr} \left((X - Y) f'(Y) \right). \tag{6.2}$$

Taking $X = \mathbf{A}$ or $X = \mathbf{B}$ and $Y = 2^{-1}(\mathbf{A} + \mathbf{B})$ and summing the two resulting inequalities shows that for any couple $\mathbf{A}, \mathbf{B}$ of $N \times N$ Hermitian matrices,

$$\operatorname{Tr} \left(f \left(\frac{1}{2} \mathbf{A} + \frac{1}{2} \mathbf{B} \right) \right) \leq \frac{1}{2} \operatorname{Tr} \left(f(\mathbf{A}) \right) + \frac{1}{2} \operatorname{Tr} \left(f(\mathbf{B}) \right)$$

which implies that $(A_{ij})_{1 \leq i \leq j \leq N} \to \operatorname{Tr}(f(\mathbf{A}))$ is convex. The result follows for general convex functions f by approximations.

To prove (6.2), let us denote by $\lambda_i(C)$ the eigenvalues of a Hermitian matrix C and by $\xi_i(C)$ the associated eigenvector and write

$$\langle \xi_i(X), (f(X) - f(Y)) \, \xi_i(X) \rangle$$

$$= f(\lambda_i(X)) - \sum_{j=1}^{N} |\langle \xi_i(X), \xi_j(Y) \rangle|^2 f(\lambda_j(Y))$$

$$= \sum_{j=1}^{N} |\langle \xi_i(X), \xi_j(Y) \rangle|^2 (f(\lambda_i(X)) - f(\lambda_j(Y)))$$

$$\geq \sum_{j=1}^{N} |\langle \xi_i(X), \xi_j(Y) \rangle|^2 (\lambda_i(X) - \lambda_j(Y)) f'(\lambda_j(Y))$$

where we have used the convexity of f to write $f(x) - f(y) \geq (x - y) f'(y)$. The right-hand side of the last inequality is equal to $\langle \xi_i(X), ((X - Y) f'(Y)) \, \xi_i(X) \rangle$ and therefore summing over i yields (6.2), which completes the first part of the proof of the lemma.

We give another proof below that also provides a lower bound of the Hessian of ψ_f. The smoothness of ψ_f is clear when f is a polynomial since then

$\psi_f((A_{ij})_{1 \le i \le j \le N})$ is a polynomial function in the entries. Let us compute its second derivative when $f(x) = x^p$. Expanding (6.1) one step further gives

$$\text{Tr}((\mathbf{A} + \epsilon\mathbf{B})^k) = \text{Tr}(\mathbf{A}^k) + \epsilon \sum_{k=0}^{p-1} \text{Tr}(\mathbf{A}^k \mathbf{B} A^{p-1-k})$$

$$+ \epsilon^2 \sum_{0 \le k+l \le p-2} \text{Tr}(\mathbf{A}^k \mathbf{B} A^l \mathbf{B} A^{p-2-k-l}) + O(\epsilon^3)$$

$$= \text{Tr}(\mathbf{A}^k) + \epsilon p \text{Tr}(A^{p-1}\mathbf{B})$$

$$+ \frac{\epsilon^2}{2} p \sum_{0 \le l \le p-2} \text{Tr}(A^l \mathbf{B} A^{p-2-l}\mathbf{B}) + O(\epsilon^3). \tag{6.3}$$

A compact way to write this formula is by defining, for two real numbers x, y,

$$K_f(x, y) := \frac{f'(x) - f'(y)}{x - y}$$

and setting for a matrix $\mathbf{A}$ with eigenvalues $\lambda_i(A)$ and eigenvector e_i, $1 \le i \le N$,

$$K_f(\mathbf{A}, \mathbf{A}) = \sum_{i,j=1}^{N} K_f(\lambda_i(\mathbf{A}), \lambda_j(\mathbf{A})) e_i e_i^* \otimes e_j e_j^*.$$

Since $K_{x^p}(x, y) = p \sum_{r=0}^{p-1} x^r y^{p-1-r}$, the last term in the r.h.s. of (6.3) reads

$$p \sum_{0 \le l \le p-1} \text{Tr}(A^l \mathbf{B} A^{p-2-l}\mathbf{B}) = \langle K_{x^p}(\mathbf{A}, \mathbf{A}), \mathbf{B} \otimes \mathbf{B} \rangle \tag{6.4}$$

where for $\mathbf{B}, \mathbf{C}, \mathbf{D}, \mathbf{E} \in M_N(\mathbb{C})$, $\langle \mathbf{B} \otimes \mathbf{C}, \mathbf{D} \otimes \mathbf{E} \rangle := \langle \mathbf{B}, \mathbf{D} \rangle_2 \langle \mathbf{C}, \mathbf{E} \rangle_2$ with $\langle \mathbf{B}, \mathbf{D} \rangle_2 = \sum_{i,j=1}^{N} B_{ij} \bar{D}_{ij}$. In particular, $\langle e_i e_i^* \otimes e_j e_j^*, \mathbf{B} \otimes \mathbf{B} \rangle = |<e_i, Be_j>|^2$ with $< u, Bv >= \sum_{i,j=1}^{N} u_i \bar{v}_j B_{ij}$. By (6.3) and (6.4), for any Hermitian matrix $\mathbf{X}$,

$$\text{Hess}(\text{Tr}(\mathbf{A}^p))[X, X] = \langle K_{x^p}(\mathbf{A}, \mathbf{A}), X \otimes X \rangle$$

$$= \sum_{r,m=1}^{N} K_{x^p}(\lambda_r(\mathbf{A}), \lambda_m(\mathbf{A}))|<e_r, Xe_m>|^2$$

Now $K_f(\mathbf{A}, \mathbf{A})$ makes sense for any twice continuously differentiable function f and by density of the polynomials in the set of twice continuously differentiable function f, we can conclude that ψ_f is twice continuously differentiable too. Moreover, for any twice continuously differentiable function f,

$$\text{Hess}(\text{Tr}(f(\mathbf{A})))[X, X] = \sum_{r,m=1}^{N} K_f(\lambda_r(\mathbf{A}), \lambda_m(\mathbf{A}))|<e_r, Xe_m>|^2.$$

Since K_f is pointwise bounded below by c when $f'' \geq c$ we finally deduce that

$$\text{Hess}(\text{Tr}(f(\mathbf{A})))[\mathbf{X}, \mathbf{X}] \geq c\text{Tr}(\mathbf{X}\mathbf{X}^*).$$

The proof is thus complete.   □

Let us also notice the following:

Lemma 6.5. *Assume* $\lambda_1(\mathbf{A}) \leq \lambda_2(\mathbf{A}) \cdots \leq \lambda_N(\mathbf{A})$. *The functions*

$$\mathbf{A} \in \mathcal{H}_N^{(2)} \to \lambda_1(\mathbf{A}) \text{ and } \mathbf{A} \in \mathcal{H}_N^{(2)} \to \lambda_N(\mathbf{A})$$

are convex. For any norm $\| \cdot \|$ *on* $\mathcal{M}_N^{(2)}$, $(A_{ij})_{1 \leq i,j \leq N} \to \|\mathbf{A}\|$ *is convex.*

Proof. The first result is clear since we have already seen that $\lambda_N(\mathbf{A} + \mathbf{B}) \leq \lambda_N(\mathbf{A}) + \lambda_N(\mathbf{B})$. Since for $\alpha \in \mathbb{R}$, $\lambda_i(\alpha\mathbf{A}) = \alpha\lambda_i(\mathbf{A})$, we conclude that $\mathbf{A} \to \lambda_N(\mathbf{A})$ is convex. The same result holds for λ_1 (by changing the sign $\mathbf{A} \to -\mathbf{A}$). The convexity of $(A_{ij})_{1 \leq i,j \leq N} \to \|\mathbf{A}\|$ is due to the definition of the norm.   □

6.2 Concentration Inequalities for the Eigenvalues of Random Matrices

We consider a Hermitian random matrix $\mathbf{A}$ whose real or complex entries have joint law μ^N that satisfies one of the two hypotheses below.
    Either the entries of $\mathbf{A}$ are independent and satisfy for some $c > 0$ the following condition:
    • (H1) $\mathbf{A} = \mathbf{X}^N/\sqrt{N} = (\mathbf{A})^*$ *with* $(\mathbf{X}_{ij}^N, 1 \leq i \leq j \leq N)$ *independent, with laws* $(\mu_{ij}^N, 1 \leq i \leq j \leq N)$, *that are probability measures on* $\mathbb{C}$ *or* $\mathbb{R}$ *satisfying a log-Sobolev inequality with constant* $c < \infty$;
or μ^N is a Gibbs measure with strictly convex potential, i.e., satisfies:
    • (H2) *there exists a strictly convex twice continuously differentiable function* $V : \mathbb{R} \to \mathbb{R}$, $V''(x) \geq \frac{1}{c} > 0$, *so that*

$$\mu^N(d\mathbf{A}) = Z_N^{-1} e^{-N\text{Tr}(V(\mathbf{A}))} d\mathbf{A}$$

with $d\mathbf{A} = \prod_{1 \leq i \leq j \leq N} d\Re(A_{ij}) \prod_{1 \leq i < j \leq N} d\Im(A_{ij})$ *for complex entries or* $d\mathbf{A} = \prod_{1 \leq i \leq j \leq N} dA_{ij}$ *for real entries.*
    Note that when $V = \frac{1}{2}x^2$, μ^N is the law of a Gaussian Wigner matrix but in any other case the entries of $\mathbf{A}$ with law μ^N are not independent.
    We can now state the following theorem.

Theorem 6.6. *Suppose there exists* $c > 0$ *so that either (H1) or (H2) holds. Then:*

1. *For any Lipschitz function f on $\mathbb{R}$, for any $\delta > 0$,*

$$\mu^N \left(|L_{\mathbf{A}}(f) - \mu^N[L_{\mathbf{A}}(f)]| \geq \delta\right) \leq 2e^{-\frac{1}{4c|f|_{\mathcal{L}}^2} N^2 \delta^2}.$$

2. *Let $\lambda_1(\mathbf{A}) \leq \lambda_2(\mathbf{A}) \cdots \leq \lambda_N(A)$ be the eigenvalues of a self-adjoint matrix $\mathbf{A}$. For any $k \in \{1, \ldots, N\}$,*

$$\mu^N \left(|\lambda_k(\mathbf{A}) - \mu^N(\lambda_k(\mathbf{A}))| \geq \delta\right) \leq 2e^{-\frac{1}{4c} N \delta^2}.$$

In particular, these results hold when the $\mathbf{X}_{ij}$ are independent Gaussian variables with uniformly bounded variances.

Proof of Theorem 6.6. For the first case, we use the product property of Lemma 4.3 which implies that $\otimes_{i \leq j} \mu_{ij}^N$ satisfies the log-Sobolev inequality with constant c. By rescaling, the law of the entries of $\mathbf{A}$ satisfies a log-Sobolev inequality with constant c/N. For the second case, the assumption $V''(x) \geq \frac{1}{c}$ implies, by Lemma 6.4, that $(A_{ij})_{1 \leq i \leq j \leq N} \in \mathcal{E}_N^{(\beta)} \rightarrow N\mathrm{Tr}(V(\mathbf{A}))$ is twice continuously differentiable with Hessian bounded below by $\frac{N}{c}$. Therefore, by Corollary 4.7, μ^N satisfies a log-Sobolev inequality with constant c/N.

Thus, to complete the proof of the first result of the theorem, we only need to recall that by Lemma 6.1, $G(A_{ij}^N, 1 \leq i \leq j \leq N) = \mathrm{Tr}(f(\mathbf{A}))$ is Lipschitz with constant bounded by $\sqrt{N}|f|_{\mathcal{L}}$ whereas $A_{ij}^N, 1 \leq i \leq j \leq N \rightarrow \lambda_k(\mathbf{A})$ is Lipschitz with constant one. For the second, we use Lemma 6.5. ☐

Exercise 6.7. *State the concentration result when the μ_{ij}^N only satisfy the Poincaré inequality.*

Exercise 6.8. *If $\mathbf{A}$ is not Hermitian but has all entries with a joint law of type μ^N as above, show that the law of the spectral radius of $\mathbf{A}$ satisfies a concentration of measure inequality.*

When the laws satisfy instead a Talagrand-type condition we state the induced concentration bounds:

Theorem 6.9. *Let $\mu^N(f(\mathbf{A})) = \int f(\mathbf{X}/\sqrt{N}) \prod d\mu_{i,j}^N(X_{ij})$ with $(\mu_{i,j}^N, i \leq j)$ compactly supported probability measures on a connected compact subset K of $\mathbb{C}$. Fix $\delta_1 = 8|K|\sqrt{\pi}$. Then, for any $\delta \geq \delta_1 N^{-1}$, for any convex function f,*

$$\mu^N \left(|\mathrm{Tr}(f(\mathbf{A})) - \mu^N[\mathrm{Tr}(f(\mathbf{A}))]| \geq N\delta|f|_{\mathcal{L}}\right) \tag{6.5}$$

$$\leq \frac{32|K|}{\delta} \exp\left(-N^2 \frac{1}{16|K|^2 a^2} \frac{(\delta - \delta_1 N^{-1})^2}{16|K|}\right).$$

If $\lambda_{\max}(\mathbf{A})$ is the largest (or smallest) eigenvalue of $\mathbf{A}$, or the spectral radius of $\mathbf{A}$, for $\delta \geq \delta_1 N^{-\frac{1}{2}}$,

$$\mu^N \left(|\lambda_{\max}(\mathbf{A}) - E^N[\lambda_{\max}(\mathbf{A})]| \geq \delta N^{-\frac{1}{2}} \right)$$

$$\leq \frac{32|K|}{\delta} \exp \left(-\frac{1}{16|K|^2 a^2} \frac{(\delta - \delta_1 N^{-\frac{1}{2}})^2}{16|K|} \right).$$

Proof. Applying Corollary 5.6, Lemmas 6.1 and 6.4 with a function $f : \mathbf{A} \to \text{Tr}(f(\mathbf{A}))$ that is Lipschitz with Lipschitz constant $|f|_{\mathcal{L}}$ provides the first bound. □

Observe that the speed of the concentration we obtained is optimal for $\text{Tr}(f(\mathbf{X}^N))$ (since it agrees with the speed of the central limit theorem). It is also optimal in view of the large deviation principle we will prove in the next section. However, it does not capture the true scale of the fluctuations of $\lambda_{\max}(\mathbf{A})$ that are of order $N^{-\frac{2}{3}}$. Improvements of concentration inequalities in that direction were obtained by M. Ledoux [139].

We emphasize that Theorem 6.6 applies also when the variance of X_{ij}^N depends on i, j. For instance, it includes the case where $X_{ij}^N = a_{ij}^N Y_{ij}^N$ with Y_{ij}^N i.i.d. with law P satisfying the log-Sobolev inequality and a_{ij} uniformly bounded (since if P satisfies the log-Sobolev inequality with constant c, the law of ax under P satisfies it also with a constant bounded by $|a|^2 c$).

6.3 Concentration Inequalities for Traces of Several Random Matrices

The previous theorems also extend to the setting of several random matrices. If we wish to consider polynomial functions of these matrices, we can use local concentration results (see Lemma 5.9). We do not need to assume the random matrices independent if they interact *via* a convex potential.

Definition 6.10. *Let V be a polynomial in m non-commutative variables. We say that V is convex iff for any $N \in \mathbb{N}$,*

$$\phi_V^N : ((A_k)_{ij})_{\substack{i \leq j \\ 1 \leq k \leq m}} \in \mathcal{E}_N^{(2)} \to \text{Tr} V(\mathbf{A}_1, \dots, \mathbf{A}_m)$$

is convex.

Exercise 6.11. • *Define $X.Y = 2^{-1} \sum_{i=1}^m (X_i Y_i + Y_i X_i)$.* *Let $D = (D_1, \dots, D_m)$ with D_i the cyclic derivative with respect to the ith variable as defined in Lemma 6.2. Show that ϕ_V^N is convex if for any $\mathbf{X} = (X_i)_{1 \leq i \leq m}$ and $\mathbf{Y} = (Y_i)_{1 \leq i \leq m}$ in $\mathcal{H}_N^{(2)}(\mathbb{C})^m$, $V(\mathbf{X})^* = V(\mathbf{X})$ and*

$$(DV(\mathbf{X}) - DV(\mathbf{Y})).(\mathbf{X} - \mathbf{Y})$$

is a non-negative matrix in $\mathcal{H}_N^{(2)}(\mathbb{C})$.

- *Show that ϕ_V^N is convex if $V(X_1, \ldots, X_m) = \sum_{i=1}^k V_i(\sum_{j=1}^m \alpha_j^i X_j)$ when α_j^i are real variables and V_i are convex functions on $\mathbb{R}$. Hint: use Klein's Lemma 6.4.*

Let c be a positive real.

$$d\mu_V^{N,\beta}(\mathbf{A}_1, \ldots, \mathbf{A}_m) := \frac{1}{Z_V^N} e^{-N\mathrm{Tr}(V(\mathbf{A}_1, \ldots, \mathbf{A}_m))} d\mu_c^{N,\beta}(\mathbf{A}_1) \cdots d\mu_c^{N,\beta}(\mathbf{A}_m)$$

with $\mu_c^{N,\beta}$ the law of an $N \times N$ Wigner matrix with complex ($\beta = 2$) or real ($\beta = 1$) Gaussian entries with variance $1/cN$, that is, the law of the self-adjoint $N \times N$ matrix A with entries with law

$$\mu_c^{N,2}(dA) = \frac{1}{Z_N^c} e^{-\frac{cN}{2} \sum_{i,j=1}^N |A_{ij}|^2} \prod_{i \leq j} d\Re A_{ij} \prod_{i < j} d\Im A_{ij}$$

and

$$\mu^{N,1}(dA) = \frac{1}{Z_N^c} e^{-\frac{cN}{4} \sum_{i,j=1}^N A_{ij}^2} \prod_{i \leq j} dA_{ij}.$$

We then have the following corollary.

Corollary 6.12. *Let $\mu_V^{N,\beta}$ be as above with V convex. Then:*

1. For any Lipschitz function f of the entries of the matrices $A_i, 1 \leq i \leq m$, for any $\delta > 0$,

$$\mu_V^{N,\beta}(|f - \mu_V^{N,\beta}(f)| > \delta) \leq 2e^{-\frac{Nc\delta^2}{2|f|_{\mathcal{L}}^2}}.$$

2. Let M be a positive real, define $\Lambda_M^N = \{\mathbf{A}_i \in \mathcal{H}_N^{(2)}; \max_{1 \leq i \leq m} \lambda_{\max}(A_i) \leq M\}$ and let P be a monomial of degree $d \in \mathbb{N}$. Then, for any $\delta > 0$

$$\mu_V^{N,\beta}\Big(\{|\mathrm{Tr}(P(\mathbf{A}_1, \ldots, \mathbf{A}_m)) - \mu_V^{N,\beta}(\mathrm{Tr}(P(\mathbf{A}_1, \ldots, \mathbf{A}_m)1_{\Lambda_M^N}))|$$

$$> \delta + \delta(M, N)\} \cap \Lambda_M^N \Big) \leq 2e^{-\frac{c\delta^2}{d^2 M^{2(d-1)}}}$$

with

$$\delta(M, N) \leq M^d \mu_V^{N,\beta}\left((1 + d\|\mathbf{A}\|_2)1_{(\Lambda_M^N)^c}\right).$$

Proof. By assumption, the law $\mu_V^{N,\beta}$ of the entries of $(\mathbf{A}_1, \ldots, \mathbf{A}_m)$ is absolutely continuous with respect to the Lebesgue measure. The Hessian of the logarithm of the density is bounded above by $-NcI$. Hence, by Corollary 4.7, $\mu_V^{N,\beta}$ satisfies a log-Sobolev inequality with constant $1/Nc$ and thus by Lemma 4.2 we find that $\mu_V^{N,\beta}$ satisfies the first statement of the corollary. We finally conclude by using Lemma 5.9 and the fact that $X_1, \ldots, X_m \to \mathrm{Tr}(P(X_1, \ldots, X_m))$ is locally Lipschitz by Lemma 6.2. $\square$

6.4 Concentration Inequalities for the Haar Measure on $O(N)$

Now, let us consider the Haar measure on the orthogonal group

$$O(N) = \{A \in \mathcal{M}_{N \times N}(\mathbb{R}); OO^T = I\}.$$

This is the unique non-negative regular Borel measure on the compact group $O(N)$ that is left-invariant (see [172, Theorem 5.14]) and with total mass one. Let us introduce

$$SO(N) = \{A \in O(N) : \det(O) = 1\}.$$

For any $A \in O(N)$, $\det(O) \in \{+1, -1\}$, so that $O(N)$ can be decomposed as two copies of $SO(N)$. One way to go from one copy of $SO(N)$ to the other is for instance to change the sign of one column vector of the matrix. Let T be such a transformation. Then, if m^N denotes the Haar measure on $O(N)$, M^N the Haar measure on $SO(N)$, and $T_\sharp M^N(.) = M^N(T.)$, we deduce that

$$m^N = \frac{1}{2} M^N + \frac{1}{2} T_\sharp M^N.$$

Note that concentration inequalities under the Haar measure on $O(N)$ do not hold in general by taking a function concentrated on only one of the copies of $SO(N)$. However, on $SO(N)$, concentration holds. Namely, endow $SO(N)$ with the Riemaniann metric given, for $M, M' \in SO(N)$, by

$$d(M, M') = \inf_{(M_t)_{t \in [0,1]}: M_0 = M, M_1 = M'} \int_0^1 \left(\frac{1}{2} \mathrm{Tr}[(\partial_t M_t)(\partial_t M_t)^*] \right)^{\frac{1}{2}} dt$$

where the infimum is taken over all differentiable paths $M_. : [0,1] \to SO(N)$. This metric is Riemannian. It is the invariant (under conjugation) metric on $SO(N)$ such that the circle consisting of the rotations around a fixed subspace $\mathbb{R}^{N-2}$ has length 2π. The later normalization can be checked by taking, if $(e_i)_{1 \leq i \leq N}$ is an orthonormal basis of $\mathbb{R}^N$, M_t to be a rotation of angle θ in the vector space generated by e_1, e_2, the identity on $e_3, \ldots, e_N$). The Ricci curvature on $SO(N)$ for this metric has been computed in [155]:

Theorem 6.13 (Gromov). *[155, p. 129]*

$$\mathrm{Ric}(SO(N)) \geq \frac{N-2}{2} I.$$

This result extends to $SU(N)$ when one replaces $O(N)$ by $U(N)$; indeed, one has (see, e.g., [118] or [6])

$$\mathrm{Ric}(SU(N)) \geq \frac{N}{2}. \tag{6.6}$$

To deduce concentration inequalities, let us compare the above metric on $SO(N)$ with the Euclidean metric on $\mathcal{M}_N(\mathbb{C})$. To do so, we observe that

$$\left(\frac{1}{2}\mathrm{Tr}((M-M')(M-M')^*)\right)^{\frac{1}{2}}$$

$$= \inf_{(M_t)_{t\in[0,1]}:M_0=M,M_1=M'} \int_0^1 \left(\frac{1}{2}\mathrm{Tr}[(\partial_t M_t)(\partial_t M_t)^*]\right)^{\frac{1}{2}} dt$$

if the infimum is taken on $\mathcal{M}_N(\mathbb{C})$ (just take $M_t = M + t(M' - M)$ to get $\geq$ and use the fact that $M \to \sqrt{\mathrm{Tr}(MM^*)}$ is convex (as a norm) with Jensen's inequality for the converse inequality). Hence, for all $M, M' \in SO(N)$,

$$\left(\frac{1}{2}\mathrm{Tr}((M-M')(M-M')^*)\right)^{\frac{1}{2}} \leq d(M, M').$$

Therefore, we deduce the following:

Corollary 6.14. *For any differentiable function* $f : SO(N) \to \mathbb{R}$ *such that, for any* $X, Y \in SO(N)$, $|f(X) - f(Y)| \leq |f|_{\mathcal{L}}\|X - Y\|_2$, *we have for all* $t \geq 0$

$$M^N \left(\left|f - \int_{SO(N)} f(O)dM^N(O)\right| \geq \delta\right) \leq 2e^{-\frac{N}{2^4|f|_{\mathcal{L}}^2}\delta^2}.$$

If f *extends to* $O(N)$ *as a Lipschitz function on* $SO(N)$ *and* $T(SO(N))$, *we have*

$$m^N \left(\left|f - \int_{O(N)} f(O)dm^N(O)\right| \geq \delta + |f|_{\mathcal{L}}\right) \leq 2e^{-\frac{N}{2^4|f|_{\mathcal{L}}^2}\delta^2}$$

Proof. The concentration result under M^N is a direct consequence of the lower bound on $\mathrm{Ric}(SO(N))$ and Corollary 5.8. Indeed, the previous comparison of the metrics shows that

$$|f(X) - f(Y)|^2 \leq |f|_{\mathcal{L}}^2 \mathrm{Tr}[(X - Y)(X - Y)^*] \leq 2|f|_{\mathcal{L}}^2 d(X, Y)^2.$$

Hence, if f is differentiable, $2|f|_{\mathcal{L}}^2 = \||\nabla f\||_2^2$ and Corollary 5.8 allows us to conclude. Concentration under m^N is based on the fact that if T is a transformation of $SO(N)$ such as a change of sign of the first column vector, then

$$\mathrm{Tr}(X - TX)(X - TX)^* = 4\sum_{i=1}^N (O_{1,i})^2 = 4.$$

Therefore,

$$\left|\int_{SO(N)} f(O)dM^N(O) - \int_{SO(N)} f(TO)dM^N(O)\right| \leq 2|f|_{\mathcal{L}},$$

and so recalling that $m^N = 2^{-1}M^N + 2^{-1}T_\#M^N$,

$$\left|\int_{SO(N)} f(O)dM^N(O) - \int_{O(N)} f(O)dm^N(O)\right| \leq |f|_{\mathcal{L}}.$$

Hence, we find that

$$m^N\left(\left|f - \int_{O(N)} f(O)dm^N(O)\right| \geq N\delta + |f|_{\mathcal{L}}\right)$$

$$\leq \frac{1}{2}M^N\left(V : \left|f(TV) - \int_{SO(N)} f(TO)dM^N(O)\right| \geq \delta\right)$$

$$+ \frac{1}{2}M^N\left(V : \left|f(V) - \int_{SO(N)} f(O)dM^N(O)\right| \geq \delta\right) \leq 2e^{-\frac{N-2}{8|f|_{\mathcal{L}}^2}\delta^2}$$

which completes the proof. $\qquad\square$

As an application, we have the following corollary.

Corollary 6.15. *Let F be a Lipschitz function on $\mathbb{R}$, and D and D' be fixed diagonal matrices (whose entries are real and uniformly bounded by $\|D\|_\infty$ and $\|D'\|_\infty$ respectively). Then, for any $\delta > 0$,*

$$M^N\left(|\mathrm{Tr}(F(D' + ODO^*)) - \mathbb{E}[\mathrm{Tr}(F(D' + ODO^*))]| \geq \delta N|F|_{\mathcal{L}}\right)$$

$$\leq 2e^{-\frac{(N-2)N}{2^5\|D\|_\infty^2}\delta^2}.$$

Proof. Put $f(O) = \mathrm{Tr}(F(D' + ODO^*))$ and note that, for any $O \in O(N)$, by Lemma 6.1,

$$|f(O) - f(\tilde{O})|^2 \leq N|F|_{\mathcal{L}}^2\|ODO^* - \tilde{O}D\tilde{O}^*\|_2^2 \leq 4N|F|_{\mathcal{L}}^2\|D\|_\infty^2\|O - \tilde{O}\|_2^2.$$

Plugging this estimate into the main result of Theorem 6.13 completes the proof. $\qquad\square$

These concentration inequalities also extend to the Haar measure on $U(N)$ even though this time $U(N)$ decomposes as a continuum of copies of $SU(N)$ (namely $SU(N)$ times a rotation). This is, however, enough to get (see, e.g., [6]) the following theorem.

Theorem 6.16. *Let $(X_1^N, \ldots, X_m^N) \in \mathcal{H}_N^{(2)}$ be a sequence such that*

$$L := \sup_{1 \leq i \leq m} \sup_{N \in \mathbb{N}} \lambda_{\max}(X_i^N) < \infty$$

and denote m^N the Haar measure on $U(N)$. Then, for any polynomial P of $m + 2$ noncommutative variables $(X_1, \ldots, X_m, U, U^*)$, there exists $c = c(L, P) > 0$ such that for N large enough

$$m^N\left(\left|\frac{1}{N}\mathrm{Tr}(P(X_1,\ldots,X_m,U,U^*))\right.\right.$$

$$\left.\left.-\int\frac{1}{N}\mathrm{Tr}(P(X_1,\ldots,X_m,V,V^*))dm^N(V)\right|>\delta\right)\leq e^{-cN^2\delta^2}.$$

Bibliographical notes. Klein's lemma can be found for instance in [173]. The idea to apply concentration of measures theory to the concentration of the spectral measure and the largest eigenvalue of random matrices started in [108], even though it was already obtained in particular cases in different papers by using for instance martingale expansions. It was generalized to the concentration of each eigenvalue around its median in [3], and to the concentration of the permanent of random matrices in [92]. In [151], the Lipschitz condition was generalized to norms different than the Euclidean norm. The applications of these ideas to the eigenvalues of Haar distributed random matrices was used in [103], based on a lower bound on the Ricci curvature of $SO(N)$ due to Gromov [155] and developed in [63] from the viewpoint of random walks on compact groups.

6.5 Brascamp–Lieb Inequalities; Applications to Random Matrices

We introduce first Brascamp–Lieb inequalities and show how they can be derived from results from optimal transport theory, following a proof of Hargé [114]. We then show how these inequalities can be used to obtain *a priori* controls for random-matrix quantities such as the spectral radius. Such controls will be particularly useful in the next chapter.

6.5.1 Brascamp–Lieb Inequalities

The Brascamp–Lieb inequalities we shall be interested in allow us to compare the expectation of convex functions under a Gaussian law and under a law with a log-concave density with respect to this Gaussian law. It is stated as follows.

Theorem 6.17. (Brascamp–Lieb [47], Hargé [114, Theorem 1.1]) *Let* $n \in \mathbb{N}$. *Let* g *be a convex function on* $\mathbb{R}^n$ *and* f *a log-concave function on* $\mathbb{R}^n$. *Let* γ *be a Gaussian measure on* $\mathbb{R}^n$. *We suppose that all the following integrals are well defined, then:*

$$\int g(x+l-m)\frac{f(x)d\gamma(x)}{\int f d\gamma}\leq\int g(x)d\gamma(x)$$

where

$$l=\int x d\gamma,\quad m=\int x\frac{f(x)d\gamma(x)}{\int f d\gamma}.$$

This theorem was proved by Brascamp and Lieb [47, Theorem 7] (case $g(x) = |x_1|^\alpha$), by Caffarelli [59, Corollary 6] (case $g(x) = g(x_1)$) and then for a general convex function g by Hargé [114]. Hargé followed the idea introduced by Caffarelli to use optimal transport of measure. Unfortunately we cannot develop the theory of optimal transport here but shall still provide Hargé's proof (which is based, as for the proof of log-Sobolev inequalities, on the use of a semi-group that interpolates between the two measures of interest) as well as the statement of the results in optimal transport theory that the proof requires. For more information on the latter, we refer the reader to the two survey books by Villani [195, 196].

We shall define $d\mu(x) = f(x)d\gamma(x)/\int f d\gamma$.

Brenier [48] (see also McCann [149]) has shown that there exists a convex function $\phi : \mathbb{R}^n \to \mathbb{R}$ such that

$$\int g(y)d\mu(y) = \int g(\nabla\phi(x))d\gamma(x).$$

In other words, μ can be realized as the image (or push forward) of μ by the map $\nabla\phi$.

Caffarelli [57, 58] then proved that if the density f is Hölder continuous with exponent $\alpha \in]0, 1[$, ϕ is $\mathcal{C}^{2,\alpha}$ for any $\alpha \in]0, 1[$ (i.e. twice continuously differentiable with a second derivative Hölder continuous with exponent α). Moreover, by Caffarelli [59, Theorem 11], we know (and here we need to have γ, μ as specified above to get the upper bound) that for any vector $e \in \mathbb{R}^n$,

$$0 \le \partial_{ee}\phi = \langle \text{Hess}(\phi)e, e \rangle \le 1.$$

We now start the proof of Theorem 6.17. Observe first that we can assume without loss of generality that γ is the law of independent centered Gaussian variables with variance one (up to a linear transformation on the x's).

We let $\psi(x) = -\phi(x) + \frac{1}{2}\|x\|_2^2$ so that $0 \le \text{Hess}(\psi) \le I$ (with I the identity matrix and where inequalities hold in the operator sense) and write

$$\int g(y)d\mu(y) = \int g(x - \nabla\psi(x))d\gamma(x).$$

The idea is then to consider the following interpolation

$$\theta(t) = \int g(x - P_t(\nabla\psi)(x))d\gamma(x)$$

with P_t the Ornstein–Uhlenbeck process given, for $h : \mathbb{R}^n \to \mathbb{R}$ by

$$P_t h(x) = \int h(e^{-\frac{t}{2}}x + \sqrt{1 - e^{-t}}y)d\gamma(y)$$

and $P_t(\nabla\psi) = (P_t(\nabla_1\psi), \ldots, P_t(\nabla_n\psi))$ with $\nabla_i\psi = \partial_{x_i}\psi$. Note that for a Lipschitz function h, for all $x \in \mathbb{R}^n$,

$$|P_t h(x) - h(x)| \leq \int |h(e^{-\frac{t}{2}}x + \sqrt{1 - e^{-t}}y) - h(x)|d\gamma(y)$$

$$\leq |h|_{\mathcal{L}}(\sqrt{1 - e^{-t}} + (1 - e^{-\frac{t}{2}}))\int (\|x\|_2 + \|y\|_2)d\gamma(y)$$

goes to zero as t goes to zero (since $\int \|x\|_2 d\gamma(x) < \infty$). Similarly, for $t > 1$, there is a finite constant C such that

$$\left|P_t h(x) - \int h d\gamma\right| \leq C|h|_{\mathcal{L}} e^{-\frac{t}{2}}\left(\|x\|_2 + \int \|y\|_2 d\gamma(y)\right)$$

which shows that $P_t h$ goes to $\int h d\gamma$ as t goes to infinity. Since ψ is twice continuously differentiable with Hessian bounded by one, each $\nabla_i \psi$, $1 \leq i \leq n$, has uniformly bounded derivatives (by one) and so is Lipschitz for the Euclidean norm (with norm bounded by $\sqrt{n}$). Hence, the above applies with $h = \nabla_i \psi$, $1 \leq i \leq n$.

Let us assume that g is smooth and ∇g is bounded. Then, we deduce from the above estimates that, again because $\int \|x\|_2 d\gamma(x)$ is finite,

$$\lim_{t \to 0} \theta(t) = \theta(0) = \int g(x - \nabla\psi(x))d\gamma(x) = \int g(x)d\mu(x),$$

$$\lim_{t \to \infty} \theta(t) = \int g(x - \int \nabla\psi d\gamma)d\gamma(x).$$

Since

$$\int \nabla\psi d\gamma = \int (\nabla\psi - x)d\gamma + \int x d\gamma = \int x d\gamma - \int x d\mu$$

we see that Theorem 6.17 is equivalent to prove that $\theta(0) \leq \theta(\infty)$ and so it is enough to show that θ is non-decreasing. But, $t \to \theta(t)$ is differentiable with derivative

$$\theta'(t) = -\int \langle \nabla g(x - P_t(\nabla\psi)(x)), \partial_t P_t(\nabla\psi)(x) \rangle \, d\gamma(x) \qquad (6.7)$$

with

$$\partial_t P_t(h)(x)$$
$$= \int \left\langle -\frac{1}{2}e^{-\frac{t}{2}}x + \frac{1}{2}e^{-t}(1 - e^{-t})^{-\frac{1}{2}}y, \nabla h(e^{-\frac{t}{2}}x + \sqrt{1 - e^{-t}}y) \right\rangle d\gamma(y)$$
$$= -\frac{1}{2}e^{-\frac{t}{2}}\langle x, P_t(\nabla h)(x)\rangle + \frac{1}{2}e^{-t}\int \Delta h(e^{-\frac{t}{2}}x + \sqrt{1 - e^{-t}}y)d\gamma(y)$$
$$= -\frac{1}{2}\langle x, \nabla P_t h(x)\rangle + \frac{1}{2}\Delta(P_t h)(x) := L(P_t h)(x)$$

where in the second line we integrated by parts under the standard Gaussian law γ. Note also, again by integration by parts, that

$$\int h_1 L h_2 d\gamma(x) = -\frac{1}{2} \int \langle \nabla h_1, \nabla h_2 \rangle d\gamma.$$

Hence, (6.7) implies

$$\theta'(t) = -\int \sum_{i=1}^{n} (\partial_i g)(x - P_t(\nabla \psi)) L P_t(\partial_i \psi) d\gamma$$

$$= \frac{1}{2} \sum_{i,j=1}^{n} \int \partial_j ((\partial_i g)(x - P_t(\nabla \psi))) \partial_j (P_t(\partial_i \psi))) d\gamma$$

$$= \frac{1}{2} \sum_{i,j,k=1}^{n} \int (1_{k=j} - \partial_j (P_t(\partial_k \psi)))(\partial_k \partial_i g)(x - P_t(\nabla \psi)) P_t(\partial_i \psi)) d\gamma.$$

Thus, if we let

$$M_{ij}(x) := \partial_j (P_t(\partial_i \psi))(x), \text{ and } C_{ij}(x) = (\partial_j \partial_i g)(x - P_t(\nabla \psi)),$$

we have written, with $I_{ij} = 1_{i=j}$ the identity matrix,

$$\theta'(t) = \frac{1}{2} \sum_{j=1}^{n} \sum_{i=1}^{n} \sum_{k=1}^{n} \int (I - M(x))_{kj} C_{ik}(x) M_{ij}(x) d\gamma(x)$$

$$= \frac{1}{2} \int \text{Tr}(C(x)(I - M(x)) M^*(x)) d\gamma(x) \geq 0$$

since by Caffarelli we know that $0 \leq M(x) \leq I$ for all x, whereas $C \geq 0$ by hypothesis.

This completes the proof for smooth g with bounded gradient. The generalization to all convex functions g is easily done by approximation. The function can indeed be assumed as smooth as wished, since we can always restrict first the integral to a large ball $B(0, R)$, then on this large ball use the Stone–Weierstrass theorem to approximate g by a smooth function, and extend again the integral. We can assume the gradient of g bounded by approximating g by

$$g_R(x) = \sup_{y \in B(0,R)} \{g(y) + \langle \nabla g(y), x - y \rangle\}.$$

g_R is convex and with bounded gradient. Moreover, since $g(x) \geq g(y) + \langle \nabla g(y), x - y \rangle$ by convexity of g, $g_R = g$ on $B(0, R)$, while $g(0) + \langle \nabla g(0), x \rangle \leq g_R(x) \leq g(x)$ shows that $g_R, R \geq 0$ is uniformly integrable so that we can use the dominated convergence theorem to show that the expectation of g_R converges to that of g.

6.5.2 Applications of Brascamp–Lieb Inequalities

We apply now Brascamp-Lieb inequalities to the setting of random matrices. To this end, we must restrict ourselves to random matrices with entries following a law that is absolutely continuous with respect to the Lebesgue measure and with strictly log-concave density. We restrict ourselves to the case of m $N \times N$ Hermitian (or symmetric) random matrices with entries following the law

$$d\mu_V^{N,\beta}(\mathbf{A}_1, \dots, \mathbf{A}_m) := \frac{1}{Z_V^N} e^{-N\operatorname{Tr}(V(\mathbf{A}_1, \dots, \mathbf{A}_m))} d\mu_c^{N,\beta}(\mathbf{A}_1) \cdots d\mu_c^{N,\beta}(\mathbf{A}_m)$$

with $\mu_c^{N,\beta}$ the law of an $N \times N$ Wigner matrix with complex ($\beta = 2$) or real ($\beta = 1$) Gaussian entries with covariance $1/cN$, that is, the law of the self-adjoint $N \times N$ matrix $\mathbf{A}$ with entries with law

$$\mu_c^{N,\beta}(d\mathbf{A}) = \frac{1}{Z_N^c} e^{-\frac{cN}{2}\operatorname{Tr}(\mathbf{A}^2)} d\mathbf{A}$$

with $d\mathbf{A} = \prod_{i \leq j} d\Re(A_{ij}) \prod_{i \leq j} d(\Im A_{ij})$ when $\beta = 2$ and $d\mathbf{A} = \prod_{i \leq j} dA_{ij}$ if $\beta = 1$.

We assume that V is convex in the sense that for any $N \in \mathbb{N}$,

$$(A_{ij})_{1 \leq i \leq j \leq N} \in \mathcal{E}_N^{(\beta)} \to \operatorname{Tr}(V(\mathbf{A}_1, \dots, \mathbf{A}_m))$$

is real valued and convex, see sufficient conditions in Exercise 6.11.

Theorem 6.17 implies that for all convex functions g on $(\mathbb{R})^{\beta m N(N-1)/2 + mN}$,

$$\int g(\mathbf{A} - \mathbf{M}) d\mu_V^{N,\beta}(\mathbf{A}) \leq \int g(\mathbf{A}) \prod_{i=1}^m d\mu_c^{N,\beta}(\mathbf{A}_i) \tag{6.8}$$

where $\mathbf{M} = \int \mathbf{A} d\mu_V^{N,\beta}(\mathbf{A})$ is the m-tuple of deterministic matrices $(\mathbf{M}_k)_{ij} = \int (\mathbf{A}_k)_{ij} d\mu_V^{N,\beta}(\mathbf{A})$. In (6.8), $g(\mathbf{A})$ is shorthand for a function of the (real and imaginary parts of the) entries of the matrices $\mathbf{A} = (\mathbf{A}_1, \dots, \mathbf{A}_m)$.

By different choices of the function g we shall now obtain some a priori bounds on the random matrices $(\mathbf{A}_1, \dots, \mathbf{A}_m)$ with law $\mu_c^{N,\beta}$.

Lemma 6.18. *Assume that the function V is convex and there exists $d > 0$ such that for some finite $c(V)$,*

$$V(X_1, \dots, X_m) \leq c(V) \left(1 + \sum_{i=1}^m X_i^{2d} \right).$$

For $c > 0$, there exists $C_0 = C_0(c, V(0), D_i V(0), c(V), d)$ finite such that for all $i \in \{1, \dots, m\}$, all $n \in \mathbb{N}$,

$$\limsup_N \mu_V^{N,\beta} \left(\frac{1}{N} \operatorname{Tr}(\mathbf{A}_i^{2n}) \right) \leq C_0^n.$$

Moreover, C_0 depends continuously on $V(0), D_i V(0), c(V)$ and in particular is uniformly bounded when these quantities are.

260

Note that this lemma shows that, for $i \in \{1, \ldots, m\}$, the spectral measure of $\mathbf{A}_i$ is asymptotically contained in the compact set $[-\sqrt{C_0}, \sqrt{C_0}]$.

Proof. Let k be in $\{1, \ldots, m\}$. As $\mathbf{A} \to \mathrm{Tr}(\mathbf{A}_k^{4d})$ is convex by Klein's lemma 6.4, Brascamp–Lieb inequality (6.8) implies that

$$\mu_V^{N,\beta} \left(\frac{1}{N} \mathrm{Tr}(\mathbf{A}_k - \mathbf{M}_k)^{4d} \right) \leq \mu_c^{N,\beta} \left(\frac{1}{N} \mathrm{Tr}(\mathbf{A}_k)^{4d} \right) = \mu_c^{N,\beta}(\mathbf{L}_{\mathbf{A}_k}(x^{4d})) \quad (6.9)$$

where $\mathbf{M}_k = \mu_V^{N,\beta}(\mathbf{A}_k)$ stands for the matrix with entries $\int (A_k)_{ij} d\mu_V^{N,\beta}(d\mathbf{A})$. Thus, since $\mu_c^{N,\beta}(\mathbf{L}_{\mathbf{A}_k}(x^{4d}))$ converges by Wigner's theorem 1.13 towards

$$c^{-2d} C_{2d} \leq (c^{-1}4)^{2d}$$

with C_{2d} the Catalan number, we only need to control $\mathbf{M}_k$. First observe that for all k the law of A_k is invariant under the multiplication by unitary matrices so that for any unitary matrices U,

$$\mathbf{M}_k = \mu_V^{N,\beta}[\mathbf{A}_k] = U\mu_V^{N,\beta}[\mathbf{A}_k]U^* \Rightarrow \mathbf{M}_k = \mu_V^{N,\beta}\left(\frac{1}{N} \mathrm{Tr}(\mathbf{A}_k) \right) I. \quad (6.10)$$

Let us bound $\mu_V^{N,\beta}(\frac{1}{N}\mathrm{Tr}(\mathbf{A}_k))$. Jensen's inequality implies

$$Z_N^V \geq e^{-N^2 \mu_c^{N,\beta}(\frac{1}{N}\mathrm{Tr}(V))} \geq e^{-N^2 c(V) \mu_c^{N,\beta}(\frac{1}{N}\mathrm{Tr}(1+\sum X_i^{2d}))}$$

By Theorem 3.3, $\mu_c^{N,\beta}(\frac{1}{N}\mathrm{Tr}(X_i^{2d}))$ converges to a finite constant and therefore we find a finite constant $C(V)$ such that $Z_N^V \geq e^{-N^2 C(V)}$.

We now use the convexity of V, to find that for all N,

$$\mathrm{Tr}(V(\mathbf{A})) \geq \mathrm{Tr}\left(V(0) + \sum_{i=1}^m D_i V(0) \mathbf{A}_i\right)$$

with D_i the cyclic derivative introduced in Lemma 6.2. By Chebyshev's inequality, we therefore obtain, for all $\lambda \geq 0$,

$$\mu_V^{N,\beta}(|\mathbf{L}_{\mathbf{A}_k}(x)| \geq y) \leq \mu_V^{N,\beta}(\mathbf{L}_{\mathbf{A}_k}(x) \geq y) + \mu_V^{N,\beta}(-\mathbf{L}_{\mathbf{A}_k}(x) \geq y)$$

$$\leq e^{N^2(C(V)-V(0)-\lambda y)} \left(\mu_c^{N,\beta}(e^{-N\mathrm{Tr}(\sum_{i=1}^m D_i V(0)\mathbf{A}_i - \lambda \mathbf{A}_k)}) \right.$$

$$\left. + \mu_c^{N,\beta}(e^{-N\mathrm{Tr}(\sum_{i=1}^m D_i V(0)\mathbf{A}_i + \lambda \mathbf{A}_k)}) \right)$$

$$= e^{N^2(C(V)-V(0)-\lambda y)} e^{\frac{N}{2c} \sum_{\ell \neq k} \mathrm{Tr}(D_i V(0)^2)}$$

$$\left(e^{\frac{N}{2c} \mathrm{Tr}((D_k V(0)-\lambda)^2)} + e^{\frac{N}{2c} \mathrm{Tr}((D_k V(0)+\lambda)^2)} \right).$$

Optimizing with respect to λ shows that there exists $B = B(V)$

$$\mu_V^N\left(|\mathbf{L}_{\mathbf{A}_k}(x)| \geq y\right) \leq e^{BN^2 - \frac{N^2 c}{4} y^2}$$

so that for N large enough,

$$\mu_V^{N,\beta}\left(|\mathbf{L}_{\mathbf{A}_k}(x)|\right) = \int_0^\infty \mu_V^N\left(|\mathbf{L}_{\mathbf{A}_k}(x)| \geq y\right) dy$$

$$\leq 4\sqrt{c^{-1}B} + \int_{y \geq 4\sqrt{c^{-1}B}} e^{-\frac{N^2 c}{4}(y^2 - 4\frac{B}{c})} dy \leq 8\sqrt{Bc^{-1}}. \quad (6.11)$$

This, with (6.9), completes the proof. $\square$

Let us derive some other useful properties due to the Brascamp–Lieb inequality. We first obtain an estimate on the spectral radius $\lambda_{\max}^N(\mathbf{A})$, defined as the maximum of the spectral radius of $\mathbf{A}_1, \ldots, \mathbf{A}_m$ under the law $\mu_V^{N,\beta}$.

Lemma 6.19. *Under the hypothesis of Lemma 6.18, there exists $\alpha = \alpha(c) > 0$ and $M_0 = M_0(V) < \infty$ such that for all $M \geq M_0$ and all integer N,*

$$\mu_V^{N,\beta}(\lambda_{max}^N(\mathbf{A}) > M) \leq e^{-\alpha MN}.$$

Moreover, $M_0(V)$ is uniformly bounded when $V(0)$, $D_iV(0)$ and $c(V)$ are.

Proof. The spectral radius $\lambda_{\max}^N(\mathbf{A}) = \max_{1 \leq i \leq m} \sup_{\|u\|_2=1} <u, \mathbf{A}_i \mathbf{A}_i^* u>^{\frac{1}{2}}$ is a convex function of the entries (see Lemma 6.5), so we can apply the Brascamp–Lieb inequality (6.8) to obtain that for all $s \in [0, \frac{c}{10}]$,

$$\int e^{sN\lambda_{\max}^N(\mathbf{A}-M)} d\mu_V^{N,\beta}(\mathbf{A}) \leq \int e^{sN\lambda_{\max}^N(\mathbf{A})} d\mu_c^{N,\beta}(\mathbf{A}).$$

But, by Theorem 6.6 applied with a quadratic potential V, we know that

$$\int e^{sN\lambda_{\max}^N(\mathbf{A})} d\mu_c^{N,\beta}(\mathbf{A})$$

$$\leq e^{sN\mu_c^{N,\beta}(\lambda_{\max}^N)} \int e^{sN(\lambda_{\max}^N - \mu_c^{N,\beta}(\lambda_{\max}^N))} d\mu_c^{N,\beta}$$

$$= sN e^{sN\mu_c^{N,\beta}(\lambda_{\max}^N)} \int_{-\infty}^\infty e^{sNy} \mu_c^{N,\beta}\left(\lambda_{\max}^N - \mu_c^{N,\beta}(\lambda_{\max}^N) \geq y\right) dy$$

$$\leq sN e^{sN\mu_c^{N,\beta}(\lambda_{\max}^N)}\left(1 + 2\int_0^\infty e^{sNy} e^{-\frac{Nc}{4}y^2} dy\right)$$

$$\leq \sqrt{2\pi} sN e^{sN\mu_c^{N,\beta}(\lambda_{\max}^N)}(1 + 2e^{\frac{2s^2N}{c}})$$

Hence, since $\mu_c^{N,\beta}(\lambda_{\max}^N)$ is uniformly bounded by Theorem 2.3, we deduce that for all $s \geq 0$, there exists a finite constant $C(s)$ such that

$$\int e^{sN\lambda_{\max}^N(\mathbf{A}-\mathbf{M})} d\mu_V^{N,\beta}(\mathbf{A}) \leq C(s)^N.$$

By (6.10) and (6.11), we know that

$$\lambda_{\max}^N(\mathbf{A}) \leq \lambda_{\max}^N(\mathbf{A}-\mathbf{M}) + \lambda_{\max}^N(\mathbf{M}) \leq \lambda_{\max}^N(\mathbf{A}-\mathbf{M}) + 8\sqrt{Bc^{-1}}$$

from which we deduce that $\int e^{sN\lambda_{\max}^N(\mathbf{A})} d\mu_V^{N,\beta}(\mathbf{A}) \leq C^N$ for a positive finite constant C. We conclude by a simple application of Chebyshev's inequality. $\square$

Lemma 6.20. *If $c > 0$, $\epsilon \in]0, \frac{1}{2}[$, then there exists $C = C(c,\epsilon) < \infty$ such that for all $d \leq N^{\frac{1}{2}-\epsilon}$,*

$$\mu_V^{N,\beta}(|\lambda_{max}^N(\mathbf{A})|^d) \leq C^d.$$

Note that this control could be generalized to $d \leq N^{2/3-\epsilon}$, by using the refinements obtained by Soshnikov in [180, Theorem 2 p. 17] but we shall not need it here.

Proof. Since $\mathbf{A} \to \lambda_{\max}^N(\mathbf{A})$ is convex, we can again use Brascamp–Lieb inequalities to insure that

$$\mu_V^{N,\beta}\left(|\lambda_{\max}^N(\mathbf{A} - \mu_V^{N,\beta}(\mathbf{A}))|^d\right) \leq \mu_c^{N,\beta}\left(|\lambda_{\max}^N(\mathbf{A} - \mu_c^{N,\beta}(\mathbf{A}))|^d\right).$$

Now, we have seen in the proof of Lemma 6.18 that $\mu_V^{N,\beta}(\mathbf{A})$ has a uniformly bounded spectral radius, say by x. Moreover, by Theorem 2.3, we find that

$$\mu_c^{N,\beta}\left(|\lambda_{\max}^N(\mathbf{A})|^{N^{\frac{1}{2}-\epsilon/2}}\right) \leq c(\epsilon)\frac{N(2c^{-1})^{N^{\frac{1}{2}-\epsilon/2}}}{\sqrt{\pi N^{3(\frac{1}{2}-\epsilon/2)}}}.$$

Applying Jensen's inequality we therefore get, for $d \leq N^{\frac{1}{2}-\epsilon}$,

$$\mu_c^{N,\beta}\left(|\lambda_{\max}^N(\mathbf{A})|^d\right) \leq c'(\epsilon)(2c^{-1})^d.$$

Hence,

$$\mu_V^{N,\beta}\left(|\lambda_{\max}^N(\mathbf{A})|^d\right)^{\frac{1}{d}} \leq x + c'(\epsilon)^{\frac{1}{d}} 2c^{-1}$$

which proves the claim. $\square$

6.5.3 Coupling Concentration Inequalities and Brascamp–Lieb Inequalities

We next turn to concentration inequalities for the trace of polynomials on the set

$$\Lambda_M^N = \{\mathbf{A} \in \mathcal{H}_N^m : \lambda_{\max}^N(\mathbf{A}) = \max_{1 \le i \le m}(\lambda_{\max}^N(\mathbf{A}_i)) \le M\} \subset \mathbb{R}^{N^2 m}.$$

We let

$$\tilde{\delta}^N(P) := \operatorname{Tr}(P(\mathbf{A}_1, \ldots, \mathbf{A}_m)) - \mu_V^{N,\beta}\left(\operatorname{Tr}(P(\mathbf{A}_1, \ldots, \mathbf{A}_m))\right).$$

Then, we have

Lemma 6.21. *For all N in $\mathbb{N}$, all $M > 0$, there exists a finite constant $C(P,M)$ and $\epsilon(P,M,N)$ such that for any $\epsilon > 0$,*

$$\mu_V^{N,\beta}\left(\{|\tilde{\delta}^N(P)| \ge \epsilon + \epsilon(P,M,N)\} \cap \Lambda_M^N\right) \le 2e^{-\frac{c\epsilon^2}{2C(P,M)}}.$$

If P is a monomial of degree d we can choose

$$C(P,M) \le d^2 M^{2(d-1)}$$

and there exists $M_0 < \infty$ so that for $M \ge M_0$, all $\epsilon \in]0, \frac{1}{2}[$, and all monomial P of degree smaller than $N^{1/2-\epsilon}$,

$$\epsilon(P,M,N) \le 3dN(CM)^{d+1} e^{-\frac{\alpha}{2}NM}$$

with C the constant of Lemma 6.20.

Proof. It is enough to consider the case where P is a monomial. By Corollary 6.12, we only need to control $\epsilon(P,M,N)$.

$$
\epsilon(P,M,N) \le \mu_V^{N,\beta}\left(1_{(\Lambda_M^N)^c}\left(|\operatorname{Tr}(P)| + dM^{d-1}\sqrt{\sum_{i=1}^m \operatorname{Tr}(\mathbf{A}_i\mathbf{A}_i^*)}\right.\right.
$$
$$
\left.\left. + \sup_{\mathbf{A} \in \Lambda_M^N} |\operatorname{Tr}(P(\mathbf{A}))|\right)\right)
$$
$$
\le N\mu_V^{N,\beta}\left(1_{(\Lambda_M^N)^c}\left(|\lambda_{\max}^N(\mathbf{A})|^d\right.\right.
$$
$$
\left.\left. + \sqrt{C(P,M)}|\lambda_{\max}^N(\mathbf{A})|^2 + M^d\right)\right).
$$

Now, by Lemmas 6.19 and 6.20, we find that

$$\mu_V^N\left(1_{(\Lambda_M^N)^c}|\lambda_{\max}^N(\mathbf{A})|^d\right) \le \mu_V^N\left(1_{(\Lambda_M^N)^c}\right)^{\frac{1}{2}} \mu_V^N\left(|\lambda_{\max}^N(\mathbf{A})|^{2d}\right)^{\frac{1}{2}} \le C^d e^{-\frac{\alpha}{2}NM}.$$

By the previous control on $C(P,M)$, we get, for $d \leq N^{\frac{1}{2}-\epsilon}$ and M large enough,

$$\epsilon(P, M, N) \leq 3dN(CM)^{d+1} e^{-\frac{\alpha}{2}NM},$$

which proves the claim. $\square$

For later purposes, we have to find a control on the variance of $\mathbf{L}$.

Lemma 6.22. *For any $c > 0$ and $\epsilon \in]0, \frac{1}{2}[$, there exists $B, C, M_0 > 0$ such that for all $\mathbf{t} \in \mathbf{B}_{\eta,\mathbf{c}}$, all $M \geq M_0$, and monomial P of degree less than $N^{\frac{1}{2}-\epsilon}$,*

$$\mu_V^{N,\beta}\left((\tilde{\delta}^N(P))^2\right) \leq BC(P,M) + C^{2d}N^4 e^{-\frac{\alpha M N}{2}}. \tag{6.12}$$

Moreover, the constants C, M_0, B depend continuously on $V(0), D_iV(0)$ and $c(V)$.

Proof. If P is a monomial of degree d, we write

$$\mu_V^{N,\beta}((\tilde{\delta}^N(P))^2) \leq \mu_V^{N,\beta}(\mathbf{1}_{\Lambda_M^N}(\tilde{\delta}^N(P))^2) + \mu_V^{N,\beta}(\mathbf{1}_{(\Lambda_M^N)^c}(\tilde{\delta}^N(P))^2) = I_1 + I_2. \tag{6.13}$$

For I_1, the previous lemma implies that, for $d \leq N$,

$$I_1 = 2\int_0^\infty x\mu_V^{N,\beta}\left(\{|\mathrm{Tr}(P) - \mu_V^{N,\beta}(\mathrm{Tr}(P))| \geq x\} \cap \Lambda_M^N\right) dx$$

$$\leq \epsilon(P, N, M)^2 + 4\int_0^\infty xe^{-\frac{cx^2}{2C(P,M)}} dx \leq BC(P,M)$$

with a constant B that depends only on c. For the second term, we take $M \geq M_0$ with M_0 as in Lemma 6.19 to get

$$I_2 \leq \mu_V^{N,\beta}[(\Lambda_M^N)^c]^{\frac{1}{2}}\mu_V^{N,\beta}((\tilde{\delta}^N(P))^4)^{\frac{1}{2}} \leq e^{-\frac{\alpha M N}{2}}\mu_V^{N,\beta}((\tilde{\delta}^N(P))^4)^{\frac{1}{2}}.$$

By the Cauchy–Schwartz inequality, we obtain the control

$$\mu_V^{N,\beta}[\tilde{\delta}^N(P)^4] \leq 2^4\mu_V^{N,\beta}((\mathrm{Tr}(P))^4).$$

Now, by non-commutative Hölder's inequality Theorem 19.5,

$$[\mathrm{Tr}(P)]^4 \leq N^4 \max_{1 \leq i \leq m} \frac{1}{N}\mathrm{Tr}(A_i^{4d})$$

so that we obtain the bound

$$\mu_V^{N,\beta}[\tilde{\delta}^N(P)^4] \leq 2^4 N^4 \max_{1 \leq i \leq m}\mu_V^{N,\beta}[\frac{1}{N}\mathrm{Tr}(A_i^{4d})].$$

By Lemma 6.20, for $d \leq N^{\frac{1}{2}-\epsilon}$,

$$\mu_V^{N,\beta} \left[\frac{1}{N} \mathrm{Tr}(A_i^{4d}) \right] \leq C^{2d}. \tag{6.14}$$

Plugging back this estimate into (6.13), we have proved that for N and M sufficiently large, all monomials P of degree $d \leq N^{\frac{1}{2}-\epsilon}$, all $\mathbf{t} \in \mathbf{B}_{\eta,\mathbf{c}}$

$$\mu_V^{N,\beta} \left((\hat{\delta}^N(P))^2 \right) \leq BC(P,M) + C^{2d} N^4 e^{-\frac{\alpha MN}{2}}$$

with a finite constant C depending only on ϵ, c and M_0. $\square$

Bibliographical Notes. Brascamp–Lieb inequalities were first introduced in [47]. The relation between FKG inequalities and optimal transportation was shown in [59], based on optimal bounds on the Hessian of the transport map. The application of this strategy to Brascamp–Lieb inequalities is due to Hargé [114]. It was used in the context of random matrices in [104, 105], following the lines that we shall develop in the next part.

Part III

Matrix Models

In this part, we study matrix models, that is, the laws of interacting Hermitian matrices of the form

$$d\mu_V^{N,2}(\mathbf{A}_1, \ldots, \mathbf{A}_m) := \frac{1}{Z_V^N} e^{-N\mathrm{Tr}(V(\mathbf{A}_1,\ldots,\mathbf{A}_m))} d\mu^{N,2}(\mathbf{A}_1) \cdots d\mu^{N,2}(\mathbf{A}_m)$$

where Z_V^N is the normalizing constant given by the matrix integral

$$Z_V^N = \int e^{-N\mathrm{Tr}(V(\mathbf{A}_1,\ldots,\mathbf{A}_m))} d\mu^{N,2}(\mathbf{A}_1) \cdots d\mu^{N,2}(\mathbf{A}_m)$$

and V is a polynomial in m non-commutative variables:

$$V(X_1, \ldots, X_m) = \sum_{i=1}^{n} t_i q_i(X_1, \ldots, X_m)$$

with q_i non-commutative monomials:

$$q_i(X_1, \ldots, X_m) = X_{j_1^i} \cdots X_{j_{r_i}^i}$$

for some $j_l^k \in \{1, \ldots, m\}$, $r_i \geq 1$. Moreover, $d\mu^{N,2}(\mathbf{A})$ denotes the standard law of the **GUE**, i.e., under $d\mu^{N,2}(\mathbf{A})$, $\mathbf{A}$ is an $N \times N$ Hermitian matrix such that

$$A(k,l) = \bar{A}(l,k) = \frac{g_{kl} + i\tilde{g}_{kl}}{\sqrt{2N}}, \ k < l, \quad A(k,k) = \frac{g_{kk}}{\sqrt{N}}$$

with independent centered standard Gaussian variables $(g_{kl}, \tilde{g}_{kl})_{k \leq l}$. In other words

$$d\mu^{N,2}(\mathbf{A}) = Z_N^{-1} 1_{\mathbf{A} \in \mathcal{H}_N^{(2)}} e^{-\frac{N}{2}\mathrm{Tr}(\mathbf{A}^2)} \prod_{1 \leq i \leq j \leq N} d\Re(A(i,j)) \prod_{1 \leq i < j \leq N} d\Im(A(i,j)).$$

Since we restrict ourselves to Hermitian matrices in this part, we shall drop the subscript $\beta = 2$ and write for short $\mu^N = \mu^{N,2}$.

Let us define by $\mathbb{C}\langle X_1, \ldots, X_m \rangle$ the set of polynomials in m non-commutative variables and, for $P \in \mathbb{C}\langle X_1, \ldots, X_m \rangle$,

$$\mathbf{L}(P) := \mathbf{L}_{\mathbf{A}_1, \ldots, \mathbf{A}_m}(P) = \frac{1}{N}\mathrm{Tr}\,(P(\mathbf{A}_1, \ldots, \mathbf{A}_m))\,.$$

When V vanishes, we have seen in Chapter 3 that for any polynomial function P, $\mathbf{L}(P)$ converges as N goes to infinity. Moreover the limit $\sigma^m(P)$ is such that if P is a monomial, $\sigma^m(P)$ is the number of non-crossing pair partitions of a set of points with m colors, or equivalently the number of planar maps with one star of type P. In this part, we shall generalize such a type of result to the case where V does not vanish but is "small" and "nice" in a sense that we shall precise.

This part is motivated by a work of Brézin, Parisi, Itzykson and Zuber [50] and large developments that occurred thereafter in theoretical physics [78]. They specialized an idea of 't Hooft [187] to show that if $V = \sum_{i=1}^{n} t_i q_i$ with fixed monomials q_i of m non-commutative variables, and if we see $Z_V^N = Z_{\mathbf{t}}^N$ as a function of $\mathbf{t} = (t_1, \ldots, t_n)$,

$$\log Z_{\mathbf{t}}^N := \sum_{g \geq 0} N^{2-2g} F_g(\mathbf{t}), \tag{III.15}$$

where

$$F_g(\mathbf{t}) := \sum_{k_1, \ldots, k_n \in \mathbb{N}^k} \prod_{i=1}^{k} \frac{(-t_i)^{k_i}}{k_i!} \mathcal{M}_g((q_i, k_i)_{1 \leq i \leq k})$$

is a generating function of integer numbers $\mathcal{M}_g((q_i, k_i)_{1 \leq i \leq k})$ that count certain graphs called maps. A map is a connected oriented graph that is embedded into a surface. Its genus g is by definition the genus of a surface in which it can be embedded in such a way that edges do not cross and the faces of the graph (that are defined by following the boundary of the graph) are homeomorphic to a disk. The vertices of the maps we shall consider will have the structure of a star, that is a vertex with colored edges embedded into a surface (that is an order on the colored edges is specified). More precisely, a star of type q, for some monomial $q = X_{\ell_1} \cdots X_{\ell_k}$, is a vertex with degree $\deg(q)$ and oriented colored half-edges with one marked half edge of color ℓ_1, the second of color ℓ_2, etc., until the last one of color ℓ_k. $\mathcal{M}_g((q_i, k_i)_{1 \leq i \leq k})$ is then the number of maps with k_i stars of type q_i, $1 \leq i \leq n$.

Adding to V a term $t\,q$ for some monomial q and identifying the first-order derivative with respect to t at $t = 0$ we derive from (III.15)

$$\int \mathbf{L}(q) d\mu_V^N = \sum_{g \geq 0} N^{-2g} \sum_{k_1, \ldots, k_n \in \mathbb{N}^k} \prod_{i=1}^{k} \frac{(-t_i)^{k_i}}{k_i!} \mathcal{M}_g((q_i, k_i)_{1 \leq i \leq k}, (q, 1))\,.$$

$$\tag{III.16}$$

The equalities (III.15) and (III.16) derived in [50] are only formal, i.e., mean that all the derivatives on both sides of the equality coincide at $\mathbf{t} = 0$. They can thus be deduced from the Wick formula (which gives the expression of arbitrary moments of Gaussian variables) or equivalently by the use of Feynman diagrams.

Even though topological expansions such as (III.15) and (III.16) were first introduced by 't Hooft in the course of computing the integrals, the natural reverse question of computing the numbers $\mathcal{M}_g((q_i, k_i)_{1 \leq i \leq k})$ by studying the associated integrals over matrices encountered a large success in theoretical physics (see, e.g., the review papers [70, 78]). In the course of doing so, one would like for instance to compute $\lim_{N \to \infty} N^{-2} \log Z_{\mathbf{t}}^N$ and claim that this limit is equal to $F_0(\mathbf{t})$. There is here the belief that one can interchange derivatives and limit, a claim that needs to be justified.

We shall indeed prove that the formal limit can be strenghtened into a large N expansion in the sense that

$$\frac{1}{N^2} \log Z_{\mathbf{t}}^N = F_0(\mathbf{t}) + \frac{1}{N^2} F_1(\mathbf{t}) + o(N^{-2})$$

where $N^2 \times o(N^{-2})$ goes to zero as N goes to infinity. This asymptotic expansion holds when V is small and satisfies some convexity hypothesis (which insures that the partition function Z_V^N is finite and the support of the limiting spectral measures of $\mathbf{A}_i$, $1 \leq i \leq m$, under μ_V^N is connected, see [106]).

This part summarizes results from [104] and [105]. The full expansion (i.e., higher-order corrections) was obtained by E. Maurel Segala [148] in the multi-matrix setting. Such expansion in the one matrix case was already derived on a physical level of rigor in [4] and then made rigorous in [2, 86]. However, in the case of one matrix, techniques based on orthogonal polynomials can be used. In the multi-matrix case this approach fails in general (or at least has not yet been extended). [104, 105, 148] take a completely different route based on the free probability setting of limiting tracial states and of the so-called Master loop or Schwinger–Dyson equations.

We start this part by introducing the combinatorial objects we shall consider and their relations with non-commutative polynomials. Then, we prove the formal expansion of Brézin, Itzykson, Parisi and Zuber. The next two chapters consider the asymptotic expansion; we first obtain the convergence of the free energy towards the expected generating function for the enumeration of planar maps, and then study the first order correction to this limit, showing it is related with the enumeration of maps with genus one.

The techniques we shall present here have the advantage to be robust. We use them here to study partition functions of Hermitian matrices, but they can be generalized to orthogonal or symplectic matrices (in a work in progress of E. Maurel Segala) or to matrices following the Haar measure on the unitary group [66]. The last extension is particularly interesting since then Gaussian calculus and Feynman diagram techniques fail (since unitary matrices have no Gaussian entries) so that the diagrammatic representation of the limit is not straightforward even on a formal level (see [65] for a formal expansion with no diagrammatic interpretation).

7

Maps and Gaussian Calculus

We start this chapter by introducing non-commutative polynomials and their relations with special vertices called stars. We then relate the enumeration of the maps buildt upon such vertices with the formal expansion of Gaussian matrix integrals.

7.1 Combinatorics of Maps and Non-commutative Polynomials

In this section, we define non-commutative polynomials and non-commutative laws such as the "empirical distribution" of matrices $\mathbf{A}_1, \ldots, \mathbf{A}_m$ which generalize the notion of probability measures and empirical measures to the case of non-commutative variables. We will then describe precisely the combinatorial objects related with matrix integrals. Recalling the bijection between non-commutative monomials and graphical objects such as stars or ordered sets of colored points, we will show how operations such as derivatives on monomials have their graphical interpretation. This will be our basis to show that some differential equations for non-commutative laws can be interpreted in terms of some surgery on maps, as introduced by Tutte [194] to prove induction relations for map enumeration (see, e.g., Bender and Canfield [29] for generalizations).

7.2 Non-Commutative Polynomials

We denote by $\mathbb{C}\langle X_1, \ldots, X_m \rangle$ the set of complex polynomials in the non-commutative unknowns $X_1, \ldots, X_m$. Let $*$ denote the linear involution such that for all complex z and all monomials

$$(zX_{i_1} \cdots X_{i_p})^* = \bar{z} X_{i_p} \cdots X_{i_1}.$$

A. Guionnet, *Large Random Matrices: Lectures on Macroscopic Asymptotics*, Lecture Notes in Mathematics 1957, DOI: 10.1007/978-3-540-69897-5_7, © 2009 Springer-Verlag Berlin Heidelberg, Reprint by Springer-Verlag Berlin Heidelberg 2012

We will say that a polynomial P is self-adjoint if $P = P^*$ and denote by $\mathbb{C}\langle X_1, \ldots, X_m \rangle_{sa}$ the set of self-adjoint elements of $\mathbb{C}\langle X_1, \ldots, X_m \rangle$.

The potential V will later on be assumed to be self-adjoint. This means that

$$V(\mathbf{A}) = \sum_{j=1}^{n} t_j q_i = \sum_{j=1}^{n} \bar{t}_j q_j^* = \sum_{j=1}^{n} \Re(t_j) \frac{q_j + q_j^*}{2} + \sum_{j=1}^{n} \Im(t_j) \frac{q_j - q_j^*}{2i}.$$

Note that the parameters $(t_j = \Re(t_j) + i\Im(t_j), 1 \le j \le n)$ may a priori be complex. This hypothesis guarantees that $\mathrm{Tr}(V(\mathbf{A}))$ is real for all $\mathbf{A} = (\mathbf{A}_1, \ldots, \mathbf{A}_m)$ in the set $\mathcal{H}_N$ of $N \times N$ Hermitian matrices.

In the sequel, the monomials $(q_i)_{1 \le i \le n}$ will be fixed and we will consider $V = V_{\mathbf{t}} = \sum_{i=1}^{n} t_i q_i$ as the parameters t_i vary in such a way that V stays self-adjoint.

7.2.1 Convexity

We shall assume hereafter that V is convex, see Definition 6.10. While it may not be the optimal hypothesis, convexity provides many simple arguments. Note that as we add a Gaussian potential $\frac{1}{2} \sum_{i=1}^{m} X_i^2$ to V we can relax the hypothesis by the notion of c-convexity.

Definition 7.1. *We say that V is c-convex if $c > 0$ and $V + \frac{1-c}{2} \sum_{1}^{m} X_i^2$ is convex. In other words, the Hessian of*

$$\phi_V^{N,c} : \begin{array}{ccc} \mathcal{E}_N^{(2)} & \longrightarrow & \mathbb{R} \\ (\Re(A_k(i,j)), \Im(A_k(i,j)))_{1 \le i \le j \le N}^{1 \le k \le m} & \longrightarrow & \mathrm{Tr}(V(\mathbf{A}_1, \ldots, \mathbf{A}_m)) \\ & & + \frac{1-c}{2} \sum_{k=1}^{m} \mathbf{A}_i^2) \end{array}$$

is non-negative. Here, for $k \in \{1, \ldots, m\}$, $\mathbf{A}_k$ is the Hermitian matrix with entries $\sqrt{2}^{-1}(A_k(p,q) + iA_k(q,p))$ above the diagonal and $A_k(i,i)$ on the diagonal.

Note that when V is c-convex, μ_V^N has a log-concave density with respect to the Lebesgue measure so that many results from the previous part will apply, in particular concentration inequalities and Brascamp–Lieb inequalities.

In the rest of this chapter, we assume that V is c-convex for some $c > 0$ fixed. Arbitrary potentials could be considered as far as first-order asymptotics are studied in [104], at the price of adding a cutoff. In fact, adding a cutoff and choosing the parameters t_i's small enough (depending eventually on this cutoff), forces the interaction to be convex so that most of the machinery we are going to describe will apply also in this context. We let $V = V_{\mathbf{t}} = \sum_{i=1}^{n} t_i q_i$ and define $U_c = \{\mathbf{t} \in \mathbb{C}^n : V_{\mathbf{t}} \text{ is } c\text{-convex}\} \subset \mathbb{C}^n$. Moreover, B_η will denote the open ball in $\mathbb{C}^n$ centered at the origin and with radius $\eta > 0$ (for instance for the metric $|\mathbf{t}| = \max_{1 \le i \le n} |t_i|$).

7.2.2 Non-commutative Derivatives

First, for $1 \leq i \leq m$, let us define the non-commutative derivatives ∂_i with respect to the variable X_i. They are linear maps from $\mathbb{C}\langle X_1, \ldots, X_m \rangle$ to $\mathbb{C}\langle X_1, \ldots, X_m \rangle^{\otimes 2}$ given by the Leibniz rule

$$\partial_i PQ = \partial_i P \times (1 \otimes Q) + (P \otimes 1) \times \partial_i Q \qquad (7.1)$$

and $\partial_i X_j = \mathbf{1}_{i=j} 1 \otimes 1$. Here, $\times$ is the multiplication on $\mathbb{C}\langle X_1, \ldots, X_m \rangle^{\otimes 2}$; $P \otimes Q \times R \otimes S = PR \otimes QS$. So, for a monomial P, the following holds:

$$\partial_i P = \sum_{P = RX_i S} R \otimes S$$

where the sum runs over all possible monomials R, S so that P decomposes into $RX_i S$. We can iterate the non-commutative derivatives; for instance ∂_i^2 : $\mathbb{C}\langle X_1, \ldots, X_m \rangle \rightarrow \mathbb{C}\langle X_1, \ldots, X_m \rangle \otimes \mathbb{C}\langle X_1, \ldots, X_m \rangle \otimes \mathbb{C}\langle X_1, \ldots, X_m \rangle$ is given for a monomial function P by

$$\partial_i^2 P = 2 \sum_{P = RX_i S X_i Q} R \otimes S \otimes Q.$$

We denote by $\sharp : \mathbb{C}\langle X_1, \ldots, X_m \rangle^{\otimes 2} \times \mathbb{C}\langle X_1, \ldots, X_m \rangle \rightarrow \mathbb{C}\langle X_1, \ldots, X_m \rangle$ the map $P \otimes Q \sharp R = PRQ$ and generalize this notation to $P \otimes Q \otimes R \sharp (S, V) = PSQVR$. So $\partial_i P \sharp R$ corresponds to the derivative of P with respect to X_i in the direction R, and similarly $2^{-1}[D_i^2 P \sharp (R, S) + D_i^2 P \sharp (S, R)]$ the second derivative of P with respect to X_i in the directions R, S.

We also define the so-called cyclic derivative D_i. If m is the map $m(A \otimes B) = BA$, let us define $D_i = m \circ \partial_i$. For a monomial P, $D_i P$ can be expressed as

$$D_i P = \sum_{P = RX_i S} SR.$$

7.2.3 Non-commutative Laws

For $(\mathbf{A}_1, \ldots, \mathbf{A}_m) \in \mathcal{H}_N^m$, let us define the linear form $\mathbf{L}_{\mathbf{A}_1, \ldots, \mathbf{A}_m}$ from $\mathbb{C}\langle X_1, \ldots, X_m \rangle$ into $\mathbb{C}$ by

$$\mathbf{L}_{\mathbf{A}_1, \ldots, \mathbf{A}_m}(P) = \frac{1}{N} \mathrm{Tr}\left(P(\mathbf{A}_1, \ldots, \mathbf{A}_m)\right)$$

where Tr is the standard trace $\mathrm{Tr}(A) = \sum_{i=1}^N A(i, i)$. $\mathbf{L}_{\mathbf{A}_1, \ldots, \mathbf{A}_m}$ will be called the empirical distribution of the matrices (note that in the case of one matrix, it is the empirical distribution of the eigenvalues of this matrix). When the matrices $\mathbf{A}_1, \ldots, \mathbf{A}_m$ are generic and distributed according to μ_V^N, we will drop the subscripts $\mathbf{A}_1, \ldots, \mathbf{A}_m$ and write for short $\hat{L}^N = \mathbf{L}_{\mathbf{A}_1, \ldots, \mathbf{A}_m}$. We define, when $V = V_{\mathbf{t}} = \sum_{i=1}^n t_i q_i$,

$$\bar{\mathbf{L}}_t^N(P) := \mu_{V_t}^N[\hat{\mathbf{L}}^N(P)].$$

$\hat{\mathbf{L}}^N, \bar{\mathbf{L}}_t^N$ will be seen as elements of the algebraic dual $\mathbb{C}\langle X_1, \ldots, X_m \rangle^{\mathcal{D}}$ of $\mathbb{C}\langle X_1, \ldots, X_m \rangle$ equipped with the involution $*$. $\mathbb{C}\langle X_1, \ldots, X_m \rangle^{\mathcal{D}}$ is equipped with its weak topology.

Definition 7.2. *A sequence $(\mu_n)_{n \in \mathbb{N}}$ in $\mathbb{C}\langle X_1, \ldots, X_m \rangle^{\mathcal{D}}$ converges weakly (or in moments) to $\mu \in \mathbb{C}\langle X_1, \ldots, X_m \rangle^{\mathcal{D}}$ iff for any $P \in \mathbb{C}\langle X_1, \ldots, X_m \rangle$,*

$$\lim_{n \to \infty} \mu_n(P) = \mu(P).$$

Lemma 7.3. *Let $C(\ell_1, \ldots, \ell_r), \ell_i \in \{1, \ldots, m\}, r \in \mathbb{N}$, be finite non-negative constants and*

$$K(C) = \{\mu \in \mathbb{C}\langle X_1, \ldots, X_m \rangle^{\mathcal{D}}; |\mu(X_{\ell_1} \cdots X_{\ell_r})| \le C(\ell_1, \ldots, \ell_r)$$
$$\forall \ell_i \in \{1, \ldots, m\}, r \in \mathbb{N}\}.$$

Then, any sequence $(\mu_n)_{n \in \mathbb{N}}$ in $K(C)$ is sequentially compact, i.e., has a subsequence $(\mu_{\phi(n)})_{n \in \mathbb{N}}$ that converges weakly (or in moments).

Proof. Since $\mu_n(X_{\ell_1} \cdots X_{\ell_r}) \in \mathbb{C}$ is uniformly bounded, it has converging subsequences. By a diagonalization procedure, since the set of monomials is countable, we can ensure that for a subsequence $(\phi(n), n \in \mathbb{N})$, the terms $\mu_{\phi(n)}(X_{\ell_1} \cdots X_{\ell_r}), \ell_i \in \{1, \ldots, m\}, r \in \mathbb{N}$ converge simultaneously. The limit defines an element of $\mathbb{C}\langle X_1, \ldots, X_m \rangle^{\mathcal{D}}$ by linearity. $\square$

The following is a triviality, that however we recall since we will use it several times.

Corollary 7.4. *Let $C(\ell_1, \ldots, \ell_r), \ell_i \in \{1, \ldots, m\}, r \in \mathbb{N}$, be finite non negative constants and $(\mu_n)_{n \in \mathbb{N}}$ a sequence in $K(C)$ that has a unique limit point. Then $(\mu_n)_{n \in \mathbb{N}}$ converges weakly (or in moments) to this limit point.*

Proof. Otherwise we could choose a subsequence that stays at positive distance of this limit point, but extracting again a converging subsequence gives a contradiction. Note as well that any limit point will belong automatically to $\mathbb{C}\langle X_1, \ldots, X_m \rangle^{\mathcal{D}}$. $\square$

Remark 7.5. *The laws $\hat{\mathbf{L}}^N, \bar{\mathbf{L}}_t^N$ are more than only linear forms on the space $\mathbb{C}\langle X_1, \ldots, X_m \rangle$; they satisfy also the properties*

$$\mu(PP^*) \ge 0, \quad \mu(PQ) = \mu(QP), \quad \mu(1) = 1 \tag{7.2}$$

for any polynomial functions P, Q. Since these conditions are closed for the weak topology, we see that any limit point of $\hat{\mathbf{L}}^N, \bar{\mathbf{L}}_t^N$ will also satisfy these properties. A linear functional on $\mathbb{C}\langle X_1, \ldots, X_m \rangle$ that satisfies such conditions are called tracial states, or non-commutative laws. This leads to the notion of C^*-algebras and representations of the laws as moments of non-commutative operators in C^*-algebras. However, we do not want to detail this point in these notes.

7.3 Maps and Polynomials

In this section, we complete Section 3.1 to describe the graphs that will be enumerated by matrix models. Let $q(X_1, \ldots, X_m) = X_{\ell_1} X_{\ell_2} \cdots X_{\ell_k}$ be a monomial in m non-commutative variables.

Hereafter, monomials $(q_i)_{1 \leq i \leq n}$ will be fixed and we will write for short, for $\mathbf{k} = (k_1, \ldots, k_n)$,

$$\mathcal{M}_{\mathbf{k}}^g = \text{card}\{ \text{ maps with genus } g$$

$$\text{and } k_i \text{ stars of type } q_i, \, 1 \leq i \leq n\},$$

and for a monomial P,

$$\mathcal{M}_{\mathbf{k}}^g(P) = \text{card}\{ \text{ maps with genus } g$$

$$k_i \text{ stars of type } q_i, \, 1 \leq i \leq n \text{ and one of type } P\}.$$

7.3.1 Maps and Polynomials

Because there is a one-to-one mapping between stars and monomials, the operations on monomials such as involution or derivatives have their graphical interpretation.

The involution comes to reverse the orientation and to shift the marked edge by one in the sense of the new orientation (see Figure 7.1). This is equivalent to considering the star in a mirror.

For derivations, the interpretation goes as follows.

Let q be a given monomial. The derivation ∂_i appears as a way to find out how to decompose a star of type q by pointing out a half-edge of color i: a star of type q can indeed be decomposed into one star of type q_1, one half-edge of color i and another star of type q_2, all sharing the same vertex, iff q can be

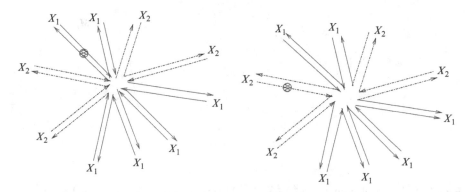

Fig. 7.1. A star of type q versus a star of type q^*

written as $q = q_1 X_i q_2$. This is particularly useful to write induction relation on the number of maps. For instance, let us consider a planar map M and the event $A_M(X_i q)$ that, inside M, a star of type $X_i q$ is such that the first marked half-edge is glued to a half-edge of q. Then, if this happens, since the map is planar, it will be decomposed into two planar maps separated by the edge between these two X_i. Such a gluing can be done only with the edges X_i appearing in the decomposition of q as $q = q_1 X_i q_2$. Moreover, the two stars of type q_1 and q_2 will belong to two "independent" planar maps. So, we can symbolically write

$$1_{A_M(X_i q)} = \sum_{q=q_1 X_i q_2} 1_{q_1 \in M_1} \otimes_{M=M_1 \otimes_i M_2} 1_{q_2 \in M_2} \tag{7.3}$$

where $M = M_1 \otimes_i M_2$ means that M decomposes into two planar maps M_1 and M_2, M_2 being surrounded by a cycle of color i that separates it from M_1 (see Figure 7.2). Note here that we forgot in some sense that these three objects were sharing the same vertex; this is somehow irrelevant here since a vertex is finally nothing but the point of junction of several edges; as long as we are concerned with the combinatorial problem of enumerating these maps, we can safely split the map M into these three objects. (7.3) is very close to the derivation operation ∂_i.

Similarly, let us consider again a planar map M containing given stars of type $X_i q$ and q' and the event $B_M(X_i q, q')$ that, inside M, the star of type

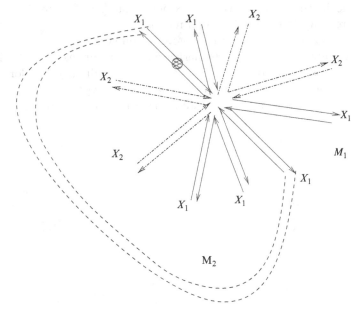

Fig. 7.2. A star of type $q = X_1^2 X_2^2 X_1^4 X_2^2$ decomposed into $X_1(X_1 X_2^2 X_1) X_1 (X_1^2 X_2^2)$

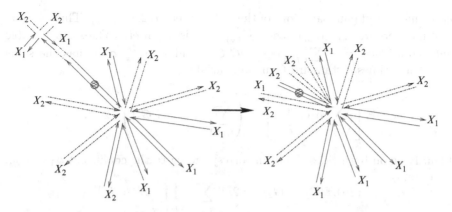

Fig. 7.3. Merging of a star of type $q = X_1^2 X_2^2 X_1^4 X_2^2$ and a star of type $X_1^2 X_2^2$

$X_i q$ is such that the first marked half-edge is glued to a half-edge of the star of type q'. Once we know that this happens, we can write

$$1_{B_M(X_i q, q')} = \sum_{q' = q_1 X_i q_2} 1_{q_2 q_1 \bullet_i q \in M}. \tag{7.4}$$

$q_2 q_1 \bullet_i q$ is a new star made of a star of type q and one of type $q_2 q_1$ with an edge of color i from one to the other just before the marked half-edges. Again, once we know that this edge of color i exists, from a combinatorial point of view, we can simply shorten it till the two stars merge into a bigger star of type $q_2 q_1 q$. This is the merging operation; it corresponds to the cyclic derivative D_i (see Figure 7.3).

7.4 Formal Expansion of Matrix Integrals

The expansion obtained by 't Hooft is based on Feynman diagrams, or equivalently on Wick's formula that can be stated as follows.

Lemma 7.6 (Wick's formula). *Let $(G_1, \ldots, G_{2n})$ be a Gaussian vector such that $\mathbb{E}[G_i] = 0$ for $i \in \{1, \ldots, 2n\}$. Then,*

$$\mathbb{E}[G_1 \cdots G_{2n}] = \sum_{\pi \in PP(2n)} \prod_{\substack{(b,b') \text{ block of } \pi, \\ b < b'}} \mathbb{E}[G_b G_{b'}]$$

where the sum runs over all pair-partitions of the ordered set $\{1, \ldots, 2n\}$.

Proof. Recall that if G is a standard Gaussian variable, for all $n \in \mathbb{N}$,

$$\mathbb{E}[G^{2n}] = 2n!! := \frac{(2n)!}{2^n n!}$$

is the number of pair-partitions of the ordered set $\{1, 2, \ldots, 2n\}$. Thus, for any real numbers $(\alpha_1, \ldots, \alpha_{2n})$, since $\sum_{i=1}^{2n} \alpha_i G_i$ is a centered Gaussian variable with covariance $\sigma^2 = \sum_{i,j=1}^{2n} \alpha_i \alpha_j \mathbb{E}[G_i G_j]$, and so $\sum_{i=1}^{2n} \alpha_i G_i$ has the same law that σ times a standard Gaussian variable,

$$\mathbb{E}\left[\left(\sum_{i=1}^{2n} \alpha_i G_i\right)^{2n}\right] = \left(\sum_{i,j=1}^{2n} \alpha_i \alpha_j \mathbb{E}[G_i G_j]\right)^n 2n!!.$$

Identifying on both sides the term corresponding to the coefficient $\alpha_1 \cdots \alpha_{2n}$, we obtain

$$(2n)! \mathbb{E}[G_1 \cdots G_{2n}] = 2n!! \sum_{\pi \in \Sigma} \prod_{(b,b') \in \pi} \mathbb{E}[G_b G_{b'}]$$

where Σ is the set of pairs of $2n$ elements. To compare this set with the collection of pairings of an ordered set, we have to order the elements of the pairs, and we have 2^n possible choices, and then order the pairs, which gives another $n!$ possible choices. Thus,

$$\sum_{\pi \in \Sigma} \prod_{(b,b') \in \pi} \mathbb{E}[G_b G_{b'}] = 2^n n! \sum_{\pi \in PP(2n)} \prod_{(i,j) \text{ block of } \pi} \mathbb{E}[G_i G_j]$$

completes the argument as $2^n n! 2n!! = (2n)!$.   □

We now consider moments of traces of Gaussian Wigner's matrices. Since we shall consider the moments of products of several traces, we shall now use the language of stars. Let us recall that a star of type $q(X) = X_{\ell_1} \cdots X_{\ell_2}$ is a vertex equipped with k colored half-edges, one marked half-edge and an orientation such that the marked half-edge is of color ℓ_1, the second (following the orientation) of color ℓ_2, etc., till the last half-edge of color ℓ_k. The graphs we shall enumerate will be obtained by gluing pairwise the half-edges.

Definition 7.7. *Let $r, m \in \mathbb{N}$. Let $q_1, \ldots, q_r$ be r monomials in m non-commutative variables. A map of genus g with a star of type q_i for $i \in \{1, \ldots, r\}$ is a connected graph embedded into a surface of genus g with r vertices so that*

1. *For $1 \leq i \leq r$, one of the vertices has degree $deg(q_i)$, and this vertex is equipped with the structure of a star of type q_i (i.e., with the corresponding colored half-edges embedded into the surface in such a way that the orientation of the star and the orientation of the surface agree). The half-edges inherit the orientation of their stars, i.e., each side of each half-edge is endowed with an opposite orientation corresponding to the orientation of a path traveling around the star by following the orientation of the star.*
2. *The half-edges of the stars are glued pair-wise and two half-edges can be glued iff they have the same color and orientation; thus edges have only one color and one orientation.*

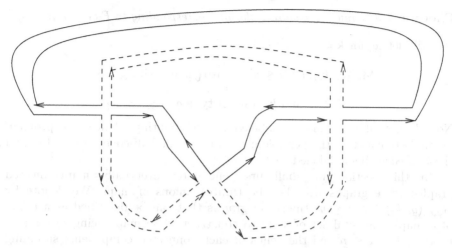

Fig. 7.4. A planar bi-colored map with stars of type $q_1 = X_1 X_2 X_1 X_2$, $q_2 = X_1^2 X_2^2$, $q_3 = X_1 X_2 X_1 X_2$

3. *A path traveling along the edges of the map following their orientation will make a loop. The surface inside this loop is homeomorphic to a disk and called a face (see Figure 7.4).*

Note that each star has a distinguished half-edge and so each half-edge of a star is labeled. Moreover, all stars are labeled. Hence, the enumeration problem we shall soon consider can be thought as the problem of matching the labeled half-edges of the stars and so we will distinguish all the maps where the gluings are not done between exactly the same set of labeled half-edges, regardless of symmetries. This is important to make clear since we shall shortly consider enumeration issues. The genus of a map is defined as in Definition 1.8. Note that since at each vertex we imposed a cyclic orientation at the ends of the edges adjacent to this vertex, there is a unique way to embed the graph drawn with stars in a surface; we have to draw the stars so that their orientation agrees with the orientation of the surface.

There is a dual way to consider maps in the spirit of Figure 1.5; as in the figure in the center of Figure 1.5, we can replace a star of type $q(X) = X_{i_1} \cdots X_{i_p}$ by a polygon (of type q) with p faces, a boundary edge of the polygon replacing an edge of the star and taking the same color as the edge, and a marked boundary edge and an orientation. A map is then a covering of a surface (with the same genus as the map) by polygonals of type $q_1, \ldots, q_r$. The constraint on the colors becomes a constraint on the colors of the sides of the polygons of the covering.

Example 1. A triangulation (resp. a quadrangulation) of a surface of genus g by F faces (the number of triangles, resp. squares) is equivalent to a map of genus g with F stars of type $q(X) = X^3$ (resp. $q(X) = X^4$).

Exercise 7.8. *Draw the quadrangulation corresponding to Figure 7.4.*

We will define, for $\mathbf{k} = (k_1, \ldots, k_n)$,

$$\mathcal{M}_g((q_i, k_i), 1 \leq i \leq n) = \text{card}\{ \text{ maps with genus } g$$

$$\text{and } k_i \text{ stars of type } q_i, 1 \leq i \leq n\}.$$

Note here that the stars are labeled in the counting. Hence, the problem amounts to counting the possible matchings of the half-edges of the stars, all the half-edges being labeled.

In this section we shall first encounter eventually non-connected graphs; these graphs will then be (finite) unions of maps. We denote by $G_{g,c}((q_i, k_i), 1 \leq i \leq n)$ the set of graphs that can be described as a union of c maps, the total set of stars to construct these maps being k_i stars of type q_i, $1 \leq i \leq n$ and the genus of each connected components summing up to g. When counting these graphs, we will also assume that all half-edges are labeled. Moreover, we shall count these graphs up to homeomorphism, that is up to continuous deformation of the surface on which the graphs are embedded. Thus, our problem is to enumerate the number of possible pairings of the half-edges (of a given color) of the stars in such a way that the resulting graph has a given genus.

We now argue that

Lemma 7.9. *Let $q_1, \ldots, q_n$ be monomials. Then,*

$$\int \prod_{i=1}^{n} (N\text{Tr}(q_i(\mathbf{A}_1, \ldots, \mathbf{A}_m)))d\mu^N(\mathbf{A}_1) \cdots d\mu^N(\mathbf{A}_m)$$

$$= \sum_{g \in \mathbb{N}} \sum_{c \geq 1} \frac{1}{N^{2g-2c}} \sharp\{G_{g,c}((q_i, 1), 1 \leq i \leq n)\}.$$

Here, $G_{g,c}((q_i, 1), 1 \leq i \leq n)$ is the set of unions of c maps drawn on the stars of type $(q_i)_{1 \leq i \leq n}$, with the sum of the genera of each map equal to g. $\sharp\{G_{g,c}((q_i, 1), 1 \leq i \leq n)\}$ is the number of graphs of the set $G_{g,c}((q_i, 1), 1 \leq i \leq n)$.

As a warm-up, let us show the following:

Lemma 7.10. *Let q be a monomial. Then, we have the following expansion*

$$\int N^{-1}\text{Tr}(q(\mathbf{A}_1, \ldots, \mathbf{A}_m))d\mu^N(\mathbf{A}_1) \cdots d\mu^N(\mathbf{A}_m) = \sum_{g \in \mathbb{N}} \frac{1}{N^{2g}} \sharp\{G_g((q, 1))\}$$

where $\sharp\{G_0((q, 1))\}$ equals $\sigma^m(q)$ as found by Voiculescu, Theorem 3.3.

Proof. As usual we expand the trace and write, if $q(X_1, \ldots, X_m) = X_{j_1} \cdots X_{j_k}$,

$$\int \mathrm{Tr}(q(\mathbf{A}_1, \ldots, \mathbf{A}_m)) d\mu^N(\mathbf{A}_1) \cdots d\mu^N(\mathbf{A}_m)$$

$$= \sum_{1 \leq r_1, \ldots, r_k \leq N} \int A_{j_1}(r_1, r_2) \cdots A_{j_k}(r_k, r_1) d\mu^N(\mathbf{A}_1) \cdots d\mu^N(\mathbf{A}_m)$$

$$= \sum_{r_1, \ldots, r_k} \sum_{\pi \in PP(k)} \prod_{\substack{(wv) \text{ block of } \pi \\ w < v}} \mathbb{E}[A_{j_w}(r_w, r_{w+1}) A_{j_v}(r_v, r_{v+1})]. \quad (7.5)$$

Note that $\prod \mathbb{E}[A_{j_w}(r_w, r_{w+1}) A_{j_v}(r_v, r_{v+1})]$ is either zero or $N^{-k/2}$. It is not zero only when $j_w = j_v$ and $r_w r_{w+1} = r_{v+1} r_v$ for all the blocks (v, w) of π. Hence, if we represent q by the star of type q, we see that all the graphs where the half-edges of the star are glued pairwise and colorwise will give a contribution. But how many indices will give the same graph? To represent the indices on the star, we fatten the half-edges as double half-edges. Thinking that each random variable sits at the end of the half-edges, we can associate to each side of the fat half-edge one of the indices of the entry (see Figure 7.5). When the fattened half-edges meet at the vertex, observe that each side of the fattened half-edges meets one side of an adjacent half-edge on which sits the same index. Hence, we can say that the index stays constant over the broken line made of the union of the two sides of the fattened half-edges.

When gluing pairwise the fattened half-edges we see that the condition $r_w r_{w+1} = r_{v+1} r_v$ means that the indices are the same on each side of the half-edge and hence stay constant on the resulting edge. The connected lines made with the sides of the fattened edges can be seen to be the boundaries of the faces of the correponding graphs. Therefore we have exactly N^F possible choices of indices for a graph with F faces. These graphs are otherwise connected, with one star of type q. (7.5) thus shows that

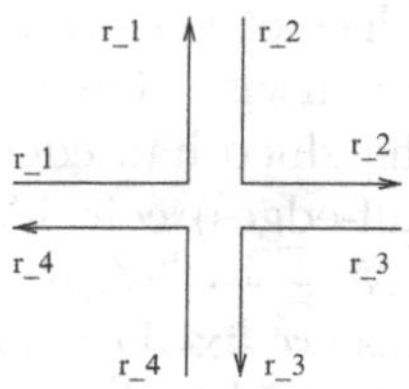

Fig. 7.5. Star of type X^4 with prescribed indices

$$\int \mathrm{Tr}(q(\mathbf{A}_1,\ldots,\mathbf{A}_m))d\mu^N(\mathbf{A}_1)\cdots d\mu^N(\mathbf{A}_m)$$

$$=\sum_{g\geq 0}\frac{N^F}{N^{\frac{k}{2}}}\sharp\{\text{maps with one star of type } q \text{ and } F \text{ faces}\}$$

Recalling that $2-2g=F+\sharp$ vertices $-\sharp$ edges $=F+1-k/2$ completes the proof. $\qquad\qquad\square$

Remark 7.11. *In the above it is important to take μ^N to be the law of the GUE (and not GOE for instance) to insure that $E[(A_k)_{ij}(A_k)_{ji}]=1/N$ but $E[((A_k)_{ij})^2]=0$. The GOE leads to the enumeration of other combinatorial objects (and in particular an expansion in N^{-1} rather than N^{-2}).*

Proof of Lemma 7.9. We let $q_i(X_1,\ldots,X_m)=X_{\ell_1^i}\cdots X_{\ell_{d_i}^i}$. As usual, we expand the traces:

$$\int \prod_{i=1}^n (N\mathrm{Tr}(q_i(\mathbf{A}_1,\ldots,\mathbf{A}_m)))d\mu^N(\mathbf{A}_1)\cdots d\mu^N(\mathbf{A}_m)$$

$$=N^n \sum_{\substack{i_1^k,\ldots,i_{d_k}^k\\1\leq k\leq n}} \mathbb{E}[\prod_{1\leq k\leq n}A_{\ell_1^k}(i_1^k i_2^k)\cdots A_{\ell_{d_k}^k}(i_{d_k}^k i_1^k)]$$

$$=N^n \sum_{\substack{i_1^k,\ldots,i_{d_k}^k\\1\leq k\leq n}} \sum_{\pi\in PP(\sum d_i)} Z(\pi,\mathbf{i})$$

where in the last line we used Wick's formula, π is a pair partition of the edges $\{(i_j^k, i_{j+1}^k)_{1\leq j\leq d_k-1}, (i_{d_k}^k, i_1^k), 1\leq k\leq n\}$ and $Z(\pi,\mathbf{i})$ is the product of the variances over the corresponding blocks of the partition. A pictorial way to represent this sum over $PP(\sum d_i)$ is to represent $X_{\ell_1^k}(i_1^k i_2^k)\cdots X_{\ell_{d_k}^k}(i_{d_k}^k i_1^k)$ by its associated star of type q_k, for $1\leq k\leq n$. Note that in the counting this star will be labeled (here by the number k). A partition π is represented by a pairwise gluing of the half-edges of the stars. $Z(\pi)$, as the product of the variances, vanishes unless each pairwise gluing is done in such a way that the indices written at the end of the glued half-edges coincide and the number of the variable (or color of the half-edges) coincide. Otherwise, each covariance being equal to N^{-1}, $Z(\pi,\mathbf{i})=N^{-\sum_{i=1}^n k_i/2}$. Note also that once the gluing is done, by construction the indices are fixed on the boundary of each face of the graph (this is due to the fact that $E[A_r(i,j)A_r(k,l)]$ is null unless $kl=ji$). Hence, there are exactly N^F possible choices of indices for a given graph, if F is the number of faces of this graph (note here that if the graph is disconnected, we count the number of faces of each connected part, including their external faces and sum the resulting numbers over all connected components). Thus, we find that

$$\sum_{\substack{i_1^k,\ldots,i_{d_k}^k \\ 1 \leq k \leq n}} \sum_{\pi \in PP(\sum d_i)} Z(\pi, \mathbf{i}) = \sum_{F \geq 0} \sum_{G \in G_F((q_i,1),1 \leq i \leq n)} N^{-\sum_{i=1}^n k_i/2} N^F$$

where G_F denotes the union of connected maps with a total number of faces equal to F (the external face of each map being counted). Note that for a connected graph, $2 - 2g = F - \sharp edges + \sharp vertices$. Because the total number of edges of the graphs is $\sharp edges = \sum_{i=1}^n k_i/2$ and the total number of vertices is $\sharp vertices = n$, we see that if $g_i, 1 \leq i \leq c$, are the genera of each connected component of our graph, we must have

$$2c - 2\sum_{i=1}^c g_i = F - \sum_{i=1}^n k_i/2 - n.$$

This completes the proof. $\square$

We then claim that we find the topological expansion of Brézin, Itzykson, Parisi and Zuber [50]:

Lemma 7.12. *Let $q_1, \ldots, q_n$ be monomials. Then, we have the following formal expansion*

$$\log\left(\int e^{\sum_{i=1}^n t_i N\mathrm{Tr}(q_i(\mathbf{A}_1,\ldots,\mathbf{A}_m))} d\mu^N(\mathbf{A}_1) \cdots d\mu^N(\mathbf{A}_m)\right)$$

$$= \sum_{g \geq 0} \frac{1}{N^{2g-2}} \sum_{k_1,\ldots,k_n \in \mathbb{N}^n \setminus \{0\}} \prod_{i=1}^n \frac{(t_i)^{k_i}}{k_i!} \mathcal{M}_g((q_i, k_i), 1 \leq i \leq n)$$

where the equality means that derivatives of all orders at $t_i = 0, 1 \leq i \leq n$, match.

Note here that the sum in the right-hand side is not absolutely convergent (in fact the left-hand side is in general infinite if the t_i's do not have the appropriate signs). However, we shall see in subsequent chapters that if we stop the expansion at $g \leq G < \infty$ (but keep the summation over all k_i's) the expansion is absolutely converging for sufficiently small t_i's.

Proof of Lemma 7.12. The idea is to expand the exponential. Again, this has no meaning in terms of convergent series (and so we do not try to justify uses of Fubini's theorem, etc.) but can be made rigorous by the fact that we only wish to identify the derivatives at $t = 0$ (and so the formal expansion is only a way to compute these derivatives). So, we find that

$$L := \int e^{\sum_{i=1}^{n} t_i N\mathrm{Tr}(q_i(\mathbf{A}_1,\ldots,\mathbf{A}_m))} d\mu^N(\mathbf{A}_1) \cdots d\mu^N(\mathbf{A}_m)$$

$$= \int \prod_{i=1}^{n} \left(e^{t_i N\mathrm{Tr}(q_i(\mathbf{A}_1,\ldots,\mathbf{A}_m))} \right) d\mu^N(\mathbf{A}_1) \cdots d\mu^N(\mathbf{A}_m)$$

$$= \int \prod_{i=1}^{n} \left(\sum_{k_i \geq 0} \frac{(t_i)^{k_i}}{k_i!} (N\mathrm{Tr}(q_i(\mathbf{A}_1,\ldots,\mathbf{A}_m)))^{k_i} \right) d\mu^N(\mathbf{A}_1) \cdots d\mu^N(\mathbf{A}_m)$$

$$= \sum_{k_1,\ldots,k_n \in \mathbb{N}} \frac{(t_1)^{k_1} \cdots (t_n)^{k_n}}{k_1! \cdots k_n!}$$

$$\int \prod_{i=1}^{n} (N\mathrm{Tr}(q_i(\mathbf{A}_1,\ldots,\mathbf{A}_m)))^{k_i} d\mu^N(\mathbf{A}_1) \cdots d\mu^N(\mathbf{A}_m)$$

$$= \sum_{k_1,\ldots,k_n \in \mathbb{N}} \frac{(t_1)^{k_1} \cdots (t_n)^{k_n}}{k_1! \cdots k_n!} \sum_{g \geq 0} \sum_{c \geq 0} \frac{1}{N^{2g-2c}} \sharp\{G_{g,c}((q_i,k_i), 1 \leq i \leq n)\}$$

$$(7.6)$$

where we finally used Lemma 7.9. Note that the case $c = 0$ is non-empty only when all the k_i's are null, and the resulting contribution is one. Now, we relate $\sharp\{G_{g,c}((q_i,k_i), 1 \leq i \leq n)\}$ with the number of maps. Since graphs in $G_{g,c}((q_i,k_i), 1 \leq i \leq n)$ can be decomposed into a union of disconnected maps, $\sharp\{G_{g,c}((q_i,k_i), 1 \leq i \leq n)\}$ is related with the ways to distribute the stars and the genus among the c maps, and the number of each of these maps. In other words, we have (since all stars are labeled)

$$\sharp\{G_{g,c}((q_i,k_i), 1 \leq i \leq n)\}$$

$$= \frac{1}{c!} \sum_{\substack{\sum_{i=1}^{c} g_i = g \\ g_i \geq 0}} \frac{g!}{g_1! \cdots g_c!} \sum_{\substack{\sum_{j=1}^{c} l_i^j = k_i \\ 1 \leq j \leq n}} \prod_{i=1}^{n} \frac{k_i!}{l_i^1! \cdots l_i^c!} \prod_{j=1}^{c} \mathcal{M}_g((q_i, l_i^j), 1 \leq i \leq n).$$

Plugging this expression into (7.6) we get

$$L := \sum_{k_1,\ldots,k_n \in \mathbb{N}} \frac{(t_1)^{k_1} \cdots (t_n)^{k_n}}{c! k_1! \cdots k_n!} \sum_{g \geq 0} \sum_{c \geq 0} \frac{1}{N^{2g-2c}} \sum_{\substack{\sum_{i=1}^{c} g_i = g \\ g_i \geq 0}} \frac{g!}{g_1! \cdots g_c!} \times$$

$$\sum_{\substack{\sum_{j=1}^{c} l_i^j = k_i \\ 1 \leq j \leq n}} \prod_{i=1}^{n} \frac{k_i!}{l_i^1! \cdots l_i^c!} \prod_{j=1}^{c} \mathcal{M}_g((q_i, l_i^j), 1 \leq i \leq n)$$

$$= \sum_{c \geq 0} \frac{1}{c!} \sum_{g = \sum_{i=1}^{c} g_i} \frac{g!}{g_1! \cdots g_c!}$$

$$\sum_{\substack{k_1,\dots,k_n \in \mathbb{N}}} \sum_{\substack{\sum_{j=1}^{c} l_i^j = k_i \\ 1 \le j \le n}} \prod_{j=1}^{c} \left(\frac{1}{N^{2g_j-2}} \prod_{i=1}^{n} \frac{(t_i)^{l_i^j}}{l_i^j!} \mathcal{M}_g((q_i, l_i^j), 1 \le i \le n) \right)$$

$$= \sum_{c \ge 0} \frac{1}{c!} \left(\sum_{g \ge 0} \frac{1}{N^{2g-2}} \sum_{l_1,\dots l_n \ge 0} \prod_{i=1}^{n} \frac{(t_i)^{l_i}}{l_i!} \mathcal{M}_g((q_i, l_i), 1 \le i \le n) \right)^{c}$$

$$= \exp \left(\sum_{g \ge 0} \frac{1}{N^{2g-2}} \sum_{l_1,\dots l_n \ge 0} \prod_{i=1}^{n} \frac{(t_i)^{l_i}}{l_i!} \mathcal{M}_g((q_i, l_i), 1 \le i \le n) \right)$$

which completes the proof.   □

The goal of subsequent chapters is to justify that this equality does not only hold formally but as a large N expansion. Instead of using Wick's formula, we shall base our analysis on differential calculus and its relations with Gaussian calculus (note here that Wick's formula might also have been proven by use of differential calculus). The point here will be that we can design an asymptotic framework for differential calculus, which will then encode the combinatorics of the first-order term in 't Hooft's expansion, that is, planar maps. To make this statement clear, we shall see that a nice set-up is when the potential $V = \sum t_i q_i$ possesses some convexity property.

Bibliographical Notes. The formal relation between Gaussian matrix integrals and the enumeration of maps first appeared in the work of 't Hooft [187] in the context of quantum chromodynamics, and soon used in many situations [32, 50] in relation with 2D gravity [51, 78, 99, 207], and with string theory [41, 79]. It was used as well in mathematics [113, 132, 133, 211].

8

First-order Expansion

At the end of this chapter (see Theorem 8.8) we will have proved that Lemma 7.12 holds as a first-order limit, i.e.,

$$\lim_{N\to\infty} \frac{1}{N^2} \log \int e^{\sum_{i=1}^{n} t_i N \mathrm{Tr}(q_i(\mathbf{A}_1,\ldots,\mathbf{A}_m))} d\mu^N(\mathbf{A}_1) \cdots d\mu^N(\mathbf{A}_m)$$

$$= \sum_{k_1,\ldots,k_n \in \mathbb{N}^n \setminus \{0\}} \prod_{i=1}^{n} \frac{(t_i)^{k_i}}{k_i!} \mathcal{M}_0((q_i, k_i), 1 \leq i \leq n)$$

provided the parameters $\mathbf{t} = (t_i)_{1\leq i\leq n}$ are sufficiently small and such that the polynomial $V = \sum t_i q_i$ is strictly convex (i.e., belong to $U_c \cap B_\eta$ for some $c > 0$ and $\eta \leq \eta(c)$ for some $\eta(c) > 0$). To prove this result we first show that, under the same assumptions, $\bar{\mathbf{L}}_t^N(q) = \mu^N_{\sum t_i q_i}(N^{-1}\mathrm{Tr}(q))$ converges as N goes to infinity to a limit that is as well related with map enumeration (see Theorem 8.4).

The central tool in our asymptotic analysis will be the so-called Schwinger–Dyson (or loop) equations. In finite dimension, they are simple emanation of the integration by parts formula (or, somewhat equivalently, of the symmetry of the Laplacian in $L^2(dx)$). As dimension goes to infinity, concentration inequalities show that $\bar{\mathbf{L}}_t^N$ approximately satisfies a closed equation that we will simply refer to as the Schwinger–Dyson equation. The limit points of $\bar{\mathbf{L}}_t^N$ will therefore satisfy this equation. We will then show that this equation has a unique solution in some small range of the parameters. As a consequence, $\bar{\mathbf{L}}_t^N$ will converge to this unique solution. Showing that an appropriate generating function of maps also satisfies the same equation will allow us to determine the limit of $\bar{\mathbf{L}}_t^N$.

8.1 Finite-dimensional Schwinger–Dyson Equations

Property 8.1. *For all $P \in \mathbb{C}\langle X_1, \ldots, X_m \rangle$, all $i \in \{1, \ldots, m\}$,*

$$\mu^N_{V_t}\left(\hat{\mathbf{L}}^N \otimes \hat{\mathbf{L}}^N(\partial_i P)\right) = \mu^N_{V_t}\left(\hat{\mathbf{L}}^N((X_i + D_i V_t)P)\right).$$

A. Guionnet, *Large Random Matrices: Lectures on Macroscopic Asymptotics*,
Lecture Notes in Mathematics 1957, DOI: 10.1007/978-3-540-69897-5_8,
© 2009 Springer-Verlag Berlin Heidelberg, Reprint by Springer-Verlag Berlin Heidelberg 2012

110   8 First-order Expansion

Proof. A simple integration by part shows that for any differentiable function f on $\mathbb{R}$ such that $fe^{-N\frac{x^2}{2}}$ goes to zero at infinity,

$$N\int f(x)xe^{-N\frac{x^2}{2}}dx = \int f'(x)e^{-N\frac{x^2}{2}}dx.$$

Such a result generalizes to a complex Gaussian by the remark that

$$N(x+iy)e^{-N\frac{|x|^2}{2}-N\frac{|y|^2}{2}} = -(\partial_x+i\partial_y)e^{-N\frac{|x|^2}{2}-N\frac{|y|^2}{2}}$$
$$= -\partial_{x-iy}e^{-N\frac{|x|^2}{2}-N\frac{|y|^2}{2}}.$$

As a consequence, applying such a remark to the entries of a Gaussian random matrix, we obtain for any differentiable function f of the entries, all $r,s \in \{1,\ldots,N\}^2$, all $r \in \{1,\ldots,m\}$,

$$N\int A_l(r,s)f(A_k(i,j),1\le i,j\le N,1\le k\le m)d\mu^N(\mathbf{A}_1)\cdots d\mu^N(\mathbf{A}_m) =$$

$$\int \partial_{A_l(s,r)}f(A_k(i,j),1\le i,j\le N,1\le k\le m)d\mu^N(\mathbf{A}_1)\cdots d\mu^N(\mathbf{A}_m).$$

Using repeatedly this equality, we arrive at

$$\int \frac{1}{N}\mathrm{Tr}(\mathbf{A}_k P)d\mu_V^N(\mathbf{A}) = \frac{1}{N^2}\sum_{i,j=1}^N \int \partial_{A_k(j,i)}(Pe^{-N\mathrm{Tr}(V)})_{ji}\prod d\mu^N(\mathbf{A}_i)$$

$$= \frac{1}{N^2}\sum_{i,j=1}^N \int \left(\sum_{P=QX_kR} Q_{ii}R_{jj}\right.$$

$$\left. -N\sum_{l=1}^n t_l \sum_{q_l=QX_kR}\sum_{h=1}^N P_{ji}Q_{hj}R_{ih}\right)d\mu_V^N(\mathbf{A})$$

$$= \int \left(\frac{1}{N^2}(\mathrm{Tr}\otimes\mathrm{Tr})(\partial_k P) - \frac{1}{N}\mathrm{Tr}(D_k VP)\right)d\mu_V^N(\mathbf{A})$$

which yields

$$\int \left(\hat{\mathbf{L}}^N((X_k+D_kV)P) - \hat{\mathbf{L}}^N\otimes\hat{\mathbf{L}}^N(\partial_k P)\right)d\mu_V^N(\mathbf{A}) = 0. \qquad (8.1)$$

□

8.2 Tightness and Limiting Schwinger–Dyson Equations

We say that $\tau \in \mathbb{C}\langle X_1,\ldots,X_m\rangle^{\mathcal{D}}$ satisfies the Schwinger–Dyson equation with potential V, denoted for short by **SD[V]**, if and only if for all $i \in \{1,\ldots,m\}$ and $P \in \mathbb{C}\langle X_1,\ldots,X_m\rangle$,

$$\tau(I) = 1, \quad \tau \otimes \tau(\partial_i P) = \tau((D_i V + X_i)P) \qquad \textbf{SD}[\textbf{V}].$$

We shall now prove the following:

Property 8.2. *Assume that V_t is c-convex. Then, $(\bar{\textbf{L}}_t^N = \mu_{V_t}^N(\hat{\textbf{L}}^N), N \in \mathbb{N})$ is tight. Its limit points satisfy* $\textbf{SD}[V_t]$ *and*

$$|\tau(X_{\ell_1} \cdots X_{\ell_r})| \leq M_0^r \tag{8.2}$$

for all $\ell_1, \ldots, \ell_r \in \mathbb{N}$, all $r \in \mathbb{N}$, with an M_0 that only depends on c.

Proof. By Lemma 6.19, we find that for all $\ell_1, \ldots, \ell_r$,

$$|\bar{\textbf{L}}_t^N(X_{\ell_1} \cdots X_{\ell_r})| \leq \mu_{V_t}^N(|\lambda_{max}(\textbf{A})|^r)$$

$$= \int_0^\infty r x^{r-1} \mu_{V_t}^N(|\lambda_{max}(\textbf{A})| \geq x) dx \tag{8.3}$$

$$\leq M_0^r + \int_{M_0}^\infty r x^{r-1} e^{-\alpha N x} dx$$

$$= M_0^r + r(\alpha N)^{-r} \int_0^\infty r x^{r-1} e^{-x} dx. \tag{8.4}$$

Hence, if $K(C)$ denotes the compact set defined in Lemma 7.4, $\bar{\textbf{L}}_t^N \in K(C)$ with $C(\ell_1, \ldots, \ell_r) = M_0^r + r\alpha^{-r} \int_0^\infty r x^{r-1} e^{-x} dx$. $(\bar{\textbf{L}}_t^N, N \in \mathbb{N})$ is therefore tight. Let us consider now its limit points; let τ be such a limit point. By (8.4), we must have

$$|\tau(X_{\ell_1} \cdots X_{\ell_r})| \leq M_0^r. \tag{8.5}$$

Moreover, by concentration inequalities (see Lemma 6.22), we find that

$$\lim_{N \to \infty} \left| \int \hat{\textbf{L}}_\textbf{A}^N \otimes \hat{\textbf{L}}_\textbf{A}^N(\partial_k P) d\mu_V^N(\textbf{A}) - \int \hat{\textbf{L}}_\textbf{A}^N d\mu_V^N(\textbf{A}) \otimes \int \hat{\textbf{L}}_\textbf{A}^N d\mu_V^N(\textbf{A})(\partial_k P) \right| = 0$$

so that Property 8.1 implies that

$$\limsup_{N \to \infty} |\bar{\textbf{L}}_t^N((X_k + D_k V_t)P) - \bar{\textbf{L}}_t^N \otimes \bar{\textbf{L}}_t^N(\partial_k P)| = 0. \tag{8.6}$$

Hence, (8.1) shows that

$$\tau((X_k + D_k V)P) = \tau \otimes \tau(\partial_k P). \tag{8.7}$$

$\square$

8.2.1 Uniqueness of the Solutions to Schwinger–Dyson's Equations for Small Parameters

Let $R \in \mathbb{R}^+$ (we will always assume $R \geq 1$ in the sequel).

(CS(R)) *An element $\tau \in \mathbb{C}\langle X_1, \ldots, X_m \rangle^{\mathcal{D}}$ satisfies* **(CS(R))** *if and only if for all $k \in \mathbb{N}$,*

$$\max_{1 \leq i_1, \ldots, i_k \leq m} |\tau(X_{i_1} \cdots X_{i_k})| \leq R^k.$$

In the sequel, we denote by D the degree of V, that is the maximal degree of the $q_i's$; $q_i(X) = X_{j_1^i} \cdots X_{j_{d_i}^i}$ with, for $1 \leq i \leq n$, $\deg(q_i) =: d_i \leq D$ and equality holds for some i.

The main result of this paragraph is:

Theorem 8.3. *For all $R \geq 1$, there exists $\epsilon > 0$ so that for $|\mathbf{t}| = \max_{1 \leq i \leq n} |t_i| < \epsilon$, there exists at most one solution $\tau_{\mathbf{t}}$ to* **SD[$V_{\mathbf{t}}$]** *that satisfies* **(CS(R))**.

Remark: Note that if $V = 0$, our equation becomes

$$\tau(X_i P) = \tau \otimes \tau(\partial_i P).$$

Because if P is a monomial, $\tau \otimes \tau(\partial_i P) = \sum_{P = P_1 X_i P_2} \tau(P_1)\tau(P_2)$ with P_1 and P_2 with degree smaller than P, we see that the equation **SD[0]** allows us to define uniquely $\tau(P)$ for all P by induction. The solution can be seen to be exactly $\tau(P) = \sigma^m(P)$, σ^m the law of m free semi-circular variables found in Theorem 3.3. When V is not zero, such an argument does not hold a priori since the right-hand side will also depend on $\tau(D_i q_j P)$, with $D_i q_j P$ of degree strictly larger than $X_i P$. However, our compactness assumption **(CS(R))** gives uniqueness because it forces the solution to be in a small neighborhood of the law $\tau_0 = \sigma^m$ of m free semi-circular variables, so that perturbation analysis applies. We shall see in Theorem 8.5 that this solution is actually the generating function for the enumeration of maps.

Proof. Let us assume we have two solutions τ and τ'. Then, by the equation **SD[V]**, for any monomial function P of degree $l - 1$, for $i \in \{1, \ldots, m\}$,

$$(\tau - \tau')(X_i P) = ((\tau - \tau') \otimes \tau)(\partial_i P) + (\tau' \otimes (\tau - \tau'))(\partial_i P) - (\tau - \tau')(D_i V P)$$

Hence, if we let, for $l \in \mathbb{N}$,

$$\Delta_l(\tau, \tau') = \sup_{\text{monomial } P \text{ of degree } l} |\tau(P) - \tau'(P)|$$

we get, since if P is of degree $l - 1$,

$$\partial_i P = \sum_{k=0}^{l-2} p_k^1 \otimes p_{l-2-k}^2$$

where p_k^i, $i = 1, 2$ are monomial of degree k or the null monomial, and $D_i V$ is a finite sum of monomials of degree smaller than $D - 1$,

$$\Delta_l(\tau, \tau') = \max_{P \text{ of degree } l-1} \max_{1 \leq i \leq m} \{|\tau(X_i P) - \tau'(X_i P)|\}$$

$$\leq 2 \sum_{k=0}^{l-2} \Delta_k(\tau, \tau') R^{l-2-k} + C|t| \sum_{p=0}^{D-1} \Delta_{l+p-1}(\tau, \tau')$$

with a finite constant C (that depends on n only). For $\gamma > 0$, we set

$$d_\gamma(\tau, \tau') = \sum_{l \geq 0} \gamma^l \Delta_l(\tau, \tau').$$

Note that under $(\mathbf{CS(R)})$, this sum is finite for $\gamma < (R)^{-1}$. Summing the two sides of the above inequality times γ^l we arrive at

$$d_\gamma(\tau, \tau') \leq 2\gamma^2 (1 - \gamma R)^{-1} d_\gamma(\tau, \tau') + C|t| \sum_{p=0}^{D-1} \gamma^{-p+1} d_\gamma(\tau, \tau').$$

We finally conclude that if $(R, |t|)$ are small enough so that we can choose $\gamma \in (0, R^{-1})$ so that

$$2\gamma^2 (1 - \gamma R)^{-1} + C|t| \sum_{p=0}^{D-1} \gamma^{-p+1} < 1$$

then $d_\gamma(\tau, \tau') = 0$ and so $\tau = \tau'$ and we have at most one solution. Taking $\gamma = (2R)^{-1}$ shows that this is possible provided

$$\frac{1}{4R^2} + C|t| \sum_{p=0}^{D-1} (2R)^{p-1} < 1$$

so that when R is large, we see that we need $|t|$ to be at most of order $|R|^{-D+2}$.
$\square$

8.3 Convergence of the Empirical Distribution

We are now in a position to state the main result of this part:

Theorem 8.4. *For all $c > 0$, there exists $\eta > 0$ and $M_0 \in \mathbb{R}^+$ (given in Lemma 6.19) so that for all $\mathbf{t} \in U_c \cap B_\eta$, $\hat{\mathbf{L}}^N$ (resp. $\bar{\mathbf{L}}_t^N$) converges almost surely (resp. everywhere) to the unique solution of $\mathbf{SD}[V_t]$ such that*

$$|\tau(X_{\ell_1} \cdots X_{\ell_r})| \leq M_0^r$$

for all choices of $\ell_1, \ldots, \ell_r$.

Proof. By Property 8.2, the limit points of $\bar{\mathbf{L}}_t^N$ satisfy $\mathbf{CS}(M_0)$ and $\mathbf{SD}[V_t]$. Since M_0 does not depend on $\mathbf{t}$, we can apply Theorem 8.3 to see that if $\mathbf{t}$ is small enough, there is only one such limit point. Thus, by Corollary 7.4 we can conclude that $(\bar{\mathbf{L}}_t^N, N \in \mathbb{N})$ converges to this limit point. From Lemma 6.22, we have that

$$\mu_V^N(|(\hat{\mathbf{L}}^N - \bar{\mathbf{L}}_t^N)(P)|^2) \leq BC(P,M)N^{-2} + C^{2d}N^2 e^{-\alpha MN/2}$$

insuring by Borel–Cantelli lemma that

$$\lim_{N \to \infty} (\hat{\mathbf{L}}^N - \bar{\mathbf{L}}_t^N)(P) = 0 \quad a.s$$

resulting with the almost sure convergence of $\hat{\mathbf{L}}^N$. $\qquad\square$

8.4 Combinatorial Interpretation of the Limit

In this part, we are going to identify the unique solution τ_t of Theorem 8.3 as a generating function for planar maps. Namely, for short, we write $\mathbf{k} = (k_1, \ldots, k_n) \in \mathbb{N}^n$ and denote by P a monomial in $\mathbb{C}\langle X_1, \ldots, X_m \rangle$,

$$\mathcal{M}_{\mathbf{k}}(P) = \text{card}\{ \text{ planar maps with } k_i \text{ labeled stars of type } q_i \text{ for } 1 \leq i \leq n$$

$$\text{and one of type } P\}.$$

This definition extends to $P \in \mathbb{C}\langle X_1, \ldots, X_m \rangle$ by linearity. By convention, $\mathcal{M}_{\mathbf{k}}(1) = 1_{\mathbf{k}=0}$. Then, we shall prove:

Theorem 8.5.

1. *The family $\{\mathcal{M}_{\mathbf{k}}(P), \mathbf{k} \in \mathbb{N}^n, P \in \mathbb{C}\langle X_1, \ldots, X_m \rangle\}$ satisfies the induction relation: for all $i \in \{1, \ldots, m\}$, all $P \in \mathbb{C}\langle X_1, \ldots, X_m \rangle$, all $\mathbf{k} \in \mathbb{N}^n$,*

$$\mathcal{M}_{\mathbf{k}}(X_i P) = \sum_{\substack{0 \leq p_j \leq k_j \\ 1 \leq j \leq n}} \prod_{j=1}^n C_{k_j}^{p_j} \sum_{P = p_1 X_i p_2} \mathcal{M}_{\mathbf{p}}(P_1) \mathcal{M}_{\mathbf{k}-\mathbf{p}}(P_2)$$

$$+ \sum_{1 \leq j \leq n} k_j \mathcal{M}_{\mathbf{k}-1_j}([D_i q_j]P) \tag{8.8}$$

   where $1_j(i) = 1_{i=j}$ and $\mathcal{M}_{\mathbf{k}}(1) = 1_{\overline{k}=0}$. (8.8) defines uniquely the family $\{\mathcal{M}_{\mathbf{k}}(P), \mathbf{k} \in \mathbb{N}^n, P \in \mathbb{C}\langle X_1, \ldots, X_m \rangle\}$.
2. *There exists A, B finite constants so that for all $\mathbf{k} \in \mathbb{N}^n$, all monomial $P \in \mathbb{C}\langle X_1, \ldots, X_m \rangle$,*

$$|\mathcal{M}_{\mathbf{k}}(P)| \leq \mathbf{k}! A^{\sum_{i=1}^n k_i} B^{deg(P)} \prod_{i=1}^n C_{k_i} C_{deg(P)} \tag{8.9}$$

   with $\mathbf{k}! := \prod_{i=1}^n k_i!$ and C_p the Catalan numbers.

3. For $\mathbf{t}$ in $B_{(4A)^{-1}}$,

$$\mathcal{M}_{\mathbf{t}}(P) = \sum_{\mathbf{k}\in\mathbb{N}^n} \prod_{i=1}^{n} \frac{(-t_i)^{k_i}}{k_i!} \mathcal{M}_{\mathbf{k}}(P)$$

is absolutely convergent. For $\mathbf{t}$ small enough, $\mathcal{M}_{\mathbf{t}}$ is the unique solution of $\mathbf{SD}[V_{\mathbf{t}}]$ that satisfies $\mathbf{CS(4B)}$.

By Theorem 8.3 and Theorem 8.4, we therefore readily obtain:

Corollary 8.6. *For all $c > 0$, there exists $\eta > 0$ so that for $\mathbf{t} \in U_c \cap B_\eta$, $\hat{\mathbf{L}}^N$ converges almost surely and in expectation to*

$$\tau_{\mathbf{t}}(P) = \mathcal{M}_{\mathbf{t}}(P) = \sum_{\mathbf{k}\in\mathbb{N}^n} \prod_{i=1}^{n} \frac{(-t_i)^{k_i}}{k_i!} \mathcal{M}_{\mathbf{k}}(P)$$

Let us remark that by definition of $\hat{\mathbf{L}}^N$, for all P, Q in $\mathbb{C}\langle X_1, \ldots, X_m \rangle$,

$$\hat{\mathbf{L}}^N(PP^*) \geq 0 \quad \text{and} \quad \hat{\mathbf{L}}^N(PQ) = \hat{\mathbf{L}}^N(QP).$$

These conditions are closed for the weak topology and hence we find:

Corollary 8.7. *There exists $\eta > 0$ $(\eta \geq (4A)^{-1})$ so that for $\mathbf{t} \in B_\eta$, $\mathcal{M}_{\mathbf{t}}$ is a linear form on $\mathbb{C}\langle X_1, \ldots, X_m \rangle$ such that for all P, Q*

$$\mathcal{M}_{\mathbf{t}}(PP^*) \geq 0 \quad \mathcal{M}_{\mathbf{t}}(PQ) = \mathcal{M}_{\mathbf{t}}(QP) \quad \mathcal{M}_{\mathbf{t}}(1) = 1.$$

Remark. This means that $\mathcal{M}_{\mathbf{t}}$ is a tracial state. The traciality property can easily be derived by symmetry properties of the maps. However, I do not know of any other way (and in particular any combinatorial way) to prove the positivity property $\mathcal{M}_{\mathbf{t}}(PP^*) \geq 0$ for all polynomial P, except by using matrix models. This property will be seen to be useful to actually solve the combinatorial problem (i.e., find an explicit formula for $\mathcal{M}_{\mathbf{t}}$), see Section 15.1.

Proof of Theorem 8.5.

1. *Proof of the induction relation* (8.8).
   - We first check them for $\mathbf{k} = 0 = (0, \ldots, 0)$. By convention, there is only one planar map with no vertex, so $\mathcal{M}_0(1) = 1$. We now check that

$$\mathcal{M}_0(X_i P) = \mathcal{M}_0 \otimes \mathcal{M}_0(\partial_i P) = \sum_{P = p_1 X_i p_2} \mathcal{M}_0(p_1)\mathcal{M}_0(p_2).$$

But this is clear from (7.3) since for any planar map with only one star of type $X_i P$, the half-edge corresponding to X_i has to be glued to another half-edge of P, hence the event $A_M(X_i P)$ must hold, and if X_i is glued to the half-edge X_i coming from the decomposition $P = p_1 X_i p_2$, the map is split into two (independent) planar maps with stars p_1 and p_2 respectively (note here that p_1 and p_2 inherits the structure of stars since they inherit the orientation from P as well as a marked half-edge corresponding to the first neighbor of the glued X_i.)

- We now proceed by induction over the **k**'s and the degree of P; we assume that (8.8) is true for $\sum k_i \leq M$ and all monomials, and for $\sum k_i = M + 1$ when $\deg(P) \leq L$. Note that $\mathcal{M}_\mathbf{k}(1) = 0$ for $|\mathbf{k}| \geq 1$ since we cannot glue a vertex with zero half-edges to any half-edge of another star. Hence, this induction can be started with $L = 0$. Now, consider $R = X_i P$ with P of degree less than L and the set of planar maps with a star of type $X_i Q$ and k_j stars of type q_j, $1 \leq j \leq n$, with $|\mathbf{k}| = \sum k_i = M + 1$. Then,

  ◇ either the half-edge corresponding to X_i is glued with an half-edge of P, say to the half-edge corresponding to the decomposition $P = p_1 X_i p_2$; we then can use (7.3) to see that this cuts the map M into two disjoint planar maps M_1 (containing the star p_1) and M_2 (resp. p_2), the stars of type q_i being distributed either in one or the other of these two planar maps; there will be $r_i \leq k_i$ stars of type q_i in M_1, the rest in M_2. Since all stars all labeled, there will be $\prod C_{k_i}^{r_i}$ ways to assign these stars in M_1 and M_2.

  Hence, the total number of planar maps with a star of type $X_i Q$ and k_i stars of type q_i, such that the marked half-edge of $X_i P$ is glued to a half-edge of P is

$$\sum_{P=p_1 X_i p_2} \sum_{\substack{0 \leq r_i \leq k_i \\ 1 \leq i \leq n}} \prod_{i=1}^{n} C_{k_i}^{r_i} \mathcal{M}_\mathbf{r}(p_1) \mathcal{M}_{\mathbf{k}-\mathbf{r}}(p_2) \qquad (8.10)$$

  ◇ Or the half-edge corresponding to X_i is glued to a half-edge of another star, say q_j; let's say to the edge coming from the decomposition of q_j into $q_j = q_1 X_i q_2$. Then, we can use (7.4) to see that once we are given this gluing of the two edges, we can replace $X_i P$ and q_j by $q_2 q_1 P$.

  We have k_j ways to choose the star of type q_j and the total number of such maps is

$$\sum_{q_j = q_1 X_i q_2} k_j \mathcal{M}_{\mathbf{k}-1_j}(q_2 q_1 P)$$

Summing over j, we obtain by linearity of $\mathcal{M}_\mathbf{k}$

$$\sum_{j=1}^{n} k_j \mathcal{M}_{\mathbf{k}-1_j}([D_i q_j] P) \qquad (8.11)$$

(8.10) and (8.11) give (8.8). Moreover, it is clear that (8.8) defines uniquely $\mathcal{M}_\mathbf{k}(P)$ by induction.

2. *Proof of* (8.9). To prove the second point, we proceed also by induction over $\mathbf{k}$ and the degree of P. First, for $\mathbf{k} = \mathbf{0}$, $\mathcal{M}_{\mathbf{0}}(P)$ is the number of colored maps with one star of type P which is smaller than the number of planar maps with one star of type $x^{\deg P}$ since colors only add constraints. Hence, we have, with C_k the Catalan numbers,

$$\mathcal{M}_{\mathbf{k}}(P) \leq C_{[\frac{\deg(P)}{2}]} \leq C_{\deg(P)}$$

showing that the induction relation is fine with $A = B = 1$ at this step. Hence, let us assume that (8.9) is true for $\sum k_i \leq M$ and all polynomials, and $\sum k_i = M+1$ for polynomials of degree less than L. Since $\mathcal{M}_{\mathbf{k}}(1) = 0$ for $\sum k_i \geq 1$ we can start this induction. Moreover, using (8.8), we get that, if we define $\mathbf{k}! = \prod_{i=1}^{n} k_i!$,

$$\frac{\mathcal{M}_{\mathbf{k}}(X_i P)}{\mathbf{k}!} = \sum_{\substack{0 \leq p_i \leq k_i \\ 1 \leq j \leq n}} \sum_{P = P_1 X_i P_2} \frac{\mathcal{M}_{\mathbf{p}}(P_1)}{\mathbf{p}!} \frac{\mathcal{M}_{\mathbf{k}-\mathbf{p}}(P_2)}{(\mathbf{k}-\mathbf{p})!}$$

$$+ \sum_{\substack{1 \leq j \leq n \\ k_j \neq 0}} \frac{\mathcal{M}_{\mathbf{k}-1_j}((D_i q_j P)}{(\mathbf{k}-1_j)!}.$$

Hence, taking P of degree less or equal to L and using our induction hypothesis, we find that

$$\left| \frac{\mathcal{M}_{\mathbf{k}}(X_i P)}{\mathbf{k}!} \right|$$

$$\leq \sum_{\substack{0 \leq p_j \leq k_j \\ 1 \leq j \leq n}} \sum_{P = P_1 X_i P_2} A^{\sum k_i} B^{\deg P - 1} \prod_{i=1}^{n} C_{p_j} C_{k_j - p_j} C_{\deg P_1} C_{\deg P_2}$$

$$+ 2 \sum_{1 \leq l \leq n} A^{\sum k_j - 1} \prod_j C_{k_j} B^{\deg P + \deg q_l - 1} C_{\deg P + \deg q_l - 1}$$

$$\leq A^{\sum k_i} B^{\deg P + 1} \prod_i C_{k_i} C_{\deg P + 1} \left(\frac{4^n}{B^2} + 2 \frac{\sum_{1 \leq j \leq n} B^{\deg q_j - 2} 4^{\deg q_j - 2}}{A} \right)$$

where we used Lemma 1.9 in the last line. It is now sufficient to choose A and B such that

$$\frac{4^n}{B^2} + 2 \frac{\sum_{1 \leq j \leq n} B^{\deg q_j - 2} 4^{\deg q_j - 2}}{A} \leq 1$$

(for instance $B = 2^{n+1}$ and $A = 4n B^{D-2} 4^{D-2}$ if D is the maximal degree of the q_j) to verify the induction hypothesis works for polynomials of all degrees (all L's).

3. *Properties of $\mathcal{M}_{\mathbf{t}}$.* From the previous considerations, we can of course define $\mathcal{M}_{\mathbf{t}}$ and the series is absolutely convergent for $|\mathbf{t}| \leq (4A)^{-1}$ since $C_k \leq 4^k$. Hence $\mathcal{M}_{\mathbf{t}}(P)$ depends analytically on $\mathbf{t} \in B_{(4A)^{-1}}$. Moreover, for all monomial P,

$$|\mathcal{M}_{\mathbf{t}}(P)| \leq \sum_{\mathbf{k}\in\mathbb{N}^n} \prod_{i=1}^{n} (4t_i A)^{k_i} (4B)^{degP} \leq \prod_{i=1}^{n} (1 - 4At_i)^{-1} (4B)^{degP}.$$

so that for small t, $\mathcal{M}_{\mathbf{t}}$ satisfies **CS(4B)**.

4. $\mathcal{M}_{\mathbf{t}}$ *satisfies* $\mathbf{SD}[V_{\mathbf{t}}]$. This is derived by summing (8.8) written for all $\mathbf{k}$ and multiplied by the factor $\prod(t_i)^{k_i}/k_i!$. From this point and the previous one (note that B is independent from $\mathbf{t}$), we deduce from Theorem 8.3 that for sufficiently small $\mathbf{t}$, $\mathcal{M}_{\mathbf{t}}$ is the unique solution of $\mathbf{SD}[V_{\mathbf{t}}]$ that satisfies **CS(4B)**.

□

8.5 Convergence of the Free Energy

Theorem 8.8. *Let $c > 0$. Then, for η small enough, for all $\mathbf{t} \in B_\eta \cap U_c$, the free energy converges towards a generating function of the numbers of certain planar maps:*

$$\lim_{N\to\infty} \frac{1}{N^2} \log \frac{Z_N^{V_{\mathbf{t}}}}{Z_N^0} = \sum_{\mathbf{k}\in\mathbb{N}^n\setminus(0,..,0)} \prod_{1\leq i\leq n} \frac{(-t_i)^{k_i}}{k_i!} \mathcal{M}_{\mathbf{k}}.$$

Moreover, the limit depends analytically on $\mathbf{t}$ in a neighborhood of the origin.

Proof. We may assume without loss of generality that $c \in (0,1]$. For $\alpha \in [0,1]$, $V_{\alpha\mathbf{t}}$ is c-convex since

$$V_{\alpha\mathbf{t}} + \frac{1}{2}\sum_{i=1}^{m} X_i^2 = \alpha(V_{\mathbf{t}}(X_1,\ldots,X_m) + \frac{1-c}{2}\sum_{i=1}^{m}X_i^2)$$

$$+ \frac{(1-\alpha)(1-c)+c}{2}\sum_{i=1}^{m}X_i^2$$

where all terms are convex (as we assumed $c \leq 1$), whereas the last one is c-convex. Set

$$F_N(\alpha) = \frac{1}{N^2}\log Z_N^{V_{\alpha\mathbf{t}}}.$$

Then, $\frac{1}{N^2}\log\frac{Z_N^{V_{\mathbf{t}}}}{Z_N^0} = F_N(1) - F_N(0)$. Moreover

$$\partial_\alpha F_N(\alpha) = -\bar{\mathbf{L}}_{\alpha t}^N(V_{\mathbf{t}}). \tag{8.12}$$

By Theorem 8.4, we know that for all $\alpha \in [0,1]$ (since $V_{\alpha t}$ is c-convex),

$$\lim_{N \to \infty} \bar{\mathbf{L}}_{\alpha t}^N(V_t) = \tau_{\alpha t}(V_t)$$

whereas by (8.4), we know that $\bar{\mathbf{L}}_{\alpha t}^N(V_t)$ stays uniformly bounded. Therefore, a simple use of dominated convergence theorem shows that

$$\lim_{N \to \infty} \frac{1}{N^2} \log \frac{Z_N^{V_t}}{Z_N^0} = -\int_0^1 \tau_{\alpha t}(V_t) d\alpha = -\sum_{i=1}^n t_i \int_0^1 \tau_{\alpha t}(q_i) d\alpha. \qquad (8.13)$$

Now, observe that by Corollary 8.6,

$$\tau_{\mathbf{t}}(q_i) = \sum_{\mathbf{k} \in \mathbb{N}^n} \prod_{1 \leq j \leq n} \frac{(-t_j)^{k_j}}{k_j!} \mathcal{M}_{\mathbf{k}+1_i}$$

$$= -\partial_{t_i} \sum_{\mathbf{k} \in \mathbb{N}^n \setminus \{0,\ldots,0\}} \prod_{1 \leq j \leq n} \frac{(-t_j)^{k_j}}{k_j!} \mathcal{M}_{\mathbf{k}}$$

so that (8.13) implies that

$$\lim_{N \to \infty} \frac{1}{N^2} \log \frac{Z_N^{V_t}}{Z_N^0} = -\int_0^1 \partial_\alpha \Big[\sum_{\mathbf{k} \in \mathbb{N}^n \setminus \{0,\ldots,0\}} \prod_{1 \leq j \leq n} \frac{(-\alpha t_j)^{k_j}}{k_j!} \mathcal{M}_{\mathbf{k}} \Big] d\alpha$$

$$= - \sum_{\mathbf{k} \in \mathbb{N}^n \setminus \{0,\ldots,0\}} \prod_{1 \leq j \leq n} \frac{(-t_j)^{k_j}}{k_j!} \mathcal{M}_{\mathbf{k}}.$$

$\square$

Bibliographical Notes. The study of matrix models in mathematics is not new. The one matrix model was already studied by Pastur [164] who derived the limiting spectral density of such measures as well as the nearest-neighbor spacing distribution. The problem of the universality of the fluctuations of the largest eigenvalue was addressed in [71, 163]. Cases where the potential is not strictly convex were studied for instance in [72, 141]. Specific two-matrix models (mainly models with quadratic interaction) were studied by Mehta and coauthors [60, 152, 154] and, on less rigorous ground, for instance in [4, 80, 87, 123, 125]. In this case, large deviations techniques [101, 109] are also available (see [146] on a less rigorous ground), yielding a non-perturbative approach. Matrix models were considered in the generality presented in this section in [104].

9

Second-order Expansion for the Free Energy

At the end of this chapter, we will have proved that Lemma 7.12 holds, up to the second-order correction in the large N limit, i.e., that

$$\frac{1}{N^2} \log \left(\int e^{\sum_{i=1}^{n} t_i N \mathrm{Tr}(q_i(X_1,...,X_m))} d\mu^N(X_1) \cdots d\mu^N(X_m) \right)$$

$$= \sum_{g=0}^{1} \frac{1}{N^{2g-2}} \sum_{k_1,...,k_n \in \mathbb{N}} \prod_{i=1}^{n} \frac{(t_i)^{k_i}}{k_i!} \mathcal{M}_g((q_i, k_i), 1 \leq i \leq n) + o\left(\frac{1}{N^2}\right)$$

when the parameters t_i are small enough and such that $\sum t_i q_i$ is c-convex. As for the first order, we shall prove first a similar large N expansion for $\bar{\mathbf{L}}_t^N$. We will first refine the arguments of the proof of Theorem 8.3 to estimate $\bar{\mathbf{L}}_t^N - \tau_t$. This will already prove that $(\bar{\mathbf{L}}_t^N - \tau_t)(P)$ is at most of order N^{-2} for any polynomial P. To get the limit of $N^2(\bar{\mathbf{L}}_t^N - \tau_t)(P)$, we will first obtain a central limit theorem for $\hat{\mathbf{L}}^N - \tau_t$ which is of independent interest. The key argument in our approach, besides further uses of integration by parts-like arguments, will be the inversion of a differential operator acting on non-commutative polynomials which can be thought as a non-commutative analog of a Laplacian operator with a drift.

We shall now estimate differences of $\hat{\mathbf{L}}^N$ and its limit. So, we set

$$\hat{\delta}_t^N = N(\hat{\mathbf{L}}^N - \tau_t)$$

$$\bar{\delta}^N = \int \hat{\delta}^N d\mu_V^N = N(\bar{\mathbf{L}}_t^N - \tau_t)$$

$$\tilde{\delta}_t^N = N(\hat{\mathbf{L}}^N - \bar{\mathbf{L}}_t^N) = \hat{\delta}_t^N - \bar{\delta}^N.$$

In order to simplify the notations, we will make $\mathbf{t}$ implicit and drop the subscript $\mathbf{t}$ in the rest of this chapter so that we will denote $\bar{\mathbf{L}}^N, \tau, \hat{\delta}^N, \bar{\delta}^N$ and $\tilde{\delta}^N$ in place of $\bar{\mathbf{L}}_t^N, \tau_t, \hat{\delta}_t^N, \bar{\delta}^N$ and $\tilde{\delta}_t^N$, as well as V in place of V_t.

A. Guionnet, *Large Random Matrices: Lectures on Macroscopic Asymptotics*,
Lecture Notes in Mathematics 1957, DOI: 10.1007/978-3-540-69897-5_9,
© 2009 Springer-Verlag Berlin Heidelberg, Reprint by Springer-Verlag Berlin Heidelberg 2012

9.1 Rough Estimates on the Size of the Correction $\tilde{\delta}_t^N$

In this section we improve on the perturbation analysis performed in Section 8.2.1 to get the order of

$$\tilde{\delta}^N(P) = N(\bar{\mathbf{L}}^N(P) - \tau)(P)$$

for all monomial P.

Proposition 9.1. *For all $c > 0, \epsilon \in]0, \frac{1}{2}[$, there exists $\eta > 0, C < +\infty$, such that for all integer number N, all $\mathbf{t} \in B_\eta \cap U_c$, and all monomial function P of degree less than $N^{\frac{1}{2}-\epsilon}$,*

$$|\tilde{\delta}^N(P)| \leq \frac{C^{deg\ (P)}}{N}.$$

Proof. The starting point is the finite dimensional Schwinger–Dyson equation of Property 8.1,

$$\mu_V^N(\hat{\mathbf{L}}^N[(X_i + D_i V)P]) = \mu_V^N\left(\hat{\mathbf{L}}^N \otimes \hat{\mathbf{L}}^N(\partial_i P)\right). \tag{9.1}$$

Therefore, since τ satisfies the Schwinger–Dyson equation $\mathbf{SD[V]}$, we get that for all polynomial P,

$$\tilde{\delta}^N(X_i P) = -\tilde{\delta}^N(D_i V P) + \tilde{\delta}^N \otimes \bar{\mathbf{L}}^N(\partial_i P) + \tau \otimes \tilde{\delta}^N(\partial_i P) + r(N, P) \tag{9.2}$$

with

$$r(N, P) := N^{-1} \mu_V^N\left(\tilde{\delta}^N \otimes \tilde{\delta}^N(\partial_i P)\right).$$

We take P as monomial of degree $d \leq N^{\frac{1}{2}-\epsilon}$ and see that

$$|r(N, P)| \leq \frac{1}{N} \sum_{P = P_1 X_i P_2} \mu_V^N\left(|\tilde{\delta}^N(P_1)|^2\right)^{\frac{1}{2}} \mu_V^N\left(|\tilde{\delta}^N(P_2)|^2\right)^{\frac{1}{2}}$$

$$\leq \frac{C}{N} \sum_{l=0}^{d-1} (Bl^2 M^{2(l-1)} + C^l N^4 e^{-\frac{\alpha M N}{2}})^{\frac{1}{2}} \times$$

$$(B(d-l-1)^2 M^{2(d-l-1)} + C^{(d-l-1)} N^4 e^{-\frac{\alpha M N}{2}})^{\frac{1}{2}}$$

$$\leq \frac{C}{N} d(B(d-1)^2 M^{2(d-2)} + C^{(d-1)} N^4 e^{-\frac{\alpha M N}{2}}) := r(N, d, M)$$

where we used in the second line Lemma 6.22 and assumed $M \geq M_0$, and $d \leq N^{\frac{1}{2}-\epsilon}$. We set

$$\Delta_d^N := \max_P \text{ monomial of degree } d |\bar{\delta}^N(P)|.$$

Observe that by (6.14), for any monomial of degree d less than $N^{\frac{1}{2}-\epsilon}$,

$$|\bar{\mathbf{L}}_t^N(P)| \leq C(\epsilon)^d, \quad |\tau(P)| \leq C_0^d \leq C(\epsilon)^d.$$

Thus, by (9.2), writing $D_i V = \sum t_j D_i q_j$, we get that for $d < N^{\frac{1}{2}-\epsilon}$

$$\Delta_{d+1}^N \leq \max_{1 \leq i \leq m} \sum_{j=1}^{n} |t_j| \Delta_{d+\deg(D_i q_j)}^N + 2 \sum_{l=0}^{d-1} C(\epsilon)^{d-l-1} \Delta_l^N + r(N, d, M).$$

We next define for $\kappa \leq 1$

$$\Delta^N(\kappa, \epsilon) = \sum_{k=1}^{N^{\frac{1}{2}-\epsilon}} \kappa^k \Delta_k^N.$$

We obtain, if D is the maximal degree of V,

$$\Delta^N(\kappa, \epsilon) \leq [C'|\mathbf{t}| + 2(1 - C(\epsilon)\kappa)^{-1}\kappa^2] \Delta^N(\kappa, \epsilon)$$
$$+ C|\mathbf{t}| \sum_{k=N^{\frac{1}{2}-\epsilon}+1}^{N^{\frac{1}{2}-\epsilon}+D} \kappa^{k-D} \Delta_k^N + \sum_{k=1}^{N^{\frac{1}{2}-\epsilon}} \kappa^k r(N, k, M)$$

where we choose κ small enough so that $C(\epsilon)\kappa < 1$. Moreover, since D is finite, bounding Δ_k^N by $2NC(\epsilon)^k$, we get

$$\sum_{k=N^{\frac{1}{2}-\epsilon}+1}^{N^{\frac{1}{2}-\epsilon}+D} \kappa^{k-D} \Delta_k^N \leq 2DN(\kappa C(\epsilon))^{N^{\frac{1}{2}-\epsilon}} \kappa^{-D}.$$

When $\kappa C(\epsilon) < 1$, as N goes to infinity, this term is negligible with respect to N^{-1} for all $\epsilon > 0$. The following estimate holds according to Lemma 6.22:

$$\sum_{k=1}^{N^{\frac{1}{2}-\epsilon}} \kappa^k r(N, k, M) \leq \frac{C}{N} \sum_{k=1}^{N^{\frac{1}{2}-\epsilon}} k\kappa^k (B(k-1)^2 M^{2(k-2)} + C^{(k-1)} N^4 e^{-\frac{\alpha NM}{2}}) \leq \frac{C''}{N}$$

if κ is small enough so that $M^2\kappa < 1$ and $C\kappa < 1$. We observed here that $N^4 e^{-\frac{\alpha NM}{2}}$ is uniformly bounded independently of $N \in \mathbb{N}$. Now, if $|\mathbf{t}|$ is small, we can choose κ so that

$$\zeta := 1 - [C'|\mathbf{t}| + 2(1 - C(\epsilon)\kappa)^{-1}\kappa^2] > 0.$$

Plugging these controls into (9.3) shows that for all $\epsilon > 0$, and for $\kappa > 0$ small enough, there exists a finite constant $C(\kappa, \epsilon)$ so that

$$\Delta^N(\kappa, \epsilon) \leq C(\kappa, \epsilon) N^{-1}$$

and so for all monomial P of degree $d \leq N^{\frac{1}{2} - \epsilon}$,

$$|\bar{\delta}^N(P)| \leq C(\kappa, \epsilon) \kappa^{-d} N^{-1}.$$

$\square$

To get the precise evaluation of $N\bar{\delta}^N(P)$, we shall first obtain a central limit theorem under μ_V^N which in turn will allow us to estimate the limit of $Nr(N, P)$.

9.2 Central Limit Theorem

We shall here prove that

$$\hat{\delta}^N(P) = N(\hat{\mathbf{L}}^N - \tau)(P)$$

satisfies a central limit theorem for all polynomial P. By Proposition 9.1, it is equivalent to prove a central limit theorem for $\tilde{\delta}^N(P)$, $P \in \mathbb{C}\langle X_1, \ldots, X_m \rangle$. We start by giving a weak form of a central limit theorem for Stieltjes-like functions. We then extend by density the result to polynomial functions in the image of some differential operator and finally to any polynomials by inverting this operator.

Until the end of this chapter, we will always assume the following hypothesis (**H**).

(**H**): Let c be a positive real number. The parameter $\mathbf{t}$ is in $B_{\eta, c}$ with η sufficiently small such that we have the convergence to the solution of $\mathbf{SD}[V_t]$ as well as the control given by Lemma 6.18, and Proposition 9.1.

Note that (**H**) implies also that the control of Lemma 6.18 is uniform, and that we can apply Lemma 6.19 and Lemma 6.21 with uniform constants.

9.2.1 Central Limit Theorem for Stieltjes Test Functions

One of the issues that one needs to address when working with polynomials is that they are not uniformly bounded. For that reason, we will prefer to work in this section with the complex vector space $\mathcal{C}_{st}^m(\mathbb{C})$ generated by the Stieltjes functionals

$$ST^m(\mathbb{C}) = \left\{ \overrightarrow{\prod_{1 \leq i \leq p}} \left(z_i - \sum_{k=1}^{m} \alpha_i^k \mathbf{X}_k \right)^{-1} \; ; \quad z_i \in \mathbb{C} \backslash \mathbb{R}, \alpha_i^k \in \mathbb{R}, p \in \mathbb{N} \right\} \quad (9.3)$$

where $\prod^{\rightarrow}$ is the non-commutative product. We can also equip $ST^m(\mathbb{C})$ with an involution

$$
\left(\overset{\rightarrow}{\prod_{1 \le k \le p}} \left(z_k - \sum_{i=1}^m \alpha_i^k \mathbf{X}_i\right)^{-1}\right)^* = \overset{\rightarrow}{\prod_{1 \le k \le p}} \left(\overline{z_{p-k}} - \sum_{i=1}^m \alpha_i^{p-k} \mathbf{X}_i\right)^{-1}.
$$

We denote by $\mathcal{C}_{st}^m(\mathbb{C})_{sa}$ the set of self-adjoint elements of $\mathcal{C}_{st}^m(\mathbb{C})$. The derivative is defined by the Leibniz rule (7.1) (taken with P, Q which are Stieltjes functionals) and

$$
\partial_i \left(z - \sum_{i=1}^m \alpha_i \mathbf{X}_i\right)^{-1} = \alpha_i \left(z - \sum_{i=1}^m \alpha_i \mathbf{X}_i\right)^{-1} \otimes \left(z - \sum_{i=1}^m \alpha_i \mathbf{X}_i\right)^{-1}.
$$

We recall two notations; first $\sharp$ is the operator

$$
(P \otimes Q)\sharp h = PhQ
$$

and

$$
(P \otimes Q \otimes R)\sharp(g, h) = PgQhR
$$

so that for a monomial q

$$
\partial_i \circ \partial_j q \#(h_i, h_j) = \sum_{q=q_0 X_i q_1 X_j q_2} q_0 h_i q_1 h_j q_2 + \sum_{q=q_0 X_j q_1 X_i q_2} q_0 h_j q_1 h_i q_2.
$$

Lemma 9.2. *Assume* **(H)** *and let* $h_1, \ldots, h_m$ *be in* $\mathcal{C}_{st}^m(\mathbb{C})_{sa}$. *Then the random variable*

$$
Y_N(h_1, \ldots, h_m) = N \sum_{k=1}^m \{\hat{\mathbf{L}}^N \otimes \hat{\mathbf{L}}^N(\partial_k h_k) - \hat{\mathbf{L}}^N[(X_k + D_k V)h_k]\}
$$

converges in law to a real centered Gaussian variable with variance

$$
C(h_1, \ldots, h_m) = \sum_{k,l=1}^m (\tau \otimes \tau[\partial_k h_l \times \partial_l h_k] + \tau(\partial_l \circ \partial_k V \sharp(h_k, h_l))) + \sum_{k=1}^m \tau(h_k^2).
$$

Proof. Define $W = \frac{1}{2}\sum_i X_i^2 + V$. Notice that $Y_N(h_1, \ldots, h_m)$ is real-valued because the $h_k's$ and W are self-adjoint. The proof of the lemma follows from a change of variable. We take $h_1, \ldots, h_m$ in $\mathcal{C}_{st}^m(\mathbb{C})_{sa}$, $\lambda \in \mathbb{R}$ and perform a change of variable $B_i = F(\mathbf{A})_i = \mathbf{A}_i + \frac{\lambda}{N}h_i(\mathbf{A})$ in Z_V^N. Note that since the h_i are $\mathcal{C}^\infty$ and uniformly bounded, this defines a bijection on $\mathcal{H}_N^m$ for N big enough. We shall compute the Jacobian of this change of variables up to its second-order correction. The Jacobian J may be seen as a matrix $(J_{i,j})_{1 \le i,j \le m}$ where the $J_{i,j}$ are in $\mathcal{L}(\mathcal{H}_N)$ the set of endomorphisms of $\mathcal{H}_N$, and we can write $J = I + \frac{\lambda}{N}\bar{J}$ with

$$\overline{J}_{i,j} : \mathcal{H}_N \longrightarrow \mathcal{H}_N$$
$$X \longrightarrow \partial_i h_j \# X.$$

Now, for $1 \leq i,j \leq m$, $X \longrightarrow \partial_i h_j \# X$ is bounded for the operator norm uniformly in N (since $h_j \in \mathcal{C}_{st}(\mathbb{C})$, $\partial_i h_j \in \mathcal{C}_{st}(\mathbb{C}) \otimes \mathcal{C}_{st}(\mathbb{C})$ is uniformly bounded) so that for sufficiently large N, the operator norm of $\frac{\lambda}{N}\overline{J}$ is less than 1. From this we deduce

$$|\det J| = \left|\det\left(I + \frac{\lambda\overline{J}}{N}\right)\right| = \exp\left(\operatorname{Tr}\log\left(I + \frac{\lambda\overline{J}}{N}\right)\right)$$

$$= \exp\left(\sum_{k\geq 1} \frac{(-1)^{k+1}\lambda^k}{kN^k}\operatorname{Tr}(\overline{J}^k)\right).$$

Observe that as $\overline{J}$ is a matrix of size $m^2 N^2$ and of uniformly bounded norm, the kth term $\frac{(-1)^{k+1}\lambda^k}{N^k} tr(\overline{J}^k)$ is of order $\frac{1}{N^{k-2}}$. Hence, only the two first terms in the expansion will contribute to the order 1. To compute them, we only have to remark that if ϕ an endomorphism of $\mathcal{H}_N$ is of the form $\phi(X) = \sum_l \mathbf{A}_l X \mathbf{B}_l$, with $N \times N$ matrices $\mathbf{A}_i, \mathbf{B}_i$ then $\operatorname{Tr}\phi = \sum_l \operatorname{Tr}\mathbf{A}_l \operatorname{Tr}\mathbf{B}_l$ (this can be checked by decomposing ϕ on the canonical basis of $\mathcal{H}_N$). Now,

$$\overline{J}_{ij}^k : X \longrightarrow \sum_{1\leq i_1,\ldots,i_{k-1}\leq m} \partial_i h_{i_2}\#(\partial_{i_2} h_{i_3}\#(\cdots(\partial_{i_{k-1}} h_j\#X)\cdots)).$$

Thus, we get

$$\operatorname{Tr}(\overline{J}) = \sum_i \operatorname{Tr}\overline{J}_{ii} = \sum_{1\leq i\leq m} tr \otimes tr(\partial_i h_i)$$

and

$$\operatorname{Tr}(\overline{J}^2) = \sum_{ij} \operatorname{Tr}(\overline{J}_{ij}\overline{J}_{ji}) = \sum_{1\leq i,j\leq m} tr \otimes tr(\partial_i h_j \times \partial_j h_i)$$

since $\overline{J}_{ij}\overline{J}_{ji}(X) = \partial_i h_j\#[\partial_j h_i\#X] = \partial_i h_j \times \partial_j h_i\#X$ where $X_i^1 \otimes Y_i^1 \times X_i^2 \otimes Y_i^2 = X_i^1 X_i^2 \otimes Y_i^2 Y_i^1$. We can now make the change of variable $\mathbf{A}_i \to \mathbf{A}_i + \frac{\lambda}{N}h_i(\mathbf{A})$ to find that

$$Z_V^N = \int e^{-NtrV}d\mu^N = \int e^{-N\operatorname{Tr}(W(\mathbf{A}_i + \frac{\lambda}{N}h_i(\mathbf{X})) - W(\mathbf{A}_i))}e^{\frac{\lambda}{N}\sum_i \operatorname{Tr}\otimes\operatorname{Tr}(\partial_i h_i)} \times$$

$$e^{-\frac{\lambda^2}{2N^2}\sum_{i,j}\operatorname{Tr}\otimes\operatorname{Tr}(\partial_i h_j \partial_j h_i)}e^{O(\frac{1}{N})}d\mu_V^N \times Z_V^N$$

where $O(\frac{1}{N})$ is a function uniformly bounded by C/N for some finite $C = C(h)$.

The first term can be expanded into

$$W\left(\mathbf{A}_i + \frac{h_i(\mathbf{A})}{N}\right) - W(\mathbf{A}_i) = \frac{1}{N}\sum_i \partial_i W \# h_i + \frac{1}{N^2}\sum_{i,j}\partial_i \circ \partial_j W \#(h_i, h_j) + \frac{R_N}{N^3}$$

where R_N is a polynomial of degree less than the degree of $\mathcal{D}V$ whose coefficients are bounded by those of a fixed polynomial R. To sum up, the following equality holds:

$$\int e^{\lambda Y_N(h_1,\ldots,h_m)-\frac{\lambda^2}{2}C_N(h_1,\ldots h_m)+\frac{1}{N}\{O(\hat{\mathbf{L}}^N(R))\}} d\mu_V^N = 1$$

with

$$C_N(h_1,\ldots,h_m) := \hat{\mathbf{L}}^N\Big(\sum_{i,j} \partial_i \circ \partial_j W\#(h_i,h_j)\Big) + \hat{\mathbf{L}}^N \otimes \hat{\mathbf{L}}^N\Big(\sum_{i,j} \partial_i h_j \partial_j h_i\Big).$$

We can decompose the previous expectation in two terms E_1 and E_2 with

$$E_1 = \mathbb{E}_V\left[1_{\Lambda_M^N} e^{\lambda Y_N(h_1,\ldots,h_m)-\frac{\lambda^2}{2}C_N(h_1,\ldots,h_m)+\frac{O(\hat{\mathbf{L}}^N(R))}{N}}\right]$$

and

$$E_2 = \mathbb{E}_V\left[1_{(\Lambda_M^N)^c} e^{\lambda Y_N(h_1,\ldots,h_m)-\frac{\lambda^2}{2}C_N(h_1,\ldots,h_m)+\frac{O(\hat{\mathbf{L}}^N(R))}{N}}\right].$$

On $\Lambda_M^N = \{\mathbf{A} : \max_i(\lambda_{\max}^N(\mathbf{A}_i)) \leq M\}$ all the quantities are Lipschitz bounded so that $\frac{O(\hat{\mathbf{L}}^N(R))}{N}$ goes uniformly to 0 and $Y_N(h_1,\ldots,h_m)$ is at most of order e^{cN}. Now, by concentration inequalities $C_N(h_1,\ldots,h_m)$ concentrates in the scale e^{-N^2} (see Lemma 6.21). Thus, in E_1, $\hat{\mathbf{L}}^N$ can be replaced by its expectation $\bar{\mathbf{L}}^N$ and then by its limit as $\bar{\mathbf{L}}^N$ converges to τ (see Theorem 8.6). This proves that we can replace C_N by C in E_1.

The aim is now to show that for M sufficiently large, E_2 vanishes when N goes to infinity. It would be an easy task if all the quantities were in $\mathcal{C}_{st}^m(\mathbb{C})$ but some derivatives of V appear so that there are polynomial terms in the exponential. The idea to pass this difficulty is to make the reverse change of variables. For N bigger than the norm of the h_i's, and with $B_i = \mathbf{A}_i + \frac{1}{N}h_i(\mathbf{A})$,

$$E_2 = \mathbb{E}_V\left[1_{\{\mathbf{A}:\max_i(\lambda_{\max}^N(\mathbf{A}_i))\geq M\}} e^{\lambda Y_N(h_1,\ldots,h_m)-\frac{\lambda^2}{2}C_N(h_1,\ldots,h_m)+\frac{O(\hat{\mathbf{L}}^N(R))}{N}}\right]$$

$$= \mu_V^N\big(\mathbf{B} : \max_i(\lambda_{\max}^N(\mathbf{A}_i)) \geq M\big) \leq \mu_V^N\big(\max_i(\lambda_{\max}^N(B_i)) \geq M - 1\big).$$

This last quantity goes exponentially fast to 0 for M sufficiently large by Lemma 6.19. Hence, we arrive, for M large enough, at

$$\lim_{N\to\infty} \int 1_{\Lambda_M^N} e^{\lambda Y_N(h_1,\ldots,h_m)} d\mu_V^N = e^{\frac{\lambda^2}{2}C(h_1,\ldots,h_m)}.$$

Because $\mu_V^N(\Lambda_M^N)$ goes to one as N goes to infinity by Lemma 6.19, we conclude that $Y_N(h_1,\ldots,h_m)$ converges in law under $1_{\Lambda_M^N} d\mu_V^N/\mu_V^N(\Lambda_M^N)$ to a centered Gaussian variable with covariance $C(h_1,\ldots,h_m)$ (since the convergence of the Laplace transforms to a Gaussian law ensures the weak convergence). But since $\mu_V^N(\Lambda_M^N)$ goes to one, this convergence in law also holds under μ_V^N (since for any bounded continuous function $\mu_V^N(f) - \int f 1_{\Lambda_M^N} d\mu_V^N/\mu_V^N(\Lambda_M^N)$ goes to zero as N goes to infinity). □

9.2.2 Central Limit Theorem for some Polynomial Functions

We now extend Lemma 9.2 to polynomial test functions.

Lemma 9.3. *Assume* **(H)**. *Then, for all* $P_1, \ldots, P_m$ *in* $\mathbb{C}\langle X_1, \ldots, X_m \rangle_{sa}$, *the variable*

$$Y_N(P_1, \ldots, P_m) = N \sum_{k=1}^{m} [\hat{\mathbf{L}}^N \otimes \hat{\mathbf{L}}^N (\partial_k P_k) - \hat{\mathbf{L}}^N [(X_k + D_k V) P_k]]$$

converges in law to a real centered Gaussian variable with variance

$$C(P_1, \cdots, P_m) = \sum_{k,l=1}^{m} (\tau_\mathbf{t} \otimes \tau_\mathbf{t} [\partial_k P_l \times \partial_l P_k] + \tau_\mathbf{t} (\partial_l \circ \partial_k V \sharp (P_k, P_l))) + \sum_{k=1}^{m} \tau_\mathbf{t}(P_k^2).$$

Proof. Let $P_1, \ldots, P_m$ be self-adjoint polynomials and $h_1^\epsilon, \ldots, h_m^\epsilon$ be Stieltjes functionals which approximate $P_1, \cdots, P_m$ such as

$$h_i^\epsilon(\mathbf{A}) = P_i \left(\frac{\mathbf{A}_1}{1 + \epsilon \mathbf{A}_1^2}, \ldots, \frac{\mathbf{A}_m}{1 + \epsilon \mathbf{A}_m^2} \right).$$

Since $E[Y_N(P_1, \ldots, P_m)] = 0$ by (9.1),

$$Y_N(P_1, \ldots, P_m) = \tilde{\delta}^N(K_N(P_1, \ldots, P_m))$$

with

$$K_N(P_1, \ldots, P_m) = \sum_{k=1}^{m} \left(\hat{\mathbf{L}}^N \otimes I(\partial_k P_k) - (X_k + D_k V) P_k \right)$$

and the same for $Y_N(h_1^\epsilon, \ldots, h_m^\epsilon)$. It is not hard to see that on Λ_M^N,

$$K_N(h_1^\epsilon, \ldots, h_m^\epsilon) - K_N(P_1, \ldots, P_m)$$

is a Lipschitz function with a constant bounded by $\epsilon C(M)$ for some finite constant $C(M)$ which depends only on M. Hence, by Lemma 6.21, we have

$$\mu_V^N \left(|\hat{\delta}^N(K_N(h_1^\epsilon, \ldots, h_k^\epsilon) - K_N(P_1, \ldots, P_k))| \geq \delta \right) \leq e^{-\alpha M N} + e^{-\frac{\delta^2}{2c\epsilon^2 C(M)^2}}$$

and so for any bounded continuous function $f : \mathbb{R} \to \mathbb{R}$, if ν_{σ^2} is the centered Gaussian law of variance σ^2, we deduce

$$\lim_{N \to \infty} \mu_V^N \left(f(\tilde{\delta}^N(K_N(P_1, \ldots, P_k))) \right)$$

$$= \lim_{\epsilon \to 0} \lim_{N \to \infty} \mu_V^N \left(f(\tilde{\delta}^N(K_N(h_1^\epsilon, \ldots, h_k^\epsilon))) \right)$$

$$= \lim_{\epsilon \to 0} \nu_{C(h_1^\epsilon, \ldots, h_m^\epsilon)}(f) = \nu_{C(P_1, \ldots, P_m)}(f)$$

where we used in the second line Lemma 9.2 and in the last line Lemma 6.18 to obtain the convergence of $C(h_1^\epsilon, \ldots, h_m^\epsilon)$ to $C(P_1, \ldots, P_m)$.   □

Y_N depends on $N\hat{\mathbf{L}}^N \otimes \hat{\mathbf{L}}^N$, in which one of the empirical distribution $\hat{\mathbf{L}}^N$ can be replaced by its deterministic limit. This is the content of the next lemma.

Lemma 9.4. *Assume* **(H)** *and let* $P_1, \ldots, P_k$ *be self-adjoint polynomial functions. Then, the variable*

$$Z_N(P_1, \ldots, P_m) = \hat{\delta}^N \left(\sum_{k=1}^{m} (X_k + D_k V) P_k - \sum_{k=1}^{m} (\tau_{\mathbf{t}} \otimes I + I \otimes \tau_{\mathbf{t}})(\partial_k P_k) \right)$$

converges in law to a centered Gaussian variable with variance

$$C(P_1, \cdots, P_m) = \sum_{k,l=1}^{m} (\tau_{\mathbf{t}} \otimes \tau_{\mathbf{t}} [\partial_k P_l \times \partial_l P_k] + \tau_{\mathbf{t}} (\partial_l \circ \partial_k V \sharp (P_k, P_l))) + \sum_{k=1}^{m} \tau_{\mathbf{t}} (P_k^2).$$

Proof. The only point is to notice that

$$Y_N(P_1, \cdots, P_m) = \sum_{k=1}^{m} \left(\hat{\delta}^N \otimes \tau_{\mathbf{t}} + \tau_{\mathbf{t}} \otimes \hat{\delta}^N \right) (\partial_k P_k) - \hat{\delta}^N ((X_k + D_k V) P_k) + r_{N,P}$$

with $r_{N,P} = N^{-1} \sum_{k=1}^{m} \hat{\delta}^N \otimes \hat{\delta}^N (\partial_k P_k)$ of order N^{-1} with probability going to 1 by Lemma 6.21 and Property 9.1. Thus

$$Y_N(P_1, \ldots, P_m)$$

$$= \hat{\delta}^N \left(\sum_{k=1}^{m} (-(X_k + D_k V) P_k + (I \otimes \tau_{\mathbf{t}} + \tau_{\mathbf{t}} \otimes I)(\partial_k P_k)) \right) + O(\frac{1}{N})$$

$$= -Z_N(P_1, \ldots, P_m) + O(\frac{1}{N}).$$

This, with the previous lemma, proves the claim.   □

9.2.3 Central Limit Theorem for all Polynomial Functions

In the previous part, we obtained the central limit theorem only for the family of random variables $\hat{\delta}^N(Q)$ with Q in

$$\mathcal{F} = \left\{ \sum_{k=1}^{m} (X_k + D_k V) P_k - \sum_{k=1}^{m} (\tau_{\mathbf{t}} \otimes I + I \otimes \tau_{\mathbf{t}})(\partial_k P_k), \forall i, P_i \text{ self-adjoint} \right\}.$$

In this section, we wish to extend this result to any self-adjoint polynomial function Q, that is, prove the following theorem:

Theorem 9.5. *Let* $\mathbf{t} \in U_c \cap B_\eta$. *There exists* $\eta_c > 0$ *so that for* $\eta < \eta_c$, *for all polynomials* $P_1, \ldots, P_k \in \mathbb{C}\langle X_1, \ldots, X_m \rangle$, $(\mathrm{Tr}(P_i) - N\tau_{\mathbf{t}}(P_i))_{1 \le i \le k}$ *converges in law to a centered Gaussian vector with covariance* $\{\sigma(P_i, P_j), 1 \le i, j \le k\}$

We shall describe σ in the course of the proof. Its interpretation as a generating function for maps is given in Section 9.3.

To prove the theorem, we have to show that the set $\mathcal{F}$ is dense for some convenient topology in $\mathbb{C}\langle X_1, \ldots, X_m \rangle$.

The strategy is to see $\mathcal{F}$ as the image of an operator that we will invert. The first operator that comes to mind is

$$\Psi : (P_1, \ldots, P_k) \to \sum_{k=1}^m (X_k + D_k V) P_k - \sum_{k=1}^m (\tau_\mathbf{t} \otimes I + I \otimes \tau_\mathbf{t})(\partial_k P_k)$$

as we immediately have $\mathcal{F} = \Psi(\mathbb{C}\langle X_1, \ldots, X_m \rangle_{sa}, \ldots, \mathbb{C}\langle X_1, \ldots, X_m \rangle_{sa})$.

In order to obtain an operator from $\mathbb{C}\langle X_1, \ldots, X_m \rangle$ to $\mathbb{C}\langle X_1, \ldots, X_m \rangle$ we will prefer to apply this with $P_k = D_k P$ for all k and for a given P; as we shall see later, $\Psi(D_1 P, \ldots, D_m P)$ is closely related with the projection on functions of the type $\mathrm{Tr} P$ of the operator on the entries $\mathcal{L} = \Delta - \nabla N tr(W).\nabla$ which is symmetric in $L^2(\mu_V^N)$ (here $W = V + \frac{1}{2} \sum \mathbf{A}_k^2$). The resulting operator is a differential operator. As such, it may be difficult to find a normed space stable for this operator (since the operator will deteriorate the smoothness of the functions) in which it is continuous and invertible.

To avoid this issue, we will first divide each monomials of P by its degree (which more or less amounts to integrate and then divide by x the function in the one variable case).

Then, we define a linear map Σ on $\mathbb{C}\langle X_1, \ldots, X_m \rangle$ such that for all monomials q of degree greater or equal to 1

$$\Sigma q = \frac{q}{\deg q}.$$

For later use, we set $\mathbb{C}\langle X_1, \ldots, X_m \rangle(0)$ to be the subset of polynomials P of $\mathbb{C}\langle X_1, \ldots, X_m \rangle_{sa}$ such that $P(0, \ldots, 0) = 0$. We let Π be the projection from $\mathbb{C}\langle X_1, \ldots, X_m \rangle_{sa}$ onto $\mathbb{C}\langle X_1, \ldots, X_m \rangle(0)$ (i.e., $\Pi(P) = P - P(0, \ldots, 0)$). We now define some operators on $\mathbb{C}\langle X_1, \ldots, X_m \rangle(0)$, i.e., from $\mathbb{C}\langle X_1, \ldots, X_m \rangle(0)$ into $\mathbb{C}\langle X_1, \ldots, X_m \rangle(0)$,

$$\Xi_1 : P \longrightarrow \Pi \left(\sum_{k=1}^m \partial_k \Sigma P \sharp D_k V \right)$$

$$\Xi_2 : P \longrightarrow \Pi \left(\sum_{k=1}^m (\mu \otimes I + I \otimes \mu)(\partial_k D_k \Sigma P) \right).$$

We define $\Xi_0 = I - \Xi_2$ and $\Xi = \Xi_0 + \Xi_1$, where I is the identity on $\mathbb{C}\langle X_1, \ldots, X_m \rangle(0)$. Note that the images of Ξ_i and Ξ are included in $\mathbb{C}\langle X_1, \ldots, X_m \rangle_{sa}$ since V is assumed self-adjoint. With these notations, Lemma 9.4, once applied to $P_i = D_i \Sigma P$, $1 \le i \le m$, reads:

Proposition 9.6. *For all P in $\mathbb{C}\langle X_1, \ldots, X_m \rangle(0)$, $\hat{\delta}^N(\Xi P)$ converges in law to a centered Gaussian variable with covariance*

$$\mathcal{C}(P) := C(D_1 \Sigma P, \ldots, D_m \Sigma P).$$

Proof. We have for all tracial states τ, $\tau(\partial_k P \sharp V) = \tau(D_k PV)$ and if P is in $\mathbb{C}\langle X_1, \ldots, X_m \rangle(0)$, we have the identity

$$P = \sum_k \partial_k \Sigma P \sharp X_k.$$

Then, as $\hat{\delta}^N$ is tracial (7.2) and vanishes on constant terms (so that the projection Π can be removed in the definition of Ξ), for all polynomial P,

$$\hat{\delta}^N(\Xi P) = \hat{\delta}^N \left(P + \sum_{k=1}^m \partial_k \Sigma P \sharp D_k V - \sum_{k=1}^m (\mu \otimes I + I \otimes \mu)(\partial_k D_k \Sigma P) \right)$$

$$= \hat{\delta}^N \left(\sum_{k=1}^m (X_k + D_k V) D_k \Sigma P - \sum_{k=1}^m (\mu \otimes I + I \otimes \mu)(\partial_k D_k \Sigma P) \right)$$

$$= Z_N(D_1 \Sigma P, \ldots, D_m \Sigma P).$$

We then use Lemma 9.4 to conclude.   □

To generalize the central limit theorem to all polynomial functions, we need to show that the image of Ξ is dense and to control approximations. If P is a polynomial and q a non-constant monomial we will denote by $\lambda_q(P)$ the coefficient of q in the decomposition of P in monomials. We can then define a norm $\|.\|_A$ on $\mathbb{C}\langle X_1, \ldots, X_m \rangle(0)$ for $A > 1$ by

$$\|P\|_A = \sum_{\deg q \neq 0} |\lambda_q(P)| A^{\deg q}.$$

In the formula above, the sum is taken over all non-constant monomials. We also define the operator norm given, for T from $\mathbb{C}\langle X_1, \ldots, X_m \rangle(0)$ to $\mathbb{C}\langle X_1, \ldots, X_m \rangle(0)$, by

$$|||T|||_A = \sup_{\|P\|_A = 1} \|T(P)\|_A.$$

Finally, let $\mathbb{C}\langle X_1, \ldots, X_m \rangle(0)_A$ be the completion of $\mathbb{C}\langle X_1, \ldots, X_m \rangle(0)$ for $\|.\|_A$. We say that T is continuous on $\mathbb{C}\langle X_1, \ldots, X_m \rangle(0)_A$ if $|||T|||_A$ is finite. We shall prove that Ξ is continuous on $\mathbb{C}\langle X_1, \ldots, X_m \rangle(0)_A$ with continuous inverse when $\mathbf{t}$ is small.

Lemma 9.7. *With the previous notations:*

1. The operator Ξ_0 is invertible on $\mathbb{C}\langle X_1, \ldots, X_m \rangle(0)$.

2. *There exists $A_0 > 0$ such that for all $A > A_0$, the operators Ξ_2, Ξ_0 and Ξ_0^{-1} are continuous on $\mathbb{C}\langle X_1, \ldots, X_m \rangle(0)_A$ and their norm are uniformly bounded for $\mathbf{t}$ in B_η.*
3. *For all $\epsilon, A > 0$, there exists $\eta_\epsilon > 0$ such for $|\mathbf{t}| < \eta_\epsilon$, Ξ_1 is continuous on $\mathbb{C}\langle X_1, \ldots, X_m \rangle(0)_A$ and $|||\Xi_1|||_A \leq \epsilon$.*
4. *For all $A > A_0$, there exists $\eta > 0$ such that for $\mathbf{t} \in B_\eta$, Ξ is continuous, invertible with a continuous inverse on $\mathbb{C}\langle X_1, \ldots, X_m \rangle(0)_A$. Besides the norms of Ξ and Ξ^{-1} are uniformly bounded for $\mathbf{t}$ in B_η.*
5. *There exists $C > 0$ such that for all $A > C$, $\mathcal{C}$ is continuous from $\mathbb{C}\langle X_1, \ldots, X_m \rangle(0)_A$ into $\mathbb{R}$.*

Proof.   1. Observe that since Ξ_2 reduces the degree of a polynomial by at least 2,

$$P \to \sum_{n \geq 0} (\Xi_2)^n(P)$$

is well defined on $\mathbb{C}\langle X_1, \ldots, X_m \rangle(0)$ as the sum is finite for any polynomial P. This gives an inverse for $\Xi_0 = I - \Xi_2$.
2. First remark that a linear operator T has a norm less than C with respect to $\|.\|_A$ if and only if for all non-constant monomial q,

$$\|T(q)\|_A \leq C A^{\deg q}.$$

Recall that μ is uniformly compactly supported (see Lemma 6.18) and let $C_0 < +\infty$ be such that $|\mu(q)| \leq C_0^{\deg q}$ for all monomial q. Take a monomial $q = X_{i_1} \cdots X_{i_p}$, and assume that $A > 2C_0$,

$$\left\| \Pi \left(\sum_k (I \otimes \mu) \partial_k D_k \Sigma q \right) \right\|_A \leq p^{-1} \sum_{\substack{k,q = q_1 X_k q_2, \\ q_2 q_1 = r_1 X_k r_2}} \|r_1 \mu(r_2)\|_A$$

$$\leq p^{-1} \sum_{\substack{k,q = q_1 X_k q_2, \\ q_2 q_1 = r_1 X_k r_2}} A^{\deg r_1} C_0^{\deg r_2} = \frac{1}{p} \sum_{n=0}^{p-1} \sum_{l=0}^{p-2} A^l C_0^{p-l-2}$$

$$\leq A^{p-2} \sum_{l=0}^{p-2} \left(\frac{C_0}{A} \right)^{p-2-l} \leq 2 A^{-2} \|q\|_A$$

where in the second line, we observed that once $\deg(q_1)$ is fixed, $q_2 q_1$ is uniquely determined and then r_1, r_2 are uniquely determined by the degree l of r_1. Thus, the factor $\frac{1}{p}$ is compensated by the number of possible decompositions of q, i.e., the choice of n, the degree of q_1. If $A > 2$, $P \to \Pi \left(\sum_k (I \otimes \mu) \partial_k D_k \Sigma P \right)$ is continuous of norm strictly less than $\frac{1}{2}$. And a similar calculus for $\Pi \left(\sum_k (\mu \otimes I) \partial_k D_k \Sigma \right)$ shows that Ξ_2 is continuous of norm strictly less than 1. It follows immediately that Ξ_0 is continuous. Since $\Xi_0^{-1} = \sum_{n \geq 0} \Xi_2^n$, Ξ_0^{-1} is continuous as soon as Ξ_2 is of norm strictly less than 1.

3. Let $q = X_{i_1} \cdots X_{i_p}$ be a monomial and let D be the degree of V and $B(\leq Dn)$ the sum of the maximum number of monomials in $D_k V$.

$$\|\Xi_1(q)\|_A \leq \frac{1}{p} \sum_{k, q = q_1 X_k q_2} \|q_1 D_k V q_2\|_A \leq \frac{1}{p} \sum_{k, q = q_1 X_k q_2} |t| B A^{p-1+D-1}$$

$$= |t| B A^{D-2} \|q\|_A.$$

It is now sufficient to take $\eta_\epsilon < (BA^{D-2})^{-1}\epsilon$.

4. We choose $\eta < (BA^{D-2})^{-1}\||\Xi_0^{-1}\||_A^{-1}$ so that when $|t| \leq \eta$,

$$\||\Xi_1\||_A \||\Xi_0^{-1}\||_A < 1.$$

By continuity, we can extend Ξ_0, Ξ_1, Ξ_2, Ξ and Ξ_0^{-1} on the space $\mathbb{C}\langle X_1, \ldots, X_m \rangle(0)_A$. The operator

$$P \to \sum_{n \geq 0} (-\Xi_0^{-1} \Xi_1)^n \Xi_0^{-1}$$

is well defined and continuous. This is an inverse of $\Xi = \Xi_0 + \Xi_1 = \Xi_0(I + \Xi_0^{-1} \Xi_1)$.

5. We finally prove that $\mathcal{C}$ is continuous from $\mathbb{C}\langle X_1, \ldots, X_m \rangle(0)_A$ into $\mathbb{R}$ where we recall that we assumed $A > C_0$. Let us consider the first term

$$\mathcal{C}_1(P) := \sum_{k, l=1}^{m} \mu \otimes \mu(\partial_k D_l \Xi P \times \partial_l D_k \Xi P).$$

Then, we obtain as in the second point of this proof

$$|\mathcal{C}_1(P)| \leq 4 \sum_{k, l=1}^{m} \sum_{q, q'} \frac{|\lambda_q(P)||\lambda_{q'}(P)|}{\deg q \deg q'} \sum_{\substack{q = q_1 X_k q_2, q' = q_1' X_l q_2' \\ q_2 q_1 = r_1 X_l r_2, q_2' q_1' = r_1' X_k r_2'}} C_0^{\deg q + \deg q' - 4}$$

$$\leq 4 \sum_{q, q'} |\lambda_q(P)||\lambda_{q'}(P)| \deg q \deg q' C_0^{\deg q + \deg q' - 4}$$

$$\leq 4(\sup_{\ell \geq 0} \ell C_0^{\ell-2} A^{-\ell})^2 \|P\|_A^2.$$

We next turn to showing that

$$\mathcal{C}_2(P) := \sum_{k, l=1}^{m} \mu\left(\partial_k \circ \partial_l V \sharp (D_k \Xi P, D_l \Xi P)\right)$$

is also continuous for $\|.\|_A$. In fact, noting that we may assume $V \in \mathbb{C}\langle X_1, \ldots, X_m \rangle(0)$ without changing $\mathcal{C}_2$, we find

$$|\mathcal{C}_2(P)| \leq \sum_{p,q,q',k,l} |\lambda_p(V)| \sum_{\substack{q,q',p=p_1 X_k p_2 X_l p_3 \\ q=q_1 X_k q_2, q'=q'_1 X_k q'_2}} \frac{|\lambda_q(P)||\lambda_{q'}(P)|C_0^{\deg p+\deg q+\deg q'-4}}{\deg q \deg q'}$$

$$\leq n|\mathbf{t}|D^2 \sum_{q,q'} |\lambda_q(P)||\lambda_{q'}(P)|C_0^{D+\deg q+\deg q'-4}$$

$$\leq n|\mathbf{t}|D^2 C_0^{D-4}\|P\|_A^2.$$

The continuity of the last term $\mathcal{C}_3(P) = \sum_{i=1}^m \mu\left((D_j\varXi P)^2\right)$ is obtained similarly.

$\square$

We can compare the norm $\|\cdot\|_A$ to a more intuitive norm, namely the Lipschitz norm $\|\cdot\|_{\mathcal{L}}^M$ defined by

$$\|P\|_{\mathcal{L}}^M = \sup_{N\in\mathbb{N}} \sup_{\substack{x_1,\dots,x_m\in\mathcal{H}_N^{(2)} \\ \forall i, \|x_i\|_\infty \leq M}} \sum_{k=1}^m (\|D_k P D_k P^*\|_A)^{\frac{1}{2}}.$$

We will say that a semi-norm $\mathcal{N}$ is weaker than a semi-norm $\mathcal{N}'$ if and only if there exists $C < +\infty$ such that for all P in $\mathbb{C}\langle X_1,\dots,X_m\rangle(0)$,

$$\mathcal{N}(P) \leq C\mathcal{N}'(P).$$

Lemma 9.8. *For $A > M$, the semi-norm $\|.\|_{\mathcal{L}}^M$ restricted to the space $\mathbb{C}\langle X_1,\dots,X_m\rangle(0)$ is weaker than the norm $\|.\|_A$.*

Proof. For all P in $\mathbb{C}\langle X_1,\dots,X_m\rangle(0)$, the following inequalities hold:

$$\|P\|_{\mathcal{L}}^M \leq \sum_q |\lambda_q(P)|\|q\|_{\mathcal{L}}^M \leq \sum_q |\lambda_q(P)| \deg q M^{\deg q} \leq \left(\sup_l l\left(\frac{M}{A}\right)^l\right) \|P\|_A.$$

$\square$

To take into account the previous results, we define a new hypothesis **(H')** stronger than **(H)**.

(H'): **(H)** is satisfied, $A - 1 > \max(A_0, M_0, C)$ for the M_0 which appear in Lemma 6.19 and the C which appear in Proposition 9.1 Besides, $|\mathbf{t}| \leq \eta$ with η as in the fourth point of Lemma 9.7 in order that $\varXi$ and $\varXi^{-1}$ are continuous on $\mathbb{C}\langle X_1,\dots,X_m\rangle(0)_A$ and $\mathbb{C}\langle X_1,\dots,X_m\rangle(0)_{A-1}$, and that $\mathcal{C}$ is also continuous for these norms.

The two main additional consequences of this hypothesis are the continuity of $\varXi$ for $\|\cdot\|_A$. The strange condition about the continuity of $\varXi$ on $\mathbb{C}\langle X_1,\dots,X_m\rangle(0)_{A-1}$ is here assumed for a technical reason which will appear only in the last section on the interpretation of the first order correction to the free energy.

While (**H'**) is full of conditions, the only important hypothesis is the c-convexity of V. Given such a V, we can always find constants A and η which satisfy the hypothesis. The only restriction will be then that $\mathbf{t}$ is sufficiently small.

We can now prove the general central limit theorem which is up to the identification of the variance equivalent to Theorem 9.5.

Theorem 9.9. *Assume* (**H'**). *For all P in $\mathbb{C}\langle X_1,\ldots,X_m\rangle_{sa}$, $\hat{\delta}^N(P)$ converges in law to a centered Gaussian variable γ_P with variance*

$$\sigma^{(2)}(P) := \mathcal{C}(\Xi^{-1}\Pi(P)) = C(D_1\Sigma\Xi^{-1}\Pi(P),\cdots,D_m\Sigma\Xi^{-1}\Pi(P)).$$

If $P \in \mathbb{C}\langle X_1,\ldots,X_m\rangle$, $\hat{\delta}^N(P)$ converges to the complex centered Gaussian variable $\gamma_{(P+P^*)/2} + i\gamma_{(P-P^*)/2i}$ (the covariance of $\gamma_{(P+P^*)/2}$ and $\gamma_{(P-P^*)/2i}$ being given by $\sigma^{(2)}((P+P^*)/2, (P-P^*)/2i)$ where $\sigma^{(2)}(\cdot,\cdot)$ is the bilinear form associated to the quadratic form $\sigma^{(2)}$).

Proof. As $\hat{\delta}^N(P)$ does not depend on constant terms, we can directly take $P = \Pi(P)$ in $\mathbb{C}\langle X_1,\ldots,X_m\rangle(0)$. Now, by Lemma 9.7. 4, we can find an element Q of $\mathbb{C}\langle X_1,\ldots,X_m\rangle(0)_A$ such that $\Xi Q = P$. But the space $\mathbb{C}\langle X_1,\ldots,X_m\rangle(0)$ is dense in $\mathbb{C}\langle X_1,\ldots,X_m\rangle(0)_A$ by construction. Thus, there exists a sequence Q_n in $\mathbb{C}\langle X_1,\ldots,X_m\rangle(0)$ such that

$$\lim_{n\to\infty} \|Q - Q_n\|_A = 0.$$

Let us define $R_n = P - \Xi Q_n$ in $\mathbb{C}\langle X_1,\ldots,X_m\rangle(0)$.

Now according to Property 9.6 for all n, $\hat{\delta}^N(\Xi Q_n)$ converges in law to a Gaussian variable γ_n of variance $\mathcal{C}(Q_n)$ with

$$\mathcal{C}(Q_n) = C(D_1\Sigma Q_n,\ldots,D_m\Sigma Q_n).$$

As $\mathcal{C}$ is continuous by Lemma 9.7.4, it can be extended to the space $\mathbb{C}\langle X_1,\ldots,X_m\rangle(0)_A$ and $\sigma^{(2)}(P) = \mathcal{C}(\Xi^{-1}P) = \mathcal{C}(Q) = \lim_n \mathcal{C}(Q_n)$ is well defined. Hence, γ_n converges weakly to γ_∞, the centered Gaussian law with covariance $\mathcal{C}(Q)$, when n goes to $+\infty$. The last step is to prove the convergence in law of $\hat{\delta}^N(P)$ to γ_∞. We will use the Dudley distance d_D. Below, as a parameter of d_D, we, for short, write $\hat{\delta}^N(P)$ for the law of $\hat{\delta}^N(P)$. We make the following decomposition:

$$d_D(\hat{\delta}^N(P),\gamma_\infty) \leq d_D(\hat{\delta}^N(P), \hat{\delta}^N(\Xi Q_n)) + d_D(\hat{\delta}^N(\Xi Q_n),\gamma_n) + d_D(\gamma_n,\gamma_\infty).$$

(9.4)

By the above remarks, $d_D(\hat{\delta}^N(\Xi Q_n),\gamma_n)$ goes to 0 when N goes to $+\infty$ and $d_D(\gamma_n,\gamma_\infty)$ goes to 0 when n goes to $+\infty$. We now use the bound on the Dudley distance

$$d_D(\hat{\delta}^N(P), \hat{\delta}^N(\Xi Q_n)) \leq E[|\hat{\delta}^N(P) - \hat{\delta}^N(\Xi Q_n)| \wedge 1] = E[|\hat{\delta}^N(R_n)| \wedge 1].$$

We control the last term by Lemmas 6.21 and 6.19 so that for $M \geq M_0$,

$$E[|\hat{\delta}^N(R_n)| \wedge 1] \leq e^{-\alpha NM} + 2\sqrt{\frac{2\pi}{c}} \|R_n\|_{\mathcal{L}}^M + \epsilon_{R_n,M}^N + |m_{R_n,M}^N|.$$

But we deduce from Lemma 9.8 that since we chose $M < A$, there exists a finite constant C such that

$$\|R_n\|_{\mathcal{L}}^M \leq C\|R_n\|_A = C\|\Xi(Q - Q_n)\|_A \leq C\|\|\Xi\|\|_A \|Q - Q_n\|_A$$

and so $\|R_n\|_{\mathcal{L}}^M$ goes to zero as n goes to infinity. And since $\|R_n\|_{\mathcal{L}}^M$ is finite, $\epsilon_{R_n,M}^N$ goes to zero. Similarly, using the bound of Lemma 6.21 on $m_{P,M}^N$ for P monomial, we find that

$$|m_{R_n,M}^N| \leq N \sum_q |\lambda_q(R_n)|\deg(q)(3M^{\deg(q)} + \deg(q)^2)e^{-\alpha MN}$$

$$\leq N \sup_{\ell \geq 0}(\ell(3M^\ell + \ell^2)A^{-\ell})\|R_n\|_A e^{-\alpha MN}$$

goes to zero as N goes to infinity. Thus, $E[|\hat{\delta}^N(R_n)| \wedge 1]$ goes to zero as n and N go to infinity. Putting things together, we obtain if we first let N go to $+\infty$ and then n, the desired convergence $\lim_N d_D(\hat{\delta}^N(P), \gamma_\infty) = 0$. □

Note that the convergence in law in Theorem 9.9 can be generalized to a convergence *in moments*:

Corollary 9.10. *Assume* **(H')**. *Let* P *be a self-adjoint polynomial, then* $\hat{\delta}^N(P)$ *converges in moments to a real centered Gaussian variable with variance* $\sigma^{(2)}(P)$, *i.e for all* k *in* $\mathbb{N}$,

$$\lim_{N \to \infty} \int (\hat{\delta}^N P)^k d\mu_V^N = \frac{1}{\sqrt{2\pi\sigma^{(2)}(P)}} \int x^k e^{-\frac{x^2}{2\sigma^{(2)}(P)}} dx.$$

Proof. Indeed, once again we decompose $\int(\hat{\delta}^N P)^k d\mu_V^N$ into $E_1^N + E_2^N$ with

$$E_1^N = \int 1_{\Lambda_M^N}(\hat{\delta}^N P)^k d\mu_V^N \quad E_2^N = \int 1_{(\Lambda_M^N)^c}(\hat{\delta}^N P)^k d\mu_V^N$$

with $M \geq M_0$. For E_1, we notice that the law of $\hat{\delta}^N P$ has a sub-Gaussian tail according to Lemma 6.21. Therefore, we can replace x^k by a bounded continuous function, producing an error independent of N. Applying Theorem 9.9 then shows that

$$\lim_{N \to \infty} \int 1_{\Lambda_M^N}(\hat{\delta}^N P)^k d\mu_V^N = \frac{1}{\sqrt{2\pi\sigma^{(2)}(P)}} \int x^k e^{-\frac{x^2}{2\sigma^{(2)}(P)}} dx.$$

For the second term, we use the trivial bound

$$|E_2^N| \leq N^k \int \mathbf{1}_{(\Lambda_M^N)^c}(|\lambda_{max}(\mathbf{A})| + |\mu|(P))^k d\mu_V^N$$

$$\leq kN^k \int_{\lambda \geq M} (\lambda + |\mu|(P))^{k-1} e^{-\alpha\lambda N} d\lambda$$

which goes to zero as N goes to infinity for all finite k.   □

9.3 Comments on the Results

There is a natural interpretation of the operator Ξ in terms of the symmetric differential operator in $L^2(\mu_V^N)$ given by

$$\mathcal{L}_N = \sum_{k=1}^m \sum_{i,j=1}^N e^{N\text{Tr}(V(\mathbf{X})+2^{-1}\sum_{i=1}^m (X^k)^2)} \partial_{X_{ij}^k} e^{-N\text{Tr}(V(\mathbf{X})+2^{-1}\sum_{i=1}^m (X^k)^2)} \partial_{X_{ji}^k}$$

$$= \sum_{k=1}^m \sum_{i,j=1}^N \left(\partial_{X_{ij}^k} \partial_{X_{ji}^k} - N(D_k V + X_k)_{ji} \partial_{X_{ji}^k} \right).$$

One checks by integration by parts that for any pair of continuously differentiable functions f, g

$$\mu_V^N (f\mathcal{L}_N g) = \mu_V^N (g\mathcal{L}_N f) = -\sum_{k=1}^m \sum_{i,j=1}^N \mu_V^N \left(\partial_{X_{ij}^k} g \partial_{X_{ij}^k} f \right). \quad (9.5)$$

Moreover, for any polynomial P, with the notation of Lemma 9.3, we find that

$$\frac{1}{N}\mathcal{L}_N \text{Tr}(\Sigma P(\mathbf{X})) = Y_N(D_1\Sigma P, \ldots, D_m\Sigma P) = \text{Tr}(\Xi P) + o(1)$$

according to Lemma 9.4. Applying (9.5) with $f = 1$ and $g = \text{Tr}(\Sigma P(\mathbf{X}))$ shows that

$$\lim_{N\to\infty} \mu_V^N (\text{Tr}(\Xi P)) = 0.$$

Furthermore, taking $f = \text{Tr} P(\mathbf{X})$ and $g = \text{Tr} Q(\mathbf{X})$ into (9.5) we deduce by (6.22) that

$$\sigma^{(2)}(\Xi P, Q) = \lim_{N\to\infty} \mu_V^N \left(\text{Tr}(\Xi P(\mathbf{X})) \left(\text{Tr} Q(\mathbf{X}) - \mu_V^N (\text{Tr} Q(\mathbf{X})) \right) \right)$$

$$= \lim_{N\to\infty} \frac{1}{N}\mu_V^N \left(\mathcal{L}_N \mathrm{Tr}(\Sigma P(\mathbf{X}))\left(\mathrm{Tr}Q(\mathbf{X}) - \mu_V^N\left(\mathrm{Tr}Q(\mathbf{X})\right)\right)\right)$$

$$= \lim_{N\to\infty} \frac{1}{N}\sum_{k=1}^{m}\sum_{i,j=1}^{N} \mu_V^N \left((D_k\Sigma P(\mathbf{X}))_{ij}(D_k Q(\mathbf{X}))_{ij}\right)$$

$$= \sum_{k=1}^{m} \tau_{\mathbf{t}} \left((D_k\Sigma P(\mathbf{X}))(D_k Q(\mathbf{X}))^*\right).$$

Thus, we have proved the following:

Lemma 9.11. *For all polynomial function P so that $P(0) = 0$,*

$$\sigma^{(2)}(\Xi P, Q) = \sum_{k=1}^{m} \tau_{\mathbf{t}} \left((D_k\Sigma P(\mathbf{X}))(D_k Q(\mathbf{X}))^*\right).$$

Let us define, for $\mathbf{k} = (k_1, \ldots, k_n)$,

$$\mathcal{M}_{\mathbf{k}}(P,Q) = \sharp\{ \text{ maps with } k_i \text{ stars of type } q_i,$$
$$\text{one of type } P \text{ and one of type } Q\}.$$

and

$$\mathcal{M}_{\mathbf{t}}(P,Q) = \sum_{k_1,\ldots,k_n} \prod_{i=1}^{n} \frac{(-t_i)^{k_i}}{k_i!} \mathcal{M}_{k_1,\ldots,k_n}(P,Q).$$

We extend $\mathcal{M}_{\mathbf{t}}$ to polynomials by linearity. Then we claim that $\sigma^2(P,Q)$ and $\mathcal{M}(P,Q)$ satisfy the same kind of induction relation.

Proposition 9.12. *For all monomials P, Q and all k,*

$$\mathcal{M}_{k_1,\ldots,k_n}(X_k P, Q)$$

$$= \sum_{0\le p_i\le k_i}\sum_{P=RX_k S}\prod_i C_{k_i}^{p_i}\mathcal{M}_{p_1,\ldots,p_n}(R,Q)\mathcal{M}_{k_1-p_1,\ldots,k_n-p_n}(S)$$

$$+ \sum_{0\le p_i\le k_i}\sum_{P=RX_k S}\prod_i C_{k_i}^{p_i}\mathcal{M}_{p_1,\ldots,p_n}(S,Q)\mathcal{M}_{k_1-p_1,\ldots,k_n-p_n}(R)$$

$$+ \sum_{0\le j\le n} k_j\mathcal{M}_{k_1,\ldots,k_j-1,\ldots,k_n}(D_k VP, Q)$$

$$+ \mathcal{M}_{k_1,\ldots,k_n}(D_k QP)$$

and

$$\mathcal{M}_{\mathbf{t}}(X_k P, Q) = \mathcal{M}_{\mathbf{t}}((I\otimes\tau_{\mathbf{t}} + \tau_{\mathbf{t}}\otimes I)D_k P) - \mathcal{M}_{\mathbf{t}}(D_k VP, Q) + \tau_{\mathbf{t}}(D_k QP). \quad (9.6)$$

Besides there exists $\eta > 0$ so that there exists $R < +\infty$ such that for all monomials P and Q, all $\mathbf{t} \in B(0, \eta)$,

$$|\mathcal{M}_{\mathbf{t}}(P,Q)| \le R^{\deg P + \deg Q}. \quad (9.7)$$

Proof. The proof is very close to that of Theorem 8.5 which explains the decomposition of planar maps with one root. First we look for a relation on $\mathcal{M}_{k_1,\ldots,k_n}(X_k P, Q)$. We look at the first half-edge associated with X_k, then three cases may occur:

1. The first possibility is that the branch is glued to another branch of $P = RX_k S$. It cuts P into two: R and S and it occurs for all decomposition of P into $P = RX_k S$, which is exactly what D does. Then either the component R is linked to Q and to p_i stars of type q_i for each i, this leads to

$$\prod C_{k_i}^{p_i} \mathcal{M}_{p_1,\ldots,p_n}(R,Q)\mathcal{M}_{k_1-p_1,\ldots,k_n-p_n}(S)$$

   possibilities or we have a symmetric case with S linked to Q in place of R.
2. The second case occurs when the branch is glued to a vertex of type q_j for a given j; first we have to choose between the k_j vertices of this type then we contract the edges arising from this gluing to form a vertex of type $D_i q_j P_1$, which creates

$$k_j \mathcal{M}_{k_1,\ldots,k_j-1,\ldots,k_n}(D_k q_j P, Q)$$

   possibilities.
3. The last case is that the branch can be glued to the star associated to $Q = RX_i S$. We then only have to count planar graphs:

$$\mathcal{M}_{k_1,\ldots,k_n}(D_k QP).$$

We can now sum on the k's to obtain the relation on $\mathcal{M}$.

Finally, to show the last point of the proposition, we only have to prove that there exist $A > 0, B > 0$ such that for all k's, for all monomials P and Q,

$$\frac{\mathcal{M}_{k_1,\ldots,k_n}(P,Q)}{\prod_i k_i!} \le A^{\sum_i k_i} B^{\deg P+\deg Q}.$$

This follows easily by induction over the degree of P with the previous relation on the $\mathcal{M}$ since we have proved such a control for $\mathcal{M}_{k_1,\ldots,k_n}(Q)$ in [104].   □

We can now prove the theorem:

Theorem 9.13. *Assume* **(H')** *with η small enough. Then, for all polynomials P, Q,*

$$\sigma^{(2)}(P,Q) = \mathcal{M}(P,Q)$$

Proof. First we transform the relation on $\mathcal{M}$. We use (9.6) with $P = D_k \Xi R$ to deduce

$$\mathcal{M}(\Xi R, Q) = \sum_k \tau_t(D_k Q D_k \Xi R).$$

Let us define $\Delta = \sigma^{(2)} - \mathcal{M}$. Then according to the previous property, Δ is compactly supported and for all polynomials P and Q,

$$\Delta(\Xi P, Q) = 0.$$

Moreover, with $\mathcal{M}(1, Q) = 0 = \sigma^{(2)}(1, Q)$,

$$\Delta(1, Q) = 0.$$

To conclude we have to invert one more time the operator Ξ. For a polynomial P we take as in the proof of the central limit theorem, a sequence of polynomial S_n which goes to $S = \Xi^{-1} P$ in $\mathbb{C}\langle X_1, \dots, X_m \rangle(0)_A$.

$$\Delta(P, Q) = \Delta(\Xi(S_n + S - S_n), Q) = \Delta(\Xi(S - S_n), Q).$$

But by continuity of Ξ, $\Xi(S - S_n)$ goes to 0 for the norm $\|.\|_A$. Moreover, because Δ is compactly supported, Δ is continuous for $\|.\|_R$, and so $\Delta(\Xi(S - S_n), Q)$ goes to zero when n goes to $+\infty$ provided $A \geq R$, which we can always assume if η is small enough. □

9.4 Second-order Correction to the Free Energy

We now deduce from the central limit theorem the precise asymptotics of $N\bar{\delta}^N(P)$ and then compute the second-order correction to the free energy.

Let ϕ_0 and ϕ be the linear forms on $\mathbb{C}\langle X_1, \dots, X_m \rangle$ which are given, if P is a monomial by

$$\phi_0(P) = \sum_{i=1}^{m} \sum_{P = P_1 X_i P_2 X_i P_3} \sigma^{(2)}(P_3 P_1, P_2).$$

and $\phi = \phi_0 \circ \Sigma$.

Proposition 9.14. *Assume* (**H'**). *Then, for any polynomial P,*

$$\lim_{N \to \infty} N\bar{\delta}^N(P) = \phi(\Xi^{-1} P).$$

Proof. Again, we base our proof on the finite-dimensional Schwinger–Dyson equation (9.1) which, after centering, reads for $i \in \{1, \dots, m\}$,

$$N^2 \mu_V^N \left((\hat{\mathbf{L}}^N - \tau_{\mathbf{t}})[(X_i + D_i V)P - (I \otimes \tau_{\mathbf{t}} + \tau_{\mathbf{t}} \otimes I)\partial_i P] \right)$$
$$= \mu_V^N \left(\hat{\delta}^N \otimes \hat{\delta}^N (\partial_i P) \right). \tag{9.8}$$

Taking $P = D_i \Sigma P$ and summing over $i \in \{1, \dots, m\}$, we thus have

$$N^2 \mu_V^N \left((\hat{\mathbf{L}}^N - \tau_{\mathbf{t}})(\Xi P) \right) = \mu_V^N \left(\hat{\delta}^N \otimes \hat{\delta}^N \left(\sum_{i=1}^{m} \partial_i \circ D_i \Sigma P \right) \right) \tag{9.9}$$

By Theorem 9.9 and Lemma 9.10 we see that

$$\lim_{N \to \infty} \mu_V^N \left(\hat{\delta}^N \otimes \hat{\delta}^N (\sum_{i=1}^{m} \partial_i \circ D_i P) \right) = \phi(P)$$

which gives the asymptotics of $N \bar{\delta}^N (\Xi P)$ for all P.

To generalize the result to arbitrary P, we proceed as in the proof of the full central limit theorem. We take a sequence of polynomials Q_n which goes to $Q = \Xi^{-1} P$ when n goes to ∞ for the norm $\|.\|_A$. We define $R_n = P - \Xi Q_n = \Xi(Q - Q_n)$. Note that as P and Q_n are polynomials then R_n is also a polynomial.

$$N \bar{\delta}^N (P) = N \bar{\delta}^N (\Xi Q_n) + N \bar{\delta}^N (R_n).$$

According to Property 9.1, for any such monomial P of degree less than $N^{\frac{1}{2} - \epsilon}$,

$$|N \bar{\delta}^N (P)| \leq C^{\deg(P)}.$$

So if we take the limit in N, for any monomial P,

$$\limsup_{N} |N \bar{\delta}^N (P)| \leq C^{\deg(P)}$$

and therefore $\limsup_N |N \bar{\delta}^N (P)| \leq \|P\|_C \leq \|P\|_A$. The last inequality comes from the hypothesis **(H')** which requires $C < A$. We now fix n and take the large N limit,

$$\limsup_{N} |N \bar{\delta}^N (P - \Xi Q_n)| \leq \limsup_{N} |N \bar{\delta}^N (R_n)| \leq \|R_n\|_A.$$

If we take the limit in n the right term vanishes and we are left with:

$$\lim_{N} N \bar{\delta}^N (P) = \lim_{n} \lim_{N} N \bar{\delta}^N (Q_n) = \lim_{n} \phi(Q_n).$$

It is now sufficient to show that ϕ is continuous for the norm $\|.\|_A$. But the map $P \to \sum_{i=1}^{m} \partial_i \circ D_i P$ is continuous from $\mathbb{C}\langle X_1, \ldots, X_m \rangle(0)_A$ to $\mathbb{C}\langle X_1, \ldots, X_m \rangle(0)_{A-1}$ and σ^2 is continuous for $\|.\|_{A-1}$ due to the technical hypothesis in **(H')**. This proves that ϕ is continuous and then can be extended on $\mathbb{C}\langle X_1, \ldots, X_m \rangle(0)_A$. Thus

$$\lim_{N} N \bar{\delta}^N (P) = \lim_{n} \phi(Q_n) = \phi(Q).$$

$\square$

Theorem 9.15. *Assume that V_t satisfies* **(H')** *with a given $c > 0$. Then*

$$\log \frac{Z_N^{V_t}}{Z_N^0} = N^2 F_{\mathbf{t}} + F_{\mathbf{t}}^1 + o(1)$$

with

$$F_{\mathbf{t}} = \int_0^1 \tau_{\alpha \mathbf{t}}(V_{\mathbf{t}})d\alpha$$

and

$$F_{\mathbf{t}}^1 = \int_0^1 \phi_{\alpha \mathbf{t}}(\Xi_{\alpha \mathbf{t}}^{-1} V_{\mathbf{t}})d\alpha.$$

Proof. As in the proof of Theorem 8.8, we note that $\alpha V_{\mathbf{t}} = V_{\alpha \mathbf{t}}$ is c-convex for all $\alpha \in [0,1]$ We use (8.12) to see that

$$\partial_\alpha \log Z_{V_{\alpha \mathbf{t}}}^N = \mu_{\alpha \mathbf{t}}^N(\hat{\mathbf{L}}^N(V_{\mathbf{t}}))$$

so that we can write

$$\log \frac{Z_{V_{\alpha \mathbf{t}}}^N}{Z_0^N} = \int_0^1 \mu_{V_{\alpha \mathbf{t}}}^N(\hat{\mathbf{L}}^N(V_{\mathbf{t}}))d\alpha$$

$$= N^2 F_{\mathbf{t}} + \int_0^1 [N\bar{\delta}_{\alpha \mathbf{t}}^N(V_{\mathbf{t}})]ds. \qquad (9.10)$$

Proposition 9.14 and (9.10) finish the proof of the theorem since by Proposition 9.1, all the $N\bar{\delta}^N(q_i)$ can be bounded independently of N and $t \in B_{\eta,c}$ so that dominated convergence theorem applies. $\qquad \square$

As for the combinatorial interpretation of the covariance we relate F^1 to a generating function of maps. This time, we will consider maps on a torus instead of a sphere. Such maps are said to be of genus 1.

$$\mathcal{M}_{k_1,\ldots,k_n}^1(P) = \sharp\{ \text{ maps of genus 1 with } k_i \text{ stars of type } q_i \text{ or } q_i^*$$
$$\text{and one of type } P\}$$

and

$$\mathcal{M}_{k_1,\ldots,k_n}^1 = \sharp\{ \text{ maps with } k_i \text{ stars of type } q_i \text{ or } q_i^*\}.$$

We also define the generating function:

$$\mathcal{M}_{\mathbf{t}}^1(P) = \sum_{k_1,\ldots,k_n} \prod_{i=1}^n \frac{(-t_i)^{k_i}}{k_i!} \mathcal{M}_{k_1,\ldots,k_n}^1(P).$$

Proposition 9.16. *For all monomials P and all k,*

$$\mathcal{M}_{k_1,\ldots,k_n}^1(X_k P)$$
$$= \sum_{0 \le p_i \le k_i} \sum_{P=RX_k S} \prod_i C_{k_i}^{p_i} \mathcal{M}_{p_1,\ldots,p_n}^1(R) \mathcal{M}_{k_1-p_1,\ldots,k_n-p_n}^1(S)$$

$$+ \sum_{0 \le p_i \le k_i} \sum_{P=RX_kS} \prod_i C_{k_i}^{p_i} \mathcal{M}_{p_1,\ldots,p_n}(R)\mathcal{M}_{k_1-p_1,\ldots,k_n-p_n}^1(S)$$

$$+ \sum_{0 \le j \le n} k_j \mathcal{M}_{k_1,\ldots,k_j-1,\ldots,k_n}^1(D_kVP,Q)$$

$$+ \sum_{P=RX_kS} \mathcal{M}_{k_1,\ldots,k_n}(R,S)$$

and

$$\mathcal{M}_{\mathbf{t}}^1(X_kP) = \mathcal{M}_{\mathbf{t}}^1((I \otimes \tau_{\mathbf{t}} + \tau_{\mathbf{t}} \otimes I)\partial_k P) - \mathcal{M}_{\mathbf{t}}^1(D_kVP) + \mathcal{M}_{\mathbf{t}} \otimes \mathcal{M}_{\mathbf{t}}(\partial_k P). \quad (9.11)$$

Besides, for η small enough, there exists $R < +\infty$ such that for all monomials P, all $\mathbf{t} \in B(0,\eta)$,

$$|\mathcal{M}^1(P)| \le R^{\deg P}.$$

Proof. We proceed as for the combinatorial interpretation of the variance. We look at the first edge which comes out of the branch X_k, then two cases may occur:

1. The first possibility is that the branch is glued to another branch of $P = RX_kS$. It forms a loop starting from P. There are two cases.
   a) The loop can be retractible. It cuts P in two: R and S and it occurs for all decomposition of P into $P = RX_kS$ which is exactly what does D. Then either the component R or the component S is of genus 1 and the other component is planar. It produces either

   $$\prod_i C_{k_i}^{p_i} \mathcal{M}_{p_1,\ldots,p_n}^1(R)\mathcal{M}_{k_1-p_1,\ldots,k_n-p_n}(S)$$

   possibilities ot the symmetric formula.
   b) The loop can also be non-trivial in the fundamental group of the surface. Then the surface is cut into two parts. We are left with a planar surface with two fixed stars R and S. This gives $\mathcal{M}_{k_1,\ldots,k_n}(R,S)$ possibilities.
2. The second possibility occurs when the branch is glued to a vertex of type q_j for a given j. First we have to choose between the k_j vertices of this type, then we contract the edges arising from this gluing to form a vertex of type $D_iq_jP_1$, which creates

   $$k_j \mathcal{M}_{k_1,\ldots,k_j-1,\ldots,k_n}^1(D_kq_jP,Q)$$

   possibilities.

We can now sum on the k's to obtain the relation on $\mathcal{M}^1$.

Finally, to show that $\mathcal{M}^1$ is compactly supported we only have to prove that there exist $A > 0, B > 0$ such that for all k's, for all monomials P,

$$\frac{\mathcal{M}^1_{k_1,\ldots,k_n}(P)}{\prod_i k_i!} \leq A^{\sum_i k_i} B^{\deg P}.$$

Again this follow easily by induction with the previous relation on the $\mathcal{M}^1$. $\square$

Proposition 9.17. *Assume* **(H')**. *There exists* $\eta > 0$ *small enough so that for* $\mathbf{t} \in B_{\eta,c}$,

1. *For all monomials* P,
$$\phi(\Xi^{-1}P) = \mathcal{M}^1(P).$$

2.

$$F_{\mathbf{t}}^1 = \sum_{\mathbf{k} \in \mathbb{N}^n \setminus \{0\}} \prod_{i=1}^n \frac{(-t_i)^{k_i}}{k_i!} \mathcal{M}^1_{k_1,\ldots,k_n}$$

Proof. We use the equation of the previous property on $\mathcal{M}^1$ with $P = D_k \Xi P$ and we sum, then

$$\mathcal{M}^1(\Xi P) = \mathcal{M}\left(\sum_k \partial_k D_k P\right) = \sum_k \sigma^2(\partial_k D_k P) = \phi(\Xi P)$$

where we have used the combinatorial interpretation of the covariance (Theorem 9.13). As $\mathcal{M}^1$ and ϕ are continuous for $\|.\|_A$ when η is small enough, we can apply this to $\Xi^{-1}P$ and conclude.

Finally, for η sufficiently small the sum is absolutely convergent so that we can interchange the integral and the sum:

$$F_{\mathbf{t}}^1 = \int_0^1 \mathcal{M}^1_{\alpha \mathbf{t}}(V_{\mathbf{t}}) d\alpha$$

$$= \sum_{i=1}^n \sum_{k_i,\ldots,k_n} \int_0^1 (-t_i) \prod_j \frac{(-\alpha t_j)^{k_j}}{k_j!} \mathcal{M}^1_{\mathbf{k}}(q_i) d\alpha.$$

$$= \int_0^1 \partial_\alpha \mathcal{M}^1_{\alpha \mathbf{t}} d\alpha = \mathcal{M}^1_{\mathbf{t}}.$$

This proves the statement. $\square$

Bibliographical notes. Central limit theorems for the trace of polynomials in independent Gaussian matrices were first considered in [54, 100], based on a dynamical approach. Similar questions were undertaken under a free probability perspective in [157] where the interpretation of the covariances appearing in the central limit theorem of independent Gaussian matrices in terms of planar diagrams is used to define a new type of freeness. The case where the potential is not convex and the limiting measure may have a disconnected support is addressed in [162]; in certain cases Pastur can compute the logarithm of the Laplace transform of linear statistics and show that it is not given by half the covariance, as it should if a central limit theorem would hold.

The asymptotic topological expansion of one-matrix integrals was studied in [2, 86], using orthogonal polynomials. The so-called Schwinger–Dyson (or loop, or Master loop) equations were already used to analyze these questions in many papers in physics, see, e.g., [31, 87, 88].

Part IV

Eigenvalues of Gaussian Wigner Matrices and Large Deviations

In this part, we consider the case where the entries of the matrix $\mathbf{X}^{N,\beta}$ are the so-called Gaussian ensembles. Moreover, since the results depend upon the fact that the entries are real or complex, we now show the difference in the notations. We consider $N \times N$ self-adjoint random matrices with entries

$$X_{kl}^{N,\beta} = \frac{\sum_{i=1}^{\beta} g_{kl}^i e_{\beta}^i}{\sqrt{\beta N}}, \quad 1 \le k < l \le N, \quad X_{kk}^{N,\beta} = \sqrt{\frac{2}{\beta N}} g_{kk} e_{\beta}^1, \quad 1 \le k \le N$$

where $(e_{\beta}^i)_{1 \le i \le \beta}$ is a basis of $\mathbb{R}^{\beta}$, that is $e_1^1 = 1$, $e_2^1 = 1, e_2^2 = i$. This definition can be extended to the case $\beta = 4$, named the Gaussian symplectic ensemble, when N is even by choosing $\mathbf{X}^{N,\beta} = \left(X_{ij}^{N,\beta}\right)_{1 \le i,j \le \frac{N}{2}}$ with $X_{kl}^{N,\beta}$ a 2×2 matrix defined as above but with $(e_{\beta}^k)_{1 \le k \le 4}$ the Pauli matrices

$$e_4^1 = \begin{pmatrix} 1 & 0 \\ 0 & 1 \end{pmatrix}, \quad e_4^2 = \begin{pmatrix} 0 & -1 \\ 1 & 0 \end{pmatrix}, \quad e_4^3 = \begin{pmatrix} 0 & -i \\ -i & 0 \end{pmatrix}, \quad e_4^4 = \begin{pmatrix} i & 0 \\ 0 & -i \end{pmatrix}.$$

$(g_{kl}^i, k \le l, 1 \le i \le \beta)$ are independent equidistributed centered Gaussian variables with variance 1. $(\mathbf{X}^{N,2}, N \in \mathbb{N})$ is commonly referred to as the Gaussian Unitary Ensemble (**GUE**), $(\mathbf{X}^{N,1}, N \in \mathbb{N})$ as the Gaussian Orthogonal Ensemble (**GOE**) and $(\mathbf{X}^{N,4}, N \in \mathbb{N})$ as the Gaussian Symplectic Ensemble (**GSE**) since they can be characterized by the fact that their laws are invariant under the action of the unitary, orthogonal and symplectic group respectively (see [153]). We denote by $P_N^{(\beta)}$ the law of $\mathbf{X}^{N,\beta}$.

The main advantage of the Gaussian ensembles is that the law of the eigenvalues of these matrices is explicit and rather simple. Namely, we now discuss the following lemma.

Lemma IV.1. *Let* $\mathbf{X} \in \mathcal{H}_N^{(\beta)}$ *be random with law* $P_N^{(\beta)}$. *The joint distribution of the eigenvalues* $\lambda_1(X) \le \cdots \le \lambda_N(X)$, *has density proportional to*

$$1_{x_1 \leq \cdots \leq x_N} \prod_{1 \leq i < j \leq N} |x_i - x_j|^\beta \prod_{i=1}^{N} e^{-\beta x_i^2/4}. \qquad (IV.12)$$

We shall prove this lemma later, when studying Dyson's Brownian motion, see Corollary 12.4. Let us, however, emphasize the ideas behind a direct proof in the case $\beta = 1$. It is simply to write the decomposition $X = UDU^*$, with the eigenvalues matrix D that is diagonal and with real entries, and with the eigenvectors matrix U (that is unitary). Suppose this map was a bijection (which it is not, at least at the matrices X that do not possess all distinct eigenvalues) and that one can parametrize the eigenvectors by $\beta N(N-1)/2$ parameters in a smooth way (which one cannot in general). Then, it is easy to deduce from the formula $X = UDU^*$ that the Jacobian of this change of variables depends polynomially on the entries of D and is of degree $\beta N(N-1)/2$ in these variables. Since the bijection must break down when $D_{ii} = D_{jj}$ for some $i \neq j$, the Jacobian must vanish on that set. When $\beta = 1$, this imposes that the polynomial must be proportional to $\prod_{1 \leq i < j \leq N}(x_i - x_j)$. Further degree and symmetry considerations allow us to generalize this to $\beta = 2$. We refer the reader to [6] for a full proof, that shows that the set of matrices for which the above manipulations are not permitted has Lebesgue measure zero.

10

Large Deviations for the Law of the Spectral Measure of Gaussian Wigner's Matrices

In this section, we consider the law of N random variables $(\lambda_1, \ldots, \lambda_N)$ with law

$$P_{V,\beta}^N(d\lambda_1, \ldots, d\lambda_N) = (Z_{V,\beta}^N)^{-1} |\Delta(\lambda)|^\beta e^{-N \sum_{i=1}^N V(\lambda_i)} \prod_{i=1}^N d\lambda_i, \qquad (10.1)$$

for a continuous function $V : \mathbb{R} \to \mathbb{R}$ such that

$$\liminf_{|x| \to \infty} \frac{V(x)}{\beta \log |x|} > 1 \qquad (10.2)$$

and a positive real number β. Here, $\Delta(\lambda) = \prod_{1 \le i < j \le N} (\lambda_i - \lambda_j)$.

When $V(x) = 4^{-1} \beta x^2$, we have seen in Lemma IV that $P_{4^{-1}\beta x^2, \beta}^N$ is the law of the eigenvalues of an $N \times N$ GOE (resp. GUE, resp GSE) matrix when $\beta = 1$ (resp. $\beta = 2$, resp. $\beta = 4$). The case $\beta = 4$ corresponds to another matrix ensemble, namely the GSE. In view of these remarks and other applications discussed in Part III, we consider in this section the slightly more general model with a potential V. We emphasize, however, that the distribution (10.1) precludes us from considering random matrices with independent non-Gaussian entries.

We have proved already at the beginning of these notes that the empirical measure

$$L_N = \frac{1}{N} \sum_{i=1}^N \delta_{\lambda_i^N}$$

converges almost surely towards the semi-circular law. Moreover, we studied its fluctuations around its mean, both by the central limit theorem and by concentration inequalities. Such results did not depend much on the Gaussian nature of the entries.

We address here a different type of question. Namely, we study the probability that L_N takes a very unlikely value. This was already considered in our discussion of concentration inequalities (cf. Part II), where the emphasis

A. Guionnet, *Large Random Matrices: Lectures on Macroscopic Asymptotics*,   149
Lecture Notes in Mathematics 1957, DOI: 10.1007/978-3-540-69897-5_10,
© 2009 Springer-Verlag Berlin Heidelberg, Reprint by Springer-Verlag Berlin Heidelberg 2012

was put on obtaining upper bounds on the probability of deviation. In contrast, the purpose of the analysis here is to exhibit a precise estimate on these probabilities, or at least on their logarithmic asymptotics. The appropriate tool for handling such questions is large deviation theory, and we present in Appendix 20.1 a concise introduction to that theory and related definitions and references.

Endow $\mathcal{P}(\mathbb{R})$ with the usual weak topology. Our goal is to estimate the probability $P_{V,\beta}^N(L_N \in A)$, for measurable sets $A \subset \mathcal{P}(\mathbb{R})$. Of particular interest is the case where A does not contain the limiting distribution of L_N. Define the *non-commutative entropy* $\Sigma : \mathcal{P}(\mathbb{R}) \to [-\infty, \infty]$, as

$$\Sigma(\mu) = \int \int \log |x - y| d\mu(x) d\mu(y) \,. \tag{10.3}$$

Set next

$$I_\beta^V(\mu) = \begin{cases} \int V(x) d\mu(x) - \frac{\beta}{2} \Sigma(\mu) - c_\beta^V \,, & \text{if } \int V(x) d\mu(x) < \infty \\ \infty, & \text{otherwise} \,, \end{cases} \tag{10.4}$$

with $c_\beta^V = \inf_{\nu \in \mathcal{P}(\mathbb{R})}\{\int V(x) d\nu(x) - \frac{\beta}{2} \Sigma(\nu)\}$.

Theorem 10.1. *Let $L_N = N^{-1} \sum_{i=1}^N \delta_{\lambda_i^N}$ be the empirical measure of the random variables $\{\lambda_i^N\}_{i=1}^N$ distributed according to the law $P_{V,\beta}^N$, see (10.1). Then, the family of random measures L_N satisfies, in $\mathcal{P}(\mathbb{R})$ equipped with the weak topology, a full large deviation principle with good rate function I_β^V in the scale N^2. That is, $I_\beta^V : \mathcal{P}(\mathbb{R}) \to [0, \infty]$ possesses compact level sets $\{\nu : I_\beta^V(\nu) \le M\}$ for all $M \in \mathbb{R}_+$, and*

For any open set $O \subset \mathcal{P}(\mathbb{R})$,
$$\liminf_{N \to \infty} \frac{1}{N^2} \log P_{\beta,V}^N (L_N \in O) \ge - \inf_O I_\beta^V \,, \tag{10.5}$$

and

For any closed set $F \subset \mathcal{P}(\mathbb{R})$,
$$\limsup_{N \to \infty} \frac{1}{N^2} \log P_{\beta,V}^N (L_N \in F) \le - \inf_F I_\beta^V \,. \tag{10.6}$$

The proof of Theorem 10.1 relies on the properties of the function I_β^V collected in Lemma 10.2 below.

Lemma 10.2.

(a) I_β^V is well defined on $\mathcal{P}(\mathbb{R})$ and takes its values in $[0, +\infty]$.

(b) I_β^V is a good rate function.

(c) I_β^V is a strictly convex function on $\mathcal{P}(\mathbb{R})$.

(d) I_β^V achieves its minimum value at a unique probability measure σ_β^V on $\mathbb{R}$ characterized, if $C_\beta^V = \inf_{\nu \in \mathcal{P}(\mathbb{R})} \left(\int (V(x) - \beta \int \log |x - y| d\sigma_\beta^V(y)) d\nu(x) \right)$, by

$$V(x) - \beta \int \log|y - x| d\sigma_\beta^V(y) = C_V^\beta, \quad \sigma_\beta^V \ a.s., \tag{10.7}$$

and, for all x outside of the support of σ_β^V,

$$V(x) - \beta \int \log|y - x| d\sigma_\beta^V(y) \geq C_\beta^V. \tag{10.8}$$

As an immediate corollary of Theorem 10.1 and of part (d) of Lemma 10.2 we have the following.

Corollary 10.3 (Second proof of Wigner's theorem). *Under $P_{V,\beta}^N$, L_N converges almost surely towards σ_β^V.*

Proof of Lemma 10.2. If $I_\beta^V(\mu) < \infty$, since V is bounded below by assumption (10.2), $\Sigma(\mu) > -\infty$ and therefore also $\int V d\mu < \infty$. This proves that $I_\beta^V(\mu)$ is well defined (and by definition non-negative), yielding point (a).
Set

$$f(x,y) = \frac{1}{2}V(x) + \frac{1}{2}V(y) - \frac{\beta}{2}\log|x - y|. \tag{10.9}$$

Note that $f(x,y)$ goes to $+\infty$ when x, y do by (10.2). Indeed, $\log|x - y| \leq \log(|x| + 1) + \log(|y| + 1)$ implies

$$f(x,y) \geq \frac{1}{2}(V(x) - \beta\log(|x| + 1)) + \frac{1}{2}(V(y) - \beta\log(|y| + 1)) \tag{10.10}$$

as well as when x, y approach the diagonal $\{x = y\}$; for all $L > 0$, there exist constants $K(L)$ (going to infinity with L) such that

$$\{(x,y) : \ f(x,y) \geq K(L)\} \subset B_L,$$

$$B_L := \{(x,y) : \ |x - y| < L^{-1}\} \cup \{(x,y) : \ |x| > L\} \cup \{(x,y) : \ |y| > L\}. \tag{10.11}$$

Since f is continuous on the compact set B_L^c, we conclude that f is bounded below, and denote $b_f > -\infty$ a lower bound.

We now show that I_V^β is a good rate function, and first that its level sets $\{I_V^\beta \leq M\}$ are closed, that is, that I_V^β is lower semi-continuous. Indeed, by the monotone convergence theorem, we have the following:

$$I_\beta^V(\mu) = \int \int f(x,y) d\mu(x) d\mu(y) - c_\beta^V$$

$$= \sup_{M \geq 0} \int \int (f(x,y) \wedge M) d\mu(x) d\mu(y) - c_\beta^V.$$

But $f^M = f \wedge M$ is bounded continuous and so for $M < \infty$,

$$I_\beta^{V,M}(\mu) = \int \int (f(x,y) \wedge M)d\mu(x)d\mu(y)$$

is bounded continuous on $\mathcal{P}(\mathbb{R})$. As a supremum of the continuous functions $I_\beta^{V,M}$, I_β^V is lower semi-continuous. Hence, by Theorem 20.11, to prove that $\{I_\beta^V \le L\}$ is compact, it is enough to show that $\{I_\beta^V \le L\}$ is included in a compact subset of $\mathcal{P}(\mathbb{R})$ of the form

$$K_\epsilon = \cap_{B \in \mathbb{N}} \{\mu \in \mathcal{P}(\mathbb{R}) : \mu([-B,B]^c) \le \epsilon(B)\}$$

with a sequence $\epsilon(B)$ going to zero as B goes to infinity.

Arguing as in (10.11), there exist constants $K'(L)$ going to infinity as L goes to infinity, such that

$$\{(x,y) : |x| > L, |y| > L\} \subset \{(x,y) : f(x,y) \ge K'(L)\}. \qquad (10.12)$$

Hence, for any $L > 0$ large,

$$
\begin{aligned}
\mu(|x| > L)^2 &= \mu \otimes \mu(|x| > L, |y| > L) \\
&\le \mu \otimes \mu(f(x,y) \ge K'(L)) \\
&\le \frac{1}{K'(L) - b_f} \int \int (f(x,y) - b_f)d\mu(x)d\mu(y) \\
&= \frac{1}{K'(L) - b_f}(I_\beta^V(\mu) + c_\beta^V - b_f)
\end{aligned}
$$

Hence, with $\epsilon(B) = [\sqrt{(M + c_\beta^V - b_f)_+}/\sqrt{(K'(B) - b_f)_+}] \wedge 1$ going to zero when B goes to infinity, one has that $\{I_\beta^V \le M\} \subset K_\epsilon$. This completes the proof of point (b).

Since I_β^V is a good rate function, it achieves its minimal value. Let σ_β^V be a minimizer. Then, for any signed measure $\bar\nu(dx) = \phi(x)\sigma_\beta^V(dx) + \psi(x)dx$ with two bounded measurable compactly supported functions (ϕ, ψ) such that $\psi \ge 0$ and $\bar\nu(\mathbb{R}) = 0$, for $\epsilon > 0$ small enough, $\sigma_\beta^V + \epsilon\bar\nu$ is a probability measure so that

$$I_\beta^V(\sigma_\beta^V + \epsilon\bar\nu) \ge I_\beta^V(\sigma_\beta^V)$$

which gives

$$\int \left(V(x) - \beta \int \log|x - y|d\sigma_\beta^V(y)\right) d\bar\nu(x) \ge 0.$$

Taking $\psi = 0$, we deduce by symmetry that there is a constant C_β^V such that

$$V(x) - \beta \int \log|x - y|d\sigma_\beta^V(y) = C_\beta^V, \quad \sigma_\beta^V \text{ a.s.,} \qquad (10.13)$$

which implies that σ_β^V is compactly supported (as $V(x) - \beta \int \log|x - y| d\sigma_\beta^V(y)$ goes to infinity when x does). Taking $\phi(x) = -\int \psi(y) dy$, we then find that

$$V(x) - \beta \int \log|x - y| d\sigma_\beta^V(y) \geq C_\beta^V \qquad (10.14)$$

Lebesgue almost surely, and then everywhere outside of the support of σ_β^V by continuity. By (10.13) and (10.14) we deduce that

$$C_\beta^V = \inf_{\nu \in \mathcal{P}(\mathbb{R})} \left\{ \int \left(V(x) - \beta \int \log|x - y| d\sigma_\beta^V(y) \right) d\nu(x) \right\}.$$

This completes the proof of (10.7) and (10.8). The claimed uniqueness of σ_β^V, and hence the completion of the proof of part (d), then follows from the strict convexity claim (point (c) of the lemma), which we turn to next.

Note first that we can rewrite I_β^V as

$$I_\beta^V(\mu) = -\frac{\beta}{2} \Sigma(\mu - \sigma_\beta^V) + \int \left(V - \beta \int \log|x - y| d\sigma_\beta^V(y) - C_\beta^V \right) d\mu(x).$$

The fact that I_β^V is strictly convex comes from the observation that Σ is strictly concave, as can be checked from the formula

$$\log|x - y| = \int_0^\infty \frac{1}{2t} \left(\exp\left\{ -\frac{1}{2t} \right\} - \exp\left\{ -\frac{|x - y|^2}{2t} \right\} \right) dt \qquad (10.15)$$

which entails that for any $\mu \in \mathcal{P}(\mathbb{R})$,

$$\Sigma(\mu - \sigma_\beta^V) = -\int_0^\infty \frac{1}{2t} \left(\int \int \exp\{ -\frac{|x - y|^2}{2t} \} d(\mu - \sigma_\beta^V)(x) d(\mu - \sigma_\beta^V)(y) \right) dt.$$

Indeed, one may apply Fubini's theorem when μ_1, μ_2 are supported in $[-\frac{1}{2}, \frac{1}{2}]$ since then $\mu_1 \otimes \mu_2 (\exp\{ -\frac{1}{2t} \} - \exp\{ -\frac{|x-y|^2}{2t} \} \leq 0) = 1$. One then deduces the claim for any compactly supported probability measures by scaling and finally for all probability measures by approximations. The fact that for all $t \geq 0$,

$$\int \int \exp\{ -\frac{|x - y|^2}{2t} \} d(\mu - \sigma_\beta^V)(x) d(\mu - \sigma_\beta^V)(y)$$

$$= \sqrt{\frac{t}{2\pi}} \int_{-\infty}^{+\infty} \left| \int \exp\{ i\lambda x \} d(\mu - \sigma_\beta^V)(x) \right|^2 \exp\{ -\frac{t\lambda^2}{2} \} d\lambda$$

therefore entails that Σ is concave since $\mu \rightarrow \left| \int \exp\{ i\lambda x \} d(\mu - \sigma_\beta^V)(x) \right|^2$ is convex for all $\lambda \in \mathbb{R}$. Strict convexity comes from Cauchy–Schwarz inequality, $\Sigma(\alpha\mu + (1 - \alpha)\nu) = \alpha\Sigma(\mu) + (1 - \alpha)\Sigma(\nu)$ if and only if $\Sigma(\nu - \mu) = 0$ which implies that all the Fourier transforms of $\nu - \mu$ are null, and hence $\mu = \nu$. This completes the proof of the lemma. □

Proof of Theorem 10.1. To begin, let us remark that with f as in (10.9),

$$P_{V,\beta}^N(d\lambda_1,\ldots,d\lambda_N) = (Z_N^{\beta,V})^{-1}e^{-N^2\int_{x\neq y}f(x,y)dL_N(x)dL_N(y)}\prod_{i=1}^N e^{-V(\lambda_i)}d\lambda_i.$$

Hence, if $\mu \to \int_{x\neq y}f(x,y)d\mu(x)d\mu(y)$ was a bounded continuous function, the proof would follow from a standard Laplace method (see Theorem 20.8 in the appendix). The main point is therefore to overcome the singularity of this function, with the most delicate part being overcoming the singularity of the logarithm.

Following Appendix 20.1 (see Corollary 20.6 and Definition 13.10), a full large deviation principle can be proved by showing that exponential tightness holds, as well as estimating the probability of small balls. We follow these steps below.

- *Exponential tightness.* Observe that by Jensen's inequality,

$$\log Z_N^{\beta,V} \geq N\log\int e^{-V(x)}dx$$

$$- N^2\int\left(\int_{x\neq y}f(x,y)dL_N(x)dL_N(y)\right)\prod_{i=1}^N\frac{e^{-V(\lambda_i)}d\lambda_i}{\int e^{-V(x)}dx} \geq -CN^2$$

with some finite constant C. Moreover, by (10.10) and (10.2), there exist constants $a > 0$ and $c > -\infty$ so that

$$f(x,y) \geq a|V(x)| + a|V(y)| + c$$

from which one concludes that for all $M \geq 0$,

$$P_{V,\beta}^N\left(\int|V(x)|dL_N \geq M\right) \leq e^{-2aN^2M+(C-c)N^2}\left(\int e^{-V(x)}dx\right)^N.$$

Since V goes to infinity at infinity, $K_M = \{\mu \in \mathcal{P}(\mathbb{R}) : \int|V|d\mu \leq M\}$ is a compact set for all $M < \infty$, so that we have proved that the law of L_N under $P_{V,\beta}^N$ is exponentially tight.

- *Large deviation upper bound.* d denotes the Dudley metric, see (0.1). We prove here that for any $\mu \in \mathcal{P}(\mathbb{R})$, if we set $\bar{P}_{V,\beta}^N = Z_N^{\beta,V}P_{V,\beta}^N$

$$\lim_{\epsilon\to 0}\limsup_{N\to\infty}\frac{1}{N^2}\log\bar{P}_{V,\beta}^N\left(d(L_N,\mu)\leq\epsilon\right) \leq -\int f(x,y)d\mu(x)d\mu(y). \qquad (10.16)$$

For any $M \geq 0$, the following bound holds:

$$\bar{P}_{V,\beta}^N\left(d(L_N,\mu)\leq\epsilon\right)$$

$$\leq \int_{d(L_N,\mu)\leq\epsilon}e^{-N^2\int_{x\neq y}f(x,y)\wedge M dL_N(x)dL_N(y)}\prod_{i=1}^N e^{-V(\lambda_i)}d\lambda_i.$$

Since under the product Lebesgue measure, the λ_i's are almost surely distinct, it holds that $L_N \otimes L_N(x = y) = N^{-1}$, $\bar{P}^N_{V,\beta}$ almost surely. Thus, we deduce for all $M \geq 0$, with $f_M(x,y) = f(x,y) \wedge M$,

$$\int f_M(x,y)dL_N(x)dL_N(y) = \int_{x \neq y} f_M(x,y)dL_N(x)dL_N(y) + MN^{-1},$$

and so

$$\bar{P}^N_{V,\beta}\left(d(L_N, \mu) \leq \epsilon\right)$$

$$\leq e^{MN} \int_{d(L_N,\mu)\leq\epsilon} e^{-N^2 \int f_M(x,y)dL_N(x)dL_N(y)} \prod_{i=1}^{N} e^{-V(\lambda_i)}d\lambda_i.$$

Since $I^{V,M}_\beta(\nu) = \int f_M(x,y)d\nu(x)d\nu(y)$ is bounded continuous, we deduce that

$$\lim_{\epsilon \to 0} \limsup_{N \to \infty} \frac{1}{N^2} \log \bar{P}^N_{V,\beta}\left(d(L_N, \mu) \leq \epsilon\right) \leq -I^{V,M}_\beta(\mu).$$

We finally let M go to infinity and conclude by the monotone convergence theorem. Note that the same argument shows that

$$\limsup_{N \to \infty} \frac{1}{N^2} \log Z^{\beta,V}_N \leq - \inf_{\mu \in \mathcal{P}(\mathbb{R})} \int f(x,y)d\mu(x)d\mu(y). \qquad (10.17)$$

- *Large deviation lower bound.* We prove here that for any $\mu \in \mathcal{P}(\mathbb{R})$

$$\lim_{\epsilon \to 0} \liminf_{N \to \infty} \frac{1}{N^2} \log \bar{P}^N_{V,\beta}\left(d(L_N, \mu) \leq \epsilon\right) \geq - \int f(x,y)d\mu(x)d\mu(y). \qquad (10.18)$$

Note that we can assume without loss of generality that $I^V_\beta(\mu) < \infty$, since otherwise the bound is trivial, and so in particular, we may and shall assume that μ has no atoms. We can also assume that μ is compactly supported since if we consider $\mu_M = \mu([-M,M])^{-1}\mathbb{1}_{|x|\leq M}d\mu(x)$, clearly μ_M converges towards μ and by the monotone convergence theorem, one checks that, since f is bounded below,

$$\lim_{M \uparrow \infty} \int f(x,y)d\mu_M(x)d\mu_M(y) = \int f(x,y)d\mu(x)d\mu(y)$$

which insures that it is enough to prove the lower bound for $(\mu_M, M \in \mathbb{R}, I^V_\beta(\mu) < \infty)$, and so for compactly supported probability measures with finite entropy.

The idea is to localize the eigenvalues $(\lambda_i)_{1 \leq i \leq N}$ in small sets and to take advantage of the fast speed N^2 of the large deviations to neglect the small volume of these sets. To do so, we first remark that for any $\nu \in \mathcal{P}(\mathbb{R})$ with no atoms, if we set

$$x^{1,N} = \inf\left\{ x \mid \nu\left(]-\infty, x]\right) \geq \frac{1}{N+1} \right\}$$

$$x^{i+1,N} = \inf\left\{ x \geq x^{i,N} \mid \nu\left(]x^{i,N}, x]\right) \geq \frac{1}{N+1} \right\} \qquad 1 \leq i \leq N-1,$$

for any real number η, there exists an integer number $N(\eta)$ such that, for any N larger than $N(\eta)$,

$$d\left(\nu, \frac{1}{N}\sum_{i=1}^{N} \delta_{x^{i,N}} \right) < \eta.$$

In particular, for $N \geq N(\frac{\delta}{2})$,

$$\left\{ (\lambda_i)_{1 \leq i \leq N} \mid |\lambda_i - x^{i,N}| < \frac{\delta}{2} \ \forall i \in [1, N] \right\} \subset \left\{ (\lambda_i)_{1 \leq i \leq N} \mid d(L_N, \nu) < \delta \right\}$$

so that we have the lower bound

$$\bar{P}_{V,\beta}^N \left(d(L_N, \mu) \leq \epsilon \right)$$

$$\geq \int_{\cap_i\{|\lambda_i - x^{i,N}| < \frac{\delta}{2}\}} e^{-N^2 \int_{x \neq y} f(x,y) dL_N(x) dL_N(y)} \prod_{i=1}^{N} e^{-V(\lambda_i)} d\lambda_i$$

$$= \int_{\cap_i\{|\lambda_i| < \frac{\delta}{2}\}} \prod_{i<j} |x^{i,N} - x^{j,N} + \lambda_i - \lambda_j|^\beta e^{-N\sum_{i=1}^N V(x^{i,N}+\lambda_i)} \prod_{i=1}^N d\lambda_i$$

$$\geq \left(\prod_{i+1<j} |x^{i,N} - x^{j,N}|^\beta \prod_i |x^{i,N} - x^{i+1,N}|^{\frac{\beta}{2}} e^{-N\sum_{i=1}^N V(x^{i,N})} \right)$$

$$\times \left(\int_{\substack{\cap_i\{|\lambda_i|<\frac{\delta}{2}\} \\ \lambda_i < \lambda_{i+1}}} \prod_i |\lambda_i - \lambda_{i+1}|^{\frac{\beta}{2}} e^{-N\sum_{i=1}^N [V(x^{i,N}+\lambda_i) - V(x^{i,N})]} \prod_{i=1}^N d\lambda_i \right)$$

$$=: P_{N,1} \times P_{N,2} \tag{10.19}$$

where we used that $|x^{i,N} - x^{j,N} + \lambda_i - \lambda_j| \geq |x^{i,N} - x^{j,N}| \vee |\lambda_i - \lambda_j|$ when $\lambda_i \geq \lambda_j$ and $x^{i,N} \geq x^{j,N}$. To estimate $P_{N,2}$, note that since we assumed that μ is compactly supported, the $(x^{i,N}, 1 \leq i \leq N)_{N \in \mathbb{N}}$ are uniformly bounded and so by continuity of V

$$\lim_{N \to \infty} \sup_{N \in \mathbb{N}} \sup_{1 \leq i \leq N} \sup_{|x| \leq \delta} |V(x^{i,N} + x) - V(x^{i,N})| = 0.$$

Moreover, writing $u_1 = \lambda_1$, $u_{i+1} = \lambda_{i+1} - \lambda_i$,

$$\int_{\substack{|\lambda_i| < \frac{\delta}{2} \ \forall i \\ \lambda_i < \lambda_{i-1}}} \prod_i |\lambda_i - \lambda_{i+1}|^{\frac{\beta}{2}} \prod_{i=1}^{N} d\lambda_i \geq \int_{0 < u_i < \frac{\delta}{2N}} \prod_{i=2}^{N} u_i^{\frac{\beta}{2}} \prod_{i=1}^{N} du_i$$

$$\geq \left(\frac{\delta}{(\beta + 2)N} \right)^{N(\frac{\beta}{2}+1)}.$$

Therefore,

$$\lim_{\delta \to 0} \liminf_{N \to \infty} \frac{1}{N^2} \log P_{N,2} \geq 0. \tag{10.20}$$

The convergence of the term $P_{N,1}$ is due to the uniform boundedness of the $x^{i,N}$'s and the convergence of their empirical measure towards μ which imply that

$$\lim_{N \to \infty} \frac{1}{N} \sum_{i=1}^{N} V(x^{i,N}) = \int V(x) d\mu(x). \tag{10.21}$$

Finally, since $x \to \log(x)$ increases on $\mathbb{R}^+$, we notice that

$$\int_{x^{1,N} \leq x < y \leq x^{N,N}} \log(y - x) d\mu(x) d\mu(y)$$

$$\leq \sum_{1 \leq i \leq j \leq N-1} \log(x^{j+1,N} - x^{i,N}) \int_{\substack{x \in [x^{i,N}, x^{i+1,N}] \\ y \in [x^{j,N}, x^{j+1,N}]}} 1_{x < y} d\mu(x) d\mu(y)$$

$$= \frac{1}{(N+1)^2} \sum_{i<j} \log |x^{i,N} - x^{j+1,N}| + \frac{1}{2(N+1)^2} \sum_{i=1}^{N-1} \log |x^{i+1,N} - x^{i,N}|.$$

Since $\log |x - y|$ is bounded when x, y are in the support of the compactly supported measure μ, the monotone convergence theorem implies that the left side in the last display converges towards $\int \int \log |x - y| d\mu(x) d\mu(x)$. Thus, with (10.21), we have proved

$$\liminf_{N \to \infty} \frac{1}{N^2} \log P_{N,1} \geq \int_{x<y} \log(y - x) d\mu(x) d\mu(y) - \int V(x) d\mu(x)$$

which concludes, with (10.19) and (10.20), the proof of (10.18).

• *Conclusion.* By (10.18), for all $\mu \in \mathcal{P}(\mathbb{R})$,

$$\liminf_{N \to \infty} \frac{1}{N^2} \log Z_{\beta,V}^N \geq \lim_{\epsilon \to 0} \liminf_{N \to \infty} \frac{1}{N^2} \log \bar{P}_{V,\beta}^N (d(L_N, \mu) \leq \epsilon)$$

$$\geq - \int f(x,y) d\mu(x) d\mu(y)$$

and so optimizing with respect to $\mu \in \mathcal{P}(\mathbb{R})$ and with (10.17),

$$\lim_{N \to \infty} \frac{1}{N^2} \log Z_{\beta,V}^N = - \inf_{\mu \in \mathcal{P}(\mathbb{R})} \left\{ \int f(x,y) d\mu(x) d\mu(y) \right\} = -c_\beta^V.$$

Thus, (10.18) and (10.16) imply the weak large deviation principle, i.e., that for all $\mu \in \mathcal{P}(\mathbb{R})$,

$$\lim_{\epsilon \to 0} \liminf_{N \to \infty} \frac{1}{N^2} \log P_{V,\beta}^N \left(d(L_N, \mu) \le \epsilon \right)$$

$$= \lim_{\epsilon \to 0} \limsup_{N \to \infty} \frac{1}{N^2} \log P_{V,\beta}^N \left(d(L_N, \mu) \le \epsilon \right) = -I_\beta^V(\mu).$$

This, together with the exponential tightness property proved above completes the proof of the full large deviation principle stated in Theorem 10.1.   □

Bibliographical Notes. The proof of Theorem 10.1 is a slight generalization of the techniques introduced in [25] to more general potentials. The ideas developed in this chapter were extended to the Ginibre ensembles in [28] and to diverse other situations, including Wishart matrices, in [117]. We discuss the generalization of large deviation principles to a multi-matrix setting in the last part of these notes.

11

Large Deviations of the Maximum Eigenvalue

We here restrict ourselves to the case where $V(x) = \beta x^2/4$ and for short denote by P_β^N the law of the eigenvalues $(\lambda_i)_{1\leq i\leq N}$:

$$P_\beta^N(d\lambda_1,\ldots,d\lambda_N) = \frac{1}{Z_\beta^N} \prod_{1\leq i<j\leq N} |\lambda_i - \lambda_j|^\beta \prod_{1\leq i\leq N} e^{-\frac{\beta N \lambda_i^2}{4}} d\lambda_i$$

with

$$Z_\beta^N = \int \prod_{1\leq i<j\leq N} |\lambda_i - \lambda_j|^\beta \prod_{1\leq i\leq N} e^{-\frac{\beta N \lambda_i^2}{4}} d\lambda_i.$$

Selberg (cf. [153, Theorem 4.1.1] or [6]) found the explicit formula for Z_β^N for any $\beta \geq 0$:

$$Z_\beta^N = (2\pi)^{\frac{N}{2}} \left(\frac{\beta N}{2}\right)^{-\beta N(N-1)/4-\frac{N}{2}} \prod_{j=1}^{N} \frac{\Gamma(\frac{j\beta}{2})}{\Gamma(\frac{\beta}{2})}. \tag{11.1}$$

The knowledge of Z_β^N up to the second order is crucial below, reason why we restrict ourselves to quadratic potentials in the next theorem (see Exercise 11.4 for a slight extension).

Theorem 11.1. *[24] The law of the maximal eigenvalue* $\lambda_N^* = \max_{i=1}^N \lambda_i$ *under* P_β^N, *with* $\beta \geq 0$, *satisfies the LDP with speed* N *and the GRF*

$$I^*(x) = \begin{cases} \beta \int_2^x \sqrt{(z/2)^2 - 1}\, dz, & x \geq 2, \\ +\infty, & \text{otherwise}. \end{cases} \tag{11.2}$$

The next estimate is key to the proof of Theorem 11.1.

Lemma 11.2. *For every M large enough and all N,*

$$P_\beta^N\left(\max_{i=1}^N |\lambda_i| \geq M\right) \leq e^{-\beta N M^2/9}.$$

A. Guionnet, *Large Random Matrices: Lectures on Macroscopic Asymptotics*, Lecture Notes in Mathematics 1957, DOI: 10.1007/978-3-540-69897-5_11,

Proof. Observe that for any $|x| \geq M \geq 8$ and $\lambda_i \in \mathbb{R}$,

$$|x - \lambda_i| e^{-\frac{\lambda_i^2}{8}} \leq (|x| + |\lambda_i|) e^{-\frac{\lambda_i^2}{8}} \leq 2|x| \leq e^{\frac{x^2}{8}}.$$

Therefore, integrating with respect to λ_1 yields, for $M \geq 8$,

$$P_\beta^N(|\lambda_1| \geq M)$$

$$= \frac{Z_\beta^{N-1}}{Z_\beta^N} \int_{|x| \geq M} dx e^{-\frac{\beta x^2}{4} \frac{(N+1)}{2}} \int \prod_{i=2}^N \left(|x - \lambda_i| e^{-\frac{\lambda_i^2}{4} - \frac{x^2}{8}}\right)^\beta dP_\beta^{N-1}(\lambda_j, j \geq 2)$$

$$\leq e^{-\frac{\beta}{8} N M^2} \frac{Z_\beta^{N-1}}{Z_\beta^N} \int_{|x| \geq M} e^{-x^2/8} dx \int \prod_{i=2}^N (|x - \lambda_i| e^{-\lambda_i^2/4} e^{-x^2/8}) dP_\beta^{N-1}$$

$$\leq e^{-\frac{\beta}{8} N M^2} \frac{Z_\beta^{N-1}}{Z_\beta^N} \int e^{-x^2/8} dx.$$

Further, following (11.1), we compute that

$$\lim_{N \to \infty} \frac{1}{N} \log \frac{Z_\beta^{N-1}}{Z_\beta^N} = -\frac{\beta}{4}. \tag{11.3}$$

Therefore, for any $M \geq 8$, for N large enough, we get

$$P_\beta^N(\max_{i=1}^N |\lambda_i| \geq M) \leq N P_\beta^N(|\lambda_1| \geq M) \leq e^{-\frac{\beta}{9} N M^2},$$

and the lemma follows. $\square$

Proof of Theorem 11.1. $I^*(x)$ is a good rate function since it is a continuous function (except at $x = 2$ where it is lower semi-continuous) and it goes to infinity at infinity. Moreover, with $I^*(x)$ continuous and strictly increasing on $[2, \infty[$ it suffices to show that for any $x < 2$,

$$\limsup_{N \to \infty} \frac{1}{N} \log P_\beta^N(\lambda_N^* \leq x) = -\infty, \tag{11.4}$$

whereas for any $x > 2$

$$\lim_{N \to \infty} \frac{1}{N} \log P_\beta^N(\lambda_N^* \geq x) = -I^*(x). \tag{11.5}$$

In fact, from these two estimates and since I^* increases on $[2, \infty[$, we find that for all $x < y$,

$$\lim_{N \to \infty} \frac{1}{N} \log P_\beta^N(\lambda_N^* \in [x, y]) = - \inf_{z \in [x,y]} I^*(z),$$

the above right-hand side being equal to $-\infty$ if $y \leq 2$, to zero if $x \leq 2 \leq y$ and to $I^*(x)$ if $x \geq 2$. By continuity of I^*, we also deduce that we have the same limits if we take (x, y) instead of $[x, y]$. Since $\mathcal{A} = \{[x, y], (x, y), x < y\}$ is a basis for the topology on $\mathbb{R}$, we conclude by Theorem 20.5.

Starting with (11.4), fix $x < 2$ and $f \in \mathcal{C}_b(\mathbb{R})$ such that $f(y) = 0$ for all $y \leq x$, whereas $\int f d\sigma > 0$. Note that $\{\lambda_N^* \leq x\} \subseteq \{\int f dL_{\mathbf{X}^N} = 0\}$, so (11.4) follows by applying the upper bound of the large deviation principle of Theorem 10.1 for the closed set $F = \{\mu : \int f d\mu = 0\}$, such that $\sigma \notin F$. Turning to the upper bound in (11.5), fix $M \geq x > 2$, noting that

$$P_\beta^N (\lambda_N^* \geq x) = P_\beta^N(\max_{i=1}^N |\lambda_i| > M) + P_\beta^N (\lambda_N^* \geq x, \max_{i=1}^N |\lambda_i| \leq M) \tag{11.6}$$

By Lemma 11.2, the first term is exponentially negligible for all M large enough. To deal with the second term, let $P_N^{N-1}(\lambda \in \cdot) = P_\beta^{N-1}((1 - N^{-1})^{1/2}\lambda \in \cdot)$, $L_{N-1} = (N - 1)^{-1} \sum_{i=2}^N \delta_{\lambda_i}$ and

$$C_N := \frac{Z_\beta^{N-1}}{Z_\beta^N}(1 - N^{-1})^{N(N-1)/4}.$$

Further, let $B(\sigma, \delta)$ denote an open ball in $\mathcal{P}(\mathbb{R})$ of radius $\delta > 0$ and center σ, and $B_M(\sigma, \delta)$ its intersection with $\mathcal{P}([-M, M])$. Observe that for any $z \in [-M, M]$ and $\mu \in \mathcal{P}([-M, M])$,

$$\Phi(z, \mu) := \beta \int \log|z - y| d\mu(y) - \frac{\beta}{4}z^2 \leq \beta \log(2M).$$

Thus, for the second term in (11.6),

$$P_\beta^N (\lambda_N^* \geq x, \max_{i=1}^N |\lambda_i| \leq M)$$

$$\leq N C_N \int_x^M d\lambda_1 \int_{[-M,M]^{N-1}} e^{(N-1)\Phi(\lambda_1, L_{N-1})} dP_N^{N-1}(\lambda_j, j \geq 2)$$

$$\leq N C_N \left(\int_x^M e^{(N-1) \sup_{\mu \in B_M(\sigma,\delta)} \Phi(z,\mu)} dz + (2M)^{\beta N} P_N^{N-1}(L_{N-1} \notin B(\sigma, \delta)) \right). \tag{11.7}$$

For any h of Lipschitz norm at most 1 and $N \geq 2$,

$$|(N - 1)^{-1} \sum_{i=2}^N (h((1 - N^{-1})^{1/2}\lambda_i) - h(\lambda_i))| \leq 3N^{-1}\max_{i=2}^N |\lambda_i|.$$

Thus, by Lemma 11.2, the spectral measures L_{N-1} under σ^{N-1} are exponentially equivalent in $\mathcal{P}(\mathbb{R})$ to the spectral measures L_{N-1} under P_N^{N-1}, so Theorem 10.1 applies also for the latter (cf. Definition 20.9 and Lemma 20.10). In particular, the second term in (11.7) is exponentially negligible as $N \to \infty$ for any $\delta > 0$ and $M < \infty$ (since it behaves like $e^{-c(\delta)N^2}$). Therefore,

$$\limsup_{N\to\infty} \frac{1}{N} \log P_\beta^N \left(\lambda_N^* \geq x, \max_{i=1}^N |\lambda_i| \leq M \right)$$

$$\leq \limsup_{N\to\infty} \frac{1}{N} \log C_N + \lim_{\delta\downarrow 0} \sup_{\substack{z\in[x,M] \\ \mu\in B_M(\sigma,\delta)}} \Phi(z,\mu) \tag{11.8}$$

Note that $\Phi(z,\mu) = \inf_{\eta>0} \Phi_\eta(z,\mu)$ with $\Phi_\eta(z,\mu) := \beta \int \log(|z-y|\vee\eta)d\mu(y) - \frac{\beta}{4}z^2$ continuous on $[-M,M] \times \mathcal{P}([-M,M])$. Thus, $(z,\mu) \mapsto \Phi(z,\mu)$ is upper semi-continuous, which implies

$$\lim_{\delta\downarrow 0} \sup_{\substack{z\in[x,M] \\ \mu\in B_M(\sigma,\delta)}} \Phi(z,\mu) = \sup_{z\in[x,M]} \Phi(z,\sigma) \tag{11.9}$$

With σ supported on $[-2,2]$, $D(z) := \frac{d}{dz}\Phi(z,\sigma)$ exists for $z \geq 2$. Moreover, $D(z) = -\beta\sqrt{(z/2)^2 - 1} \leq 0$. It is shown in [25, Lemma 2.7] that $\Phi(2,\sigma) = -\beta/2$. Hence, for $x > 2$,

$$\sup_{z\geq x} \Phi(z,\sigma) = \Phi(x,\sigma) = -\frac{1}{2} - I^*(x). \tag{11.10}$$

By (11.3), we deduce that

$$\lim_{N\to\infty} N^{-1} \log C_N = \frac{\beta}{2}.$$

Combining this with (11.8)–(11.10) completes the proof of the upper bound for (11.5). To prove the complementary lower bound, fix $y > x > r > 2$ and $\delta > 0$, noting that for all N,

$$P_\beta^N (\lambda_N^* \geq x)$$
$$\geq P_\beta^N \left(\lambda_1 \in [x,y], \max_{i=2}^N |\lambda_i| \leq r \right)$$
$$= C_N \int_x^y e^{-\lambda_1^2/4} d\lambda_1 \int_{[-r,r]^{N-1}} e^{(N-1)\Phi(\lambda_1,L_{N-1})} dP_N^{N-1}(\lambda_j, j \geq 2)$$
$$\geq k C_N \exp\left((N-1) \inf_{\substack{z\in[x,y] \\ \mu\in B_r(\sigma,\delta)}} \Phi(z,\mu) \right) P_N^{N-1}(L_{N-1} \in B_r(\sigma,\delta))$$

with $k = k(x,y) > 0$. Recall that the large deviation principle with speed N^2 and good rate function $I(\cdot)$ applies for the measures L_{N-1} under P_N^{N-1}. It follows by this LDP's upper bound that $P_N^{N-1}(L_{N-1} \notin B(\sigma,\delta)) \to 0$, whereas by the symmetry of $P_\beta^N(\cdot)$ and the upper bound of (11.5),

$$P_N^{N-1} (L_{N-1} \notin \mathcal{P}([-r,r])) \leq 2P_\beta^{N-1}(\lambda_N^* \geq r) \to 0$$

as $N \to \infty$. Consequently,

$$\liminf_{N\to\infty} \frac{1}{N} \log P_\beta^N (\lambda_N^* \geq x) \geq \frac{1}{2} + \beta \inf_{\substack{z\in[x,y] \\ \mu\in B_r(\sigma,\delta)}} \Phi(z,\mu)$$

Observe that $(z,\mu) \mapsto \Phi(z,\mu)$ is continuous on $[x,y] \times \mathcal{P}([-r,r])$, for $y > x > r > 2$. Hence, considering $\delta \downarrow 0$ followed by $y \downarrow x$ results in the required lower bound

$$\liminf_{N\to\infty} \frac{1}{N} \log P_\beta^N (\lambda_N^* \geq x) \geq \frac{\beta}{2} + \beta\Phi(x,\sigma) \ .$$

$\square$

Exercise 11.3 (suggested by B. Collins). *Generalize the proof to obtain the large deviation principle for the joint law of the kth largest eigenvalues (k finite) with good rate function given by*

$$I^*(x_1,\ldots,x_k) = \sum_{l=1}^{k} I^*(x_k) - \beta \sum_{1\leq \ell \leq p \leq k} \log(x_\ell - x_k) + constant.$$

if $x_1 \geq x_2 \cdots \geq x_k \geq 2$ and $+\infty$ otherwise.

Exercise 11.4. *Consider*

$$P_{\alpha V}^N(d\lambda_1,\ldots,d\lambda_N) = e^{-N\alpha \sum_{i=1}^{N} V(\lambda_i)} dP_2^N(d\lambda_1,\ldots,d\lambda_N)/Z_{\alpha V}^N$$

with V a polynomial such that $V''(x) \geq 0$ for $|x|$ large enough. Show that for α positive small enough, the law of λ_N^* under $P_{\alpha V}^N$ satisfies a large deviation principle with rate function

$$I_{\alpha V}^*(x) = \begin{cases} \Phi_\alpha(x) - \inf_y \Phi_\alpha(y), & x \geq x_V \ , \\ +\infty, & otherwise . \end{cases}$$

with $\Phi_\alpha(x) = 2\int \log|x-y| d\mu_{\alpha V}(y) - \frac{1}{2}x^2 - \alpha V(x)$ and μ_V the unique solution of the Schwinger–Dyson equation of Theorem 8.3.

Hint: Observe that we are in the situation of Part III so that we know that $\frac{1}{N}\sum \delta_{\lambda_i}$ converges almost surely to $\mu_{\alpha V}$ and $Z_{\alpha V}^N = e^{N^2 I_{\alpha V}} C_{\alpha V}(1 + o(1))$. Then, show that the proof of Theorem 11.1 extends.

Bibliographical Notes. This proof is taken from [24]. It was generalized to the case of a deformed Gaussian ensemble in [143].

Part V

Stochastic Calculus

We shall now study the Hermitian Brownian motion. It is a matrix-valued process $(H_t^N)_{t\geq0}$ constructed as Gaussian Wigner matrices but with Brownian motion entries instead of Gaussian entries. We shall describe below the symmetric and the Hermitian Brownian motions, leaving the generalization to the symplectic Brownian motions as exercises. We define the symmetric (resp. Hermitian) Brownian motion $H^{N,\beta}$ for $\beta = 1$ (resp. $\beta = 2$) as a process with values in the set of $N \times N$ symmetric (resp. Hermitian) matrices with entries $\{H_{i,j}^{N,\beta}(t), t \geq 0, i \leq j\}$ constructed via independent real-valued Brownian motions $(B_{i,j}, \tilde{B}_{i,j}, 1 \leq i \leq j \leq N)$ by

$$H_{k,l}^{N,\beta}(t) = \begin{cases} \frac{1}{\sqrt{\beta N}}(B_{k,l}(t) + i(\beta - 1)\tilde{B}_{k,l}(t)), & \text{if } k < l \\ \frac{\sqrt{2}}{\sqrt{\beta N}} B_{l,l}(t), & \text{if } k = l. \end{cases} \tag{V.1}$$

Considering the matrix-valued processes, and the associated dynamics, has the advantage to allow us not only to consider one Gaussian Wigner matrix $X^N = H^{N,\beta}(1)$ but also, if $X^N(0)$ is some Hermitian Wigner matrix, the sum $X^N(1) = H^{N,\beta}(1) + X^N(0)$ seen as the matrix at time one of the matrix-valued process $X^N(t) = H^{N,\beta}(t) + X^N(0)$. Studying the evolution of the eigenvalues of $X^N(t)$ allows us to prove the law of large numbers for the spectral measure of $X^N(1)$ (see Lemma 12.5) as well as large deviation principles (see Theorem 13.1). The latter large deviations estimates result in the asymptotics for the spherical or Itzykson–Zuber–Harich–Chandra integrals (see Theorem 14.1) that in turn will give us the value of free energies for diverse two matrices matrix models (see Theorem 15.1) as well as estimates on Schur functions (see Corollary 14.2).

12

Stochastic Analysis for Random Matrices

12.1 Dyson's Brownian Motion

Let $X^N(0)$ be a symmetric (resp. Hermitian) matrix with eigenvalues $(\lambda_N^1(0), \ldots, \lambda_N^N(0))$. Let, for $t \geq 0$, $\lambda_N(t) = (\lambda_N^1(t), \ldots, \lambda_N^N(t))$ denote the (real) eigenvalues of $X^N(t) = X^N(0) + H^{N,\beta}(t)$ for $t \geq 0$. We shall prove that $(\lambda_N(t))_{t \geq 0}$ is a semi-martingale with respect to the filtration $\mathcal{F}_t = \sigma(B_{i,j}(s), \tilde{B}_{ij}(s), 1 \leq i, j \leq N, s \leq t)$ whose evolution is described by a stochastic differential system. This result was first stated by Dyson [85], and $(\lambda_N(t))_{t \geq 0}$, when $X^N(0) = 0$, has since then been called Dyson's Brownian motion. To begin with, let us describe the stochastic differential system that governs the evolution of $(\lambda_N(t))_{t \geq 0}$ and show that it is well defined.

Lemma 12.1. *Let $(W^1, \ldots, W^N)$ be an N-dimensional Brownian motion in a probability space (Ω, P) equipped with a filtration $\mathcal{F} = \{\mathcal{F}_t, t \geq 0\}$. Let Δ_N be the simplex $\Delta_N = \{(x_i)_{1 \leq i \leq N} \in \mathbb{R}^N : x_1 < x_2 < \cdots < x_{N-1} < x_N\}$ and take $\lambda_N(0) = (\lambda_N^1(0), \ldots, \lambda_N^N(0)) \in \Delta_N$. Let $\beta \geq 1$. Let $T \in \mathbb{R}^+$. There exists a unique strong solution (see Definition 20.13) to the stochastic differential system*

$$d\lambda_N^i(t) = \frac{\sqrt{2}}{\sqrt{\beta N}} dW_t^i + \frac{1}{N} \sum_{j \neq i} \frac{1}{\lambda_N^i(t) - \lambda_N^j(t)} dt \tag{12.1}$$

with initial condition $\lambda_N(0)$ such that $\lambda_N(t) \in \Delta_N$ for all $t \geq 0$. We denote by $P_{T,\lambda_N(0)}^N$ its law in $\mathcal{P}(\mathcal{C}([0,T], \Delta_N))$. It is called Dyson's Brownian motion. This weak solution (see Definition 20.14) is as well unique.

For any $\beta \geq 1$, the Dyson Brownian motion can be defined from general initial conditions, thus extending Lemma 12.1 to $\lambda_N(0) \in \overline{\Delta}_N$ (cf. [6]). We shall take this generalization for granted in the sequel.

A. Guionnet, *Large Random Matrices: Lectures on Macroscopic Asymptotics*,   167
Lecture Notes in Mathematics 1957, DOI: 10.1007/978-3-540-69897-5_12,
© 2009 Springer-Verlag Berlin Heidelberg 2009, Reprint by Springer-Verlag Berlin Heidelberg 2012

Proof. To prove the claim, let us introduce, for $R > 0$, the auxiliary system

$$d\lambda_{N,R}^i(t) = \sqrt{\frac{2}{\beta N}} dW_t^i + \frac{1}{N} \sum_{j \neq i} \phi_R(\lambda_{N,R}^i(t) - \lambda_{N,R}^j(t)) dt, \qquad (12.2)$$

with $\phi_R(x) = x^{-1}$ if $|x| \geq R^{-1}$ and $\phi_R(x) = R\,\mathrm{sgn}(x)$ if $|x| < (R)^{-1}$. We take $\lambda_{N,R}^i(0) = \lambda_N^i(0)$ for $i \in \{1,\ldots,N\}$. Since ϕ_R is uniformly Lipschitz, it is known (cf. Theorem 20.16) that this system admits a unique strong solution as well as a unique weak solution $P_{T,\lambda_N(0)}^{N,R}$, that is a probability measure on $\mathcal{C}([0,T], \mathbb{R}^N)$. Moreover, this strong solution is adapted to the filtration $\mathcal{F}$. We can now construct a solution to (12.1) by putting $\lambda_N(t) = \lambda_{N,R}(t)$ on $|\lambda_N^i(s) - \lambda_N^j(s)| \geq R^{-1}$ for all $s \leq t$ and all $i \neq j$. To prove that this construction is possible, we need to show that almost surely $\lambda_{N,R}(t) = \lambda_{N,R'}(t)$ for $R > R'$ for some (random) R' and for all t's in a compact set. To do so, we want to prove that for all times $t \geq 0$, $(\lambda_N^1(t), \ldots, \lambda_N^N(t))$ stay sufficiently apart. To prove this fact, let us consider the Lyapounov function

$$f(x_1, \ldots, x_N) = \frac{1}{N} \sum_{i=1}^N x_i^2 - \frac{1}{N^2} \sum_{i \neq j} \log |x_i - x_j|$$

and for $M > 0$ set

$$T_M = \inf\{t \geq 0 : f(\lambda_N(t)) \geq M\}.$$

Since f is $\mathcal{C}^\infty(\Delta_N, \mathbb{R})$ on sets where it is uniformly bounded (note here that f is bounded below uniformly), we also deduce that $\{T_M > T\}$ is in $\mathcal{F}_T$ for all $T \geq 0$. Thus, T_M is a stopping time. Moreover, using that $\log|x - y| \leq \log(|x| + 1) + \log(|y| + 1)$ and $x^2 - 2\log(|x| + 1) \geq c$ for some finite constant, we find that for all $i \neq j$,

$$-\frac{1}{N^2} \log |x_i - x_j| \leq f(x_1, \ldots, x_N) - c.$$

Thus, on $\{T_M > T\}$, for all $t \leq T$,

$$|\lambda_i(t) - \lambda_j(t)| \geq e^{N^2(-M+c)} =: R^{-1}$$

so that λ_N coincides with $\lambda_{N,R}$ and is therefore adapted. Itô's calculus (see Theorem 20.18) gives

$$df(\lambda_N(t))$$

$$= \frac{2}{N} \sum_{i=1}^N \left(\lambda_N^i(t) - \frac{1}{N} \sum_{k \neq i} \frac{1}{\lambda_N^i(t) - \lambda_N^k(t)} \right) \frac{1}{N} \sum_{l \neq i} \frac{1}{\lambda_N^i(t) - \lambda_N^l(t)} dt$$

$$+ \frac{2}{\beta N} \sum_{i=1}^N \left(1 + \frac{1}{N} \sum_{k \neq i} \frac{1}{(\lambda_N^i(t) - \lambda_N^k(t))^2} \right) dt + dM_N(t)$$

with M_N the local martingale

$$dM_N(t) = \frac{2^{\frac{3}{2}}}{\beta^{\frac{1}{2}}N^{\frac{3}{2}}}\sum_{i=1}^{N}\left(\lambda_N^i(t) - \frac{1}{N}\sum_{k\neq i}\frac{1}{\lambda_N^i(t) - \lambda_N^k(t)}\right)dW_t^i.$$

Observing that for all $i \in \{1,\ldots,N\}$,

$$\sum_{i=1}^{N}\left(\left(\sum_{k\neq i}\frac{1}{\lambda_N^i(t) - \lambda_N^k(t)}\right)\left(\sum_{l\neq i}\frac{1}{\lambda_N^i(t) - \lambda_N^l(t)}\right) - \sum_{k\neq i}\frac{1}{(\lambda_N^i(t) - \lambda_N^k(t))^2}\right)$$

$$= \sum_{\substack{k\neq i,l\neq i\\k\neq l}}\frac{1}{\lambda_N^i(t) - \lambda_N^k(t)}\frac{1}{\lambda_N^i(t) - \lambda_N^l(t)}$$

$$= \sum_{\substack{k\neq i,l\neq i\\k\neq l}}\frac{1}{\lambda_N^k(t) - \lambda_N^l(t)}\left(\frac{1}{\lambda_N^i(t) - \lambda_N^k(t)} - \frac{1}{\lambda_N^i(t) - \lambda_N^l(t)}\right)$$

$$= -2\sum_{\substack{k\neq i,l\neq i\\k\neq l}}\frac{1}{\lambda_N^i(t) - \lambda_N^k(t)}\frac{1}{\lambda_N^i(t) - \lambda_N^l(t)} = 0$$

and

$$\sum_{i=1}^{N}\lambda_N^i(t)\sum_{k\neq i}\frac{1}{\lambda_N^i(t) - \lambda_N^k(t)} = \frac{N(N-1)}{2},$$

we obtain that

$$df(\lambda_N(t)) = \frac{\beta(N+1)}{\beta N}dt + \frac{(2-2\beta)}{\beta N^2}\sum_{k,i,k\neq i}\frac{1}{(\lambda_N^i(t) - \lambda_N^k(t))^2}dt + dM_N(t).$$

Thus, for all $\beta \geq 1$, for all $M < \infty$, since $(M_N(t\wedge T_M), t \geq 0)$ is a martingale with vanishing expectation,

$$\mathbb{E}[f(\lambda_N(t\wedge T_M))] \leq 3\mathbb{E}[t\wedge T_M] + f(\lambda_N(0)).$$

Therefore, if $c = -\inf\{f(x_1,\ldots,x_N); (x_i)_{1\leq i\leq N} \in \mathbb{R}^N\}$,

$$(M+c)\mathbb{P}(t \geq T_M) \leq \mathbb{E}[(f(\lambda_N(t\wedge T_M)) + c)\,1_{t\geq T_M}]$$
$$\leq \mathbb{E}[f(\lambda_N(t\wedge T_M)) + c]$$
$$\leq 3\mathbb{E}[t\wedge T_M] + c + f(\lambda_N(0))$$
$$\leq 3t + c + f(\lambda_N(0))$$

which proves that for $M + c > 0$,

$$\mathbb{P}(t \geq T_M) \leq \frac{3t + c + f(\lambda_N(0))}{M+c}.$$

Hence, the Borel–Cantelli lemma implies that for all $t \in \mathbb{R}^+$,

$$\mathbb{P}\left(\cup_{M_0} \cap_{M \geq M_0} \{T_{M^2} \geq t\}\right) = 1,$$

and so in particular, $\sum_{i \neq j} \log |\lambda_N^i(t) - \lambda_N^j(t)| > -\infty$ almost surely for all times. Since $(\lambda_N(t \wedge T_M))_{t \geq 0}$ are continuous for all $M < \infty$ (as bounded perturbations of Brownian motions), we conclude that $\lambda_N(t \wedge T_M) \in \Delta_N$ for all $t \geq 0$ and all $M > 0$. As T_M goes to infinity almost surely as M goes to infinity, we conclude that $\lambda_N(t) \in \Delta_N$ for all $t \geq 0$.

Now, to prove uniqueness of the weak solution, let us consider, for $R \in \mathbb{R}^+$, the auxiliary system (12.2) with strong solution $(\lambda_{N,R}^j(t), 1 \leq j \leq N)_{t \geq 0}$. Remark that for all R, there exists $M = M(R, N) < \infty$ so that

$$\{T \leq T_M\} \subset \cap_{t \leq T} \cap_{i \neq j} \{|\lambda_N^i(t) - \lambda_N^j(t)| \geq R^{-1}\}.$$

Hence, on $\{T \leq T_M\}$, $(\lambda_N(t))_{0 \leq t \leq T}$ satisfies (12.2) and therefore $\lambda_N^i(t) = \lambda_{N,R}^i(t)$ is uniquely determined for all $i \in \{1, \ldots, N\}$ and all $t \leq T$. Since we have seen that T_{M^2} goes to infinity almost surely, we conclude that there exists a unique strong solution $(\lambda_N(t))_{0 \leq t \leq T}$ to (12.1); it coincides with $(\lambda_{N,R}^j(t), 1 \leq j \leq N)_{0 \leq t \leq T}$ for some R sufficiently large (and random). Its weak solution $P_{T,\lambda_N(0)}^N$ is also unique since its restriction to $T \leq T_M$ is uniquely determined for all $M < \infty$. □

Let $\beta = 1$ or 2 and $X^{N,\beta}(0) \in \mathcal{H}_N^{(\beta)}$ with eigenvalues $\lambda_N(0) \in \mathbb{R}^N$ and set

$$X^{N,\beta}(t) = X^{N,\beta}(0) + H^{N,\beta}(t).$$

Theorem 12.2 (Dyson). *[85] Let $\beta = 1$ or 2 and $\lambda_N(0)$ be in Δ_N. Then, the eigenvalues $(\lambda_N(t))_{t \geq 0}$ of $\left(X^{N,\beta}(t)\right)_{t \geq 0}$ are semi-martingales. Their joint law is the weak solution to (12.1).*

The proof we present goes "backward" by proposing a way to construct the matrix $X^N(t)$ from the solution of (12.1) and a Brownian motion on the orthogonal group. Its advantage with respect to a "forward" proof is that we do not need to care about justifying that certain quantities defined from X^N are semi-martingales so that Itô's calculus applies.

Proof of Theorem 12.2. We present the proof in the case $\beta = 1$ and leave the generalization to $\beta = 2$ as an exercise.

We can assume without loss of generality that $X^N(0)$ is the diagonal matrix $\mathrm{diag}(\lambda_N^1(0), \ldots, \lambda_N^N(0))$ since otherwise if O is an orthogonal matrix so that $X^N(0) = ODO^T$,

$$X^N(t) = ODO^* + H^N(t) = O(D + \tilde{H}^N(t))O^T = O\tilde{X}^N(t)O^T$$

with $\tilde{H}^N(t) = O^T H^N(t) O^T$ another Hermitian Brownian motion, independent from O. Since $\tilde{X}^N(t)$ has the same eigenvalues than $X^N(t)$ and the same law, we can assume without loss of generality that $X^N(0)$ is diagonal.

Let $M > 0$ be fixed. We consider the strong solution of (12.1) till the random time T_M. We let $w_{ij}, 1 \leq i < j \leq N$ be independent Brownian motions. Hereafter, all solutions will be equipped with the natural filtration $\mathcal{F}_t = \sigma((w_{ij}(s), W_i(s), s \leq t \wedge T_M)$ with W_i the Brownian motions of (12.1), independent of $w_{ij}, 1 \leq i < j \leq N$. We set for $i < j$

$$dR_{ij}^N(t) = \frac{1}{\sqrt{N}} \frac{1}{\lambda_N^j(t) - \lambda_N^i(t)} dw_{ij}(t).$$

We let $R^N(t)$ be the skew-symmetric matrix (i.e., $R^N(t) = -R^N(t)^T$) with such entries above the diagonal and set O^N to be the strong solution of

$$dO^N(t) = O^N(t)dR^N(t) - \frac{1}{2}O^N(t)d\langle (R^N)^T R^N \rangle_t \qquad (12.3)$$

with $O^N(0) = I$. Here, for semi-martingales A, B with values in $\mathcal{M}_N(\mathbb{R})$, $\langle A, B \rangle_t = (\sum_{k=1}^N \langle A_{ik}, B_{kj} \rangle_t)_{1 \leq i,j \leq N}$ is the martingale bracket of A and B and $\langle A \rangle_t$ is the finite variation part of A at time t. Existence and uniqueness of strong solutions of (12.3) for the filtration $\mathcal{F}_t$ are given till the random time T_M since (12.3) has bounded Lipschitz coefficients (see Theorem 20.16). Note that since the martingale bracket of a semi-martingale is given by the bracket of its martingale part,

$$d\langle (O^N)^T O^N \rangle_t = \langle [d(R^N)^T](O^N)^T, O^N dR^N \rangle_t = d\langle (R^N)^T R^N \rangle_t.$$

Hence

$$dO^N(t)^T O^N(t) = O^N(t)^T dO^N(t) + (dO^N(t)^T)O^N(t) + d\langle (O^N)^T O^N \rangle_t.$$

If $O^N(t)^T O^N(t) = I$, we deduce that $dO^N(t)^T O^N(t)$ is equal to $dR^N(t) + dR^N(t)^T = 0$ as $R^N(t)$ is skew-symmetric, from which it can be guessed that $O^N(t)^T O^N(t) = I$ at all times. This can be in fact proved, see [6], by showing that $dO^N(t)^T O^N(t)$ is linear in $O^N(t)^T O^N(t) - I$ with uniformly bounded coefficients on $\{T_M \geq t\}$. Thus, $(O^N)^T(t)O^N(t) = I$ at all times. We now show that $Y^N(t) := O^N(t)^T D(\lambda_N(t))O^N(t)$ has the same law than $X^N(t)$ which will prove the claim. By construction, $Y^N(0) := \text{diag}(\lambda_N(0)) = X^N(0)$. Moreover,

$$dY^N(t) = dO^N(t)D(\lambda_N(t))O^N(t)^T + O^N(t)D(\lambda_N(t))dO^N(t)^T$$
$$+ O^N(t)dD(\lambda_N(t))O^N(t)^T + d\langle O^N D(\lambda_N)(O^N)^T \rangle(t) \qquad (12.4)$$

where for all $i, j \in \{1, \ldots, N\}$, we adopted the notation

$$\left(d\langle O^N D(\lambda_N)(O^N)^T\rangle_t\right)_{ij}$$

$$= \sum_{k=1}^{N} \left(\frac{1}{2}O_{ik}^N(t)d\langle \lambda_N^k, O_{jk}^N\rangle_t + \lambda_N^k(t)d\langle O_{ik}^N, O_{jk}^N\rangle_t \right.$$

$$\left. + \frac{1}{2}O_{jk}^N(t)d\langle \lambda_N^k, O_{ik}^N\rangle_t \right)$$

$$= \sum_{k=1}^{N} \lambda_N^k(t)d\langle O_{ik}^N, O_{jk}^N\rangle_t.$$

The last equality is due to the independence of $(W_i)_{1\leq i\leq N}$ and $(w_{ij})_{1\leq i<j\leq N}$ which results in $\langle \lambda_N^k, O_{ik}^N\rangle_t \equiv 0$. By left multiplication by $(O^N(t))^T$ and right multiplication by $O^N(t)$ of (12.4) we arrive at

$$dW_N(t) = (O^N(t))^T dO^N(t)D(\lambda_N(t)) + D(\lambda_N(t))dO^N(t)^T O^N(t) \quad (12.5)$$
$$+ dD(\lambda_N(t)) + (O^N(t))^T d\langle O^N D(\lambda_N)(O^N)^T\rangle(t)O^N(t)$$

with $dW_N(t) = (O^N(t))^T dY^N(t)O^N(t)$. Let us compute the last term in the right-hand side of (12.5). For all $i,j \in \{1,\ldots,N\}^2$, we have

$$d\langle O^N D(\lambda_N)(O^N)^T\rangle_t^{ij} = \sum_{k=1}^{N} \lambda_N^k(t)d\langle O_{ik}^N, O_{jk}^N\rangle_t$$

$$= \sum_{k,l,m=1}^{N} \lambda_N^k(t)O_{il}^N(t)O_{jm}^N(t)d\langle R_{lk}^N, R_{mk}^N\rangle_t$$

$$= \frac{1}{N}\sum_{k\neq l} \frac{\lambda_N^k(t)}{(\lambda_N^k(t) - \lambda_N^l(t))^2}O_{il}^N(t)O_{jl}^N(t)dt$$

where we finally used the definition of R^N to compute the martingale brackets that gives

$$d\langle R_{lk}^N, R_{mk}^N\rangle_t = 1_{l=m\neq k}\frac{1}{N(\lambda_N^k(t) - \lambda_N^l(t))^2}dt.$$

Hence, for all $i,j \in \{1,\ldots,N\}^2$, we get

$$[(O^N)^T(t)d\langle O^N D(\lambda_N)(O^N)^T\rangle_t O^N(t)]_{ij} = 1_{i=j}\sum_{k\neq i}^{N} \frac{\lambda_N^k(t)}{N(\lambda_N^i(t) - \lambda_N^k(t))^2}dt.$$

Similarly, recall that

$$(O^N)^T(t)dO^N(t) = dR^N(t) - 2^{-1}d\langle (R^N)^T R^N\rangle_t$$

with for all $i, j \in \{1, \ldots, N\}^2$,

$$d\langle (R^N)^T R^N \rangle_t^{ij} = \sum_{k=1}^{N} d\langle R_{ki}^N, R_{kj}^N \rangle_t$$

$$= \frac{1_{i=j}}{N} \sum_{k \neq i} (\lambda_N^k(t) - \lambda_N^i(t))^{-2} dt.$$

Therefore, identifying the terms on the diagonal in (12.5) and recalling that R^N is null on the diagonal, we find that

$$dW_N^{ii}(t) = \sqrt{\frac{2}{N}} dW_t^i.$$

Outside the diagonal, for $i \neq j$, we get

$$dW_N^{ij}(t) = [dR^N(t)D(\lambda_N(t)) + D(\lambda_N(t))dR^N(t)^T]_{ij}$$

$$= \frac{1}{\sqrt{N}} dw_{ij}(t).$$

Hence, $W_N(t)$ has the law of a symmetric Brownian motion. Thus, since $(O^N(t), t \geq 0)$ is adapted, $dY^N(t) = O^N(t)dW_N(t)(O^N(t))^T$ is a continuous matrix-valued martingale whose quadratic variation is given by

$$\langle Y_{ij}^N, Y_{kl}^N \rangle_t = 1_{ij=kl \text{ or } lk} N^{-1} t \quad i \neq j, \langle Y_{ii}^N, Y_{kl}^N \rangle_t = 1_{i=k=l} 2N^{-1} t.$$

Therefore, by Levy's theorem (cf. [122, p. 157]), $(Y^N(t) - Y^N(0), t \geq 0)$ is a symmetric Brownian motion, and so $(Y^N(t), t \geq 0)$ has the same law than $(X^N(t), t \geq 0)$ since $X^N(0) = Y^N(0)$.   □

Corollary 12.3 (Dyson). *Let $\beta = 1$ or 2 and $\lambda_N(0)$ in $\overline{\Delta}_N$. Then, the eigenvalues $(\lambda_N(t))_{t\geq 0}$ of $(X^{N,\beta}(t))_{t\geq 0}$ are continuous semi-martingales with values in Δ_N for all $t > 0$. The joint law of $(\lambda_N(t))_{t\geq \epsilon}$ is the weak solution to (12.1) starting from $\lambda_N(\epsilon) \in \Delta_N$ for any $\epsilon > 0$. $\lambda_N(\epsilon)$ converges to $\lambda_N(0)$ as ϵ goes to zero.*

Proof. To remove the hypothesis that the eigenvalues of $X_N(0)$ belong to Δ_N, note that for all $t > 0$, $\lambda_N(t)$ belongs to Δ_N almost surely. Indeed, the set of symmetric matrices with at least one double eigenvalue can be characterized by the fact that the discriminant of their characteristic polynomial vanishes, and so the entries of such matrices belong to a submanifold with codimension greater than one. Since the law of the entries of $X^N(t)$ is absolutely continuous with respect to the Lebesgue measure, this set has measure zero. Therefore, we can represent the eigenvalues of $(X^N(t), t \geq \epsilon)$ as solution of (12.1) for any $\epsilon > 0$. By using Lemma 1.16, we see that for all $s, t \in \mathbb{R}$

$$\sum_{i=1}^{N}(\lambda_N^i(t) - \lambda_N^i(s))^2 \leq \frac{1}{N}\sum_{i,j=1}^{N}(B_{ij}(t) - B_{ij}(s))^2$$

so that the continuity of the Brownian motions paths results in the continuity of $t \to \lambda_N(t)$ for any given N. Hence, the eigenvalues of $(X^N(t))_{t \leq T}$ are strong solutions of (12.1) for all $t > 0$ and are continuous at the origin. $\qquad\square$

Exercise 2. Let $X^{N,4} = \left(X_{ij}^{N,4}\right)$ be a $2N \times 2N$ complex matrix defined as the $N \times N$ self-adjoint random matrices with entries

$$X_{kl}^{N,4} = \frac{\sum_{i=1}^{\beta} g_{kl}^i e_4^i}{\sqrt{\beta N}}, \quad 1 \leq k < l \leq N, \quad X_{kk}^{N,4} = \sqrt{\frac{2}{\beta N}} g_{kk} e_4^1, \quad 1 \leq k \leq N$$

where $(e_4^i)_{1 \leq i \leq 4}$ are the Pauli matrices

$$e_4^1 = \begin{pmatrix} 1 & 0 \\ 0 & 1 \end{pmatrix}, \, e_4^2 = \begin{pmatrix} 0 & -1 \\ 1 & 0 \end{pmatrix}, \, e_4^3 = \begin{pmatrix} 0 & -i \\ -i & 0 \end{pmatrix}, \, e_4^4 = \begin{pmatrix} i & 0 \\ 0 & -i \end{pmatrix}.$$

Define $H^{N,4}$ similarly by replacing the Gaussian entries by Brownian motions. Show that if $X^N(0)$ a Hermitian matrix with eigenvalues $\lambda_{2N}(0)$, the eigenvalues $\lambda_{2N}(t)$ of $X^N(0) + H^{N,4}$ satisfy the stochastic differential system

$$d\lambda_{2N}^i(t) = \frac{1}{\sqrt{4N}}dW_t^i + \frac{1}{2N}\sum_{j \neq i}\frac{1}{\lambda_N^i(t) - \lambda_N^j(t)}dt. \qquad (12.6)$$

Corollary 12.4. *For $\beta = 1$ or 2, the law of the eigenvalues of the Gaussian Wigner matrix $X^{N,\beta}$ is given by*

$$P_\beta^N(dx_1, \dots, dx_N) = \frac{1}{Z_N}1_{x_1 \leq \cdots \leq x_N}\prod_{1 \leq i < j \leq N}|x_i - x_j|^\beta \prod_{i=1}^{N}e^{-\beta x_i^2/4}. \qquad (12.7)$$

Proof. In Dyson's theorem, we can easily replace the Hermitian Brownian motion $(H^N(t))_{t \in \mathbb{R}^+}$ by the Hermitian Ornstein–Uhlenbeck process $(\tilde{H}^N(t))_{t \in \mathbb{R}^+}$ whose entries are solutions of

$$d\tilde{H}_{k,l}(t) = dH_{k,l}(t) - \frac{1}{2}\tilde{H}_{k,l}(t)dt.$$

$\tilde{H}_{k,l}(t)$ converges as t goes to infinity towards a centered Gaussian variable with covariance N^{-1}, independently of the initial condition $X^N(0)$. Hence, Wigner matrices appear as the large time limit of $\tilde{H}$ and in particular their law is invariant for the dynamics of the Ornstein–Uhlenbeck process. On the other hand, a slight modification of the proof of Dyson's Theorem 12.2 (we can here assume that $X^N(0)$ has eigenvalues in the simplex) shows that the eigenvalues of $\tilde{H}$ follow the SDE

$$d\lambda_N^i(t) = \frac{\sqrt{2}}{\sqrt{\beta N}}dW_t^i - \frac{1}{2}\lambda_N^i(t)dt + \frac{1}{N}\sum_{j\neq i}\frac{1}{\lambda_N^i(t) - \lambda_N^j(t)}dt.$$

Hence, the law $P_N^{(\beta)}$, as the large time limit of the law of $\lambda_N(t)$, must be invariant under the above dynamics. Itô's calculus shows that the infinitesimal generator of these dynamics is

$$\mathcal{L} = \frac{1}{\beta N}\sum_{i=1}^{N}\partial_i^2 + \sum_{i=1}^{N}\left(\frac{1}{N}\sum_{j\neq i}\frac{1}{\lambda^i - \lambda^j} - \frac{1}{2}\lambda_i\right)\partial_i$$

and therefore we must have, for any twice continuously differentiable function f on $\mathbb{R}^N$,

$$\int \mathcal{L}f(\lambda_1,\ldots,\lambda_N)dP_N^{(\beta)}(\lambda_1,\ldots,\lambda_N) = 0.$$

Some elementary algebra shows that the choice proposed in (12.7) fulfills this requirement. Furthermore it is the unique such probability measure on the simplex since if there was another invariant distribution Q_N for $\mathcal{L}$, we could follow the proof of Theorem 12.2 to reconstruct a Hermitian Ornstein-Uhlenbeck process $\tilde{H}^N(t)$ and a matrix $X^N(0)$ whose eigenvalues would follow Q_N so that $\tilde{H}_{k,l}(0) = X^N(0)_{k,l}$ and

$$d\tilde{H}_{k,l}(t) = dH_{k,l}(t) - \frac{1}{2}\tilde{H}_{k,l}(t)dt.$$

But this gives a contradiction since as time goes to infinity, the law of $\tilde{H}_{k,l}$ is a Gaussian law, independently of the law Q_N. $\square$

12.2 Itô's Calculus

Let $(W^1,\ldots,W^N)$ be independent Brownian motions and $(\lambda_N^1(0),\ldots,\lambda_N^N(0))$ be real numbers. Let β be a real number greater than one and let $(\lambda_N(t))_{t\geq 0}$ be the unique strong solution to (12.1). We denote by

$$L_N(t,dx) := \frac{1}{N}\sum_{i=1}^{N}\delta_{\lambda_N^i(t)} \in \mathcal{P}(\mathbb{R})$$

the empirical measure of $\lambda_N(t)$. We shall sometimes use the short notation for bounded measurable functions f on $\mathbb{R}$,

$$\int f dL_N(t) := \int f(x)L_N(t,dx) = \frac{1}{N}\sum_{i=1}^{N}f(\lambda_N^i(t)).$$

Then, by Itô's calculus Theorem 20.18, we know that for all $f \in C^2([0,T] \times \mathbb{R}, \mathbb{R})$,

$$\int f(t,x)L_N(t,dx) = \int f(0,x)L_N(0,dx) + \int_0^t \int \partial_s f(s,x)L_N(s,dx)ds$$

$$\tag{12.8}$$

$$+ \frac{1}{2}\int_0^t \int \frac{\partial_x f(s,x) - \partial_x f(s,y)}{x-y} L_N(s,dx)L_N(s,dy)ds$$

$$+ \left(\frac{2}{\beta}-1\right)\frac{1}{2N}\int_0^t \int \partial_x^2 f(s,x)L_N(s,dx)ds + M_N^f(s)$$

with M_N^f the martingale given for $s \leq T$ by

$$M_N^f(t) = \frac{\sqrt{2}}{\sqrt{\beta}N^{\frac{3}{2}}} \sum_{i=1}^N \int_0^t \partial_x f(s,\lambda_N^i(s))dB_s^i.$$

Note that M_N^f is a martingale with bracket

$$\langle M_N^f \rangle_t = \frac{2}{\beta N^2}\int_0^t \int (\partial_x f(s,x))^2 L_N(s,dx)du \leq \frac{2\|\partial_x f\|_\infty^2 t}{\beta N^2}.$$

12.3 A Dynamical Proof of Wigner's Theorem 1.13

In this section, we shall give a dynamical proof of Theorem 1.13; it is restricted to Gaussian entries but generalized in the sense that we can study the asymptotic behavior of the spectral measure of any sum of two independent symmetric matrices, one being a Gaussian Wigner matrix, the second being deterministic with a converging spectral distribution. Moreover, our proof only relies on (12.1) and thus our result generalizes to any $\beta \geq 1$, and in particular to the Hermitian and the symplectic case too (that corresponds to $\beta = 2$ and 4).

For $T > 0$, we denote by $C([0,T],\mathcal{P}(\mathbb{R}))$ the space of continuous processes from $[0,T]$ into $\mathcal{P}(\mathbb{R})$ equipped with its weak topology. We have

Lemma 12.5. *Let* $\lambda_N(0) \in \mathbb{R}^N$ *so that* $L_N(0) = \frac{1}{N}\sum_{i=1}^N \delta_{\lambda_N^k(0)}$ *converges as* N *goes to infinity towards* $\mu \in \mathcal{P}(\mathbb{R})$. *We assume*

$$C_0 := \sup_{N \geq 0} \int \log(x^2+1)dL_N(0)(x) < \infty. \tag{12.9}$$

Let $(\lambda_N^1(t),\ldots,\lambda_N^N(t))_{t \geq 0}$ *be the solution to* (12.1) *and set*

$$L_N(t) = \frac{1}{N}\sum_{i=1}^N \delta_{\lambda_N^i(t)}.$$

Then, for any finite time T, $(L_N(t), t \in [0, T])$ converges almost surely in $\mathcal{C}([0, T], \mathcal{P}(\mathbb{R}))$. Its limit is the unique measure-valued process $(\mu_t, t \in [0, T])$ so that $\mu_0 = \mu$ and for all $z \in \mathbb{C}\backslash\mathbb{R}$

$$G_t(z) = \int (z - x)^{-1} d\mu_t(x)$$

satisfies the complex Burgers equation

$$G_t(z) = G_0(z) - \int_0^t G_s(z) \partial_z G_s(z) ds. \qquad (12.10)$$

with given initial condition G_0.

We begin the proof by showing that $(L_N(t), t \in [0, T])$ is almost surely tight in $\mathcal{C}([0, T], \mathcal{P}(\mathbb{R}))$ and then show that it has a unique limit point characterized by (12.10).

We first describe compact sets of $\mathcal{C}([0, T], \mathcal{P}(\mathbb{R}))$; they are of the form

$$\mathcal{K} = \{\forall t \in [0, T], \mu_t \in K\} \cap_{i \geq 0} \{t \to \mu_t(f_i) \in C_i\} \qquad (12.11)$$

where

- K is a compact set of $\mathcal{P}(\mathbb{R})$ such as

$$K_{\epsilon,m} := \cap_{m \geq 0} \{\mu([-m, m]) \leq \epsilon_m\} \qquad (12.12)$$

for a sequence $\{\epsilon_m, m \geq 0\}$ of positive real numbers going to zero as m goes to infinity.

- $(f_i)_{i \geq 0}$ is a sequence of bounded continuous functions dense in $\mathcal{C}_0(\mathbb{R})$ and C_i are compact sets of $\mathcal{C}([0, T], \mathbb{R})$. By the Arzela–Ascoli theorem, it is known that the latter are of the form

$$C_{\epsilon,M} := \{g : [0, T] \to \mathbb{R}, \sup_{\substack{t,s \in [0,T] \\ |t-s| \leq \eta_n}} |g(t) - g(s)| \leq \epsilon_n, \sup_{t \in [0,T]} |g(t)| \leq M\} \quad (12.13)$$

with sequences $\{\epsilon_n, n \geq 0\}$ and $\{\eta_n, n \geq 0\}$ of positive real numbers going to zero as n goes to infinity.

In fact, if we take a sequence μ^n in $\mathcal{K}$, for all $i \in \mathbb{N}$, we can find a subsequence such that $\mu^{\phi_i(n)}(f_i)$ converges as a bounded continuous function on $[0, T]$. By a diagonalization procedure, we can find ϕ so that $\mu^{\phi(n)}(f_i)$ converges simultaneously towards some $\mu.(f_i)$ for all $i \in \mathbb{N}$. Note at this point that since the f_i have compact support, the limit $\mu.$ might not have mass one. This is dealt with by the second condition. Indeed, since the time marginals of $\mu_t^{\phi(n)}$ are tight for all t we can, again up to take another subsequence, insure that $\mu_t \in \mathcal{P}(\mathbb{R})$, at least for a countable number of times. The continuity of $t \to \mu_t(f_i)$ and the density of the family f_i then shows that $\mu_t \in \mathcal{P}(\mathbb{R})$ for all t. Hence, we have proved that μ^n is sequentially compact. Further, the limit μ also belongs to $\mathcal{K}$, which finishes to show that $\mathcal{K}$ is compact.

We shall prove below that $\hat{\mathbf{L}}^N$ is almost surely tight. For later purposes (namely the study of large deviation properties), we next prove a slightly stronger result.

Lemma 12.6. *Let $T \in \mathbb{R}^+$. Assume (12.9). Then, there exists $a = a(T) > 0$ and $M(T) < \infty$ so that:*

1. For $M \geq M(T)$

$$P\left(\sup_{t \in [0,T]} \int \log(x^2 + 1) L_N(t, dx) \geq M\right) \leq e^{-a(T)MN^2}.$$

2. For any $\delta > 0$ and $M > 0$, for any twice continuously differentiable function f so that $\|f''\|_\infty \leq 2^{-3} M \delta^{-\frac{3}{4}}$,

$$P\left(\sup_{\substack{t,s \in [0,T] \\ |t-s| \leq \delta}} \left|\int f(x) L_N(t, dx) - \int f(x) L_N(s, dx)\right|\right.$$
$$\left. \geq M\delta^{\frac{1}{4}}\right) \leq 2(T\delta^{-1} + 1) e^{-\frac{\beta N^2 M^2}{2^8 \|f'\|_\infty^2 \delta^{\frac{1}{2}}}}.$$

3. For all $T \in \mathbb{R}^+$, all $L \in \mathbb{N}$, there exists a compact set $\mathcal{K}(L)$ of the set $\mathcal{C}([0,T], \mathcal{P}(\mathbb{R}))$ of continuous probability measures-valued processes so that

$$P\left(L_N(.) \in \mathcal{K}(L)^c\right) \leq e^{-N^2 L}.$$

In particular, the law of $(L_N(s), s \in [0,T])$ is almost surely tight in $\mathcal{C}([0,T], \mathcal{P}(\mathbb{R}))$.

Proof. We base our proof on (12.1). Using Section 12.2 with $f(x) = \log(x^2 + 1)$, we get since

$$\left|\frac{f'(x) - f'(y)}{x - y}\right| = \left|\int_0^1 f''(\alpha x + (1 - \alpha)y) d\alpha\right| \leq \|f''\|_\infty < \infty,$$

for all $s \geq 0$,

$$\left|\int \log(x^2 + 1) dL_N(s)\right| \leq \left|\int \log(x^2 + 1) dL_N(0)\right| + 2\|f''\|_\infty s + |M_s^N| \quad (12.14)$$

with M_s^N the martingale

$$M_s^N = \frac{2\sqrt{2}}{\sqrt{\beta} N^{\frac{3}{2}}} \sum_{i=1}^N \int_0^s \frac{\lambda_N^i(u)}{\lambda_N^i(u)^2 + 1} dW_u^i.$$

Note that

$$\langle M^N \rangle_s = \frac{8}{\beta N^3} \sum_{i=1}^{N} \int_0^s \frac{\lambda_N^i(u)^2}{(\lambda_N^i(u)^2 + 1)^2} du \le \frac{8s}{\beta N^2}.$$

Hence, we can use Corollary 20.23 to obtain for all $L \ge 0$

$$P\left(\sup_{s \le T} |M_s^N| \ge L \right) \le 2e^{-\frac{\beta N^2 L^2}{16T}}.$$

Thus, (12.14) shows that for $M \ge C_0 + 2\|f''\|_\infty T$,

$$P\left(\sup_{s \le T} \left| \int \log(x^2 + 1) dL_N(s) \right| \ge M \right) \le 2e^{-\frac{\beta N^2 (M - C_0 + 2\|f''\|_\infty T)^2}{16T}} \quad (12.15)$$

which proves the first point. For the second point we proceed similarly by first noticing that if $t_i = i\delta$ for $i \in [0, [T/\delta] + 1]$,

$$\left\{ \sup_{\substack{t,s \in [0,T] \\ |t-s| \le \delta}} \left| \int f dL_N(t) - \int f dL_N(s) \right| \ge M\delta^{\frac{1}{4}} \right\}$$

$$\subset \cup_{1 \le i \le [T/\delta]+1} \left\{ \sup_{t_i \le s \le t_{i+1}} \left| \int f dL_N(s) - \int f dL_N(t_i) \right| \ge 2^{-1} M\delta^{\frac{1}{4}} \right\}.$$

Now, for $s \in [t_i, t_{i+1}]$, we write by a further use of Section 12.2,

$$\left| \int f dL_N(s) - \int f dL_N(t_i) \right| \le 2\|f''\|_\infty \delta + |M_N^f(s)|$$

with

$$\langle M_N^f \rangle_s = \frac{8}{\beta N^3} \sum_{i=1}^{N} \int_{t_i}^s (f'(\lambda_N^i(u)))^2 du \le \frac{8\delta \|f'\|_\infty^2}{\beta N^2}.$$

Thus Corollary 20.23 shows that

$$\sup_{t_i \le s \le t_{i+1}} \left| \int f dL_N(s) - \int f dL_N(t_i) \right| \le 2\|f''\|_\infty \delta + \epsilon$$

with probability greater than $1 - 2e^{-\frac{\beta N^2 (\epsilon)^2}{16\delta \|f'\|_\infty^2}}$. As a conclusion, for $\epsilon = 2^{-1} M\delta^{\frac{1}{4}} - 2\|f''\|_\infty \delta \ge 2^{-2} M\delta^{\frac{1}{4}}$ we have

$$P\left(\sup_{\substack{t,s \in [0,T] \\ |t-s| \le \delta}} \left| \int f dL_N(t) - \int f dL_N(s) \right| \ge M\delta^{\frac{1}{4}} \right) \le \sum_{i=1}^{[T/\delta]+1} 2e^{-\frac{\beta N^2 M^2}{2^8 \|f'\|_\infty^2 \delta^{\frac{1}{2}}}}$$

which proves the second claim.

To conclude our proof, let us notice that:

- The set

$$K_M = \{\mu \in \mathcal{P}(\mathbb{R}) : \int \log(1 + x^2)d\mu(x) \leq M\}$$

is compact and by Borel–Cantelli lemma, we deduce from point 1 that

$$P(\{L_N(t) \in K_M \ \forall \ t \in [0,T]\}^c) \leq e^{-a(T)MN^2}.$$

- Take f twice continuously differentiable and consider the compact subset of $\mathcal{C}([0,T], \mathcal{P}(\mathbb{R}))$

$$C_T(f, M) := \{\mu \in \mathcal{C}([0,T], \mathcal{P}(\mathbb{R})) :$$

$$\sup_{|t-s| \leq n^{-2}} |\mu_t(f) - \mu_s(f)| \leq M\sqrt{n}^{-1} \quad \forall n \in \mathbb{N}^*\}.$$

The previous estimates imply that if $\|f''\|_\infty \leq Mn$ for all n

$$P(L_N \in C_T(f, M)^c) \leq \sum_{n \geq 1} 2(Tn^2 + 1)e^{-\frac{\beta N^2 n M^2}{2^{16}\|f'\|_\infty^2}} \leq C(T)e^{-\frac{\beta N^2 M^2}{2^{16}\|f'\|_\infty^2}}$$

with some finite constant $C(T)$ that only depends on T. Choosing a countable family f_i of twice continuously differentiable functions dense in $\mathcal{C}_0(\mathbb{R})$ and such that $\|f''\|_\infty \leq i$ and $\|f'\|_\infty \leq \sqrt{i}$ for all i, we obtain

$$P(\{L_N \in (\cap_{i \geq 0} C_T(f_i, Mi))^c\}) \leq C(T)\sum_{i \geq 1} e^{-\frac{\beta N^2 i^2 M^2}{2^{16}\|f_i'\|_\infty^2}} \leq C'(T)e^{-\frac{\beta N^2 M^2}{2^{16}}}.$$

- Hence, we conclude that the compact set

$$\mathcal{K}(M) = K_M \cap \cap_{i \geq 0} C_T(f_i, Mi)$$

of $\mathcal{C}([0,T], \mathcal{P}(\mathbb{R}))$ is such that $P(L_N \in \mathcal{K}(M)^c) \leq C(T)e^{-\frac{\beta N^2 M^2}{2^{16}}}$ and thus by Borel–Cantelli lemma

$$P(\cup_{N_0} \cap_{N \geq N_0} \{L_N \in \mathcal{K}(M)\}) = 1.$$

$\square$

To characterize the limit points of L_N, let us use also Itô's calculus of Section 12.2 with $f(t,x) = f(x) = (z-x)^{-1}$ for $z \in \mathbb{C}\backslash\mathbb{R}$ (or separately for its real and imaginary parts). Again by Corollary 20.23, M_f^N goes almost surely to zero and therefore, any limit point $(\mu_t, t \in [0,T])$ of $(L_N(t), t \in [0,T])$ satisfies the equation

$$\int f(x)d\mu_t(x) = \int f(x)d\mu_0(x) + \int_0^t \int \partial_s f(x)d\mu_s(x)ds \qquad (12.16)$$

$$+\frac{1}{2}\int_0^t\int\int\frac{\partial_x f(x)-\partial_x f(y)}{x-y}d\mu_s(x)d\mu_s(y)ds.$$

Thus, $G_t(z)=\int(z-x)^{-1}d\mu_t(x)$ satisfies (12.10).

To conclude our proof, we show that (12.10) has a unique solution.

Lemma 12.7. *For any $t>0$ and z with modulus $|z|$ large enough, $z+tG_0(z)$ is invertible with inverse H_t. The solution of (12.10) is the unique analytic function on $\mathbb{C}\backslash\mathbb{R}$ such that for any $t>0$ and z with large enough modulus*

$$G_t(z)=G_0\left(H_t(z)\right).$$

Comments. This result is a particular case of free convolution (see Section 17.3.2) and of the notion of subordination (cf. [35]).

Exercise 3. Take $\mu=\delta_0$ and prove that μ_t is the semicircular law with variance t. *Hint.* Use that by scaling property $G_t(z)=t^{-\frac{1}{2}}G_1(t^{-\frac{1}{2}}z)$ for all $t>0$ and deduce a formula for G_1.

Proof. We use the characteristic method. Let us associate to z the solution $\{z_t, t\ge 0\}$ of the equation

$$\partial_t z_t=G_t(z_t),\ z_0=z.$$

Such a solution exists at least up to time $(\Im z)^2/2$ since if $\Im(z)>0$, $\Im(G_t(z))\in[-\frac{1}{\Im(z)},0]$ implies that we can construct a unique solution z_t with $\Im(z_t)>0$ up to that time, a domain on which G_t is Lipschitz. Now, $\partial_t G_t(z_t)=0$ implies

$$z_t=tG_0(z)+z,\ G_t(z+tG_0(z))=G_0(z)$$

from which the conclusion follows. □

Bibliographical Notes. The previous arguments on Dyson's Brownian motion are inspired by [150, p.123], where the density of the eigenvalues of symmetric Brownian motions are discussed, [170] where the stochastic differential equation (12.1) is studied, [160] where Brownian motions of ellipsoids is considered as well as [165] where decompositions of Brownian motions on certain manifolds of matrices are analyzed. Theorem 12.2 is also stated in Mehta's book [153, Theorem 8.2.1](see also Chan [61]). Similar results can be obtained for Wishart processes, see [52]. The interpretation of Dyson's Brownian motion as a Brownian motion in a Weyl chamber was used in [36, 40].

13

Large Deviation Principle for the Law of the Spectral Measure of Shifted Wigner Matrices

The goal of this section is to prove the following theorem.

Theorem 13.1. *Assume that D_N is uniformly bounded with spectral measure converging to μ_D. Let $X^{\beta,N}$ be a Gaussian symmetric (resp. Hermitian) Wigner matrix when $\beta = 1$ (resp. $\beta = 2$). Then the law of the spectral measure $L_{Y^{N,\beta}}$ of the Wigner matrix $Y^{N,\beta} = D_N + X^{\beta,N}$ satisfies a large deviation principle in the scale N^2 with a certain good rate function $J_\beta(\mu_D, .)$.*

We shall base our approach on Bryc's theorem, that says that the above large deviation principle statement is equivalent to the fact that for any bounded continuous function f on $\mathcal{P}(\mathbb{R})$,

$$\Lambda(f) = \lim_{N \to \infty} \frac{1}{N^2} \log \int e^{N^2 f(L_{Y^{N,\beta}})} d\mathbb{P}$$

exists and is given by $-\inf\{J_\beta(\mu_D, \nu) - f(\nu)\}$. It is not clear how one could a priori study such limits, except for very trivial functions f. However, if we consider the matrix-valued process $Y^{N,\beta}(t) = D^N + H^{N,\beta}(t)$ with Brownian motion $H^{N,\beta}$ described in (V.1) and its spectral measure process

$$L_{N,\beta}(t) := L_{Y^{N,\beta}(t)} = \frac{1}{N} \sum_{i=1}^{N} \delta_{\lambda_i(Y^{N,\beta}(t))} \in \mathcal{P}(\mathbb{R}),$$

we may construct martingales by use of Itô's calculus. Indeed, continuous martingales lead to exponential martingales, which have constant expectation, and therefore allow one to compute the exponential moments of a whole family of functionals of $L_N(t)$. This idea gives easily a large deviation upper bound for the law of $(L_{N,\beta}(t), t \in [0,1])$, and therefore for the law of $L_{Y^{N,\beta}}$, that is, the law of $L_{N,\beta}(1)$. The difficult point here is to check that this bound is sharp, i.e., it is enough to compute the exponential moments of this family of functionals in order to obtain the large deviation lower bound.

A. Guionnet, *Large Random Matrices: Lectures on Macroscopic Asymptotics*, 183
Lecture Notes in Mathematics 1957, DOI: 10.1007/978-3-540-69897-5_13,
© 2009 Springer-Verlag Berlin Heidelberg, Reprint by Springer-Verlag Berlin Heidelberg 2012

An alternative tempting way to prove this large deviation lower bound would be, as for the proof of Theorem 10.1, to force the paths of the eigenvalues to be in small tubes around the quantiles of their limiting law. However, these tubes would need to be very small, with width of order $\delta \approx N^{-1}$ and the probability $P(\sup_{0 \leq s \leq 1} |B_s| \leq \delta) \approx e^{-\frac{c}{\delta^2}}$ is now giving a contribution on the scale e^{N^2}.

Let us now state more precisely our result. We shall consider $\{L_{N,\beta}(t), t \in [0,1]\}$ as an element of the set $\mathcal{C}([0,1], \mathcal{P}(\mathbb{R}))$ of continuous processes with values in $\mathcal{P}(\mathbb{R})$. The rate function for these deviations shall be given as follows. For any $f, g \in \mathcal{C}_b^{2,1}(\mathbb{R} \times [0,1])$, any $s \leq t \in [0,1]$, and any $\nu. \in \mathcal{C}([0,1], \mathcal{P}(\mathbb{R}))$, we let

$$
S^{s,t}(\nu, f) = \int f(x,t) d\nu_t(x) - \int f(x,s) d\nu_s(x)
$$

$$
- \int_s^t \int \partial_u f(x,u) d\nu_u(x) du
$$

$$
- \frac{1}{2} \int_s^t \int \int \frac{\partial_x f(x,u) - \partial_x f(y,u)}{x-y} d\nu_u(x) d\nu_u(y) du, \quad (13.1)
$$

$$
\langle f, g \rangle_\nu^{s,t} = \int_s^t \int \partial_x f(x,u) \partial_x g(x,u) d\nu_u(x) du, \quad (13.2)
$$

and

$$
\bar{S}_\beta^{s,t}(\nu, f) = S^{s,t}(\nu, f) - \frac{1}{\beta} \langle f, f \rangle_{s,t}^\nu. \quad (13.3)
$$

Set, for any probability measure $\mu \in \mathcal{P}(\mathbb{R})$,

$$
S_{\mu,\beta}(\nu) := \begin{cases} +\infty, & \text{if } \nu_0 \neq \mu, \\ S_\beta^{0,1}(\nu) := \sup_{f \in \mathcal{C}_b^{2,1}(\mathbb{R} \times [0,1])} \sup_{0 \leq s \leq t \leq 1} \bar{S}_\beta^{s,t}(\nu, f), & \text{otherwise.} \end{cases}
$$

$$(13.4)$$

Then, the main theorem of this section is the following:

Theorem 13.2. Let $\beta = 1$ or 2. (1) For any $\mu \in \mathcal{P}(\mathbb{R})$, $S_{\mu,\beta}$ is a good rate function on $\mathcal{C}([0,1], \mathcal{P}(\mathbb{R}))$, i.e. $\{\nu \in \mathcal{C}([0,1], \mathcal{P}(\mathbb{R})); S_\mu(\nu) \leq M\}$ is compact for any $M \in \mathbb{R}^+$.

(2) Assume that

$$
\sup_N L_{D_N}(|x|^4) < \infty, \quad L_{D_N} \text{ converges to } \mu_D, \quad (13.5)
$$

then the law of $(L_{N,\beta}(t), t \in [0,1])$ satisfies a large deviation upper-bound in the scale N^2 with good rate function $S_{\mu_D,\beta}$. The large deviation lower bound holds around any measure-valued path with uniformly bounded fourth moment, i.e., for all μ. so that $\sup_{t \in [0,1]} \mu_t(x^4)$ is finite:

$$\liminf_{\delta \to 0} \liminf_{N \to \infty} \frac{1}{N^2} \log P \left(\sup_{t \in [0,1]} d(L_{N,\beta}(t), \mu_t) < \delta \right) \geq -S_{\mu_D,\beta}(\mu_.).$$

In [110, Theorem 3.3] O. Zeitouni and I proved the following:

Theorem 13.3. *Take* $\beta = 1$ *or* 2 *and take*

$$\mathcal{A} = \{\mu \in \mathcal{P}(\mathbb{R}) : \text{ there exists } \epsilon > 0, \int |x|^{5+\epsilon} d\mu(x) < \infty\}.$$

Assume $\mu_D \in \mathcal{A}$. *Then, for any* ν *such that* $S_{\mu_D,\beta}(\nu) < \infty$, *there exists a sequence* $\nu^n : [0,1] \to \mathcal{A}$ *of measure-valued paths with uniformly bounded fourth moment such that*

$$\lim_{n \to \infty} \nu_n = \nu \quad \lim_{n \to \infty} S_{\mu_D,\beta}(\nu_n) = S_{\mu_D,\beta}(\nu).$$

This result could be extended by replacing $5 + \epsilon$ by $4 + \epsilon$ (which we needed first, since Theorem 13.2 was originally obtained under these 5^+ moment conditions) but we decided not to enter into this proof here since it is purely analytical. We shall, however, provide here a complete proof of Theorem 13.2 that slightly simplifies that given in [109].

Theorems 13.2 and 13.3 imply the following:

Theorem 13.4. *Assume that* (13.5) *holds and* $\mu_D \in \mathcal{A}$. *Then the law of* $(L_{N,\beta}(t), t \in [0,1])$ *satisfies a large deviation principle in the scale* N^2 *with good rate function* $S_{\mu_D,\beta}$.

Note that the application $(\mu_t, t \in [0,1]) \to \mu_1$ is continuous from $\mathcal{C}([0,1], \mathcal{P}(\mathbb{R}))$ into $\mathcal{P}(\mathbb{R})$, so that Theorem 13.2 and the contraction principle Theorem 20.7 imply that the law of $L_N(1)$ satisfies a large deviation principle.

Theorem 13.5. *Under assumption* (13.5) *and* $\mu_D \in \mathcal{A}$, *Theorem 13.1 is true with*

$$J_\beta(\mu_D, \mu_E) = \frac{\beta}{2} \inf\{S_{\mu_D}(\nu.); \nu_1 = \mu\}.$$

Remark 4. Remark that without using Theorem 13.3, we could still get a large deviation lower bound for the law of $L_{N,\beta}(1)$ around measure with fourth moments; this is due to the fact that the optimal paths in the above infimum have fourth moments (as can be guessed from the remark that optimal paths have to be Brownian bridges, cf., e.g., [55]).

In [101], the infimum in Theorem 13.5 was studied. It was shown that it is achieved and that, if $\int \log |x - y| d\mu_D(x) d\mu_D(y) > -\infty$, the minimizer $\mu^* \in (C([0,1], \mathcal{P}(\mathbb{R}))$ is such that $\mu_t^*(dx) = \rho_t^*(x) dx$ is absolutely continuous with respect to Lebesgue measure for all $t \in (0,1)$ and there exists a measurable function u. such that $f_t(x) = u_t(x) + i\pi\rho_t(x)$ is solution (at least in a weak sense) of the complex Burgers equation

$$\partial_t f_t(x) = -\frac{1}{2}\partial_x f_t(x)^2$$

with boundary conditions given by the imaginary part of f at $t = 0$ and $t = 1$.

This result was stated first by Matytsin [146]. Interestingly, the complex Burgers equation also describes limit shapes of plane partitions and dimers, see [127].

The main point to prove Theorem 13.2 is to observe that the evolution of L_N is described, thanks to Itô's calculus 12.2, by an autonomous differential equation. This is the starting point to use the ideas of Kipnis–Olla–Varadhan papers [130, 131]. These papers concern the case where the diffusive term is not vanishing (βN is of order one). The large deviations for the law of the empirical measure of the particles following (12.1) in such a scaling have been studied by Fontbona [91] in the context of McKean–Vlasov diffusion with singular interaction. We shall first recall for the reader the techniques of [130,131] applied to the empirical measures of independent Brownian motions as presented in [130]. We will then describe the necessary changes to adapt this strategy to our setting.

13.1 Large Deviations from the Hydrodynamical Limit for a System of Independent Brownian Particles

Note that the deviations of the law of the empirical measure of independent Brownian motions on path space

$$L_N = \frac{1}{N}\sum_{i=1}^{N}\delta_{B^i_{[0,1]}} \in \mathcal{P}(\mathcal{C}([0,1],\mathbb{R}))$$

are well known by Sanov's theorem which yields (cf. [74, Section 6.2]):

Theorem 13.6. *Let* $\mathcal{W}$ *be the Wiener law. Then, the law* $(L_N)_{\#}\mathcal{W}^{\otimes N}$ *of* L_N *under* $\mathcal{W}^{\otimes N}$ *satisfies a large deviation principle in the scale* N *with rate function given, for* $\mu \in \mathcal{P}(\mathcal{C}([0,1],\mathbb{R}))$, *by* $I(\mu|\mathcal{W})$ *that is infinite if* μ *is not absolutely continuous with respect to Wiener measure and otherwise given by*

$$I(\mu|\mathcal{W}) = \int \log\frac{d\mu}{d\mathcal{W}}\log\frac{d\mu}{d\mathcal{W}}d\mathcal{W}.$$

Thus, if we consider

$$L_N(t) = \frac{1}{N}\sum_{i=1}^{N}\delta_{B^i_t}, \quad t \in [0,1],$$

since $L_N \rightarrow (L_N(t), t \in [0,1])$ is continuous from $\mathcal{P}(\mathcal{C}([0,1], \mathbb{R}))$ into $\mathcal{C}([0,1], \mathcal{P}(\mathbb{R}))$, the law of $(L_N(t), t \in [0,1])$ under $\mathcal{W}^{\otimes N}$ satisfies a large deviation principle by the contraction principle Theorem 20.7. Its rate function is given, for $p \in \mathcal{C}([0,1], \mathcal{P}(\mathbb{R}))$, by

$$S(p) = \inf\{I(\mu|\mathcal{W}) \quad : \quad (x_t)_{\#}\mu = p_t \;\; \forall t \in [0,1]\}.$$

Here, $(x_t)_{\#}\mu$ denotes the law of x_t under μ. It was shown by Föllmer [90] that in fact $S(p)$ is infinite unless there exists $k \in L^2(p_t(dx)dt)$ such that

$$\inf_{f \in \mathcal{C}^{1,1}(\mathbb{R} \times [0,1])} \int_0^1 \int (\partial_x f(x,t) - k(x,t))^2 p_t(dx)dt = 0, \qquad (13.6)$$

and for all $f \in \mathcal{C}^{2,1}(\mathbb{R} \times [0,1])$,

$$\partial_t p_t(f_t) = p_t(\partial_t f_t) + \frac{1}{2}p_t(\partial_x^2 f_t) + p_t(\partial_x f_t k_t).$$

Moreover, we then have

$$S(p) = \frac{1}{2}\int_0^1 p_t(k_t^2)dt. \qquad (13.7)$$

Kipnis and Olla [130] proposed a direct approach to obtain this result based on exponential martingales. Its advantage is to be much more robust and to adapt to many complicated settings encountered in hydrodynamics (cf. [129]). Let us now summarize it. It follows the following scheme:

- *Exponential tightness and study of the rate function S.* Since the rate function S is the contraction of the relative entropy $I(.|\mathcal{W})$, it is clearly a good rate function. This can be proved directly from formula (13.7) as we shall detail it in the context of the eigenvalues of large random matrices. Similarly, we shall not detail here the proof that $L_{N\#}\mathcal{W}^{\otimes N}$ is exponentially tight, a property that reduces the proof of the large deviation principle to the proof of a weak large deviation principle and thus to estimate the probability of deviations into small open balls (cf. Theorem 20.4). We will now concentrate on this last point.
- *Itô's calculus.* Itô's calculus (cf. Theorem 20.18) implies that for any function F in $\mathcal{C}_b^{2,1}(\mathbb{R}^N \times [0,1])$, any $t \in [0,1]$

$$F(B_t^1, \ldots, B_t^N, t) = F(0, \ldots, 0) + \int_0^t \partial_s F(B_s^1, \ldots, B_s^N, s)ds$$

$$+ \sum_{i=1}^N \int_0^t \partial_{x_i} F(B_s^1, \ldots, B_s^N, s)dB_s^i$$

$$+ \frac{1}{2} \sum_{1 \leq i,j \leq N} \int_0^t \partial_{x_i}\partial_{x_j} F(B_s^1, \ldots, B_s^N, s)ds.$$

Moreover, $M_t^F = \sum_{i=1}^N \int_0^t \partial_{x_i} F(B_s^1, \ldots, B_s^N, s) dB_s^i$ is a martingale with respect to the filtration of the Brownian motion, with bracket

$$\langle M^F \rangle_t = \sum_{i=1}^N \int_0^t [\partial_{x_i} F(B_s^1, \ldots, B_s^N, s)]^2 ds.$$

Taking $F(x^1, \ldots, x^N, t) = N^{-1} \sum_{i=1}^N f(B_t^i, t) = \int f(x, t) L_N(t, dx) = \int f_t dL_N(t)$, we deduce that for any $f \in \mathcal{C}_b^{2,1}(\mathbb{R} \times [0,1])$,

$$M_f^N(t) = \int f_t dL_N(t) - \int f_0 dL_N(0) - \int_0^t \int \partial_s f_s dL_{N_s} ds$$
$$- \int_0^t \frac{1}{2} \int \partial_x^2 f_s dL_N(s) ds$$

is a martingale with bracket

$$\langle M_f^N \rangle_t = \frac{1}{N} \int_0^t \int (\partial_x f_s)^2 dL_N(s) ds.$$

The last ingredient of stochastic calculus we want to use is that (cf. Theorem 20.20) for any bounded continuous martingale m_t with bracket $\langle m \rangle_t$, any $\lambda \in \mathbb{R}$,

$$\left\{ \exp(\lambda m_t - \frac{\lambda^2}{2} \langle m \rangle_t), t \in [0,1] \right\}$$

is a martingale. In particular, it has constant expectation. Thus, we deduce that for all $f \in \mathcal{C}_b^{2,1}(\mathbb{R} \times [0,1])$, all $t \in [0,1]$,

$$\mathbb{E}[\exp\{N(M_f^N(t) - \frac{1}{2}\langle M_f^N \rangle_t)\}] = 1. \tag{13.8}$$

- *Weak large deviation upper bound.*
  We equip $\mathcal{C}([0,1], \mathcal{P}(\mathbb{R}))$ with the weak topology on $\mathcal{P}(\mathbb{R})$ and the uniform topology on the time variable. It is then a Polish space. A distance compatible with such a topology is given, for any $\mu, \nu \in \mathcal{C}([0,1], \mathcal{P}(\mathbb{R}))$, by

$$D(\mu, \nu) = \sup_{t \in [0,1]} d(\mu_t, \nu_t)$$

with a distance d on $\mathcal{P}(\mathbb{R})$ compatible with the weak topology such as the Dudley distance (0.1).

Lemma 13.7. *For any* $p \in \mathcal{C}([0,1], \mathcal{P}(\mathbb{R}))$,

$$\limsup_{\delta \to 0} \limsup_{N \to \infty} \frac{1}{N} \log \mathcal{W}^{\otimes N}(D(L_N, p) \leq \delta) \leq -S(p).$$

Proof. Let $p \in \mathcal{C}([0,1], \mathcal{P}(\mathbb{R}))$. Observe first that if $p_0 \neq \delta_0$, since $L_N(0) = \delta_0$ almost surely,

$$\limsup_{\delta \to 0} \limsup_{N \to \infty} \frac{1}{N} \log \mathcal{W}^{\otimes N} \left(\sup_{t \in [0,1]} d(L_N(t), p_t) \leq \delta \right) = -\infty.$$

Therefore, let us assume that $p_0 = \delta_0$. We set

$$B(p, \delta) = \{\mu \in \mathcal{C}([0,1], \mathcal{P}(\mathbb{R})) : D(\mu, p) \leq \delta\}.$$

Let us define, for $f, g \in \mathcal{C}_b^{2,1}(\mathbb{R} \times [0,1])$, $\mu \in \mathcal{C}([0,1], \mathcal{P}(\mathbb{R}))$, $0 \leq t \leq 1$,

$$T^{0,t}(f, \mu) = \mu_t(f_t) - \mu_0(f_0) - \int_0^t \mu_s(\partial_s f_s) ds - \int_0^t \mu_s \left(\frac{1}{2} \partial_x^2 f_s \right) ds$$

and

$$\langle f, g \rangle_\mu^{0,t} := \int_0^t \mu_s(\partial_x f_s \partial_x g_s) ds.$$

Then, by (13.8), for any $t \leq 1$,

$$\mathbb{E} \left[\exp \left\{ N \left(T^{0,t}(f, L_N) - \frac{1}{2} \langle f, f \rangle_{L_N}^{0,t} \right) \right\} \right] = 1.$$

Therefore, if we write for short $T(f, \mu) = T^{0,1}(f, \mu) - \frac{1}{2} \langle f, f \rangle_\mu^{0,1}$,

$$\mathcal{W}^{\otimes N} (D(L_N, p) \leq \delta)$$

$$= \mathcal{W}^{\otimes N} \left(1_{D(L_N, p) \leq \delta} \frac{e^{NT(f, L_N)}}{e^{NT(f, L_N)}} \right)$$

$$\leq \exp\{-N \inf_{B(p, \delta)} T(f, \cdot)\} \mathcal{W}^{\otimes N} \left(1_{D(L_N, p) \leq \delta} e^{NT(f, L_N)} \right)$$

$$\leq \exp\{-N \inf_{B(p, \delta)} T(f, \cdot)\} \mathcal{W}^{\otimes N} \left(e^{NT(f, L_N)} \right) \qquad (13.9)$$

$$= \exp\{-N \inf_{\mu \in B(p, \delta)} T(f, \mu)\}.$$

Since $\mu \to T(f, \mu)$ is continuous when $f \in \mathcal{C}_b^{2,1}(\mathbb{R} \times [0,1])$, we arrive at

$$\limsup_{\delta \to 0} \limsup_{N \to \infty} \frac{1}{N} \log \mathcal{W}^{\otimes N} \left(\sup_{t \in [0,1]} d(L_N(t), p_t) \leq \delta \right) \leq -T(f, p).$$

We now optimize over f to obtain a weak large deviation upper bound with rate function

$$S(p) = \sup_{f \in \mathcal{C}_b^{2,1}(\mathbb{R} \times [0,1])} \left(T^{0,1}(f,p) - \frac{1}{2} \langle f, f \rangle_p^{0,1} \right)$$

$$= \sup_{f \in \mathcal{C}_b^{2,1}(\mathbb{R} \times [0,1])} \sup_{\lambda \in \mathbb{R}} (\lambda T^{0,1}(f,p) - \frac{\lambda^2}{2} \langle f, f \rangle_p^{0,1})$$

$$= \frac{1}{2} \sup_{f \in \mathcal{C}_b^{2,1}(\mathbb{R} \times [0,1])} \frac{T^{0,1}(f,p)^2}{\langle f, f \rangle_p^{0,1}} \tag{13.10}$$

From the last formula, one sees that any p such that $S(p) < \infty$ is such that $f \to T_f(p)$ is a linear map that is continuous with respect to the norm $\|f\|_p^{0,1} = (\langle f, f \rangle_p^{0,1})^{\frac{1}{2}}$. Hence, Riesz's theorem asserts that there exists a function k verifying (13.6, 13.7).

- *Large deviation lower bound.* The derivation of the large deviation upper bound was thus fairly easy. The lower bound is a bit more sophisticated and relies on the proof of the following points:
(a) The solutions to the heat equations with a smooth drift are unique.
(b) The set described by these solutions is dense in $\mathcal{C}([0,1], \mathcal{P}(\mathbb{R}))$.
(c) The entropy behaves continuously with respect to some approximation by elements of this dense set.
We now describe more precisely these ideas. In the previous section (see (13.9)), we have merely obtained the large deviation upper bound from the observation that for all $\nu \in \mathcal{C}([0,1], \mathcal{P}(\mathbb{R}))$, all $\delta > 0$ and any $f \in \mathcal{C}_b^{2,1}([0,1], \mathbb{R})$,

$$\mathbb{E}\left[1_{L_N \in B(\nu, \delta)} \exp\left(N \left(T^{0,1}(L_N, f) - \frac{1}{2} \langle f, f \rangle_{L_N}^{0,1} \right) \right) \right]$$

$$\leq \mathbb{E}\left[\exp\left(N \left(T^{0,1}(L_N, f) - \frac{1}{2} \langle f, f \rangle_{L_N}^{0,1} \right) \right) \right] = 1.$$

To make sure that this upper bound is sharp, we need to check that for any $\nu \in \mathcal{C}([0,1], \mathcal{P}(\mathbb{R}))$ and $\delta > 0$, this inequality is almost an equality for some function $f = k$, i.e., there exists $k \in \mathcal{C}_b^{2,1}([0,1], \mathbb{R})$,

$$\liminf_{N \to \infty} \frac{1}{N} \log \frac{\mathbb{E}\left[1_{L_N \in B(\nu, \delta)} \exp\left(N(T^{0,1}(L_N, k) - \frac{1}{2} \langle k, k \rangle_{L_N}^{0,1}) \right) \right]}{\mathbb{E}\left[\exp\left(N(T^{0,1}(L_N, k) - \frac{1}{2} \langle k, k \rangle_{L_N}^{0,1}) \right) \right]} \geq 0.$$

In other words that we can find a k such that the probability that $L_N(.)$ belongs to a small neighborhood of ν under the shifted probability measure

$$\mathbb{P}^{N,k} = \frac{\exp\left(N(T^{0,1}(L_N, k) - \frac{1}{2} \langle k, k \rangle_{L_N}^{0,1}) \right)}{\mathbb{E}[\exp\left(N(T^{0,1}(L_N, k) - \frac{1}{2} \langle k, k \rangle_{L_N}^{0,1}) \right)]}$$

is not too small. In fact, we shall prove that for good processes ν, we can find k such that this probability goes to one by the following argument. Take $k \in C_b^{2,1}(\mathbb{R} \times [0,1])$. Under the shifted probability measure $\mathbb{P}^{N,k}$, it is not hard to see that $L_N(.)$ is exponentially tight (indeed, for $k \in C_b^{2,1}(\mathbb{R} \times [0,1])$, the density of $\mathbb{P}^{N,k}$ with respect to $\mathbb{P}$ is uniformly bounded by $e^{C(k)N}$ with a finite constant $C(k)$ so that $\mathbb{P}^{N,k} \circ (L_N(.))^{-1}$ is exponentially tight since $\mathbb{P} \circ (L_N(.))^{-1}$ is). As a consequence, $L_N(.)$ is almost surely tight. We let $\mu_.$ be a limit point. Now, by Itô's calculus, for any $f \in C_b^{2,1}(\mathbb{R} \times [0,1])$, any $0 \le t \le 1$,

$$T^{0,t}(L_N, f) = \int_0^t \int \partial_x f_u(x) \partial_x k_u(x) dL_N(u)(x) du + M_t^N(f)$$

with a martingale $(M_t^N(f), t \in [0,1])$ with bracket

$$\left(N^{-1} \int_0^t \int (\partial_x f(x))^2 dL_N(s)(x) ds, t \in [0,1] \right).$$

Since the bracket of $M_t^N(f)$ goes to zero, the martingale $(M_t^N(f), t \in [0,1])$ goes to zero uniformly almost surely. Hence, any limit point $\mu_.$ of $L_N(.)$ under $\mathbb{P}^{N,k}$ must satisfy

$$T^{0,1}(\mu, f) = \int_0^1 \int \partial_x f_u(x) \partial_x k_u(x) d\mu_u(x) du \qquad (13.11)$$

for any $f \in C_b^{2,1}(\mathbb{R} \times [0,1])$.

When (μ, k) satisfies (13.11) for all $f \in C_b^{2,1}(\mathbb{R} \times [0,1])$, we say that k is the field associated with μ.

Therefore, if we can prove that there exists a unique solution ν to (13.11), we see that $L_N(.)$ converges almost surely under $\mathbb{P}^{N,k}$ to this solution. This proves the lower bound at any measure-valued path $\nu_.$ that is the unique solution of (13.11), namely for any $k \in C_b^{2,1}(\mathbb{R} \times [0,1])$ such that there exists a unique solution ν_k to (13.11),

$$\liminf_{\delta \to 0} \liminf_{N \to \infty} \frac{1}{N} \log \mathcal{W}^{\otimes N} \left(\sup_{t \in [0,1]} d(L_N(t), \nu_k) < \delta \right)$$

$$= \liminf_{\delta \to 0} \liminf_{N \to \infty} \frac{1}{N} \log \mathbb{P}^{N,k} \left(1_{\sup_{t \in [0,1]} d(L_N(t), \nu_k) < \delta} e^{-NT(k, L_N)} \right)$$

$$\ge -T(k, \nu_k) + \liminf_{\delta \to 0} \liminf_{N \to \infty} \frac{1}{N} \log \mathbb{P}^{N,k} \left(\sup_{t \in [0,1]} d(L_N(t), \nu_k) < \delta \right)$$

$$\ge -S(\nu_k). \qquad (13.12)$$

where we used in the second line the continuity of $\mu \to T(\mu, k)$ due to our assumption that $k \in C_b^{2,1}(\mathbb{R} \times [0,1])$ and the fact that

$$\mathbb{P}^{N,k}\left(\sup_{t \in [0,1]} d(L_N(t), \nu_k) < \delta\right)$$

goes to one in the third line. Hence, the question boils down to uniqueness of the weak solutions of the heat equation with a drift. This problem is not too difficult to solve here and one can see that for instance for fields k that are analytic within a neighborhood of the real line, there is at most one solution to this equation. To generalize (13.12) to any $\nu \in \{S < \infty\}$, it is not hard to see that it is enough to find, for any such ν, a sequence ν_{k_n} for which (13.12) holds and such that

$$\lim_{n \to \infty} \nu_{k_n} = \nu, \quad \lim_{n \to \infty} S(\nu_{k_n}) = S(\nu). \tag{13.13}$$

Now, observe that S is a convex function so that for any probability measure p_ϵ,

$$S(\mu * p_\epsilon) \leq \int S((.-x)_\# \mu)p_\epsilon(dx) = S(\mu) \tag{13.14}$$

where in the last inequality we neglected the condition at the initial time to say that $S((.-x)_\# \mu) = S(\mu)$ for all x. Hence, since S is also lower semicontinuous, one sees that $S(\mu * p_\epsilon)$ will converge to $S(\mu)$ for any μ with finite entropy S. Performing also a regularization with respect to time and taking care of the initial conditions allows us to construct a sequence ν_n with analytic fields satisfying (13.13). This point is quite technical but still manageable in this context. Since it will be done quite explicitly in the case we are interested in, we shall not detail it here.

13.2 Large Deviations for the Law of the Spectral Measure of a Non-centered Large Dimensional Matrix-valued Brownian Motion

To prove a large deviation principle for the law of the spectral measure of Hermitian Brownian motions, the first natural idea would be, following (12.1), to prove a large deviation principle for the law of the spectral measure of $\tilde{L}^N : t \to N^{-1}\sum_{i=1}^N \delta_{\sqrt{N}^{-1}B^i(t)}$, to use Girsanov's theorem to show that the law we are considering is absolutely continuous with respect to the law of independent Brownian motions, with a density that only depends on $\tilde{L}^N$ and conclude by Laplace's method (cf. Theorem 20.8). However, this approach presents difficulties due to the singularity of the interacting potential, and thus of the density. Here, the techniques developed in [130] will, however, be very efficient because they only rely on smooth functions of the empirical measure since the empirical measures are taken as distributions so that the interacting potential is smoothed by the test functions. (Note, however, that this strategy would not have worked with more singular potentials.) According to (12.1), we can in fact follow the very same approach.

Itô's Calculus

With the notations of (13.1) and (13.2), we have by Section 12.2:

Theorem 13.8. *For all $\beta \geq 1$, for any $N \in \mathbb{N}$, any $f \in C_b^{2,1}(\mathbb{R} \times [0,1])$ and any $s \in [0,1)$, $\left(S^{s,t}(L_{N,\beta}, f) + \frac{\beta/2-1}{2N} \int_s^t \int \partial_x^2 f(y,s) dL_{N,\beta}(s)(y) ds, s \leq t \leq 1 \right)$ is a bounded martingale with quadratic variation*

$$\langle S^{s,\cdot}(L_{N,\beta}, f) \rangle_t = \frac{2}{\beta N^2} \langle f, f \rangle_{L_{N,\beta}}^{s,t}.$$

Remark 5. Observe that if the entries were not Brownian motions but diffusions described for instance as solution of a stochastic differential equation

$$dx_t = dB_t + U(x_t)dt,$$

then the evolution of the spectral measure of the matrix would no longer be autonomous. In fact, our strategy is strongly based on the fact that the variations of the spectral measure under small changes of time only depends on the spectral measure, allowing us to construct exponential martingales that are functions of the process of the spectral measure only. It is easy to see that if the entries of the matrix are not Gaussian, the variations of the spectral measures will depend on much more general functions of the entries than those of the spectral measure.

However, this strategy can also be used to study the spectral measure of other Gaussian matrices as emphasized in [55, 101].

From now on, we shall consider the case where $\beta = 2$ and drop the subscript β in $\mathbf{H}^{N,\beta}$, $L_{N,\beta}$, etc. This case is slightly easier to write down since there are no error terms in Itô's formula, but everything extends readily to the cases $\beta \geq 1$. One also needs to notice that

$$S_\beta(\mu) = \sup_{\substack{f \in C_\beta^{2,1}(\mathbb{R} \times [0,1]) \\ 0 \leq s \leq t \leq 1}} \left\{ S^{s,t}(\mu, f) - \frac{1}{\beta} \langle f, f \rangle_\mu^{s,t} \right\} = \frac{\beta}{2} S_2(\mu)$$

where the last equality is obtained by changing f into $2^{-1}\beta f$.

Large Deviation Upper Bound

From the previous Itô's formula, one can deduce by following the ideas of [131] (see Section 13.1) a large deviation upper bound for the measure-valued process $L_N(.) \in \mathcal{C}([0,1], \mathcal{P}(\mathbb{R})))$. To this end, we shall make the following assumption on the initial condition D_N:

$$(\mathrm{H}) \qquad C_D := \sup_{N \in \mathbb{N}} L_{D_N}(\log(1 + |x|^2)) < \infty,$$

implying that $(L_{D_N}, N \in \mathbb{N})$ is tight. Moreover, L_{D_N} converges weakly, as N goes to infinity, to a probability measure μ_D.
Then, we shall prove, with the notations of (13.1)–(13.3), the following:

Theorem 13.9. Assume (H). Then:
(1) S_{μ_D} is a good rate function on $\mathcal{C}([0,1], \mathcal{P}(\mathbb{R}))$.
(2) For any closed set F of $\mathcal{C}([0,1], \mathcal{P}(\mathbb{R}))$,

$$\limsup_{N \to \infty} \frac{1}{N^2} \log \mathbb{P}\left(L_N(.) \in F\right) \leq -\inf_{\nu \in F} S_{\mu_D}(\nu).$$

We first prove that S_{μ_D} is a good rate function. Then, we show that exponential tightness holds and finally obtain a weak large deviation upper bound, these two arguments yielding (2) (cf. Theorem 20.4).
    (a) Let us first observe that $S_{\mu_D}(\nu)$ is also given, when $\nu_0 = \mu_D$, by

$$S_{\mu_D}(\nu) = \frac{1}{2} \sup_{f \in \mathcal{C}_b^{2,1}(\mathbb{R} \times [0,1])} \sup_{0 \leq s \leq t \leq 1} \frac{S^{s,t}(\nu, f)^2}{\langle f, f \rangle_\nu^{s,t}}. \tag{13.15}$$

Consequently, S_{μ_D} is non-negative. Moreover, S_{μ_D} is obviously lower semi-continuous as a supremum of continuous functions.
    Hence, we merely need to check that its level sets are contained in relatively compact sets. By (12.11), it is enough to show that, for any $M > 0$,
    (1) For any integer m, there is a positive real number L_m^M so that for any $\nu \in \{S_{\mu_D} \leq M\}$,

$$\sup_{0 \leq s \leq 1} \nu_s(|x| \geq L_m^M) \leq \frac{1}{m}, \tag{13.16}$$

proving that $\nu_s \in K_{\frac{1}{m}, L_m^M}$ defined in (12.12) for all $s \in [0,1]$.
    (2) For any integer m and $f \in \mathcal{C}_b^2(\mathbb{R})$, there exists a positive real number δ_m^M so that for any $\nu \in \{S_{\mu_D} \leq M\}$,

$$\sup_{|t-s| \leq \delta_m^M} |\nu_t(f) - \nu_s(f)| \leq \frac{1}{m}, \tag{13.17}$$

showing that $s \to \nu_s(f)$ belongs to the compact set $C_{\delta^M, \|f\|_\infty}$ as defined in (12.13).

To prove (13.16), we consider, for $\delta > 0$, $f_\delta(x) = \log\left(x^2(1 + \delta x^2)^{-1} + 1\right) \in \mathcal{C}_b^{2,1}(\mathbb{R} \times [0,1])$. We observe that

$$C := \sup_{0 < \delta \leq 1} ||\partial_x f_\delta||_\infty + \sup_{0 < \delta \leq 1} ||\partial_x^2 f_\delta||_\infty$$

is finite and, for $\delta \in (0,1]$,

$$\left| \frac{\partial_x f_\delta(x) - \partial_x f_\delta(y)}{x - y} \right| \leq C.$$

Hence, (13.15) implies, by taking $f = f_\delta$ in the supremum, that for any $\delta \in (0,1]$, any $t \in [0,1]$, any $\mu. \in \{S_{\mu_D} \leq M\}$,

$$\mu_t(f_\delta) \leq \mu_0(f_\delta) + 2Ct + 2C\sqrt{Mt}.$$

Consequently, we deduce by the monotone convergence theorem and letting δ decrease to zero that for any $\mu. \in \{S_{\mu_D} \leq M\}$,

$$\sup_{t \in [0,1]} \mu_t(\log(x^2 + 1)) \leq \mu_D(\log(x^2 + 1)) + 2C(1 + \sqrt{M}).$$

Chebycheff's inequality and hypothesis (H) thus imply that for any $\mu. \in \{S_{\mu_D} \leq M\}$ and any $K \in \mathbb{R}^+$,

$$\sup_{t \in [0,1]} \mu_t(|x| \geq K) \leq \frac{C_D + 2C(1 + \sqrt{M})}{\log(K^2 + 1)}$$

which finishes the proof of (13.16).

The proof of (13.17) again relies on (13.15), which implies that for any $f \in \mathcal{C}_b^2(\mathbb{R})$, any $\mu. \in \{S_{\mu_D} \leq M\}$ and any $0 \leq s \leq t \leq 1$,

$$|\mu_t(f) - \mu_s(f)| \leq ||\partial_x^2 f||_\infty |t - s| + 2||\partial_x f||_\infty \sqrt{M}\sqrt{|t - s|}. \tag{13.18}$$

(b) Exponential tightness. By Lemma 12.6, we have:

Lemma 13.10. *For any integer number L, there exists a finite integer number $N_0 \in \mathbb{N}$ and a compact set $\mathcal{K}_L$ in $\mathcal{C}([0,1], \mathcal{P}(\mathbb{R}))$ such that $\forall N \geq N_0$,*

$$\mathbb{P}(L_N \in \mathcal{K}_L^c) \leq \exp\{-LN^2\}.$$

(c) Weak large deviation upper bound. Following the arguments of Section 13.1, we readily get:

Lemma 13.11. *For every process ν in $\mathcal{C}([0,1], \mathcal{P}(\mathbb{R}))$, if $B_\delta(\nu)$ denotes the open ball with center ν and radius δ for the distance D, then*

$$\lim_{\delta \to 0} \limsup_{N \to \infty} \frac{1}{N^2} \log \mathbb{P}(L_N \in B_\delta(\nu)) \leq -S_{\mu_D}(\nu).$$

Moreover, processes with finite entropy are characterized as follows.

Lemma 13.12. *For any $\mu \in \{S_{\mu_D} < \infty\}$, there exists a measurable function k such that for any $f \in \mathcal{C}_b^2([0,1] \times \mathbb{R}, \mathbb{R})$*

$$S^{s,t}(\mu, f) = \int_s^t \int k_u(x) \partial_x f(x,u) d\mu_u(x) du. \tag{13.19}$$

Moreover,

$$S_{\mu_D}(\mu.) = \frac{1}{2} \int_0^1 \int k_u(x)^2 d\mu_u(x) du.$$

Proof. By (13.15), if $S_{\mu_D}(\mu.) < \infty$, for any $f \in \mathcal{C}_b^2([0,1] \times \mathbb{R}, \mathbb{R})$, all $s \leq t$

$$\left| \int_s^t \int k_u(x) \partial_x f(x,u) d\mu_u(x) du \right| \leq 2 S_{\mu_D}(\mu.)^{\frac{1}{2}} \left(\int_0^1 \int f(x,u)^2 d\mu_u(x) du \right)^{\frac{1}{2}}.$$

Hence, $f \to \int_0^1 \int k_u(x) \partial_x f(x,u) d\mu_u(x)$ is linear, bounded in the Hilbert space obtained by completing and separating $\mathcal{C}_b^2([0,1] \times \mathbb{R}, \mathbb{R})$ for the norm

$$\|f\|_2 = \left(\int_0^1 \int f(x,u)^2 d\mu_u(x) du \right)^{\frac{1}{2}}.$$

Riesz's theorem (cf. [172]) allows us to conclude for $s = 0, t = 1$. We get the general case by taking $f = 0$ outside $[s - \epsilon, t + \epsilon]$ and a smooth interpolation in the time variable in $[s - \epsilon, t + \epsilon] \backslash [s, t]$.   □

Large Deviation Lower Bound

We shall prove at the end of this section the following:

Lemma 13.13. *Let*

$$\mathcal{MF}_\infty = \left\{ h \in \mathcal{C}_b^\infty(\mathbb{R} \times [0,1]); \exists \epsilon > 0, C. \in L^2([0,1], dt) \text{ so that} \right.$$

$$h_t(x) = C_t + \int e^{i\lambda x} \hat{h}_t(\lambda) d\lambda \text{ with } \int_0^1 \max_{\lambda \in \mathbb{R}} (e^{2\epsilon|\lambda|} |\hat{h}_t(\lambda)|^2) dt < \infty. \right\}$$

For any field k in $\mathcal{MF}_\infty$, there exists a unique solution ν_k to

$$S^{s,t}(f, \nu) = \langle f, k \rangle_\nu^{s,t} \tag{13.20}$$

for any $f \in \mathcal{C}_b^{2,1}(\mathbb{R} \times [0,1])$. We set $\mathcal{MC}([0,1], \mathcal{P}(\mathbb{R}))$ to be the subset of $\mathcal{C}([0,1], \mathcal{P}(\mathbb{R}))$ consisting of such solutions.

Note that h belongs to $\mathcal{MF}_\infty$ implies that it can be extended analytically to $\{z : |\Im(z)| < \epsilon\}$ for almost all $t \in [0, 1]$.

As a consequence of Lemma 13.13, if we take $\nu \in \mathcal{MC}([0, 1], \mathcal{P}(\mathbb{R}))$ associated with a field k, and if we define

$$\mathbb{P}^{N,k} = \exp\left\{N^2\left(S^{0,1}(L_N(.), k) - \frac{1}{2}\langle k, k\rangle_{L_N(.)}^{0,1}\right)\right\}\mathbb{P}^N_{T,\lambda_N(0)},$$

the limit points of $L_N(.)$ under $\mathbb{P}^{N,k}$ coincide with ν (see the classical analog (13.11)). Thus, for any open subset $O \in \mathcal{C}([0, 1], \mathcal{P}(\mathbb{R}))$, any $\nu \in O \cap \mathcal{MC}([0, 1], \mathcal{P}(\mathbb{R}))$, for δ small enough,

$$\mathbb{P}(L_N(.) \in O) \geq \mathbb{P}(D(L_N(.), \nu) < \delta)$$

$$= \mathbb{P}^{N,k}\left(1_{D(L_N(.),\nu)<\delta}e^{-N^2(S^{0,1}(L_N,k)-\frac{1}{2}\langle k,k\rangle_{L_N}^{0,1})}\right)$$

$$\geq e^{-N^2(S^{0,1}(\nu,k)-\frac{1}{2}\langle k,k\rangle_\nu^{0,1})-g(\delta)N^2}\mathbb{P}^{N,k}(D(L_N(.),\nu) < \delta)$$

with a function g vanishing at the origin. Hence, for any $\nu \in O \cap \mathcal{MC}([0, 1], \mathcal{P}(\mathbb{R}))$

$$\liminf_{N\to\infty}\frac{1}{N^2}\log\mathbb{P}(L_N(.) \in O) \geq -\left(S^{0,1}(\nu, k) - \frac{1}{2}\langle k, k\rangle_\nu^{0,1}\right) = -S_{\mu_D}(\nu)$$

and therefore

$$\liminf_{N\to\infty}\frac{1}{N^2}\log\mathbb{P}(L_N(.) \in O) \geq -\inf_{O\cap\mathcal{MC}([0,1],\mathcal{P}(\mathbb{R}))}S_{\mu_D}. \tag{13.21}$$

To complete the lower bound, it is therefore sufficient to prove that for any $\nu \in \mathcal{C}([0, 1], \mathcal{P}(\mathbb{R}))$ with uniformly bounded fourth moment, there exists a sequence $\nu^n \in \mathcal{MC}([0, 1], \mathcal{P}(\mathbb{R}))$ such that

$$\lim_{n\to\infty}\nu^n = \nu \text{ and } \lim_{n\to\infty}S_{\mu_D}(\nu^n) = S_{\mu_D}(\nu). \tag{13.22}$$

The rate function S_{μ_D} is not convex a priori since it is the supremum of **quadratic** functions of the measure-valued path ν so that there is no reason why it should be reduced by standard convolution as in the classical setting (cf. Section 13.1). Thus, it is now unclear how we can construct the sequence ν^n satisfying (13.22). Further, we begin with a degenerate rate function that is infinite unless $\nu_0 = \mu_D$.

To overcome the lack of convexity, we shall remember the origin of the problem; in fact, we have been considering the spectral measure of matrices and should not forget the special features of operators due to the matrices structure. By definition, the differential equation satisfied by a Hermitian Brownian motion should be invariant if we translate the entries, that is, translate the Hermitian Brownian motion by a self-adjoint matrix. The natural limiting framework of large random matrices is free probability, and the limiting

spectral measure of the sum of a Hermitian Brownian motion and a deterministic self-adjoint matrix converges to the free convolution of their respective
limiting spectral measure. Intuitively, we shall therefore expect (and in fact
we will show in the specific case of Cauchy laws) that the rate function $S^{0,1}$
decreases by free convolution, generalizing the fact that standard convolution
was decreasing the Brownian motion rate function (cf. (13.14)). However, because free convolution by a Cauchy law is equal to the standard convolution
by a Cauchy law, we shall regularize our laws by convolution by Cauchy laws.
We now prove the large deviation lower bound of Theorem 13.2.

- *Regularization by Cauchy laws.* We prove below that convolution by
Cauchy laws reduces the entropy, a point analogous to (13.14). This result is
a special case of Theorem 4.1 in [56] where it is shown that any free convolution reduces the entropy. This generalization was in fact used with the free
convolution with respect to the semi-circular in [110] to prove Theorem 13.3.

Lemma 13.14. *Let $\mu.$ satisfy (13.19) for some measurable function $k \in L^2(d\mu_t(x)dt)$. Let p_ϵ be the Cauchy law*

$$dp_\epsilon(c) = \frac{\epsilon}{\pi(\epsilon^2 + c^2)}dc.$$

Then, $(\mu_t * p_\epsilon)_{t \in [0,1]}$ satisfies (13.19) with field k^ϵ given by

$$k_t^\epsilon(c) = \frac{\int \frac{k_t(x)}{\epsilon^2 + (x-c)^2}d\mu_t(x)}{\int \frac{1}{\epsilon^2 + (x-c)^2}d\mu_t(x)}.$$

Moreover

$$S_{\mu_D}(\mu.) \geq S_{\mu_D * p_\epsilon}(\mu. * p_\epsilon).$$

Proof. We verify (13.19) for Stieltjes transform, i.e., functions of the form
$f(x) = (z - x)^{-1}$ and any $[s,t]$. Note that if we do that, we can then use
the density of these functions in $C_b^2(\mathbb{R}, \mathbb{R})$ (again based on the Weierstrass
theorem) and then play on the parameters s, t to get the equation for all
$f \in C_b^2([0, 1] \times \mathbb{R}, \mathbb{R})$. Now, for $f = (z - x)^{-1}$ we find that

$$p_\epsilon * \mu_t((z - x)^{-1})$$
$$= \int \mu_t((z - c - x)^{-1})dp_\epsilon(c)$$
$$= p_\epsilon * \mu_s((z - x)^{-1})$$
$$+ \int \int_s^t \mu_u((z - c - x)^{-1})\mu_u((z - c - x)^{-2})dudp_\epsilon(c)$$
$$+ \int \int_s^t \mu_u((z - c - x)^{-2}k_u(x))dudp_\epsilon(c)$$

where we have used Fubini and equation (13.19) for $\mu_.$. Observe that if $z \in \mathbb{C}^+ = \{z : \Im(z) > 0\}$, as $c \to (z-c-x)^{-1}(z-c-x')^{-2}$ is analytic on $\mathbb{C}^-$, the residue theorem implies that, since $dp_\epsilon(c)/dc$ has a unique pole at $-i\epsilon$ in $\mathbb{C}^-$,

$$\int (z-c-x)^{-1}(z-c-x')^{-2} dp_\epsilon(c)$$

$$= (z+i\epsilon-x)^{-1}(z+i\epsilon-x')^{-2}$$

$$= \int (z-c-x)^{-1} dp_\epsilon(c) \int (z-c-x')^{-2} dp_\epsilon(c)$$

and therefore we deduce

$$p_\epsilon * \mu_t((z-x)^{-1}) = p_\epsilon * \mu_s((z-x)^{-1})$$

$$+ \int_s^t p_\epsilon * \mu_u((z-x)^{-1}) p_\epsilon * \mu_u((z-x)^{-2}) du$$

$$+ \int_s^t p_\epsilon * \mu_u((z-x)^{-2} k_u^\epsilon(x)) du$$

where we finally used

$$\int \mu_u((z-c-x)^{-2} k_u(x)) dp_\epsilon(c)$$

$$= \int (z-c)^{-2} \left(\int \frac{\epsilon k_u(x)}{\pi((x-c)^2+\epsilon^2)} d\mu_u(x) \right) dc$$

$$= \int (z-c)^{-2} k_u^\epsilon(c) dp_\epsilon * \mu_u(c).$$

This is (13.20) with the dense set of functions $f = (z-x)^{-1}$. For the last point notice that in fact

$$k_u^\epsilon(x) = E[k_u(x)|x + C_\epsilon]$$

when C_ϵ is a Cauchy variable independent of x and x has law μ_u. Hence,

$$S_{\mu_D * p_\epsilon}(\mu_. * p_\epsilon) = \frac{1}{2} \int_0^1 \int E[(E[k_u(x_u)|x_u + C_\epsilon])^2] du$$

$$\leq \frac{1}{2} \int_0^1 \int E[k_u(x_u)^2] du = S_{\mu_D}(\mu_.).$$

$\square$

Thus, we find that convolution by Cauchy laws $(p_\epsilon)_{\epsilon>0}$ decreases the entropy. Since the entropy is lower semicontinuous (as a supremum of continuous functions), we deduce that

$$\lim_{\epsilon \to 0} S_{\mu_D * p_\epsilon}(\mu_. * p_\epsilon) = S_{\mu_D}(\mu_.). \tag{13.23}$$

Note that $x \to k_u^\epsilon(x)$ is analytic in the strip $|\Im(z)| \leq \epsilon$ so that there is some good chance that $p_\epsilon * \mu$. will be in $\mathcal{MC}([0,1], \mathcal{P}(\mathbb{R}))$. We shall see this point below for $\mu. \in \mathcal{A}$ with

$$\mathcal{A} = \{\mu. : [0,1] \to \mathcal{P}(\mathbb{R}) : \sup_{t \in [0,1]} \mu_t(x^4) < \infty\}.$$

Namely, we prove:

Lemma 13.15. *For* $\mu. \in \mathcal{A} \cap \{S_{\mu_D} < \infty\}$, *for all* $\epsilon > 0$, $p_\epsilon * \mu$. *belongs to* $\mathcal{MC}([0,1], \mathcal{P}(\mathbb{R}))$.

Proof. The strategy is to show that $k_t^\epsilon \in L^2(dx) \cap L^1(dx)$ in order to use Plancherel representation. Since $k_t^\epsilon(x)$ goes to $\int k_t(y) d\mu_t(y)$ as x goes to infinity, we need to substract this quantity to make sure this can happen. For further use, we need to consider $k_t^\epsilon(x + i\delta)$ for $\delta < \epsilon$, say $\delta = \epsilon/2$. We then write

$$k_t^\epsilon(x + i\delta) = \frac{\int \frac{k_t(y)}{(x+i\delta-y)^2+\epsilon^2} d\mu_t(y)}{\int \frac{1}{(x+i\delta-y)^2+\epsilon^2} d\mu_t(y)}$$

Note that since $\mu_t(x^2)$ is uniformly bounded by say C, $\mu_t([-M,M]^c) \leq CM^{-2}$ for all t by Chebyshev inequality. Thus, for $x \in [-M, M]$, with M sufficiently large

$$\left| \int \frac{1}{(x + i\delta - y)^2 + \epsilon^2} d\mu_t(y) \right| \geq \frac{1}{(2M + \delta)^2 + \epsilon^2}$$

and therefore there exists a finite constant $C = C(\epsilon, M)$ so that

$$|k_t^\epsilon(x + i\delta)| \leq C \left(\int k_t(y)^2 d\mu_t(y) \right)^{\frac{1}{2}}.$$

We choose below M large. For $x \in [-M, M]^c$, we have that

$$\int_{[-2|x|,2|x|]} \frac{((x + i\delta)^2 + \epsilon^2) k_t(y)}{(x + i\delta - y)^2 + \epsilon^2} d\mu_t(y)$$

$$= \int_{[-2|x|,2|x|]} k_t(y) \left(1 + 2\frac{(x + i\delta)y}{(x + i\delta)^2 + \epsilon^2} \right.$$

$$\left. + O\left(\frac{y^2}{(x + i\delta)^2 + \epsilon^2} \right) \right) d\mu_t(y)$$

$$\left| \int_{[-2|x|,2|x|]^c} \frac{((x + i\delta)^2 + \epsilon^2) k_t(y)}{(x + i\delta - y)^2 + \epsilon^2} d\mu_t(y) \right|$$

$$\leq \frac{1}{\epsilon^2 - \delta^2} \int_{[-2|x|,2|x|]^c} |k_t(y)| d\mu_t(y)$$

$$\leq \frac{1}{\epsilon^2 - \delta^2} \left(\int k_t(y)^2 d\mu_t(y) \right)^{\frac{1}{2}} \left(\int \frac{y^4}{x^4} d\mu_t(y) \right)^{\frac{1}{2}}.$$

Letting $C_t^1(k) = \int k_t(y)d\mu_t(y)$, $C_t^2(k) = \int y k_t(y)d\mu_t(y)$ and $C_t^3(k) = (\int k_t(y)^2 d\mu_t(y))^{\frac{1}{2}}$, we get

$$\int \frac{k_t(y)}{(x + i\delta - y)^2 + \epsilon^2} d\mu_t(y)$$
$$= C_t^1(k) + (2C_t^2(k))(x + i\delta)((x + i\delta)^2 + \epsilon^2)^{-1}) + h_t^\epsilon(x + i\delta)$$

with some function h_t^ϵ so that

$$|h_t^\epsilon(x + i\delta)| \le cC_t^3(k)\frac{1}{|x|^2 + 1}$$

with a constant c that only depends on $\sup_t \mu_t(y^4)$. Doing the same for the denominator, we conclude that

$$k_t^\epsilon(x+i\delta) = C_t^1 + 2[C_t^2(k) - C_t^1(k)C_t^2(1)](x + i\delta)((x + i\delta)^2 + \epsilon^2)^{-1} + \bar{k}_t^\epsilon(x + i\delta) \tag{13.24}$$

with some bounded function $\bar{k}_t^\epsilon(x + i\delta)$ such that

$$|\bar{k}_t^\epsilon(x + i\delta)| \le cC_t^3(k)\frac{x^2}{(x^2 - \delta^2 + \epsilon^2)^2}.$$

Applying this result to $\delta = 0$, we see that $\bar{k}_t^\epsilon(x) \in L^1(dx) \cap L^2(dx)$ extends to $|\Im(x)| \le \delta$ analytically while staying in L^1. This implies in particular that for $\lambda > 0$

$$|\hat{k}_t^\epsilon(\lambda)| = \pi^{-1}\left|\int e^{i\lambda x}\bar{k}_t^\epsilon(x)dx\right| = \pi^{-1}\left|\int e^{i\lambda(x+i\delta)}\bar{k}_t^\epsilon(x + i\delta)dx\right|$$
$$\le e^{-\delta\lambda}\int |\bar{k}_t^\epsilon(x + i\delta)|dx \le c'C_t^3(k)e^{-\delta\lambda}$$

and the same bound for $\lambda < 0$. Hence, by the Plancherel formula

$$\bar{k}_t^\epsilon(x) = \int e^{i\lambda x}\hat{k}_t^\epsilon(\lambda)d\lambda,$$

with $\hat{k}_t^\epsilon(\lambda)$ satisfying the required property that

$$|\hat{k}_t^\epsilon(\lambda)| \le c'C_t^3(k)e^{-\delta|\lambda|}$$

with $C_t^3(k)$ in $L^2([0, 1], dt)$ since μ has finite entropy. Moreover, Note that

$$\frac{x}{x^2 + \epsilon^2} = \Re(x + i\epsilon)^{-1}$$

can be written for $\epsilon > 0$ as

$$\frac{x}{x^2 + \epsilon^2} = \Re\left(-i\int_0^\infty e^{i\xi(x+i\epsilon)}d\xi\right) = \int e^{i\xi x}\left(\sin(\xi)e^{-\epsilon|\xi|}\right)d\xi. \tag{13.25}$$

Hence, we have written by (13.24)

$$k_t^\epsilon(x + i\delta) = C_t^1 + \int e^{i\lambda x} \hat{k}_t^{\epsilon,2}(\lambda) d\lambda \tag{13.26}$$

with

$$|\hat{k}_t^{\epsilon,2}(\lambda) e^{\delta|\lambda|}| \leq 2[C_t^2(k) + C_t^1(k)C_t^2(1)] + c' C_t^3(k)$$

for all $\lambda \in \mathbb{R}$. The proof of the lemma is thus complete since $C_t^2(k)$, $C_t^2(1)$, $C_t^3(k)$ are in $L^2([0,1], dt)$, whereas $C_t^2(1) = \int y d\mu_t(x)$ is uniformly bounded as $\int y^2 d\mu_t(y)$ is.   □

• *Large deviations lower bound for processes regularized by Cauchy distribution.* Everything looks nice except that we modified the initial condition from μ_D into $\mu_D * p_\epsilon$, so that in fact $S_{\mu_D}(\nu^{\epsilon,\Delta}) = +\infty$! and moreover, the empirical measure-valued process cannot deviate toward processes of the form $\nu^{\epsilon,\Delta}$ even after some time because these processes do not have a finite second moment (it can indeed be checked that if $\mu_D(x^2) < \infty$, $S_{\mu_D}(\mu.) < \infty$ implies that $\sup_t \mu_t(x^2) < \infty$). To overcome this problem, we first note that this result will still give us a large deviation lower bound if we change the initial data of our matrices. Namely, let, for $\epsilon > 0$, C_ϵ^N be an $N \times N$ diagonal matrix with spectral measure converging to the Cauchy law p_ϵ and consider the matrix-valued process

$$\mathbf{X}_t^{N,\epsilon} = \mathbf{U}_N C_\epsilon^N \mathbf{U}_N^* + D_N + \mathbf{H}^N(t)$$

with $\mathbf{U}_N$ a $N \times N$ unitary measure following the Haar measure m_2^N on $U(N)$. Then, it is well known [30] that the spectral distribution of $\mathbf{U}_N C_\epsilon^N \mathbf{U}_N^* + D_N$ converges to $= p_\epsilon * \mu_D$. We choose C_ϵ^N satisfying (H).

Hence, we can proceed as before to obtain the following large deviation estimates on the law of the spectral measure $L_{N,\epsilon}(t)t = L_{X_t^{N,\epsilon}}$. We define

$$\mathcal{A}^p = \{\mu \in \mathcal{C}([0,1], \mathcal{P}(\mathbb{R})); \mu_t \in \mathcal{A} \quad \forall t \in [0,1]\}$$

Corollary 13.16. *Assume (H). For any $\epsilon > 0$, for any closed subset F of $\mathcal{C}([0,1], \mathcal{P}(\mathbb{R}))$,*

$$\text{*} \quad \limsup_{N \to \infty} \frac{1}{N^2} \log \mathbb{P}\left(L_{N,\epsilon}(.) \in F\right) \leq -\inf\{S_{P_\epsilon * \mu_D}(\nu), \nu \in F\}.$$

Further, for any open set O of $\mathcal{C}([0,1], \mathcal{P}(\mathbb{R}))$,

$$\liminf_{N \to \infty} \frac{1}{N^2} \log \mathbb{P}\left(L_{N,\epsilon}(.) \in O\right)$$
$$\geq -\inf\{S_{P_\epsilon * \mu_D}(\nu), \nu \in O, \nu = P_\epsilon * \mu, \mu \in \mathcal{A}^p \cap \{S_{\mu_D} < \infty\}\}.$$

The only point is to prove the lower bound (see Theorem 13.11 for the upper bound). In [109] we proceed by an extra regularization in time. We will bypass this argument here.

1. *Time discretization of $k^\epsilon \in \mathcal{MF}_\infty$.* We note that since $\int_0^1 \int k_t^\epsilon(x)^2 dp_\epsilon * d\mu_t(x)dt$ is finite, we can approximate $k_t^\epsilon(x)$ by

$$k_t^{\epsilon,p}(x) = k_{t_n}^\epsilon(x) + \frac{t - t_n}{t_{n+1} - t_n}(k_{t_{n+1}}^\epsilon(x) - k_{t_n}^\epsilon(x)) \quad \text{for } t \in [t_n, t_{n+1}[$$

for some time discretization $0 = t_0 < t_1 < \cdots t_p = 1$ in such a way that

$$\lim_{p \to \infty} \int_0^1 \int (k_t^\epsilon(x) - k_t^{\epsilon,p}(x))^2 \, dp_\epsilon * \mu_t(x)dt = 0.$$

Indeed, this is clearly true if $t \to k_t^\epsilon(x)$ is continuous (since $k_t^\epsilon(x)$ is uniformly bounded, see Lemma 13.15 so·that bounded convergence theorem applies) and then generalizes to all uniformly bounded field by density of continuous functions in L^2.

2. *Change of measure and convergence under the shifted probability measure.* We let $\mathbb{P}^{N,k,\epsilon}$ be the law with density

$$\Lambda^{N,k} = \exp\left\{\sqrt{N} \sum_{i=1}^N \int_0^T k_t^\epsilon(\lambda_N^i(t))dW_t^i - \frac{N}{2} \sum_{i=1}^N \int_0^T k_t^\epsilon(\lambda_N^i(t))^2 dt\right\}$$

with respect to the law $P_{T,\lambda_N^\epsilon(0)}^N$ of the eigenvalues of $X^{N,\epsilon}(t), t \in [0,1]$ (W being the d dimensional Brownian motion appearing in (12.1)). By Girsanov's theorem 20.21, since $k_t^\epsilon(x)$ is uniformly bounded, $\mathbb{P}^{N,k,\epsilon}$ is the law of

$$d\lambda_N^i(t) = \frac{1}{N} \sum_{j=1}^N \frac{1}{\lambda_N^i(t) - \lambda_N^j(t)} dt + k_t^\epsilon(\lambda_N^i(t))dt + \frac{1}{\sqrt{N}}d\tilde{W}_t^i$$

with an N-dimensional Brownian motion $\tilde{W}$ under $\mathbb{P}^{N,k,\epsilon}$. Applying Itô's calculus and exactly the same argument than in Lemma 12.5 (we leave this as an exercise, and note that it is important that k_t^ϵ is uniformly bounded and continuous for all $t \in [0,1]$), we find that the limit point of $L_N(.)$ under $\mathbb{P}^{N,k,\epsilon}$ are solution of (13.20) with $k = k^\epsilon$ and $\mu_0 = \mu_D * p_\epsilon$. Hence, by Lemma 13.13,

$$\lim_{N \to \infty} \mathbb{P}^{N,k,\epsilon}(L_N(.) \in B(p_\epsilon * \mu_., \delta)) = 1 \tag{13.27}$$

for all $\delta > 0$.

3. *Approximating the shifted law.* Because we did not regularize k^ϵ in time, $\frac{1}{N^2} \log \Lambda^{N,k}$ is not necessarily a continuous function of $L_N(.)$; indeed, we cannot use Itô's calculus to transform $\sum_{i=1}^N \int_0^T k_t^\epsilon(\lambda_N^i(t))dW_t^i$ into an integral over dt since $t \to k_t^\epsilon$ may not be differentiable. To circumvent this problem, we consider the law $\mathbb{P}^{N,k^p,\epsilon}$ corresponding to the discretized field $k^{p,\epsilon}$. Then, since $k^{\epsilon,p}$ is continuously differentiable in time, if Λ^{N,k^p} is the density of $\mathbb{P}^{N,k^p,\epsilon}$ with respect to $\mathbb{P}_{T,\lambda_N(0)}^N$, we have

$$\frac{1}{N^2} \log \Lambda^{N,k^{p,\epsilon}} = N^2 \left(S^{0,1}(L_N(.), K^{p,\epsilon}) - \frac{1}{2} \langle K^{p,\epsilon}, K^{p,\epsilon} \rangle_{L_N}^{0,1} \right)$$

$$= N^2 \bar{S}^{0,1}(L_N(.), K^{p,\epsilon})$$

with $K_t^{p,\epsilon} = \int_{-\infty}^x k_t^{p,\epsilon}(y)dy$. Hence, $\frac{1}{N^2} \log \Lambda^{N,k^{p,\epsilon}}$ is a smooth function of $L_N(.)$ since $K_t^{p,\epsilon}$ and its time derivative are C^∞ with uniformly bounded derivatives. Therefore, for a fixed δ, we find a $\kappa(\epsilon, p)$ vanishing as δ goes to zero for any $p \in \mathbb{N}$, such that

$$\mathbb{P}^{N,\epsilon}(B(p_\epsilon * \mu., \delta)) \geq e^{-N^2(\bar{S}^{0,1}(p_\epsilon * \mu., K^p) + \kappa(\epsilon,p))} \mathbb{P}^{N,k^p,\epsilon}(B(p_\epsilon * \mu., \delta)) \tag{13.28}$$

To replace $\mathbb{P}^{N,k^p,\epsilon}$ by $\mathbb{P}^{N,k,\epsilon}$ and conclude, note that by Girsanov's theorem 20.21, if $\Delta k_t^{\epsilon,p} = (k_t^{\epsilon,p} - k_t^\epsilon)$,

$$\Lambda_N := \log \frac{d\mathbb{P}^{N,k^p,\epsilon}}{d\mathbb{P}^{N,k,\epsilon}} = M_1^N - \frac{1}{2} \langle M^N \rangle_1$$

with the martingale $(M_s^N)_{0 \leq s \leq 1}$ given by

$$M_s^N = \sqrt{N} \sum_{i=1}^N \int_0^s \Delta k_t^{\epsilon,p}(\lambda_N^i(t))dW_t^i$$

for an N-dimensional Brownian motion $(W^i)_{1 \leq i \leq N}$ under $\mathbb{P}^{N,k,\epsilon}$. We have

$$\mathbb{P}^{N,k^p,\epsilon}(B(p_\epsilon * \mu., \delta)) \geq e^{-\kappa N^2} \mathbb{P}^{N,k,\epsilon}\left(\{\Lambda_N \geq e^{-\kappa N^2}\} \cap B(p_\epsilon * \mu., \delta) \right)$$

$$\geq e^{-\kappa N^2} \left(\mathbb{P}^{N,k,\epsilon}(B(p_\epsilon * \mu., \delta)) \right.$$

$$\left. - \mathbb{P}^{N,k,\epsilon}(\{\Lambda_N \leq e^{-\kappa N^2}\} \cap B(p_\epsilon * \mu., \delta)) \right). \tag{13.29}$$

Since $x \to k_t^\epsilon(x)$ is uniformly Lipschitz, we find that on $B(p_\epsilon * \mu., \delta)$,

$$\frac{1}{N^2} \langle M^N \rangle_1 = \frac{1}{N} \sum_{i=1}^N \int_0^1 (\Delta k_t^{\epsilon,p}(\lambda_N^i(t)))^2 dt$$

$$= \int_0^1 \int (\Delta k_t^{\epsilon,p}(x))^2 dp_\epsilon * \mu_t(x)dt + O(\delta) = o(p, \delta)$$

by our choice of $k^{\epsilon,p}$, and with $o(p, \delta) = o(p) + o(\delta)$ going to zero as p goes to infinity and δ to zero. Thus, we obtain that

$$\mathbb{P}^{N,k,\epsilon}\left(\{\Lambda_N \leq e^{-\kappa N^2}\} \cap B(p_\epsilon * \mu., \delta) \right)$$

$$\leq \mathbb{P}^{N,k,\epsilon}\left(\{M_1^N \leq -(\kappa - o(\delta, p))N^2\} \cap B(p_\epsilon * \mu., \delta) \right)$$

$$\leq \mathbb{P}^{N,k,\epsilon}\left(e^{-\lambda M_1^N - \frac{\lambda^2}{2} \langle M^N \rangle_1} \geq e^{[\lambda(\kappa - o(\delta,p)) - \frac{\lambda^2}{2} o(\delta,p)]N^2} \right)$$

$$\leq e^{-[\lambda(\kappa - o(\delta,p)) - \frac{\lambda^2}{2} o(\delta,p)]N^2}$$

for any $\lambda > 0$. Hence, taking $\lambda = (\kappa - o(\delta, p))/o(\delta, p)$ we conclude that for any $\kappa > 0$,

$$\limsup_{\substack{\delta \to 0 \\ p \to \infty}} \limsup_{N \to \infty} \frac{1}{N^2} \log \mathbb{P}^{N,k,\epsilon} \left(\{ \Lambda_N \leq e^{-\kappa N^2} \} \cap B(p_\epsilon * \mu., \delta) \right) = -\infty.$$

(13.30)

By (13.28), (13.29) and (13.27) we thus conclude that for $\kappa > 0$, δ small enough and p large enough

$$\liminf_{N \to \infty} \frac{1}{N^2} \log \mathbb{P}^{N,\epsilon} \left(B(p_\epsilon * \mu., \delta) \right) \geq -\bar{S}^{0,1}(p_\epsilon * \mu., K^p) - \kappa + o(\epsilon, p, \delta)$$

(13.31)

with $o(\epsilon, p, \delta)$ going to zero as p goes to infinity and δ to zero. Finally, our choice of $k^{p,\epsilon}$ shows that

$$\lim_{p \to \infty} \bar{S}^{0,1}(p_\epsilon * \mu., K^p) = \bar{S}^{0,1}(p_\epsilon * \mu., K)$$

completing the proof of the lower bound by letting δ going to zero, p going to infinity and then κ to zero.

• *Large deviation lower bound for processes in $\mathcal{A}^p$.* To deduce our result for the case $\epsilon = 0$, we proceed by exponential approximation. In fact, we have the following lemma:

Lemma 13.17. *Consider, for $L \in \mathbb{R}^+$, the compact set K_L of $\mathcal{P}(\mathbb{R})$ given by*

$$K_L = \{ \mu \in \mathcal{P}(\mathbb{R}); \mu(\log(x^2 + 1)) \leq L \}.$$

Then, on $\mathcal{K}_\epsilon^N(K_L) := \bigcap_{t \in [0,1]} \{ \{L_{N,\epsilon}(t) \in K_L\} \cap \{L_N(t) \in K_L\} \}$, with d the Duddley distance (0.1),

$$D(L_{N,\epsilon}(.), L_N(.)) \leq f(N, \epsilon)$$

where

$$\limsup_{\epsilon \to 0} \limsup_{N \to \infty} f(N, \epsilon) = 0.$$

Proof. *Step 1: compactly supported measure approximation.* Write $C_\epsilon^N = \sqrt{\epsilon} C_N$ with a matrix C_N whose spectral measure converges to a standard Cauchy law. Denote by $(c_i)_{1 \leq i \leq N}$ the eigenvalues of C_N. For $M > 0$, we set

$$B_M := \{ i : |c_i| > M \} := \{ j_1, \ldots, j_{|B_M|} \}.$$

Define

$$C_{N,M}(i,i) = \begin{cases} c_i & \text{if } i \notin B_M \\ 0 & \text{otherwise.} \end{cases}$$

Let

$$X_N^{\epsilon, M}(t) = H_N(t) + D_N + \sqrt{\epsilon} U_N C_{N,M} U_N^*$$

and denote by $L_{N,\epsilon,M}(t)$ its spectral measure. Then,

$$D(L_{N,\epsilon,M}(.), L_N(.)) \leq \epsilon M.$$

In fact, for any continuously differentiable function f, any $t \in [0,1]$,

$$|\hat{\mu}_t^{N,\epsilon,M}(f) - L_{Nt}(f)|$$

$$= \epsilon \left| \int_0^1 \frac{1}{N} \text{Tr} \left(f'(X_N(t) + \alpha\sqrt{\epsilon}U_N C_{N,M} U_N^*) U_N C_{N,M} U_N \right) d\alpha \right|$$

$$\leq \frac{\epsilon}{N} \sum_{i \in B_M} |c_i| \int_0^1 |\langle e_i, f'(X_N(t) + \alpha\sqrt{\epsilon}U_N C_{N,M} U_N^*) e_i \rangle| \, d\alpha$$

$$\leq \epsilon M \int_0^1 \left(\frac{1}{N} \text{Tr}(f'(X_N(t) + \alpha\epsilon C_{N,M})^2) \right)^{\frac{1}{2}} d\alpha.$$

Extending this inequality to Lipschitz functions, we deduce that

$$\left| \int f dL_{N,\epsilon,M}(t) - \int f dL_N(t) \right| \leq ||f||_{\mathcal{L}} \epsilon M$$

which gives the desired estimate on $D(\hat{\mu}^{N,\epsilon,M}, L_N.)$.

Step 2: small rank perturbation approximation. On the compact set K_L, the Duddley distance is equivalent to the distance

$$d_1(\mu, \nu) = \sup_{||f||_{\mathcal{L}} \leq 1, f \uparrow} \left| \int f d\nu - \int f d\mu \right|.$$

Write

$$X_N^\epsilon(t) = X_N^{\epsilon,M}(t) + \epsilon \sum_{i=1}^{|B_M|} c_{j_i} e_{j_i} e_{j_i}^T.$$

Following Lidskii's theorem (see (1.18)), we find that

$$d_1(L_{N,\epsilon,M}(t), L_{N,\epsilon}(t)) \leq \frac{4|B_M|}{N}. \tag{13.32}$$

But Chebycheff's inequality yields

$$\frac{|B_M|}{N} = \int 1_{|x| \geq M} dL_{C_N}(x)$$

$$\leq \frac{1}{(\log(M^2+1))^2} \sup_N \int (\log(x^2+1))^2 dL_{C_N}(x)$$

giving finally, according to condition (13.32), a finite constant C such that

$$d_1(L_{N,\epsilon,M}(t), L_{N,\epsilon}(t)) \leq \frac{CL}{(\log(M^2+1))^2}.$$

Now, since d_1 and d are equivalent on K_L, the proof of the lemma is complete.
□

We then can prove the following:

Theorem 13.18. *Assume that L_{D_N} converges to μ_D while*

$$\sup_{N \in \mathbb{N}} L_{D_N}(x^4) < \infty.$$

Then, for any $\mu_{\cdot} \in \mathcal{A}$

$$\lim_{\delta \to 0} \liminf_{N \to \infty} \frac{1}{N^2} \log \mathbb{P}\left(D(L_N(.), \mu_{\cdot}) \leq \delta\right) \geq -\overset{\bullet}{S}_{\mu_D}(\mu_{\cdot}).$$

so that for any open subset $O \in \mathcal{C}([0,1], \mathcal{P}(\mathcal{P}(\mathbb{R})))$,

$$\liminf_{N \to \infty} \frac{1}{N^2} \log \mathbb{P}\left(L_N(.) \in O\right) \geq - \inf_{O \cap \mathcal{A}} S_{\mu_D}$$

Proof of Theorem 13.18. Following Lemma 13.10, we deduce that for any $M \in \mathbb{R}^+$, we can find $L_M \in \mathbb{R}^+$ such that for any $L \geq L_M$,

$$\sup_{0 \leq \epsilon \leq 1} \mathbb{P}(\mathcal{K}_\epsilon^N(K_L)^c) \leq e^{-MN^2}. \tag{13.33}$$

Fix $M > S_{\mu_D}(\mu) + 1$ and $L \geq L_M$. Let $\delta > 0$ be given. Next, observe that $P_\epsilon * \mu_{\cdot}$ converges weakly to $\mu_{\cdot}$ as ϵ goes to zero and choose consequently ϵ small enough so that $D(P_\epsilon * \mu_{\cdot}, \mu_{\cdot}) < \frac{\delta}{3}$. Then, write

$$\mathbb{P}\left(L_N(.) \in B(\mu_{\cdot}, \delta)\right)$$

$$\geq \mathbb{P}\left(D(L_N(.), \mu_{\cdot}) < \frac{\delta}{3}, L_{N,\epsilon_{\cdot}} \in B(P_\epsilon * \mu_{\cdot}, \frac{\delta}{3}), \mathcal{K}_\epsilon^N(K_L)\right)$$

$$\geq \mathbb{P}\left(L_{N,\epsilon_{\cdot}} \in B(P_\epsilon * \mu_{\cdot}, \frac{\delta}{3})\right) - \mathbb{P}(\mathcal{K}_\epsilon^N(K_L)^c)$$

$$- \mathbb{P}\left(D(L_{N,\epsilon}(.), L_N(.)) \geq \frac{\delta}{3}, \mathcal{K}_\epsilon^N(K_L)\right) = I - II - III.$$

(13.33) implies, up to terms of smaller order, that

$$II \leq e^{-N^2(S_{\mu_D}(\mu)+1)}.$$

Lemma 13.17 shows that $III = 0$ for ϵ small enough and N large, while Corollary 13.16 implies that for any $\eta > 0$, N large and $\epsilon > 0$

$$I \geq e^{-N^2 S_{P_\epsilon * \mu_D}(P_\epsilon * \mu) - N^2 \eta} \geq e^{-N^2 S_{\mu_D}(\mu) - N^2 \eta}.$$

Theorem 13.18 is proved. □

Proof of Lemma 13.13. Following [55], we take $f(x,t) := e^{i\lambda x}$ for some $\lambda \in \mathbb{R}$ in (13.20) and denote by $\mathcal{L}_t(\lambda) = \int e^{i\lambda x} d\nu_t(x)$ the Fourier transform of ν_t. $\nu \in \mathcal{MC}([0,1], \mathcal{P}(\mathbb{R}))$ implies that if k is the field associated with ν,

$$k_t(x) = C_t + \int e^{i\lambda x} \hat{k}_t(\lambda) d\lambda$$

with $\int_0^1 \max_\lambda e^{-2\epsilon|\lambda|} |\hat{k}_t(\lambda)|^2 dt \leq C$ for a given $\epsilon > 0$. Then, we find that for $t \in [0,1]$,

$$\mathcal{L}_t(\lambda) = \mathcal{L}_0(\lambda) - \frac{\lambda^2}{2} \int_0^t \int_0^1 \mathcal{L}_s(\alpha\lambda)\mathcal{L}_s((1-\alpha)\lambda)d\alpha ds$$

$$+ i\lambda \int_0^t \int \mathcal{L}_s(\lambda + \lambda')\hat{k}(\lambda', s)d\lambda' ds + i\lambda \int_0^t \mathcal{L}_s(\lambda)C_s ds. \tag{13.34}$$

Multiplying both sides of this equality by $e^{-\frac{\epsilon}{4}|\lambda|}$ gives, with $\mathcal{L}_t^\epsilon(\lambda) = e^{-\frac{\epsilon}{4}|\lambda|}\mathcal{L}_t(\lambda)$,

$$\mathcal{L}_t^\epsilon(\lambda) = \mathcal{L}_0^\epsilon(\lambda) - \frac{\lambda^2}{2} \int_0^t \int_0^1 \mathcal{L}_s^\epsilon(\alpha\lambda)\mathcal{L}_s^\epsilon((1-\alpha)\lambda)d\alpha ds$$

$$+ i\lambda \int_0^t \int \mathcal{L}_s^\epsilon(\lambda + \lambda')e^{\frac{\epsilon}{4}|\lambda+\lambda'|-\frac{\epsilon}{4}|\lambda|}\hat{k}(\lambda', s)d\lambda' ds + i\lambda \int_0^t \mathcal{L}_s^\epsilon(\lambda)C_s ds. \tag{13.35}$$

Therefore, if $\nu, \tilde{\nu}'$ are two solutions with Fourier transforms $\mathcal{L}$ and $\tilde{\mathcal{L}}$ respectively and if we set $\Delta_t^\epsilon(\lambda) = |\mathcal{L}_t^\epsilon(\lambda) - \tilde{\mathcal{L}}_t^\epsilon(\lambda)|$, we deduce from (13.35) that if we denote $D_s = \sup_{\lambda \in \mathbb{R}} e^{\epsilon|\lambda|}|\hat{k}(\lambda, s)|$,

$$\Delta_t^\epsilon(\lambda) \leq \lambda^2 \int_0^t \int_0^1 \Delta_s^\epsilon(\alpha\lambda)e^{-\frac{1}{4}(1-\alpha)\epsilon\lambda}d\alpha ds + |\lambda| \int_0^t \Delta_s^\epsilon(\lambda)C_s ds$$

$$+ |\lambda| \int_0^t \int \Delta_s^\epsilon(\lambda + \lambda')D_s e^{\frac{\epsilon}{4}|\lambda+\lambda'|-\frac{\epsilon}{4}|\lambda|-\epsilon|\lambda'|}d\lambda' ds$$

$$\leq \frac{4|\lambda|}{\epsilon} \int_0^t \sup_{|\lambda'| \leq |\lambda|} \Delta_s^\epsilon(\lambda')ds + |\lambda| \int_0^t \Delta_s^\epsilon(\lambda)C_s ds$$

$$+ |\lambda| \int_0^t D_s [\sup_{|\lambda'| \leq R} \Delta_s^\epsilon(\lambda') + 2e^{-\frac{\epsilon}{4}R}] \int e^{\frac{\epsilon}{4}|\lambda+\lambda'|-\frac{\epsilon}{4}|\lambda|-\epsilon|\lambda'|}d\lambda' ds$$

where R is any positive constant and we used that $\Delta_t^\epsilon(\lambda) \leq 2e^{-\frac{\epsilon}{4}|\lambda|}$. Considering $\bar{\Delta}_t^\epsilon(R) = \sup_{|\lambda'| \leq R} \Delta_s^\epsilon(\lambda')$, we therefore obtain, since $|\lambda| + |\lambda'| \geq |\lambda + \lambda'|$,

$$\bar{\Delta}_t^\epsilon(R) \leq \frac{R}{\epsilon} \int_0^t (D_s + C_s + 4)\bar{\Delta}_s^\epsilon(R)ds + 2\frac{R}{\epsilon}e^{-\frac{\epsilon}{4}R} \int_0^t D_s ds$$

By Gronwall's lemma, we deduce that

$$\bar{\Delta}^\epsilon_t(R) \leq 2\frac{R}{\epsilon}e^{-\frac{\epsilon}{4}R}\int_0^t D_s e^{\frac{R}{\epsilon}\int_s^t(D_u+C_u+4)du}ds.$$

Now, since we assumed $D^2 := \int_0^1 D_s^2 ds < \infty$ and $C^2 := \int_0^1 C_s^2 ds < \infty$, $\int_0^t(D_s + C_s)ds \leq (C+D)\sqrt{t}$, we have

$$\bar{\Delta}^\epsilon_t(R) \leq 2\frac{R}{\epsilon}De^{-\frac{\epsilon}{4}R}e^{\frac{R}{\epsilon}[(C+D)\sqrt{t}+4t]}.$$

Thus $\bar{\Delta}^\epsilon_t(\infty) = 0$ for $t < \tau \equiv (\frac{\epsilon^2}{4(C+D+4)})^2$. By induction over the time, we conclude that $\bar{\Delta}^\epsilon_t(\infty) = 0$ for any time $t \leq 1$, and therefore that $\nu = \tilde{\nu}$. □

Bibliographical Notes. The previous large deviation estimates were proved in [109]. Further analysis of the rate function was performed in [101], and of related PDE questions in [5, 134, 142]. On less rigorous ground, we refer to [146]. In particular, the rate function $S_{\mu_D}(\mu)$ is given as an infimum achieved at the solution of a complex Burgers equation [101]. By completely different techniques, Kenyon, Okounkov, Sheffield [128] obtained large deviation principles for the law of the random surfaces given by dimers. Interestingly, the limiting shapes are also described by the complex Burgers equation [127]. This point may indicate that there is a link between random matrices and random partitions at the level of large deviations, generalizing the well-known local connection [17, 161].

14

Asymptotics
of Harish–Chandra–Itzykson–Zuber Integrals
and of Schur Polynomials

Let $Y^{N,\beta}$ be the random matrix $\mathbf{D}^N + X^{N,\beta}$ with a deterministic diagonal matrix $\mathbf{D}^N$ and $X^{N,\beta}$ a Gaussian Wigner matrix. We now show how the deviations of the law of the spectral measure of $Y^{N,\beta}$ are related to the asymptotics of the Harish–Chandra–Itzykson–Zuber (or spherical) integrals

$$I_N(A, B) = \int e^{N\operatorname{Tr}(AUBU^*)} dm_N^\beta(U)$$

where $m_N^\beta(U)$ is the Haar measure on $U(N)$ when $\beta = 2$ and $O(N)$ when $\beta = 1$. m_N will stand for m_N^2 to simplify the notations. Here, $I_N(A, B)$ makes sense for any $A, B \in \mathcal{M}_n(\mathbb{C})$, but we shall consider asymptotics only when $A, B \in \mathcal{H}_N^{(\beta)}(\mathbb{C})$ (the extension of our results to non-self-adjoint matrices is still open). To this end, we shall make the following hypothesis:

Assumption 1. 1. There exists $d_{max} \in \mathbb{R}^+$ such that for any integer number N, $L_{D_N}(\{|x| \geq d_{max}\}) = 0$. Moreover, L_{D_N} converges weakly to $\mu_D \in \mathcal{P}(\mathbb{R})$.

2. L_{E_N} converges to $\mu_E \in \mathcal{P}(\mathbb{R})$ while $L_{E_N}(x^2)$ stays uniformly bounded.

Theorem 14.1. *Under Assumption 1:*

1) There exists a function $g : [0, 1] \times \mathbb{R}^+ \mapsto \mathbb{R}^+$, *depending on* μ_E *only, such that* $g(\delta, L) \to_{\delta \to 0} 0$ *for any* $L \in \mathbb{R}^+$, *and, for* $\hat{E}_N, \bar{E}_N$ *such that*

$$d(L_{\hat{E}_N}, \mu_E) + d(L_{\bar{E}_N}, \mu_E) \leq \delta/2 \tag{14.1}$$

and

$$\int x^2 dL_{\bar{E}_N}(x) + \int x^2 dL_{\hat{E}_N}(x) \leq L, \tag{14.2}$$

and it holds that

$$\limsup_{N \to \infty} \left| \frac{1}{N^2} \log \frac{I_N^{(\beta)}(D_N, \hat{E}_N)}{I_N^{(\beta)}(D_N, \bar{E}_N)} \right| \leq g(\delta, L).$$

A. Guionnet, *Large Random Matrices: Lectures on Macroscopic Asymptotics*, 211
Lecture Notes in Mathematics 1957, DOI: 10.1007/978-3-540-69897-5_14,
© 2009 Springer-Verlag Berlin Heidelberg, Reprint by Springer-Verlag Berlin Heidelberg 2012

We define

$$\bar{I}^{(\beta)}(\mu_D, \mu_E) = \limsup_{N \uparrow \infty} \frac{1}{N^2} \log I_N^{(\beta)}(D_N, E_N)$$

$$\underline{I}^{(\beta)}(\mu_D, \mu_E) = \liminf_{N \uparrow \infty} \frac{1}{N^2} \log I_N^{(\beta)}(D_N, E_N),$$

$\bar{I}^{(\beta)}(\mu_D, \mu_E)$ *and* $\underline{I}^{(\beta)}(\mu_D, \mu_E)$ *are continuous functions on* $\{(\mu_E, \mu_D) \in \mathcal{P}(\mathbb{R})^2 : \int x^2 d\mu_E(x) + \int x^2 d\mu_D(x) \leq L\}$ *for any* $L < \infty$.

2) For any probability measure $\mu \in \mathcal{P}(\mathbb{R})$,

$$\inf_{\delta \to 0} \liminf_{N \to \infty} \frac{1}{N^2} \log \mathbb{P}\left(d(L_{Y^{N,\beta}}, \mu) < \delta\right)$$

$$= \inf_{\delta \to 0} \limsup_{N \to \infty} \frac{1}{N^2} \log \mathbb{P}\left(d(L_{Y^N,\beta}, \mu) < \delta\right)$$

$$:= -J_\beta(\mu_D, \mu).$$

3) We let, for any $\mu \in \mathcal{P}(\mathbb{R})$,

$$I_\beta(\mu) = \frac{\beta}{4} \int x^2 d\mu(x) - \frac{\beta}{2} \int \int \log|x - y| d\mu(x) d\mu(y).$$

If L_{E_N} *converges to* $\mu_E \in \mathcal{P}(\mathbb{R})$ *with* $I_\beta(\mu_E) < \infty$, *we have*

$$I^{(\beta)}(\mu_D, \mu_E) := \bar{I}^{(\beta)}(\mu_D, \mu_E) = \underline{I}^{(\beta)}(\mu_D, \mu_E)$$

$$= -J_\beta(\mu_D, \mu_E) + I_\beta(\mu_E) - \inf_{\mu \in \mathcal{P}(\mathbb{R})} I_\beta(\mu) + \frac{\beta}{4} \int x^2 d\mu_D(x).$$

Before going any further, let us point out that these results give interesting asymptotics for Schur polynomials that are defined as follows.

- A Young shape λ is a finite sequence of non-negative integers $(\lambda_1, \lambda_2, \ldots, \lambda_l)$ written in non-increasing order. One should think of it as a diagram whose ith line is made of λ_i empty boxes: for example,

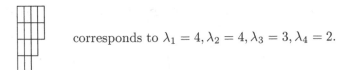

corresponds to $\lambda_1 = 4, \lambda_2 = 4, \lambda_3 = 3, \lambda_4 = 2$.

We denote by $|\lambda| = \sum_i \lambda_i$ the total number of boxes of the shape λ.
In the sequel, when we have a shape $\lambda = (\lambda_1, \lambda_2, \ldots)$ and an integer N greater than the number of lines of λ having a strictly positive length, we will define a sequence l associated to λ and N, that is an N-tuple of integers $l_i = \lambda_i + N - i$. In particular we have that $l_1 > l_2 > \cdots > l_N \geq 0$ and $l_i - l_{i+1} \geq 1$.

- For some fixed $N \in \mathbb{N}$, a Young tableau will be any filling of the Young shape with integers from 1 to N that is increasing on each line and (strictly) increasing on each column. For each such filling, we define the content of a Young tableau as the N-tuple $(\mu_1, \ldots, \mu_N)$ where μ_i is the number of i's written in the tableau.

For example,

$$\begin{array}{|c|c|c|}\hline 1 & 1 & 2 \\\hline 2 & 3 \\\cline{1-2} 3 \\\cline{1-1}\end{array}$$

is allowed (and has content $(2, 2, 2)$),

whereas

$$\begin{array}{|c|c|c|}\hline 1 & 1 & 2 \\\hline 1 & 3 \\\cline{1-2} 3 \\\cline{1-1}\end{array}$$

is not.

Notice that, for $N \in \mathbb{N}$, a Young shape can be filled with integers from 1 to N if and only if $\lambda_i = 0$ for $i > N$.

- For a Young shape λ and an integer N, the Schur polynomial s_λ is an element of $\mathbb{C}[x_1, \ldots, x_N]$ defined by

$$s_\lambda(x_1, \ldots, x_N) = \sum_T x_1^{\mu_1} \ldots x_N^{\mu_N}, \tag{14.3}$$

where the sum is taken over all Young tableaux T of fixed shape λ and $(\mu_1, \ldots, \mu_N)$ is the content of T. On a statistical point of view, one can think of the filling as the heights of a surface sitting on the tableau λ, μ_i being the height of the surface at i. s_λ is then a generating function for these heights when one considers the surfaces uniformly distributed under the constraints prescribed for the filling. Note that s_λ is positive whenever the x_i's are and, although it is not obvious from this definition (cf. for example [175] for a proof), s_λ is a symmetric function of the x_i's and actually (s_λ, λ) form a basis of symmetric functions and hence play a key role in representation theory of the symmetric group. If A is a matrix in $\mathcal{M}_N(\mathbb{C})$, then define $s_\lambda(A) \equiv s_\lambda(A_1, \ldots, A_N)$, where the A_i's are the eigenvalues of A. Then, by Weyl's formula (cf. [204, Theorem 7.5.B]), for any matrices V, W,

$$\int s_\lambda(UVU^*W)dm_N(U) = \frac{1}{d_\lambda}s_\lambda(V)s_\lambda(W), \tag{14.4}$$

with $d_\lambda = s_\lambda(I) = \prod_{i<j}(l_i - l_j)/\prod_{i=1}^{N-1} i!$ with $l_i = \lambda_i - i + N$. s_λ can also be seen as a generating function for the number of surfaces constructed on the Young shape with prescribed level areas.

The Schur function s_λ has a determinantal formula (cf. [175] and [98]);

$$s_\lambda(x) = \frac{\det(x_i^{l_j})_{1 \leq i,j \leq N}}{\Delta(x)}$$

with $\Delta(x)$ the Vandermonde determinant $\Delta(x) = \prod_{i<j}(x_i - x_j)$. Since also the spherical integral $I_N^{(2)}$ has a determinantal expression for $A = \mathrm{diag}(a_1, \ldots, a_N)$ and $B = \mathrm{diag}(b_1, \ldots, b_N)$,

$$I_N^{(2)}(A, B) = \frac{\det(e^{a_i b_j})_{1 \le i,j \le N}}{\Delta(a)\Delta(B)},$$

with $\Delta(A) = \Delta(a), \Delta(B) = \Delta(a)$, we deduce

$$s_\lambda(M) = I_N^{(2)}\left(\log M, \frac{l}{N}\right)\Delta\left(\frac{l}{N}\right)\frac{\Delta(\log M)}{\Delta(M)}, \tag{14.5}$$

where $\frac{l}{N}$ denotes the diagonal matrix with entries $N^{-1}(\lambda_i - i + N)$. Therefore, we have the following immediate corollary to Theorem 14.1:

Corollary 14.2. *Let λ^N be a sequence of Young shapes and set $D_N = (N^{-1}(\lambda_i^N - i + N))_{1 \le i \le N}$. We pick a sequence of Hermitian matrices $(E_N)_{N \ge 0}$ and assume that $(D_N, E_N)_{N \in \mathbb{N}}$ satisfy Assumption 1 and that $\Sigma(\mu_D) > -\infty$. Then,*

$$\lim_{N \to \infty} \frac{1}{N^2} \log s_{\lambda^N}(e^{E_N})$$

$$= I^{(2)}(\mu_E, \mu_D) - \frac{1}{2}\int \log\left[\int_0^1 e^{\alpha x + (1-\alpha)y} d\alpha\right] d\mu_E(x)d\mu_E(y) + \frac{1}{2}\Sigma(\mu_D).$$

Proof of Theorem 14.1:. To simplify, let us assume that E_N and $\hat{E}_N$ are uniformly bounded by a constant M. Let $\delta' > 0$ and $\{A_j\}_{j \in \mathcal{J}}$ be a partition of $[-M, M]$ such that $|A_j| \in [\delta', 2\delta']$ and the end points of A_j are continuity points of μ_E. Define

$$\hat{I}_j = \{i : \hat{E}_N(ii) \in A_j\}, \quad \bar{I}_j = \{i : \bar{E}_N(ii) \in A_j\}.$$

By (14.1),

$$|\mu_E(A_j) - |\hat{I}_j|/N| + |\mu_E(A_j) - |\bar{I}_j|/N| \le \delta.$$

We construct a permutation σ_N so that $|\hat{E}(ii) - \bar{E}(\sigma_N(i), \sigma_N(i))| < 2\delta$ except possibly for very few i's as follows. First, if $|\bar{I}_j| \le |\hat{I}_j|$ then $\tilde{I}_j := \bar{I}_j$, whether if $|\bar{I}_j| > |\hat{I}_j|$ then $|\tilde{I}_j| = |\hat{I}_j|$ while $\tilde{I}_j \subset \bar{I}_j$. Then, choose and fix a permutation σ_N such that $\sigma_N(\tilde{I}_j) \subset \hat{I}_j$. Then, one can check that if $\mathcal{J}_0 = \{i : |\hat{E}(ii) - \bar{E}(\sigma_N(i), \sigma_N(i))| < 2\delta\}$,

$$|\mathcal{J}_0| \ge |\cup_j \sigma_N(\tilde{I}_j)| = \sum_j |\sigma_N(\tilde{I}_j)| \ge N - \sum_j |\bar{I}_j \setminus \tilde{I}_j|$$

$$\ge N - \max_j(|\bar{I}_j| - |\tilde{I}_j|)|\mathcal{J}| \ge N - 2\delta N\frac{M}{\delta'}.$$

Next, note the invariance of $I_N^{(\beta)}(D_N, E_N)$ to permutations of the matrix elements of D_N. That is,

$$I_N^{(\beta)}(D_N, \bar{E}_N) = \int \exp\left\{ N\frac{\beta}{2}\mathrm{Tr}(UD_N U^* \bar{E}_N) \right\} dm_N^\beta(U)$$

$$= \int \exp\left\{ N\frac{\beta}{2} \sum_{i,k} u_{ik}^2 D_N(kk)\bar{E}_N(ii) \right\} dm_N^\beta(U)$$

$$= \int \exp\left\{ N \sum_{i,k} u_{ik}^2 D_N(kk)\bar{E}_N(\sigma_N(i)\sigma_N(i)) \right\} dm_N^\beta(U).$$

But, with $d_{\max} = \max_k |D_N(kk)|$ bounded uniformly in N,

$$N^{-1} \sum_{i,k} u_{ik}^2 D_N(kk)\bar{E}_N(\sigma_N(i)\sigma_N(i))$$

$$= N^{-1} \sum_{i\in\mathcal{J}_0} \sum_k u_{ik}^2 D_N(kk)\bar{E}_N(\sigma_N(i)\sigma_N(i))$$

$$\quad + N^{-1} \sum_{i\notin\mathcal{J}_0} \sum_k u_{ik}^2 D_N(kk)\bar{E}_N(\sigma_N(i)\sigma_N(i))$$

$$\leq N^{-1} \sum_{i,k} u_{ik}^2 D_N(kk)(\hat{E}_N(ii) + 2\delta) + N^{-1}d_{\max}M|\mathcal{J}_0^c|$$

$$\leq N^{-1} \sum_{i,k} u_{ik}^2 D_N(kk)\hat{E}_N(ii) + d_{\max}\frac{M^2\delta}{\delta'}.$$

Hence, we obtain, taking $d_{\max}\frac{M^2\delta}{\delta'} = \sqrt{\delta}$,

$$I_N^{(\beta)}(D_N, \bar{E}_N) \leq e^{N\sqrt{\delta}} I_N^{(\beta)}(D_N, \hat{E}_N)$$

and the reverse inequality by symmetry. This proves the first point of the theorem when $(\bar{E}_N, \hat{E}_N)$ are uniformly bounded. The general case (which is not much more complicated) is proved in [109] and follows from first approximating $\bar{E}_N$ and $\hat{E}_N$ by bounded operators using (14.2).

The second and the third points are proved simultaneously: in fact, writing

$$\mathbb{P}\left(d(L_{Y^{N,\beta}}, \mu) < \delta \right)$$

$$= \frac{1}{Z_N^\beta} \int_{d(L_{Y^{N,\beta}},\mu)<\delta} e^{-\frac{N\beta}{4}\mathrm{Tr}((Y^{N,\beta}-D_N)^2)} dY^{N,\beta}$$

$$= \frac{e^{-\frac{N\beta}{4}\mathrm{Tr}(D_N^2)}}{Z_N^\beta} \int_{d(\frac{1}{N}\sum_{i=1}^N \delta_{\lambda_i},\mu)<\delta} I_N^{(\beta)}(D(\lambda), D_N) e^{-\frac{N\beta}{4}\sum_{i=1}^N \lambda_i^2} \Delta(\lambda)^\beta \prod_{i=1}^N d\lambda_i$$

with Z_N^β the normalizing constant

$$Z_N^\beta = \int e^{-\frac{N}{2}\mathrm{Tr}((Y^{N,\beta}-D_N)^2)} dY^{N,\beta} = \int e^{-\frac{N}{2}\mathrm{Tr}((Y^{N,\beta})^2)} dY^{N,\beta},$$

we see that the first point gives, since $I_N^{(\beta)}(D(\lambda), D_N)$ is approximately constant on $\{d(N^{-1} \sum \delta_{\lambda_i}, \mu) < \delta\} \cap \{d(L_{D_N}, \mu_D) < \delta\}$,

$$
\begin{aligned}
&\mathbb{P}\left(d(L_{Y^{N,\beta}}, \mu) < \delta\right) \\
&\approx \frac{e^{N^2(I^{(\beta)}(\mu_D, \mu) - \frac{\beta}{4} L_{D_N}(x^2))}}{Z_N^\beta} \int_{d(\frac{1}{N} \sum_{i=1}^N \delta_{\lambda_i}, \mu) < \delta} e^{-\frac{N\beta}{4} \sum_{i=1}^N \lambda_i^2} \Delta(\lambda)^\beta \prod_{i=1}^N d\lambda_i \\
&= e^{-\frac{N^2 \beta}{4} L_{D_N}(x^2) + N^2 I^{(\beta)}(\mu_D, \mu)} \mathbb{P}\left(d(L_{X^{N,\beta}}, \mu) < \delta\right)
\end{aligned}
$$

where $A_{N,\delta} \approx B_{N,\delta}$ means that $N^{-2} \log A_{N,\delta} B_{N,\delta}^{-1}$ goes to zero as N goes to infinity first and then δ goes to zero. When $\beta = 2$ (resp. $\beta = 1$), $X^{N,\beta}$ is an $N \times N$ Hermitian (resp. symmetric) matrix with centered Gaussian entries.

The large deviation principle proved in the Chapter 10 shows 2) and 3). □

Note for 3) that if $I_\beta(\mu_E) = +\infty$, $J(\mu_D, \mu_E) = +\infty$ so that in this case the result is empty since it leads to an indetermination. Still, if $I_\beta(\mu_D) < \infty$, by symmetry of $I^{(\beta)}$, we obtain a formula by exchanging μ_D and μ_E. If both $I_\beta(\mu_D)$ and $I_\beta(\mu_E)$ are infinite, we can only argue, by continuity of $I^{(\beta)}$, that for any sequence $(\mu_E^\epsilon)_{\epsilon > 0}$ of probability measures with uniformly bounded variance and finite entropy I_β converging to μ_E,

$$
I^{(\beta)}(\mu_D, \mu_E) = \lim_{\epsilon \to \infty} \{-J_\beta(\mu_D, \mu_E^\epsilon) + I_\beta(\mu_E^\epsilon)\} - \inf I_\beta + \frac{\beta}{4} \int x^2 d\mu_D(x).
$$

A more explicit formula is not yet available.

Bibliographical Notes. Note that the convergence of the spherical integral is in fact not obvious and is given by the large deviation principle for the law of the spectral measure of non-centered Wigner matrices. Such types of convergence were shown to hold for more general integrals, but in a small-parameters region, in [66].

Harish–Chandra–Itzykson–Zuber integral was studied intensively [49, 65, 89, 210]. An important tool, when the integral holds over the unitary group, is the use of the Harish–Chandra formula that expresses this integral as a determinant [49, 115, 116]. Our approach is based on its relation with the large deviation principle for the law of the Hermitian Brownian motion. A parallel approach uses the heat kernel [146]. The asymptotics of the Harish–Chandra–Itzykson–Zuber integral when one of the matrices has a rank that is negligible with respect to N were studied in [67, 103]; they are given by the so-called R-transform.

15

Asymptotics of Some Matrix Integrals

We would like to consider integrals of more than one matrix. The simplest interaction that one can think of is the quadratic one. Such an interaction describes already several classical models in random-matrix theory; We refer here to the works of M. Mehta, A. Matytsin, A. Migdal, V. Kazakov, P. Zinn Justin and B. Eynard for instance. We list below a few models that were studied.

- The random Ising model on random graphs is described by the Gibbs measure

$$\mu_{Ising}^{N}(dA, dB) = \frac{1}{Z_{Ising}^{N}} e^{N\operatorname{Tr}(AB) - N\operatorname{Tr}(P_1(A)) - N\operatorname{Tr}(P_2(B))} dAdB$$

with Z_{Ising}^{N} the partition function

$$Z_{Ising}^{N} = \int e^{N\operatorname{Tr}(AB) - N\operatorname{Tr}(P_1(A)) - N\operatorname{Tr}(P_2(B))} dAdB$$

and two polynomial functions P_1, P_2. The limiting free energy for this model was calculated by M. Mehta [153] in the case $P_1(x) = P_2(x) = x^2 + gx^4$ and integration holds over $\mathcal{H}_N$. The limit was studied in [43]. However, the limiting spectral measures of A and B under μ_{Ising}^{N} were not considered in these papers. A discussion about this problem can be found in P. Zinn Justin [209].

- One can also define the $q - 1$ Potts model on random graphs described by the Gibbs measure

$$\mu_{Potts}^{N}(d\mathbf{A}_1, ..., d\mathbf{A}_q)$$

$$= \frac{1}{Z_{Potts}^{N}} \prod_{i=2}^{q} e^{N\operatorname{Tr}(\mathbf{A}_1\mathbf{A}_i) - N\operatorname{Tr}(P_i(\mathbf{A}_i))} d\mathbf{A}_i e^{-N\operatorname{Tr}(P_1(\mathbf{A}_1))} d\mathbf{A}_1.$$

The limiting spectral measures of $(\mathbf{A}_1, \ldots, \mathbf{A}_q)$ were first discussed in [80].

A. Guionnet, *Large Random Matrices: Lectures on Macroscopic Asymptotics*, 217
Lecture Notes in Mathematics 1957, DOI: 10.1007/978-3-540-69897-5_15,
© 2009 Springer-Verlag Berlin Heidelberg, Reprint by Springer-Verlag Berlin Heidelberg 2012

- As a straightforward generalization, one can consider matrices coupled in chain following S. Chadha, G. Mahoux and M. Mehta [60] given by

$$\mu_{chain}^N(d\mathbf{A}_1, ..., d\mathbf{A}_q)$$

$$= \frac{1}{Z_{chain}^N} \prod_{i=2}^{q} e^{N\text{Tr}(\mathbf{A}_{i-1}\mathbf{A}_i) - N\text{Tr}(P_i(\mathbf{A}_i))} d\mathbf{A}_i e^{-N\text{Tr}(P_1(\mathbf{A}_1))} d\mathbf{A}_1.$$

q can eventually go to infinity as in [147].

The first-order asymptotics of these models can be studied thanks to the control of spherical integrals obtained in the last chapter.

Theorem 15.1. *Assume that* $P_i(x) \geq c_i x^4 + d_i$ *with* $c_i > 0$ *and some finite constants* d_i. *Hereafter,* $\beta = 1$ *(resp.* $\beta = 2$, *resp.* $\beta = 4$) *when* dA *denotes the Lebesgue measure on* $\mathcal{S}_N$ *(resp.* $\mathcal{H}_N$, *resp.* $\mathcal{H}_N$ *with* N *even). Then, with* $c = \inf_{\nu \in \mathcal{P}(\mathbb{R})} I_\beta(\nu)$,

$$F_{Ising} = \lim_{N \to \infty} \frac{1}{N^2} \log Z_{Ising}^N$$

$$= -\inf \left\{ \mu(P) + \nu(Q) - I^{(\beta)}(\mu, \nu) - \frac{\beta}{2}\Sigma(\mu) - \frac{\beta}{2}\Sigma(\nu) \right\} - 2c \quad (15.1)$$

$$F_{Potts} = \lim_{N \to \infty} \frac{1}{N^2} \log Z_{Potts}^N$$

$$= -\inf \left\{ \sum_{i=1}^{q} \mu_i(P_i) - \sum_{i=2}^{q} I^{(\beta)}(\mu_1, \mu_i) - \frac{\beta}{2} \sum_{i=1}^{q} \Sigma(\mu_i) \right\} - qc \quad (15.2)$$

$$F_{chain} = \lim_{N \to \infty} \frac{1}{N^2} \log Z_{chain}^N$$

$$= -\inf \left\{ \sum_{i=1}^{q} \mu_i(P_i) - \sum_{i=2}^{q} I^{(\beta)}(\mu_{i-1}, \mu_i) - \frac{\beta}{2} \sum_{i=1}^{q} \Sigma(\mu_i) \right\} - qc \quad (15.3)$$

Remark 6. (1) The above theorem actually extends to polynomial functions going to infinity like x^2. However, the case of quadratic polynomials is trivial since it boils down to the Gaussian case and therefore the next interesting case is the quartic polynomial as above. Moreover, Theorem 15.2 fails in the case where P, Q go to infinity only like x^2. However, all our proofs would extend easily for any continuous functions $P_i's$ such that $P_i(x) \geq a|x|^{2+\epsilon} + b$ with some $a > 0$ and $\epsilon > 0$.

(2) Note that we did not assume here that potentials are small perturbations of the Gaussian potential as in Part III.

(3) The above free energies are not very explicit and not easy to analyze. To give a taste of the kind of information we have been able to establish so far, we state below a result about the Ising model.

Proof of Theorem 15.1. It is enough to notice that, when diagonalizing the matrices $\mathbf{A}_i$'s, the interaction in the models under consideration is expressed in terms of spherical integrals since, under dA, $A = UD_AU^*$, with D_A diagonal, U independent from D_A following the Haar measure on $U(N)$ when $\beta = 2$ and $O(N)$ when $\beta = 1$, so that

$$\mathbb{E}[e^{N\mathrm{Tr}(AB)}|\lambda_N^i(A), \ldots, \lambda_N^i(B), 1 \le i \le N] = I_N^{(\beta)}(A_N, B_N).$$

Laplace's (or saddle point) method then gives the result (up to the boundedness of the matrices A_i's in the spherical integrals, that can be obtained by approximation). We shall not detail it here and refer the reader to [101]. □

We shall then study the variational problems for the above energies; indeed, by standard large deviation considerations, it is clear that the spectral measures of the matrices $(\mathbf{A}_i)_{1\le i\le d}$ will concentrate on the set of the minimizers defining the free energies, and in particular converge to these minimizers when they are unique. We prove in [101] the following for the Ising model.

Theorem 15.2. *Assume* $P_1(x) \ge ax^4 + b, P_2(x) \ge ax^4 + b$ *for some positive constant* a. *Then:*

(0) The infimum in F_{Ising} *is achieved at a unique couple* (μ_A, μ_B) *of probability measures.*

(1) (L_A, L_B) *converges* μ_{Ising}^N- *almost surely to* (μ_A, μ_B).

(2) (μ_A, μ_B) *are compactly supported with finite non-commutative entropy*

$$\Sigma(\mu) = \int\int \log|x - y|d\mu(x)d\mu(y).$$

(3) There exists a couple $(\rho^{A\to B}, u^{A\to B})$ *of measurable functions on* $\mathbb{R} \times (0, 1)$ *such that* $\rho_t^{A\to B}(x)dx$ *is a probability measure on* $\mathbb{R}$ *for all* $t \in (0, 1)$ *and* $(\mu_A, \mu_B, \rho^{A\to B}, u^{A\to B})$ *are characterized uniquely as the minimizer of a strictly convex function under a linear constraint.*

In particular, $(\rho^{A\to B}, u^{A\to B})$ *are solution of the Euler equation for isentropic flow with negative pressure* $p(\rho) = -\frac{\pi^2}{3}\rho^3$ *such that, for all* (x, t) *in the interior of* $\Omega = \{(x, t) \in \mathbb{R} \times [0, 1]; \rho_t^{A\to B}(x) \ne 0\}$,

$$\begin{cases} \partial_t\rho_t^{A\to B} + \partial_x(\rho_t^{A\to B}u_t^{A\to B}) = 0 \\ \partial_t(\rho_t^{A\to B}u_t^{A\to B}) + \partial_x(\rho_t^{A\to B}(u_t^{A\to B})^2 - \frac{\pi^2}{3}(\rho_t^{A\to B})^3) = 0 \end{cases} \tag{15.4}$$

with the probability measure $\rho_t^{A\to B}(x)dx$ *weakly converging to* $\mu_A(dx)$ *(resp.* $\mu_B(dx)$*) as* t *goes to zero (resp. one). Moreover, we have*

$$P'(x) - x - \frac{\beta}{2}u_0^{A\to B}(x) - \frac{\beta}{2}H\mu_A(x) = 0 \quad \mu_A - a.s$$

and $\quad Q'(x) - x + \frac{\beta}{2}u_1^{A\to B}(x) - \frac{\beta}{2}H\mu_B(x) = 0 \quad \mu_B - a.s.$

For the other models, uniqueness of the minimizers is not always clear. For instance, we obtain uniqueness of the minimizers for the q-Potts models only for $q \leq 2$, whereas it is also expected for $q = 3$ (when the potential is convex, uniqueness is, however, always true, see [106]). For the description of these minimizers, I refer the reader to [101].

15.1 Enumeration of Maps from Matrix Models

As we have seen in Part III, the enumeration of maps with one color is related with matrix integrals of the form

$$Z_{\mathbf{t}}^N = \int e^{-N\mathrm{Tr}(V_{\mathbf{t}}(X))} d\mu^N(X)$$

with $V_{\mathbf{t}} = \sum_{i=1}^{n} t_i x^i$ a polynomial function depending on parameters $\mathbf{t} = (t_1, \ldots, t_n)$. When n is even and $t_n > 0$, the above matrix integral is finite and we can apply the results of Theorem 10.1 to see that

$$\lim_{N \to \infty} \frac{1}{N} \log Z_{\mathbf{t}}^N = \sup_{\mu \in \mathcal{P}(\mathbb{R})} \left\{ \int \int \log |x - y| d\mu(x) d\mu(y) - \int V_{\mathbf{t}}(x) d\mu(x) \right\} - c.$$

By Lemma 10.2, there is a unique optimizer $\mu_{\mathbf{t}}$ to the above supremum and it is characterized by the fact that there exists a constant ℓ such that

$$\ell = -2 \int \log |x - y| d\mu_{\mathbf{t}}(y) + V_{\mathbf{t}}(x) + \frac{1}{2}x^2 \quad \mu_{\mathbf{t}} \text{ a.s.} \tag{15.5}$$

$$\ell \leq -2 \int \log |x - y| d\mu_{\mathbf{t}}(y) + V_{\mathbf{t}}(x) + \frac{1}{2}x^2 \quad \mu_{\mathbf{t}} \text{ everywhere.} \tag{15.6}$$

The large deviation principle of Theorem 10.1 as well as the uniqueness of the minimizers assert that under the Gibbs measure

$$d\mu_{\mathbf{t}}^N(X) = (Z_{\mathbf{t}}^N)^{-1} e^{-N\mathrm{Tr}(V_{\mathbf{t}}(X))} d\mu^N(X)$$

$\hat{\mathbf{L}}^N$ converges almost surely towards $\mu_{\mathbf{t}}$. Assume now that there exists $c > 0$ such that $V_{\mathbf{t}}$ is c-convex. Then, if the parameters $(t_i)_{1 \leq i \leq n}$ are small enough, Theorem 8.4 and Corollary 8.6 assert that the limit $\mu_{\mathbf{t}}$ is also a generating function for planar maps;

$$\mu_{\mathbf{t}}(x^p) = \sum_{k \in \mathbb{N}^n} \prod_{i=1}^{n} \frac{(-t_i)^{k_i}}{k_i!} \mathcal{M}_{\mathbf{k}}(x^p)$$

with $\mathcal{M}_{\mathbf{k}}(x^p)$ the number of planar maps with k_i stars of type x^i and one star of type x^p.

Let us show how to deduce formulae for $\mathcal{M}_k(x^p)$ when $V_t(x) = tx^4$ from the above large deviation result, i.e., count quadrangulations and recover the result of Tutte [194] from the matrix-model approach. The analysis below is inspired from [32]. As can be guessed, formulae become more complicated as V_t becomes more complex (see [72] for a more general treatment).

To find an explicit formula for μ_t from (15.5) and (15.6), let us observe first that differentiating (15.5) (or using directly the Schwinger–Dyson's equation) and integrating with respect to $(z - x)^{-1} d\mu_t(x)$ gives

$$G\mu_t(z)^2 = -4t(\alpha_t + z^2) - 1 + 4tz^3 G\mu_t(z) + z G\mu_t(z)$$

with $G\mu_t(z) = \int (z - x)^{-1} d\mu_t(x)$, $z \in \mathbb{C}\backslash\mathbb{R}$ and $\alpha_t = \int x^2 d\mu_t(x)$. Solving this equation yields

$$G\mu_t(z) = \frac{1}{2}\left(4tz^3 + z - \sqrt{(4tz^3 + z)^2 - 4(4t(\alpha_t + z^2) + 1)}\right)$$

where we have chosen the solution so that $G\mu_t(z) \approx z^{-1}$ as $|z| \to \infty$. The square root is chosen as the analytic continuation in $\mathbb{C}\backslash\mathbb{R}^-$ of the square root on $\mathbb{R}^+$. Recall that if p_ϵ is the Cauchy law with parameter $\epsilon > 0$, for $x \in \mathbb{R}$,

$$-\Im(G\mu_t(x + i\epsilon)) = \int \frac{\epsilon}{(x - y)^2 + \epsilon^2} d\mu_t(y) = \pi p_\epsilon * \mu_t(x).$$

Hence, if $-\Im(G\mu_t(x + i\epsilon))$ converges as ϵ decreases towards zero, its limit is the density of μ_t. Thus, we in fact have

$$\frac{d\mu_t}{dx} = -\frac{1}{\pi}\lim_{\epsilon\downarrow 0}\Im\left(\sqrt{(4t(x + i\epsilon)^3 + (x + i\epsilon))^2 - 4(4t(\alpha_t + (x + i\epsilon)^2) + 1)}\right).$$

To analyze this limit, we write

$$(4tz^3 + z)^2 - 4(4t(\alpha_t + z^2) + 1) = (4t)^2(z^2 - a_1)(z^2 - a_2)(z^2 - a_3)$$

for some $a_1, a_2, a_3 \in \mathbb{C}$. Note that since $G\mu_t$ is analytic on $\mathbb{C}\backslash\mathbb{R}$, either we have a double root and a real non-negative root, or three real non-negative roots. We now argue that when V_t is convex, $a_1 = a_2$ and $a_3 \in \mathbb{R}^+$. In fact, the function

$$f(x) := -2\int \log|x - y| d\mu_t(y) + V_t(x) + \frac{1}{2}x^2$$

is strictly convex on $\mathbb{R}\backslash\mathrm{support}(\mu_t)$ and it is continuous at the boundaries of the support of μ_t since μ_t as a bounded density. Since f equals ℓ on the support of μ_t and is greater or equal to ℓ outside, we deduce that if there is a hole in the support of μ_t, f must also be constant equal to ℓ on this hole. This contradicts the strict convexity of f outside the support. Hence, the support S of μ_t must be an interval and $G\mu_t$ must be analytic outside S. Thus, we must have $a_1 = a_2 := b \in \mathbb{R}$ and $a_3 := a \in \mathbb{R}^+$ and $S = [-\sqrt{a}, +\sqrt{a}]$. Plugging back this equality gives the system of equations

$$a + 2b = -\frac{1}{2t}, \quad -2ba + b^2 = \frac{1}{4t^2} - \frac{1}{t}, \quad 4t^2 ab^2 = -(4t\alpha_t + 1),$$

which has a unique solution $(a, b) \in \mathbb{R}^+ \times \mathbb{R}$, which in turn prescribes α_t uniquely. Thus, we now have

$$\frac{d\mu_t}{dx}(x) = c_t 1_{[-\sqrt{a}, \sqrt{a}]}(x^2 - b)\sqrt{a - x^2}$$

$$c_t^{-1} = \int_{[-\sqrt{a}, \sqrt{a}]} (x^2 - b)\sqrt{a - x^2}dx = \frac{\pi}{2}a\left[\frac{a}{4} - b\right].$$

In particular, this expression allows us to write all moments of μ_t in terms of Catalan numbers since

$$\int x^{2p}d\mu_t(x) = c_t \int_{-\sqrt{a}}^{\sqrt{a}} x^{2p}(x^2 - b)\sqrt{a - x^2}dx$$

$$= \frac{4}{a[\frac{a}{4} - b]}(a/4)^p \int [ax^2 - b]x^{2p}d\sigma(x)$$

$$= \frac{4}{a[\frac{a}{4} - b]}(a/4)^p[aC_{p+1} - bC_p]$$

where we finally used Property 1.11. Thus, we have found exact formulae for $\mathcal{M}((x^4, k), (x^p, 1))$.

Remark. Note that the connectivity argument for the support of the optimizing measure is valid for any c-convex potential, $c > 0$. It was shown in [72] that the optimal measure has always the form, in the small-parameters region,

$$d\mu_t(x) = ch(x)\sqrt{(x - a_1)(a_2 - x)}dx$$

with h a polynomial. However, as the degree of V_t grows, the equations for the parameters of h become more and more complicated.

15.2 Enumeration of Colored Maps from Matrix Models

In the context of colored maps, exact computations are much more scarce. However, for the Ising model, some results can again be obtained (it corresponds to maps with colored vertices of a given degree, say p, corresponding to monomials $V_1(A) = A^p$ and $V_2(B) = B^p$) that can be glued together by a bi-colored straight line (corresponding to the monomial AB). For the Ising model with quartic polynomial (the case $p = 4$), M. Mehta [152] obtained an explicit expression for the free energy corresponding to the potential

$V(A, B) = V_{Ising}(A, B) = -cAB + V_1(A) + V_2(B)$ when $V_1 = V_2 = (g/4)x^4$. The corresponding results for the enumeration of colored quadrangulation was recently recovered by M. Bousquet-Melou and G. Scheaffer. Let us emphasize that there is a general approach also based on Schwinger–Dyson equations that should allow us to understand these results, see B. Eynard [87]. The remark is that by the Schwinger-Dyson equation we know that the limiting state μ_t satisfies

$$\mu_t((W_1'(A) - B)P) = \mu_t \otimes \mu_t(\partial_A P),$$

$$\mu_t((W_2'(B) - A)P) = \mu_t \otimes \mu_t(\partial_B P).$$

Taking $P = P(A) = (x - A)^{-1}$ in the second equation and $P(A, B) = \frac{1}{(x-A)} \frac{(W_2'(y) - W_2'(B))}{(y-B)}$ we find that if

$$E(x, y) = (y - W_1'(x))(x - W_2'(y)) + 1 - Q(x, y)$$

with

$$Q(x, y) = \mu_t \left(\frac{W_1'(x) - W_1'(A)}{(x - A)} \frac{W_2'(y) - W_2'(B)}{(y - B)} \right),$$

$$E(x, W_1'(x) - G\mu_A(x)) = 0$$

where $G\mu_A(x) = \mu_t((x - A)^{-1})$. Hence, as for one-matrix models, $G\mu_A$ is solution to an algebraic equation, with some unknown coefficients $\mu_t(A^i B^j)$ with i, j smaller are equal to the degree of W_1' (resp. W_2') minus one. The large deviations theorem 15.1 should now show us (but we have not yet been able to prove it) that for small enough parameters, the support of μ_A and μ_B are connected. Connectivity of the support was in fact proved by using dynamics in a general convex potential setting, including the Ising model, in [106]. This information should also prescribe uniquely the solution.

Bibliographical Notes. The question of enumerating maps was first tackled by Tutte [192–194] who enumerated rooted planar triangulations and quadrangulations (see, e.g., E. Bender and E. Canfield [29] for generalizations). In general, the equations obtained by Tutte's approach are not exactly solvable; their analysis was the subject of subsequent developments (see [97]). Because this last problem is in general difficult, a bijective approach was developed after the work of Cori and Vauquelin [69] and Schaeffer's thesis (see e.g [176]). It was shown that planar triangulations and quadrangulations can be encoded by labeled trees, which are much easier to count. This idea proved to be very fruitful in many respects and was generalized in many ways [24, 46]. It allows us not only to study the number of maps but also part of their geometry; P. Chassaing and G. Schaeffer [62] could prove that the diameter of uniformly distributed quadrangulations with n vertices behaves like $n^{\frac{1}{4}}$. This in particular allowed a limiting object for random planar maps to be defined [135, 144]. Such results seem to be out of reach of random-matrix techniques. The case of planar bi-colored maps related to the so-called Ising model on random planar graphs could also be studied [44]. However, there are still many several-colors problems that could be solved by using random matrices but not on a direct combinatorial approach, see e.g the Potts model [80] or the dually weighted graph model [102, 124].

Part VI

Free Probability

Free probability is a probability theory for non-commutative variables. In this field, random variables are usually bounded operators on a Hilbert space. The law of a self-adjoint operator T is given as the evaluation $(\langle \zeta, T^n \zeta \rangle)_{n \geq 0}$ of its moments in the direction of a fixed vector ζ of this Hilbert space. Large $N \times N$ matrices can be seen to fit in this framework as bounded operators on the Hilbert space $\mathbb{C}^N$ equipped for instance with the Euclidean scalar product. We will see in fact that free probability is the right framework to consider random matrices as their size goes to infinity.

For the sake of completeness, but actually not needed for our purpose, we shall recall some notions of operator algebras. We shall then describe free probability as a probability theory on non-commutative functionals (a point of view that forgets the space of realizations of the laws) equipped with the notion of freeness that generalizes the idea of independence to this non-commutative setting. We will then focus on large random matrices and show that their asymptotics are related with freeness. In particular, independent Wigner's matrices converge to free semi-circular operators and the Hermitian Brownian motion converges to the free Brownian motion. Conversely, large random matrices can be seen as an approximation to a large class of (and maybe all) operators. In particular, ideas from classical probability, once applied to large random matrices, can be imported to operator algebra theory via such an approximating scheme. In this part, we shall emphasize the uses of stochastic dynamics, as applied to the Hermitian and the free Brownian motions, to obtain large deviations estimates and study free entropies.

16

Free Probability Setting

16.1 A Few Notions about Algebras and Tracial States

Definition 16.1. *A C^*-algebra $(\mathcal{A}, *)$ is a complex algebra equipped with an involution $*$ and a norm $||.||_{\mathcal{A}}$ such that $\mathcal{A}$ is complete for the norm $|| \cdot ||_{\mathcal{A}}$ and, for any $X, Y \in \mathcal{A}$,*

$$\|XY\|_{\mathcal{A}} \leq \|X\|_{\mathcal{A}} \|Y\|_{\mathcal{A}}, \quad \|X^*\|_{\mathcal{A}} = \|X\|_{\mathcal{A}}, \quad \|XX^*\|_{\mathcal{A}} = \|X\|_{\mathcal{A}}^2.$$

$X \in \mathcal{A}$ is self-adjoint iff $X^* = X$. $\mathcal{A}_{sa}$ denote the set of self-adjoint elements of $\mathcal{A}$. A C^*-algebra $(\mathcal{A}, *)$ is unital if it contains a neutral element I.

$\mathcal{A}$ can always be realized as a sub-C^*-algebra of the space $B(H)$ of bounded linear operators on a Hilbert space H. For instance, if $\mathcal{A}$ is a unital C^*-algebra furnished with a positive linear form τ, one can always construct such a Hilbert space H by completing and separating $L^2(\tau)$ (this is the Gelfand–Neumark–Segal construction, see [186, Theorem 2.2.1]). We shall restrict ourselves to this case in the sequel and denote by H a Hilbert space equipped with a scalar product $\langle ., . \rangle_H$ such that $\mathcal{A} \subset B(H)$.

Definition 16.2. *If $\mathcal{A}$ is a sub-C^*-algebra of $B(H)$, $\mathcal{A}$ is a von Neumann algebra iff it is closed for the weak topology, generated by the semi-norms $\{p_{\xi,\eta}(X) = \langle X\xi, \eta \rangle_H, \xi, \eta \in H\}$.*

Let us notice that by definition, a von Neumann algebra contains only bounded operators. The theory nevertheless allows us to consider unbounded operators thanks to the notion of affiliated operators. A densely defined self-adjoint operator X on H is said to be affiliated to $\mathcal{A}$ iff for any Borel function f on the spectrum of X, $f(X) \in \mathcal{A}$ (see [167, p.164]). Here, $f(X)$ is well defined for any operator X as the operator with the same eigenvectors as X and eigenvalues given by the image of those of X by the map f. Murray and von Neumann have proved that if X and Y are affiliated with $\mathcal{A}$, $aX + bY$ is also affiliated with $\mathcal{A}$ for any $a, b \in \mathbb{C}$.

A. Guionnet, *Large Random Matrices: Lectures on Macroscopic Asymptotics*, Lecture Notes in Mathematics 1957, DOI: 10.1007/978-3-540-69897-5_16, © 2009 Springer-Verlag Berlin Heidelberg, Reprint by Springer-Verlag Berlin Heidelberg 2012

A state τ on a unital von Neumann algebra $(\mathcal{A}, *)$ is a linear form on $\mathcal{A}$ such that $\tau(\mathcal{A}_{sa}) \subset \mathbb{R}$ and:

1. **Positivity** $\tau(\mathbf{A}\mathbf{A}^*) \geq 0$, for any $\mathbf{A} \in \mathcal{A}$.
2. **Total mass** $\tau(I) = 1$.

A tracial state satisfies the additional hypothesis:

3. **Traciality** $\tau(\mathbf{A}\mathbf{B}) = \tau(\mathbf{B}\mathbf{A})$ for any $\mathbf{A}, \mathbf{B} \in \mathcal{A}$.

The couple $(\mathcal{A}, \tau)$ of a von Neumann algebra equipped with a state τ is called a W^*- probability space.

Exercise 16.3.   *1. Let $n \in \mathbb{N}$, and consider $\mathcal{A} = M_n(\mathbb{C})$ as the set of bounded linear operators on $\mathbb{C}^n$. For any $v \in \mathbb{C}^n$, $\langle v, v \rangle_{\mathbb{C}^n} = \sum_{i=1}^{n} |v_i|^2 = \|v\|_{\mathbb{C}^n}^2 = 1$,*

$$\tau_v(M) = \langle v, Mv \rangle_{\mathbb{C}^n}$$

is a state. There is a unique tracial state on $M_n(\mathbb{C})$ that is the normalized trace

$$\frac{1}{n}\mathrm{Tr}(M) = \frac{1}{n}\sum_{i=1}^{n} M_{ii}.$$

2. Let $(X, \Sigma, d\mu)$ be a classical probability space. Then $\mathcal{A} = L^\infty(X, \Sigma, d\mu)$ equipped with the expectation $\tau(f) = \int f d\mu$ is a (non-)commutative probability space. Here, $L^\infty(X, \Sigma, d\mu)$ is identified with the set of bounded linear operators on the Hilbert space H obtained by separating $L^2(X, \Sigma, d\mu)$ (by the equivalence relation $f \simeq g$ iff $\mu((f - g)^2) = 0$). The identification follows from the multiplication operator $M(f)g = fg$. Observe that $\mathcal{A}$ is weakly closed for the semi-norms $(\langle f, .g \rangle_H, f, g \in L^2(\mu))$ as $L^\infty(X, \Sigma, d\mu)$ is the dual of $L^1(X, \Sigma, d\mu)$.

3. Let G be a discrete group, and $(e_h)_{h \in G}$ be a basis of $\ell^2(G)$. Let $\lambda(h)e_g = e_{hg}$. Then, we take $\mathcal{A}$ to be the von Neumann algebra generated by the linear span of $\lambda(G)$. The (tracial) state is the linear form such that $\tau(\lambda(g)) = 1_{g=e}$ (e = neutral element).

We refer to [200] for further examples and details.

The notion of law $\tau_{X_1,\dots,X_m}$ of m operators $(X_1, \dots, X_m)$ in a W^*-probability space $(\mathcal{A}, \tau)$ is simply given by the restriction of the trace τ to the algebra generated by $(X_1, \dots, X_m)$, that is by the values

$$\tau_{X_1,\dots,X_m}(P) := \tau(P(X_1, \dots, X_m)), \quad P \in \mathbb{C}\langle X_1, \dots X_m \rangle$$

where $\mathbb{C}\langle X_1, \dots X_m \rangle$ denotes the set of polynomial functions of m non-commutative variables.

16.2 Space of Laws of m Non-commutative Self-adjoint Variables

Following the above description, laws of m non-commutative self-adjoint variables can be seen as elements of the set $\mathcal{M}^{(m)}$ of linear forms τ on the set of polynomial functions of m non-commutative variables $\mathbb{C}\langle X_1, \ldots X_m \rangle$ furnished with the involution

$$(X_{i_1} X_{i_2} \cdots X_{i_n})^* = X_{i_n} X_{i_{n-1}} \cdots X_{i_1}$$

and such that:

1. **Positivity** $\tau(PP^*) \geq 0$, for any $P \in \mathbb{C}\langle X_1, \ldots X_m \rangle$,
2. **Traciality** $\tau(PQ) = \tau(QP)$ for any $P, Q \in \mathbb{C}\langle X_1, \ldots X_m \rangle$,
3. **Total mass** $\tau(I) = 1$.

This point of view is identical to the previous one. Indeed, being given $\mu \in \mathcal{M}^{(m)}$ such that

$$|\mu(X_{i_1} \cdots X_{i_k})| \leq R^k$$

for any choices of $i_1, \cdots, i_k \in \{1, \cdots, m\}$ and k and some finite constant R, we can construct a W^*-probability space $(\mathcal{A}, \tau)$ and operators $(X_1, \ldots, X_m)$ such that

$$\mu = \tau_{X_1, \ldots, X_m}. \tag{16.1}$$

We refer to [6, 167, 200] for such a construction.

The topology under consideration is usually in free probability the $\mathbb{C}\langle X_1, \ldots X_m \rangle$-* topology that is $\{\tau_{X_1^n, \ldots, X_m^n}\}_{n \in \mathbb{N}}$ converges to $\tau_{X_1, \ldots, X_m}$ iff for every $P \in \mathbb{C}\langle X_1, \ldots X_m \rangle$,

$$\lim_{n \to \infty} \tau_{X_1^n, \ldots, X_m^n}(P) = \tau_{X_1, \ldots, X_m}(P).$$

If $(X_1^n, \ldots, X_m^n)_{n \in \mathbb{N}}$ are non-commutative variables whose law $\tau_{X_1^n, \ldots, X_m^n}$ converges to $\tau_{X_1, \ldots, X_m}$, then we shall also say that $(X_1^n, \ldots, X_m^n)_{n \in \mathbb{N}}$ converges in law (or in distribution) to $(X_1, \ldots, X_m)$.

Such a topology is reasonable when one deals with uniformly bounded non-commutative variables. In fact, if we consider for $R \in \mathbb{R}^+$,

$$\mathcal{M}_R^{(m)} := \{\mu \in \mathcal{M}^{(m)} : \mu(X_i^{2p}) \leq R^p, \ \forall p \in \mathbb{N}, \ 1 \leq i \leq m\},$$

then $\mathcal{M}_R^{(m)}$, equipped with this $\mathbb{C}\langle X_1, \ldots X_m \rangle$-* topology, is a Polish space (i.e a complete metric space). In fact, $\mathcal{M}_R^{(m)}$ is compact by the Banach–Alaoglu theorem. A distance is for instance given by

$$d(\mu, \nu) = \sum_{n \geq 0} \frac{1}{2^n} |\mu(P_n) - \nu(P_n)|$$

where $\{P_n\}_{n \in \mathbb{N}}$ is a dense sequence of polynomials with operator norm bounded by one when evaluated at any set of self-adjoint operators with operator norms bounded by R.

This notion is the generalization of laws of m real-valued variables bounded by a given finite constant R, in which case the $\mathbb{C}\langle X_1, \ldots X_m \rangle$-* topology driven by polynomial functions is the same as the standard weak topology. Actually, it is not hard to check that $\mathcal{M}_R^{(1)} = \mathcal{P}([-R, R])$. However, it may be useful to consider more general topologies, compatible with the existence of unbounded operators, as might be encountered for instance when considering the deviations of large random matrices. One way to do that is to change the set of test functions (as one does in the case $m = 1$ where bounded continuous test functions are often chosen to define the standard weak topology). In [55], the set of test functions was chosen to be the complex vector space $CC_{st}^m(\mathbb{C})$ generated by the Stieltjes functionals

$$ST^m(\mathbb{C}) = \left\{ \overrightarrow{\prod_{1 \le i \le n}} \left(z_i - \sum_{k=1}^{m} \alpha_i^k X_k \right)^{-1} ; \quad z_i \in \mathbb{C}\backslash\mathbb{R}, \alpha_i^k \in \mathbb{Q}, n \in \mathbb{N} \right\}$$

(16.2)

where $\overrightarrow{\prod}$ denotes the non-commutative product. It can be checked easily that, with such type of test functions, $\mathcal{M}^{(m)}$ is again a Polish space.

A particular important example of non-commutative laws is given by the empirical distribution of matrices.

Definition 16.4. *Let $N \in \mathbb{N}$ and consider m Hermitian matrices $A_1^N, \ldots, A_m^N \in \mathcal{H}_N^m$. Then, the empirical distribution of the matrices $(A_1^N, \ldots, A_m^N)$ is given by*

$$\mathbf{L}_{A_1^N, \ldots, A_m^N}(P) := \frac{1}{N} \mathrm{Tr}\left(P(A_1^N, \ldots, A_m^N) \right), \quad \forall P \in \mathbb{C}\langle X_1, \cdots X_m \rangle.$$

Exercise 16.5. *Show that if the spectral radius of $(A_1^N, \ldots, A_m^N)$ is uniformly bounded by R, $\mathbf{L}_{A_1^N, \ldots, A_m^N}$ is an element of the set $\mathcal{M}_R^{(m)}$ of non-commutative laws. Moreover, if $(A_1^N, \ldots, A_m^N)_{N \in \mathbb{N}}$ is a sequence such that*

$$\lim_{N \to \infty} \mathbf{L}_{A_1^N, \ldots, A_m^N}(P) = \tau(P), \quad \forall P \in \mathbb{C}\langle X_1, \cdots X_m \rangle,$$

show that $\tau \in \mathcal{M}_R^{(m)}$.

It is actually a long-standing question posed by A. Connes to know whether all $\tau \in \mathcal{M}^{(m)}$ can be approximated in such a way. In the case $m = 1$, the question amounts to asking if for all $\mu \in \mathcal{P}([-R, R])$, there exists a sequence $(\lambda_1^N, \ldots, \lambda_N^N)_{N \in \mathbb{N}}$ such that

$$\lim_{N \to \infty} \frac{1}{N} \sum_{i=1}^{N} \delta_{\lambda_i^N} = \mu.$$

This is well known to be true by the Birkhoff's theorem (which is based on the Krein–Milman theorem), but still an open question when $m \ge 2$.

Bibliographical Notes. Introductory notes to free probability can be found in [118, 200, 202, 203]. Basics on operator algebra theory are taken from [83, 167].

17

Freeness

In this chapter we first define freeness as a non-commutative analog of independence. We then show how independent matrices, as their size goes to
infinity, converge to free variables.

17.1 Definition of Freeness

Free probability is not only a theory of probability for non-commutative variables; it contains also the central notion of freeness, that is, the analog of
independence in standard probability.

Definition 17.1. *The variables* $(X_1, \ldots, X_m)$ *and* $(Y_1, \ldots, Y_n)$ *are said to be
free iff for any* $(P_i, Q_i)_{1 \le i \le p} \in (\mathbb{C}\langle X_1, \ldots, X_m\rangle \times \mathbb{C}\langle X_1, \ldots, X_n\rangle)^p$,

$$\tau\left(\overrightarrow{\prod_{1 \le i \le p}} P_i(X_1, \ldots, X_m)Q_i(Y_1, \ldots, Y_n)\right) = 0 \qquad (17.1)$$

as soon as

$$\tau\left(P_i(X_1, \ldots, X_m)\right) = 0, \qquad \tau\left(Q_i(Y_1, \ldots, Y_n)\right) = 0, \qquad \forall i \in \{1, \ldots, p\}.$$

More generally, let $(\mathcal{A}, \phi)$ *be a non-commutative probability space and consider
unital subalgebras* $\mathcal{A}_1, \ldots, \mathcal{A}_m \subset \mathcal{A}$. *Then,* $\mathcal{A}_1, \ldots, \mathcal{A}_m$ *are free if and only if
for any* $(a_1, \ldots, a_n) \in \mathcal{A}$, $a_j \in \mathcal{A}_{i_j}$

$$\phi(a_1 \cdots a_n) = 0$$

as soon as $i_j \ne i_{j+1}$ *for* $1 \le j \le n - 1$ *and* $\phi(a_i) = 0$ *for all* $i \in \{1, \ldots, n-1\}$.

Observe that the assumption that $\tau(Q_p(Y_1, \ldots, Y_n)) = 0$ can be removed by
linearity.

A. Guionnet, *Large Random Matrices: Lectures on Macroscopic Asymptotics*,
Lecture Notes in Mathematics 1957, DOI: 10.1007/978-3-540-69897-5_17,
© 2009 Springer-Verlag Berlin Heidelberg, Reprint by Springer-Verlag Berlin Heidelberg 2012

Remark 17.2. *(1) The notion of freeness defines uniquely the law of*

$$\{X_1, \ldots, X_m, Y_1, \ldots, Y_n\}$$

once the laws of $(X_1, \ldots, X_m)$ and $(Y_1, \ldots, Y_n)$ are given (in fact, check that every expectation of any polynomial is given uniquely by induction over the degree of this polynomial).

(2) If X and Y are free variables with joint law τ, and $P, Q \in \mathbb{C}\langle X \rangle$ such that $\tau(P(X)) = 0$ and $\tau(Q(Y)) = 0$, it is clear that $\tau(P(X)Q(Y)) = 0$ as it should for independent variables, but also $\tau(P(X)Q(Y)P(X)Q(Y)) = 0$ which is very different from what happens with usual independent commutative variables where $\mu(P(X)Q(Y)P(X)Q(Y)) = \mu(P(X)^2 Q(Y)^2) > 0$.

(3) The above notion of freeness is related with the usual notion of freeness in groups as follows. Let $(x_1, \ldots, x_m, y_1, \ldots, y_n)$ be elements of a group. Then, $(x_1, \ldots, x_m)$ is said to be free from $(y_1, \ldots, y_n)$ if any non-trivial words in these elements is not the neutral element of the group, i.e., that for every monomials $P_1, \ldots, P_k \in \mathbb{C}\langle X_1, \ldots, X_m \rangle$ and $Q_1, \ldots, Q_k \in \mathbb{C}\langle X_1, \ldots, X_n \rangle$, $P_1(x)Q_1(y)P_2(x) \cdots Q_k(y)$ is not the neutral element as soon as the $Q_k(y)$ and the $P_i(x)$ are not the neutral element. If we consider, following example 16.3.3), the map that is one on trivial words and zero otherwise and extend it by linearity to polynomials, we see that this defines a tracial state on the operators of left multiplication by the elements of the group and that the two notions of freeness coincide.

(4) We shall see below that examples of free variables naturally show up when considering random matrices with size going to infinity.

17.2 Asymptotic Freeness

Definition 17.3. *Let $(X_1^N, \ldots, X_m^N)_{N \geq 0}$ and $(Y_1^N, \ldots, Y_n^N)_{N \geq 0}$ be two families of $N \times N$ random Hermitian matrices on a probability space (Ω, P). $(X_1^N, \ldots, X_m^N)_{N \geq 0}$ and $(Y_1^N, \ldots, Y_n^N)_{N \geq 0}$ are asymptotically free almost surely (respectively in expectation) iff $\mathbf{L}_{X_1^N, \ldots, X_m^N, Y_1^N, \ldots, Y_n^N}(Q)$ (respectively $\mathbb{E}[\mathbf{L}_{X_1^N, \cdots X_m^N, Y_1^N, \cdots Y_n^N}(Q)])$ converges for all polynomial Q to $\tau(Q)$ and τ satisfies (17.1).*

We claim that:

Lemma 17.4. *Let $(X_1^N, \ldots, X_m^N)_{N \in \mathbb{N}}$ be m independent matrices taken from the GUE. Then, $(X_1^N, \ldots, X_{m-1}^N)_{N \in \mathbb{N}}$ and $(X_m^N)_{N \in \mathbb{N}}$ are asymptotically free almost surely (and in expectation).*

Proof. By Theorem 8.4 with $V = 0$, $\mathbf{L}_{X_1^N, \cdots X_m^N}$ converges almost surely and in expectation as N goes to infinity. Moreover, the limit satisfies the Schwinger–Dyson equation **SD**$[0]$

$$\tau(X_i P) = \tau \otimes \tau(\partial_i P).$$

We check by induction over the degree of P that $\mathbf{SD}[0]$ implies (17.1). Indeed, (17.1) is true if the total degree of $R = \prod^{\rightarrow} P_i Q_i$ is one by definition. Let us assume this relation is true for any monomial $R = \prod^{\rightarrow} P_j Q_j$ of degree less than k with $\tau(P_j(X_1,\ldots,X_{m-1})) = 0$ and $\tau(Q_j(X_m)) = 0$ except possibly for the first P_j. If now $R' = X_i R$ for some $i \leq m-1$, the Schwinger–Dyson equation $\mathbf{SD}[0]$ gives

$$\tau(R') = \sum_{R=R_1 X_i R_2} \tau(R_1)\tau(R_2)$$

where R_1 and R_2 are polynomials of degree less than k which are products of centered polynomials P_i and Q_i as above, except possibly for the first and the last terms in this alternating product which may be not centered. Thus, $\tau(R_2) = 0$ by induction unless $R_2 = 1$ in which case $\tau(R_1)$ vanishes since by traciality it can be written as $\tau(P_1' R_1')$ with R_1' an alternated product of centered monomials. Hence, $\tau(R') = 0$. We can prove similarly that $\tau(X_m \prod^{\rightarrow} Q_j P_j)$ vanishes. This proves the induction. $\qquad\square$

We next show that Lemma 17.4 can be generalized to laws that are invariant by multiplication by unitary matrices. Let $(A_1^N,\ldots,A_m^N)_{N\geq 0}$ be self-adjoint matrices with operator norm bounded independently of N. Finally, suppose that

$$\lim_{N\to\infty} \mathbf{L}_{A_1^N,\ldots,A_m^N} = \mu.$$

Let $U_1^N,\ldots,U_m^N$ be m independent unitary matrices, independent of the A_N^i's, following the Haar measure on $\mathcal{U}(N)$. Then:

Theorem 17.5. $\{A_i^N\}_{1\leq i\leq m}$ and $\{U_i^N,(U_i^N)^{-1}\}_{1\leq i\leq m}$ are asymptotically free almost surely and in expectation. Moreover, the variables $\{U_i^N, (U_i^N)^{-1}\}_{1\leq i\leq m}$ are asymptotically free almost surely with limit law τ such that $\tau(U_i^n) = 1_{n=0}$ for all $n \in \mathbb{Z}$. In particular, $(U_i^N A_i^N U_i^*)_{1\leq i\leq m}$ are asymptotically free variables.

Exercise 17.6. Extend the theorem to the case where the Haar measure on $\mathcal{U}(N)$ is replaced by the Haar measure on $\mathcal{O}(N)$ Hint: extend the proof below.

The proof we shall provide here follows the change of variable trick that we used to analyze matrix models. It can be found in [199] (and is taken, in the form below, from a work with B. Collins and E. Maurel Segala [66]).

Proof. 1. *Corresponding Schwinger–Dyson equation.* We denote by m_N the Haar measure on $\mathcal{U}(N)$. By definition, m_N is invariant under left multiplication by a unitary matrix. In particular, if $P \in \mathbb{C}\langle A_i, U_i, U_i^{-1}, 1 \leq i \leq m\rangle$, we have for all $kl \in \{1,\ldots,N\}^2$,

$$\partial_t \int \left(P(A_i, e^{tB_i}U_i, U_i^* e^{-tB_i})\right)_{kl} dm_N(U_1)\cdots dm_N(U_m) = 0$$

for any antihermitian matrices B_i ($B_i^* = -B_i$). Taking B_i null except for $i = i_0$ and B_{i_0} null except at the entries qr and rq, we find that

$$\int [\partial_{i_0} P]_{kr,ql}(A_i, U_i, U_i^*) dm_N(U_1) \cdots dm_N(U_m) = 0$$

with ∂_i the derivative that obeys Leibniz's rule

$$\partial_i(PQ) = \partial_i P \times 1 \otimes Q + P \otimes 1 \times \partial_i Q \qquad (17.2)$$

so that

$$\partial_i U_j = 1_{j=i} U_j \otimes 1, \partial_i U_j^* = -1_{j=i} 1 \otimes U_j^* \qquad (17.3)$$

and $[A \otimes B]_{kr,ql} := A_{kr} B_{ql}$. Taking $k = r$ and $q = l$ and summing over r, q gives

$$\mathbb{E}\left[\mathbf{L}_{A_i^N, U_i^N, (U_i^N)^*, 1 \leq i \leq m} \otimes \mathbf{L}_{A_i^N, U_i^N, (U_i^N)^*, 1 \leq i \leq m} \right] = 0.$$

Observe that $\mathbf{L}_{A_i^N, U_i^N, (U_i^N)^*, 1 \leq i \leq m}(P)$ is a Lipschitz function of the unitary matrices $\{U_i^N, (U_i^N)^*\}_{1 \leq i \leq p}$ with uniformly bounded constant since we assumed that the A_i^N's are uniformly bounded for the operator norm. Thus, we can use the concentration result of Theorem 6.16 to deduce that for all $P \in \mathbb{C}\langle A_i, U_i, U_i^{-1}, 1 \leq i \leq m \rangle$

$$\lim_{N \to \infty} \mathbb{E}[\mathbf{L}_{A_i^N, U_i^N, (U_i^N)^*, 1 \leq i \leq m}] \otimes \mathbb{E}[\mathbf{L}_{A_i^N, U_i^N, (U_i^N)^*, 1 \leq i \leq m}] = 0. \qquad (17.4)$$

Observe that $\mathbb{E}[\mathbf{L}_{A_i^N, U_i^N, (U_i^N)^*, 1 \leq i \leq m}]$ can be identified with the non-commutative law of self-adjoint variables $\mathbb{E}[\mathbf{L}_{A_i^N, V_i^N, W_i^N, 1 \leq i \leq m}]$ where $V_i^N = U_i^N + (U_i^N)^*$ and $W_i^N = (U_i^N - (U_i^N)^*)/\sqrt{-1}$ up to an obvious change of variables. If A denotes a uniform bound on the spectral radius of the $\{A_i^N\}_{1 \leq i \leq m}$, $\mathbb{E}[\mathbf{L}_{A_i^N, V_i^N, W_i^N, 1 \leq i \leq m}]$ belongs to the compact set $\mathcal{M}_{A \vee 2}^{(3m)}$. Thus, we can consider limit points of $\mathbb{E}[\mathbf{L}_{A_i^N, V_i^N, W_i^N, 1 \leq i \leq m}]$, and therefore of $\mathbb{E}[\mathbf{L}_{A_i^N, U_i^N, (U_i^N)^*, 1 \leq i \leq m}]$. If τ is such a limit point, we deduce from (17.4) that it must satisfy the Schwinger–Dyson equation

$$\tau \otimes \tau(\partial_i P) = 0 \qquad (17.5)$$

for all $i \in \{1, \ldots, m\}$ and $P \in \mathbb{C}\langle A_i, U_i, U_i^{-1}, 1 \leq i \leq m \rangle$. Note here that we can bypass the concentration argument by using the change of variable trick that shows that the above equation is satisfied almost surely asymptotically (see [66]).

2. *Uniqueness of the solution to* (17.5).

Let τ be a tracial solution to (17.5) and P be a monomial. Note that if P depends only on the A_i's, $\tau(P) = \mu(P)$ is uniquely determined. If P belongs to $\mathbb{C}\langle A_i, U_i, U_i^{-1}, 1 \leq i \leq m \rangle \backslash \mathbb{C}\langle A_i, 1 \leq i \leq m \rangle$, we can always write $\tau(P) = \tau(QU_i)$ or $\tau(P) = \tau(U_i^{-1}Q)$ for some monomial Q. Let us consider the first case (the second can then be deduced from the fact that $\tau(U_i^{-1}Q) = \tau(Q^*U_i)$). Then, we have

$$\partial_i(QU_i) = \partial_i Q \times 1 \otimes U_i + (QU_i) \otimes 1$$

and so (17.5) gives

$$\tau(QU_i) = -\tau \otimes \tau(\partial_i Q \times 1 \otimes U_i)$$

$$= - \sum_{Q=Q_1 U_i Q_2} \tau(Q_1 U_i)\tau(Q_2 U_i) + \sum_{Q=Q_1 U_i^* Q_2} \tau(Q_1)\tau(Q_2)$$

where we used $\tau(U_i^{-1}Q_2 U_i) = \tau(Q_2)$ by traciality. Each term in the above right-hand side is the expectation under τ of a polynomial of degree strictly smaller in U_i and U_i^{-1} than QU_i. Hence, this relation defines uniquely τ by induction.

3. *The solution is the law of free variables.* It is enough to show by the previous point that if the algebra generated by $\{U_i, U_i^{-1}, 1 \leq i \leq m\}$ is free from the $A_i's$, then the corresponding tracial state on $\mathbb{C}\langle A_i, U_i, U_i^{-1}, 1 \leq i \leq m\rangle$ satisfies (17.5). So take $P = U_{i_1}^{n_1} B_1 \cdots U_{i_p}^{n_p} B_p$ with some B_k's in the algebra generated by $(A_i, 1 \leq i \leq m)$. We wish to show that for all $i \in \{1, \ldots, m\}$,

$$\mu \otimes \mu(\partial_i P) = 0.$$

By linearity, it is enough to prove this equality when $\mu(B_j) = 0$ for all j. Using repeatedly (17.2) and (17.3), we find that

$$\partial_i P = \sum_{k:i_k=i, n_k>0} \sum_{l=1}^{n_k} U_{i_1}^{n_1} B_1 \cdots B_{k-1} U_i^l \otimes U_i^{n_k-l} B_k \cdots U_{i_p}^{n_p} B_p$$

$$- \sum_{k:i_k=i, n_k<0} \sum_{l=0}^{n_k-1} U_{i_1}^{n_1} B_1 \cdots B_{k-1} U_i^{-l} \otimes U_i^{n_k+l} B_k \cdots U_{i_p}^{n_p} B_p$$

Taking the expectation on both sides, since $\mu(U_i^i) = 0$ and $\mu(B_j) = 0$ for all $i \neq 0$ and j, we see that freeness implies that the right-hand side vanishes (recall here that in the definition of freeness, two consecutive elements have to be in free algebras but the first and the last element can be in the same algebra). Thus, $\mu \otimes \mu(\partial_i P)$ vanishes, which proves the claim.

The last point of the theorem is a direct consequence of the asymptotic freeness of the algebra which implies that for all $B_i \in \mathbb{C}\langle A_i, 1 \leq i \leq m\rangle$ such that $\mu(B_i) = \tau(U_j B_i U_j^{-1}) = 0$

$$\tau(B_1 U_{i_1} B_2 U_{i_1}^{-1} B_3 \cdots U_{i_p}^{-1}) = 0$$

and therefore $(U_i^N A_i^N (U_i^N)^{-1})_{1 \leq i \leq m}$ are asymptotically free.

$\square$

17.3 The Combinatorics of Freeness

It is a natural question to wonder, if a, b are two free bounded variables in a non-commutative probability space, what is the law of their sum $a + b$ or of their product ab. The study of this question is the object of this section. We restrict ourselves to bounded variables.

17.3.1 Free Cumulants

In practice, the notion of cumulants often appears to be easier to work with to compute laws than the direct use of (17.1). We recall below the definition of free cumulants that is based on non-crossing partitions.

Definition 17.7. • *A partition of a the set $S := \{1, \ldots, n\}$ is a decomposition*

$$\pi = \{V_1, \ldots, V_r\}$$

of S into disjoint and non empty sets V_i.
- *The set of all partitions of S is denoted by $\mathcal{P}(S)$, and for short by $\mathcal{P}(n)$ if $S := \{1, \ldots, n\}$.*
- *The $V_i, 1 \le i \le r$, are called the blocks of the partition and we say that $p \sim_\pi q$ if p, q belong to the same block of the partition π.*

The central result of this section is that freeness is related with *non-crossing* partitions:

Definition 17.8. • *A partition π of $\{1, \ldots, n\}$ is said to be crossing if there exists $1 \le p_1 < q_1 < p_2 < q_2 \le n$ with*

$$p_1 \sim_\pi p_2 \nsim_\pi q_1 \sim_\pi q_2.$$

It is non- crossing otherwise.
- *The set of non-crossing partitions of $\{1, \ldots, n\}$ is denoted by $NC(n)$.*
- *We let $\le$ be the partial order on $NC(n)$; if $\pi, \pi' \in NC(n)$, $\pi \le \pi'$ iff every block of π is included into a block of π'. For this partial order, $\mathbf{0_n} := \{(1), \ldots, (n)\}$ (resp. $\mathbf{1_n} := \{(1, \ldots, n)\}$) is the "smallest" (resp. "largest") element of $NC(n)$.*
- *We write $NC(S_1) \equiv NC(S_2)$ iff S_1 and S_2 have the same number of elements.*

A pictural description of non-crossing versus crossing partitions was given in Figure 1.4. In practice, the following recursive definition of non-crossing partition shall be used.

Property 17.9. *A partition π of $S = \{1, \ldots, n\}$ is non-crossing iff there is at least one block V of π that is an interval (i.e., of the form $\{p, p+1, \ldots, p+q\}$ for some $q \ge 0$ and $1 \le p \le p + q \le n$) and the restriction of π to $S \backslash V$ is non-crossing.*

Proof. If π is non-crossing, we consider the block V that contains 1. If it is an interval, we are done, and otherwise we consider a block S' in $S\backslash V$ that is contained in between the first and the last elements of V (here, elements of a block are ordered) and whose first element is the smaller between the possible choices of S'. Since the restriction of π to this block is non-crossing, we can reiterate this procedure. Since the cardinality of elements of S' is strictly less than the cardinality of S, the iteration of this decomposition will lead us to a block W of the partition that is of the form $\{p, p+1, \ldots, p+q\}$ with $q \geq 0$. Moreover, the restriction of π to $S\backslash W$ is non-crossing. Reciprocally, since the restriction of π to an interval is non-crossing and we assumed π non-crossing once restricted to $S\backslash V$, π is non-crossing.   □

Definition 17.10. *Let $(\mathcal{A}, \phi)$ be a non-commutative probability space. The free cumulants are defined as a collection of multi-linear functionals*

$$k_n : \mathcal{A}^n \to \mathbb{C} \quad (n \in \mathbb{N})$$

by the following system of equations:

$$\phi(a_1 \cdots a_n) = \sum_{\pi \in NC(n)} k_\pi(a_1, \ldots, a_n)$$

with, if $\pi = (V_1, \ldots, V_r)$ with $V_i = \{v_1^i, \cdots, \ldots, v_{l_i}^i\}$ for $1 \leq i \leq r$,

$$k_\pi(a_1, \ldots, a_n) = k_{l_1}(a_{v_1^1}, \ldots, a_{v_{l_1}^1}) k_{l_2}(a_{v_1^2}, \ldots, a_{v_{l_2}^2}) \cdots k_{l_r}(a_{v_1^r}, \ldots, a_{v_{l_r}^r}).$$

Observe that the above system of equations is implicit but defines uniquely the cumulants k_n since

$$k_n(a_1, \ldots, a_n) = \phi(a_1 \cdots a_n) - \sum_{\substack{\pi \in NC(n) \\ \pi \neq 1_n}} k_\pi(a_1, \ldots, a_n)$$

where the last term in the right-hand side only depends on $(k_l, l \leq n - 1)$. This last definition of k_n also shows its existence by induction over n (remark that k_π is multilinear).

Example 17.11. • $n = 1$, $k_1(a_1) = \phi(a_1)$.
• $n = 2$, $\phi(a_1 a_2) = k_2(a_1, a_2) + k_1(a_1)k_1(a_2)$ *and so*

$$k_2(a_1, a_2) = \phi(a_1 a_2) - \phi(a_1)\phi(a_2).$$

• $n = 3$,

$$k_3(a_1, a_2, a_3) = \phi(a_1 a_2 a_3) - \phi(a_1)\phi(a_2 a_3) - \phi(a_1 a_3)\phi(a_2)$$
$$- \phi(a_1 a_2)\phi(a_3) + 2\phi(a_1)\phi(a_2)\phi(a_3).$$

We now turn to the description of freeness in terms of cumulants.

Theorem 17.12. *Let $(\mathcal{A}, \phi)$ be a non-commutative probability space and consider unital subalgebras $\mathcal{A}_1, \ldots, \mathcal{A}_m \subset \mathcal{A}$. Then, $\mathcal{A}_1, \ldots, \mathcal{A}_m$ are free if and only if for all $n \geq 2$ and for all $a_i \in \mathcal{A}_{j(i)}$ with $1 \leq j(1), \ldots, j(n) \leq m$,*

$$k_n(a_1, \ldots, a_n) = 0 \quad \text{if there exists } 1 \leq l, k \leq n \text{ with } j(l) \neq j(k). \quad (17.6)$$

Observe here that the description of freeness by cumulants does not require any centering of the variables; all the questions of centering concern only the cumulant k_1. In fact, we have:

Property 17.13. *Let $(\mathcal{A}, \phi)$ be a non-commutative probability space and $a_1, \ldots, a_n$ be elements of $\mathcal{A}$. Assume $n \geq 2$. If there is $i \in \{1, \ldots, n\}$ so that $a_i = 1$, then*

$$k_n(a_1, \ldots, a_n) = 0.$$

As a consequence, for $n \geq 2$, any $a_1, \ldots, a_n \in \mathcal{A}$,

$$k_n(a_1, \ldots, a_n) = k_n(a_1 - \phi(a_1), a_2 - \phi(a_2), \ldots, a_n - \phi(a_n)).$$

Proof. We prove this result by induction over $n \geq 2$. First, for $n = 2$ we have, since $k_1(a) = \phi(a)$

$$\phi(a_1 a_2) = k_2(a_1, a_2) + \phi(a_1)\phi(a_2)$$

and so if $a_1 = 1$, we deduce, since $\phi(1) = 1$, that

$$\phi(a_2) = \phi(1 a_2) = k_2(a_1, a_2) + \phi(a_2) \Rightarrow k_2(a_1, a_2) = 0.$$

The same argument holds when $a_2 = 1$. Let us assume that for $p \leq n - 1$, $k_p(a_1, \ldots, a_p) = 0$ if one of the a_p is the neutral element. Consider the step n with $a_i = 1$. Then

$$\phi(a_1 \cdots a_n) = k_n(a_1, \ldots, a_n) + \sum_{\substack{\pi \in NC(n) \\ \pi \neq 1_n}} k_\pi(a_1, \ldots, a_n) \quad (17.7)$$

where by our induction hypothesis all the partitions π in the above sum where the element i is not a block of the partition do not contribute. But then

$$\sum_{\substack{\pi \in NC(n) \\ \pi \neq 1_n}} k_\pi(a_1, \ldots, a_n) = \sum_{\pi \in NC(n-1)} k_\pi(a_1, \ldots, a_{i-1}, a_{i+1}, \ldots, a_n)$$

$$= \phi(a_1 \cdots a_{i-1} a_{i+1} \cdots a_n) = \phi(a_1 \cdots a_n)$$

which proves $k_n(a_1, \ldots, a_n) = 0$ with (17.7). □

Proof of Theorem 17.12. Assume that the cumulants vanish when evaluated at elements of different algebras $\mathcal{A}_1, \ldots, \mathcal{A}_m$ and consider, for $a_i \in \mathcal{A}_{j(i)}$,

$$\phi((a_1 - \phi(a_1)) \cdots (a_n - \phi(a_n))) = \sum_{\pi \in NC(n)} k_\pi(a_1, \ldots, a_n).$$

By our hypothesis, k_π vanishes as soon as a block of π contains $p, q \in \{1, \ldots, n\}$ so that $j(p) \neq j(q)$. Therefore, if we assume $j(p) \neq j(p+1)$, we see that the contribution in the above sum comes from partitions π whose blocks cannot contain two nearest neighbors $\{p, p+1\}$. By Property 17.9, this implies that π contains an interval of the form $V = \{p\}$. But then k_π also vanishes since $k_1 = 0$ by centering of the variables. Therefore, if for $1 \leq p \leq n-1$ $j(p) \neq j(p+1)$, we get

$$\phi((a_1 - \phi(a_1)) \cdots (a_n - \phi(a_n))) = 0,$$

that is ϕ satisfies (17.1).

Reciprocally, let us assume that ϕ satisfies (17.1). We prove that (17.6) is satisfied by induction over n. It is clear for $n = 2$ since then we saw that $k_2(a_1, a_2) = \phi(a_1 a_2) - \phi(a_1)\phi(a_2)$. Let us assume it is true for $p \leq n - 1$, $n \geq 3$.

We first prove that $k_n(a_1, \ldots, a_n) = 0$ when $a_i \in \mathcal{A}_{j(i)}, 1 \leq i \leq n$ with $j(i) \neq j(i+1)$ for all $1 \leq i \leq n - 1$. Indeed, for any $\pi \in NC(n)$, $\pi \neq 1_n$, k_π will vanish as soon as it contains two nearest neighbors by our induction hypothesis. But again by property 17.9, this implies that π contains a singleton. Thus $k_\pi(a_1 - \phi(a_1), \ldots, a_n - \phi(a_n)) = 0$ since $k_1(a_i - \phi(a_i)) = 0$ and since also $\phi((a_1 - \phi(a_1)) \cdots (a_n - \phi(a_n))$ vanish, we deduce that $k_n(a_1 - \phi(a_1), \ldots, a_n - \phi(a_n))$ vanishes. Since k_n does not depend on the centering by the previous property, we have shown that $k_n(a_1, \ldots, a_n) = 0$ when $j(i) \neq j(i+1)$ for all $1 \leq i \leq n - 1$, $j(n) \neq j(1)$.

To prove that $k_n(a_1, \ldots, a_n)$ vanishes as soon as a couple of a_i's belong to different subalgebras, we shall show how to come back to the situation of alternating moments by the next lemma.

Lemma 17.14. *Consider $n \geq 2$, $a_1, \ldots, a_n \in \mathcal{A}$ and $1 \leq p \leq n - 1$. Then,*

$$k_{n-1}(a_1, \ldots, a_{p-1}, a_p a_{p+1}, a_{p+2}, \ldots, a_n)$$
$$= k_n(a_1, \cdots a_p, a_{p+1}, \ldots, a_n)$$
$$+ \sum_{\substack{\pi \in NC(n) \\ \sharp \pi = 2, p \not\sim_\pi p+1}} k_\pi(a_1, \ldots, a_p, a_{p+1}, \ldots, a_n)$$

where $\sharp \pi = 2$ means that π has exactly two blocks.

We complete the proof of the theorem before proving the lemma. We have $a_i \in \mathcal{A}_{j(i)}$ with some $j(l) \neq j(p)$. If $j(p) \neq j(p+1)$ for all $1 \leq p \leq n - 1$ then we are done by the previous consideration. Otherwise, there is a p so that $j(p) = j(p+1)$. We can then use Lemma 17.14 to reduce the number of variables. By our induction hypothesis $k_{n-1}(a_1, \ldots, a_{p-1}, a_p a_{p+1}, a_{p+2}, \ldots, a_n)$ vanishes, but also $k_\pi(a_1, \ldots, a_p, a_{p+1}, \ldots, a_n)$ since π decomposes into two blocks of size strictly smaller than n, one of which containing an element of a subalgebra free with $\mathcal{A}_{j(p)}$. Therefore, $k_n(a_1, \ldots, a_p, a_{p+1}, \ldots, a_n)$ vanishes and the theorem is proved. □

Proof of Lemma 17.14. Again, we prove it by induction over n. For $n = 2$, the equality reads

$$k_1(a_1 a_2) = k_2(a_1, a_2) + k_1(a_1)k_1(a_2) \Leftrightarrow k_2(a_1, a_2) = \phi(a_1 a_2) - \phi(a_1)\phi(a_2)$$

that we have already seen. We thus assume the equality true for $p \leq n - 1$ for some $n \geq 3$. For $\pi \in NC(n)$, let us denote by $\pi|_{p=p+1}$ the partition obtained by identifying p and $p + 1$, namely if $\pi = \{V_1, \ldots, V_r\}$ with $p \in V_j, p+1 \in V_l$,

$$\pi|_{p=p+1} = \{V_1, \ldots, V_j \cup V_l \backslash \{p + 1\}, \ldots, V_r\}.$$

Note that $\pi|_{p=p+1} \in NC(n - 1)$. In terms of such a partition, the equality of the lemma can be restated as

$$k_{\mathbf{1}_{n-1}}(a_1, \ldots, a_p a_{p+1}, \ldots, a_n) = \sum_{\substack{\pi \in NC(n) \\ \pi|_{p=p+1} = \mathbf{1}_{n-1}}} k_\pi(a_1, \ldots, a_p, a_{p+1}, \ldots, a_n)$$

Since we assumed that this equality is true for $l < n$, we deduce that for any $\sigma \in NC(n - 1)$, $\sigma \neq \mathbf{1}_{n-1}$ so that the block containing $a_p a_{p+1}$ has length strictly smaller than $n - 1$,

$$k_\sigma(a_1, \ldots, a_p a_{p+1}, \ldots, a_n) = \sum_{\substack{\pi \in NC(n) \\ \pi|_{p=p+1} = \sigma}} k_\pi(a_1, \ldots, a_p, a_{p+1}, \ldots, a_n).$$

Therefore, we have

$$k_{\mathbf{1}_{n-1}}(a_1, \ldots, a_{p-1}, a_p a_{p+1}, a_{p+2}, \ldots, a_n)$$

$$= \phi(a_1 \cdots a_n) - \sum_{\substack{\sigma \in NC(n-1) \\ \sigma \neq \mathbf{1}_n}} k_\sigma(a_1, \ldots, a_p a_{p+1}, \ldots, a_n)$$

$$= \phi(a_1 \cdots a_n) - \sum_{\substack{\sigma \in NC(n-1) \\ \sigma \neq \mathbf{1}_{n-1}}} \sum_{\substack{\pi \in NC(n) \\ \pi|_{p=p+1} = \sigma}} k_\pi(a_1, \ldots, a_p, a_{p+1}, \ldots, a_n)$$

$$= \sum_{\sigma \in NC(n)} k_\sigma(a_1, \ldots, a_n) - \sum_{\substack{\pi \in NC(n) \\ \pi|_{p=p+1} \neq \mathbf{1}_{n-1}}} k_\pi(a_1, \ldots, a_n)$$

$$= \sum_{\substack{\sigma \in NC(n) \\ \sigma|_{p=p+1} = \mathbf{1}_{n-1}}} k_\sigma(a_1, \ldots, a_n)$$

which proves the claim. □

Bibliographical notes. This section followed quite closely R. Speicher [185]. Note that in classical probability, cumulants play also a similar role but then partition can be crossing (e.g., Shiryaev [178, p. 290]).

We next exhibit a consequence of free independence, namely free harmonic analysis. The problem of interest is to determine the law of $a + b$ when a, b are free. Since the law of (a, b) with a, b free is uniquely determined by the laws μ_a of a and μ_b of b, the law of their sum is a function of μ_a and μ_b denoted by $\mu_a \boxplus \mu_b$. There are several approaches to the problem; we shall present the combinatorial approach based on free cumulants and refer the interested reader to [159] for more details.

17.3.2 Free Additive Convolution

Definition 17.15. *Let a, b be two operators in a non-commutative probability space $(\mathcal{A}, \phi)$ with law μ_a, μ_b respectively. If a, b are free, the law of $a + b$ is denoted by $\mu_a \boxplus \mu_b$.*

We write for short $k_n(a) = k_n(a, \ldots, a)$ as the nth cumulant of the variable a.

Lemma 17.16. *Let a, b be two bounded operators in a non-commutative probability space $(\mathcal{A}, \phi)$. If a and b are free, for all $n \geq 1$*

$$k_n(a + b) = k_n(a) + k_n(b).$$

Proof. The result is obvious for $n = 1$ by linearity of k_1. Moreover, for all $n \geq 2$, by multilinearity of the cumulants,

$$k_n(a + b) = \sum_{\epsilon_i = 0, 1} k_n(\epsilon_1 a + (1 - \epsilon_1)b, \ldots, \epsilon_n a + (1 - \epsilon_n)b) = k_n(a) + k_n(b)$$

where the second line is a direct consequence of Theorem 17.12. □

As a consequence, let us define the following generating function of cumulants:

Definition 17.17. *For a bounded operator a the formal power series*

$$R_a(z) = \sum_{n \geq 0} k_{n+1}(a) z^n$$

is called the R-transform of the law μ_a of the operator a. We also write $R_{\mu_a} := R_a$ since R only depends on the law μ_a.

Then, the R-transform is to free probability what the log-Fourier transform is to classical probability in the sense that it is linear for free convolution.

Corollary 17.18. *Let a, b be two bounded operators in a non-commutative probability space $(\mathcal{A}, \phi)$. If a and b are free,*

$$R_{a+b} = R_a + R_b \Leftrightarrow R_{\mu_a \boxplus \mu_b} = R_{\mu_a} + R_{\mu_b}.$$

We next provide a more tractable definition of the R-transform in terms of the Cauchy transform. Suppose that $\mu : \mathbb{C}[X] \to \mathbb{C}$ is a distribution with all moments. Then we may define G_μ as the formal series

$$G_\mu(z) = \sum_{n \geq 0} \mu(X^n) z^{-(n+1)}. \tag{17.8}$$

Let $K_\mu(z)$ be the formal inverse of G_μ, i.e., $G_\mu(K_\mu(z)) = z$. The *formal* power series expansion of K_μ is

$$K_\mu(z) = \frac{1}{z} + \sum_{n=1}^{\infty} C_n z^{n-1}.$$

Then, we shall prove the following:

Lemma 17.19. *Let μ be a compactly supported probability measure. For all $n \in \mathbb{N}$, $C_n = k_{n+1}$. Therefore, $R_\mu(z) = K_\mu(z) - 1/z$.*

Proof. To prove this lemma, we compare the generating function of the cumulants as the formal power series

$$C_a(z) = 1 + \sum_{n=1}^{\infty} k_n(a) z^n$$

and the generating function of the moments as the formal power series

$$M_a(z) = 1 + \sum_{n=1}^{\infty} m_n(a) z^n$$

with $m_n(a) := \mu(a^n)$. Then, we shall prove that

$$C_a(z M_a(z)) = M_a(z). \tag{17.9}$$

The rest of the proof is pure algebra since

$$G_a(z) := G_{\mu_a}(z) = z^{-1} M_a(z^{-1}), R_a(z) := z^{-1}(C_a(z) - 1)$$

then gives $C_a(G_a(z)) = z G_a(z)$ and so by composition by K_a

$$z R_a(z) + 1 = C_a(z) = z K_a(z).$$

This equality is formal and only proves $k_{n+1} = C_n$. We thus need to derive (17.9). To do so, we show that

$$m_n(a) = \sum_{s=1}^{n} \sum_{\substack{i_1, \dots, i_s \in \{0,1 \cdots, n-s\} \\ i_1 + \cdots + i_s = n - s}} k_s(a) m_{i_1}(a) \cdots m_{i_s}(a). \tag{17.10}$$

Once (17.10) holds, (17.9) follows readily since

$$M_a(z) = 1 + \sum_{n=1}^{\infty} z^n m_n(a)$$

$$= 1 + \sum_{n=1}^{\infty} \sum_{s=1}^{n} \sum_{\substack{i_1,\ldots,i_s \in \{0,1\cdots,n-s\} \\ i_1+\cdots+i_s=n-s}} k_s(a) z^s m_{i_1}(a) z^{i_1} \cdots m_{i_s}(a) z^{i_s}$$

$$= 1 + \sum_{s=1}^{\infty} k_s(z) z^s \left(\sum_{i=0}^{\infty} z^i m_i(a) \right)^s = C_a(z M_a(z)).$$

To prove (17.10), recall that by definition of the cumulants,

$$m_n(a) = \sum_{\pi \in NC(n)} k_\pi(a).$$

Let $\pi = \{V_1, \ldots, V_r\} \in NC(n)$ be given, and let us fix its first block $V_1 = (1, v_2, \ldots, v_s)$ with $s = |V_1| \in \{1, \ldots, n\}$. Being given V_1, since π is non-crossing, we see that for any $l \in \{2, \ldots, r\}$, there exists $k \in \{1, \ldots, s\}$ so that the elements of V_l lies between v_k and v_{k+1}. Here $v_{s+1} = n + 1$ by convention. This means that π decomposes into V_1 and at most s other partitions $\tilde{\pi}_1, \ldots, \tilde{\pi}_s$. Therefore

$$k_\pi = k_s k_{\tilde{\pi}_1} \cdots k_{\tilde{\pi}_s}.$$

If we denote by i_k the number of elements in $\tilde{\pi}_k$, we thus have proved that

$$m_n(a) = \sum_{s=1}^{n} k_s(a) \sum_{\substack{\tilde{\pi}_k \in NC(i_k), \\ i_1+\cdots+i_s=n-s}} k_{\tilde{\pi}_1}(a) \cdots k_{\tilde{\pi}_s}(a)$$

$$= \sum_{s=1}^{n} k_s(a) \sum_{\substack{i_1+\cdots+i_s=n-s \\ i_k \geq 0}} m_{i_1}(a) \cdots m_{i_s}(a)$$

where we used again the relation between cumulants and moments. The proof is thus complete.   □

Example 17.20. *Let $\nu_a = \sigma(x)dx$ be the standard semicircle, and note that $G_{\nu_a}(z) = -S_{\nu_a}(z)$. We saw in Corollary 1.12 that $G_{\nu_a}(1/\sqrt{z}) = (1 - \sqrt{1 - 4z})/2\sqrt{z}$. Thus, $G_{\nu_a}(z) = (z \pm \sqrt{z^2 - 4})/2$, and the correct choice of the branch of the square-root, dictated by the fact that $\Im z > 0$ implies $\Im G_{\nu_a}(z) < 0$, leads to the formula*

$$G_{\nu_a}(z) = \frac{z - \sqrt{z^2 - 4}}{2}.$$

Thus,

$$z = \frac{K_{\nu_a}(z) - \sqrt{K_{\nu_a}^2(z) - 4}}{2}$$

with solution $K_{\nu_a}(z) = z^{-1} + z$. *In particular, the R-transform of the semicircle is the linear function* z, *and summing two (freely independent) semicircles yields again a semicircle with a different variance. Indeed, repeating the computation above, the R-transform of a semicircle with support* $[-\alpha, \alpha]$ *(or equivalently with covariance* $\alpha^2/2$) *is* $\alpha^2 z/2$. *Note here that the linearity of the R-transform is equivalent to* $k_n(a) = 0$ *except if* $n = 2$, *and* $k_2(a) = \alpha^2/2 = \phi(a^2)$.

Exercise 17.21. *Let* $\mu = \frac{1}{2}(\delta_{+1} + \delta_{-1})$. *Then,* $G_\mu(z) = (z^2 - 1)^{-1} z$ *and*

$$R_\mu(z) = \frac{\sqrt{1 + 4z^2} - 1}{2z}.$$

Show that $\mu \boxplus \mu$ *is absolutely continuous with respect to the Lebesgue measure and with density const.*$/\sqrt{4 - x^2}$.

Bibliographical Notes. In these notes, we only considered the R-transform as formal series. It is, however, possible to see it as an analytic function once restricted to an appropriate subset of the complex plane. The study of multiplicative convolution can be performed similarly, see, e.g., [159]. This section closely follows the lecture notes of Roland Speicher [185]. Lots of refinements of the relation between free cumulants and freeness can be found for instance in the book by Nica and Speicher [159]. Here, we only considered free convolution of bounded operators; the generalization holds for unbounded operators and can be found in [30]. Our last example is a particularly simple example of infinite divisibility; the theory of free infinite divisibility parallels the classical one (in particular, a Levy–Khitchine formula does exist to characterize them) (see cf. [30], [20]). Related with free convolution come natural questions such as the regularizing effect of free convolution. A detailed study of free convolution by a semi-circular variable was done by P. Biane [34].

18

Free Entropy

Free entropy was defined by Voiculescu as a generalization of classical entropy to the non-commutative context. There are several definitions of free entropy; we shall concentrate on two of them. The first is the so-called microstates entropy that measures a volume of matrices with empirical distribution approximating a given law. The second, called the microstates-free entropy, is defined via a non-commutative version of Fisher information. The classical analog of these definitions is, on one hand, the definition of the entropy of a measure μ as the volume of points whose empirical distribution approximates μ, and, on the other hand, the well-known entropy $\int \frac{d\mu}{dx} \log \frac{d\mu}{dx} dx$. In this classical setting, Sanov's theorem shows that these two entropies are equal. The free analog statement is still open but we shall give in this section bounds to compare the microstates and the microstates-free entropies. The ideas come from [37, 55, 56] but we shall try to simplify the proof to hopefully make it more accessible to non-probabilists (the original proof uses Malliavin calculus but we shall here give an elementary version of the few properties of Malliavin calculus we need). In the following, we consider only laws of self-adjoint variables (i.e., $A_i^* = A_i$ for $1 \le i \le m$). We do not loose generality since any operator can be decomposed as the sum of two self-adjoint operators.

Definition 18.1. *Let $\tau \in \mathcal{M}^m$. Let $R \in \mathbb{R}^+$, $\epsilon > 0$ and $k, N \in \mathbb{N}$. We then define the microstate*

$$\Gamma_N(\tau; \epsilon, k, R) = \{A_1, \ldots, A_m \in \mathcal{H}_N^m : \|A_i\|_\infty \le R,$$
$$|\mathbf{L}_{A_1, \ldots, A_m}(X_{i_1} \cdots X_{i_p}) - \tau(X_{i_1} \cdots X_{i_p})| \le \epsilon$$
$$\text{for all } i_j \in \{1, \ldots, m\}, p \le k\}.$$

We then define the microstates entropy of τ) by

$$\chi(\tau) = \limsup_{\substack{\epsilon \to 0, L \to \infty \\ k \to \infty}} \limsup_{N \to \infty} \frac{1}{N^2} \log \mu_N^{\otimes m} \left(\Gamma_N(\tau; \epsilon, k, L) \right).$$

A. Guionnet, *Large Random Matrices: Lectures on Macroscopic Asymptotics*, Lecture Notes in Mathematics 1957, DOI: 10.1007/978-3-540-69897-5_18, © 2009 Springer-Verlag Berlin Heidelberg, Reprint by Springer-Verlag Berlin Heidelberg 2012

424

Remark 18.2. • *Voiculescu's original definition $\chi_{original}$ of the entropy consists in taking the Lebesgue measure over $\mathcal{H}_N^m$ rather than the Gaussian measure $\mu_N^{\otimes m}$. However, both definitions are equivalent up to a quadratic weight since as soon as $k \geq 2$, the Gaussian weight is almost constant on a small microstate. Hence, we have (see [56])*

$$\chi(\tau) = \chi_{original}(\tau) - \frac{1}{2}\sum_{i=1}^{m} \tau(X_i^2) - mc$$

with

$$c = \lim_{N\to\infty} \frac{1}{N^2}\log\int_{\mathcal{H}_N} e^{-\frac{N}{2}\mathrm{Tr}(A^2)} dA = \sup_{\mu\in\mathcal{P}(\mathbb{R})}\left\{\Sigma(\mu) - \frac{1}{2}\mu(x^2)\right\}.$$

- *It was proved by Belinschi and Bercovici [23] that in the definition of χ, one does not need to take L going to infinity but rather any L fixed greater than R if $\tau \in \mathcal{M}_R^m$. For the same reason, χ can be defined as the asymptotic volume of $\Gamma_N(\tau; \epsilon, k, \infty)$.*
- *The classical analog is, if γ is the standard Gaussian law, to take a probability measure μ on $\mathbb{R}$ and define, if d is a distance on $\mathcal{P}(\mathbb{R})$ compatible with the weak topology,*

$$S(\mu) = \limsup_{\epsilon\to\infty}\limsup_{N\to\infty} \frac{1}{N^2}\log\gamma^{\otimes N}\left(d(\frac{1}{N}\sum_{i=1}^{N}\delta_{x_i}, \mu) < \epsilon\right).$$

the main difference is here that we take bounded continuous test functions, instead of polynomials, and so do not need the cutoff $\cap_i\{\|A_i\|_\infty \leq L\}$. We shall later on also adopt this point of view in the proofs, to avoid dealing with the cutoff.
- *It is an open problem whether one can replace the limsup by a liminf in the definition of χ without changing its value. This question can be seen to be equivalent to the convergence of $N^{-2}\log\int_{\|A_i\|_\infty\leq R} e^{\mathrm{Tr}\otimes\mathrm{Tr}(V)} d\mu_N^{\otimes m}$ as N goes to infinity for any polynomial V in $\mathbb{C}\langle X_1,\ldots,X_m\rangle^{\otimes 2}$ and all $R \in \mathbb{R}^+$ large enough (see Property 18.3).*

Hereafter, when no confusion is possible, $\mathbf{L}^N$ will write for short $\mathbf{L}_{A_1,\ldots,A_m}$ with $A_1,\ldots,A_m$ generic Hermitian $N \times N$ matrices.

There is a dual definition to the microstates entropy χ, namely:

Property 18.3. *Let $F \in \mathbb{C}\langle X_1,\ldots,X_m\rangle \otimes \mathbb{C}\langle X_1,\ldots,X_m\rangle$ and define, for $L \in \mathbb{R}^+$, its Legendre transform by*

$$\Lambda_L(F) = \limsup_{N\to\infty} \frac{1}{N^2}\log\int_{\|A_i\|_\infty\leq L} e^{N^2\mathbf{L}^N\otimes\mathbf{L}^N(F)} d\mu_N^{\otimes m}.$$

Then, if $\tau \in \mathcal{M}_R^m$ is such that $\chi(\tau) > -\infty$, for any $L > R$,

$$\chi(\tau) = \inf_F\{-\tau\otimes\tau(F) + \Lambda_L(F)\}.$$

Proof. Clearly, for all $F \in \mathbb{C}\langle X_1, \ldots, X_m \rangle \otimes \mathbb{C}\langle X_1, \ldots, X_m \rangle$,

$$\mu_N^{\otimes m}\left(\Gamma_N(\tau; \epsilon, k, L)\right)$$

$$= \mu_N^{\otimes m}\left(1_{\Gamma_N(\tau;\epsilon,k,L)} e^{N^2 \mathbf{L}^{\mathbf{N}} \otimes \mathbf{L}^{\mathbf{N}}(F)}\right)$$

$$\leq e^{-N^2(\tau \otimes \tau(F) + \delta(F,\epsilon))} \mu_N^{\otimes m}\left(1_{\Gamma_N(\tau;\epsilon,k,L)} e^{N^2 \mathbf{L}^{\mathbf{N}} \otimes \mathbf{L}^{\mathbf{N}}(F)}\right) \qquad (18.1)$$

$$\leq e^{-N^2(\tau \otimes \tau(F) + \delta(F,\epsilon))} \mu_N^{\otimes m}\left(1_{\max_i \|A_i\|_\infty \leq L} e^{N^2 \mathbf{L}^{\mathbf{N}} \otimes \mathbf{L}^{\mathbf{N}}(F)}\right) \qquad (18.2)$$

where we assumed in (18.1) that k is larger than the degree of all monomials in F so that $\delta(F, \epsilon)$ goes to zero with ϵ. Taking the logarithm and the large N limit and then the small ϵ limit (together with the remark of Belinschi and Bercovici) we conclude that for L sufficiently large,

$$\chi(\tau) \leq -\tau \otimes \tau(F) + \Lambda_L(F),$$

which gives the upper bound by optimizing over F. For the lower bound, remark that we basically need to show that the inequalities in (18.1) and (18.2) are almost equalities on the large deviation scale for some F. The candidate for F will be given as a multiple of

$$F := \sum_{\ell=1}^{k} (m+1)^{-\ell} \sum_{i_1,\ldots,i_\ell=1}^{m} (X_{i_1} \cdots X_{i_\ell} - \tau(X_{i_1} \cdots X_{i_\ell}))$$

$$\otimes (X_{i_1} \cdots X_{i_\ell} - \tau(X_{i_1} \cdots X_{i_\ell}))$$

so that

$$\mu \otimes \mu(F) = \sum_{\ell=1}^{k} (m+1)^{-\ell} \sum_{i_1,\ldots,i_\ell=1}^{m} (\mu(X_{i_1} \cdots X_{i_\ell}) - \tau(X_{i_1} \cdots X_{i_\ell}))^2.$$

Note that for matrices bounded by L, $A_1, \ldots, A_m \in \Gamma_N(\tau, \epsilon, k, L)$ if

$$0 \leq \mathbf{L}_{A_1,\ldots,A_m} \otimes \mathbf{L}_{A_1,\ldots,A_m}(F) \leq (m+1)^{-k} \epsilon^2$$

so that for all $L \geq 0$, if we set $F_N := \mathbf{L}^N \otimes \mathbf{L}^N(F)$,

$$\lim_{\epsilon \downarrow 0} \limsup_{N \to \infty} \frac{1}{N^2} \log \mu_N^{\otimes m}\left(\Gamma_N(\tau; \epsilon, k, L)\right)$$

$$\geq \lim_{\epsilon \downarrow 0} \limsup_{N \to \infty} \frac{1}{N^2} \log \mu_N^{\otimes m}\left(\max_i \|A_i\|_\infty \leq L, F_N \leq \epsilon\right). \qquad (18.3)$$

But, since $F_N \geq 0$, for any $\gamma > 0$,

$$\mu_N^{\otimes m}\left(\max_i \|A_i\|_\infty \leq L, F_N \leq \epsilon\right) \geq \mu_N^{\otimes m}\left(1_{\max_i \|A_i\|_\infty \leq L, F_N \leq \epsilon} e^{-\gamma N^2 F_N}\right)$$

$$= \mu_N^{\otimes m}\left(1_{\max_i \|A_i\|_\infty \leq L} e^{-\gamma N^2 F_N}\right) - \mu_N^{\otimes m}\left(1_{\max_i \|A_i\|_\infty \leq L, F_N > \epsilon} e^{-\gamma N^2 F_N}\right)$$

$$=: I_N^1 - I_N^2$$

For the first term, for any ϵ', k' and L we have

$$I_N^1 \geq \mu_N^{\otimes m} \left(1_{\Gamma_N(\tau;\epsilon',k',L)} e^{-\gamma N^2 F_N}\right) \geq e^{-\gamma N^2 (m+1)(\epsilon')^2} \mu_N^{\otimes m} \left(\Gamma_N(\tau;\epsilon',k',L)\right).$$

Therefore, for L large enough (but finite according to [23])

$$\limsup_{N\to\infty} \frac{1}{N^2} \log I_N^1 \geq \chi(\tau) > -\infty.$$

On the other hand, $I_N^2 \leq e^{-\gamma N^2 \epsilon}$ is negligible with respect to I_N^1 if $\gamma\epsilon > -\chi(\tau)$. Thus, we conclude by (18.3) that

$$\lim_{\epsilon\downarrow 0}\limsup_{N\to\infty} \frac{1}{N^2} \log \mu_N^{\otimes m}\left(\Gamma_N(\tau;\epsilon,k,L)\right) \geq \limsup_{N\to\infty}\frac{1}{N^2}\log I_N^1 = \Lambda_L(-\gamma F) \tag{18.4}$$

and therefore that, with $G = -\gamma F$,

$$\lim_{\epsilon\downarrow 0}\limsup_{N\to\infty}\frac{1}{N^2}\log\mu_N^{\otimes m}\left(\Gamma_N(\tau;\epsilon,k,L)\right) \geq -\tau\otimes\tau(G) + \Lambda_L(G)$$

$$\geq \inf_F\{-\tau\otimes\tau(F) + \Lambda_L(F)\}.$$

We finally take the limit k, L going to infinity to conclude. □

Remark 18.4. *In the classical case, it is enough to take a linear function of* $\mathbf{L}^N$. *This in particular implies that the rate function (corresponding to* $-\chi$) *is convex. This cannot be the case in the non-commutative case since Voiculescu (see [198]) proved that if* $\chi(\tau_1) > -\infty$ *and* $\chi(\tau_2) > -\infty$, $\chi(\alpha\tau_1 + (1-\alpha)\tau_2) = -\infty$ *for all* $\alpha \in]0,1[.$

Let us now introduce the microstates-free entropy. Its definition is based on the notion of free Fisher information which is given, for a tracial state τ, by

$$\Phi^*(\tau) = 2\sum_{i=1}^m \sup_{P\in\mathbb{C}\langle X_1,\dots,X_m\rangle}\left\{\tau\otimes\tau(\partial_i P) - \frac{1}{2}\tau(P^2)\right\}$$

with ∂_i the non-commutative derivative defined in section 7.2.2. Then, we define the microstates-free entropy χ^* by

$$\chi^*(\tau) = -\frac{1}{2}\int_0^1 \Phi^*\left(\tau_{tX+\sqrt{t(1-t)}S}\right)dt$$

with S an m-dimensional semicircular law, free from X.

Theorem 18.5. *There exists* $\chi^{**} \leq 0$ *so that for all* $\tau \in \mathcal{M}_R^m,$

$$\chi^{**}(\tau) \leq \chi(\tau) \leq \chi^*(\tau).$$

Remark 18.6. *Many questions around microstates-free entropy remain open and important. It would be very interesting to show that the limsup in its definition can be replaced by liminf. The two bounds above still hold in fact if we perform this change. It would be great to prove that $\chi = \chi^*$ at least on $\chi < \infty$. It is not clear at all that $\chi^{**} = \chi^*$ in general (since this amounts to saying that any law can be achieved (at least approximatively) as the marginal at time one of a very smooth process) but should be expected for instance for laws such as those encountered in Part III).*

The idea is to try to compute the Legendre transforms $\Lambda_\infty(F)$ for F in $\mathbb{C}\langle X_1, \ldots, X_m \rangle^{\otimes 2}$. However, this is not easy directly. We shall follow a standard path of thoughts in large deviation theory and consider the problem in a bigger space; namely instead of random Wigner matrices, we shall consider Hermitian Brownian motions (which, at time one, have the same law than Gaussian Wigner matrices) and generalize the ideas of Part V to the multi-matrix setting. We then study the deviations of the empirical distributions of m independent Hermitian Brownian motions. It is simply defined as a linear form on the set of polynomials of the indeterminates $(X_{t_j}^{i_j}, 1 \le j \le n)$ for any choices of times $(t_j, 1 \le j \le n)$. It might be possible to argue that we can use the polynomial topology, but it is in fact easier to use a topology of bounded functions at this point, since it is easier for us to estimate Laplace transforms without cutoff, a cutoff that is necessary in general to insure that the polynomial topology is good. The space of test functions we shall use is the set of Stieltjes functionals as introduced in (16.2). To go from processes with continuous-time parameters to finite-time marginals, a standard way is to consider *continuous processes* and to make sure that large deviations hold in this space of continuous processes. A final point is that changing the topology (from polynomials to Stieltjes functionals) does not change the entropy of laws in $\mathcal{M}_R^m$, as was proved in Lemma 7.1 of [37]. We next describe the setting of continuous processes and show how Laplace transforms of time marginals can be computed.

1. *Space of laws of continuous processes.* We let $\mathcal{F}$ be the space of functions on processes indexed by $t \in [0, 1]$ such that $F \in \mathcal{F}$ iff there exists $n \in \mathbb{N}$, $t_1, \ldots, t_n \in [0, 1]^n$, $i_1, \ldots, i_n \in \{1, \ldots, m\}^n$ and $S \in ST^n(\mathbb{C})$ such that

$$F(X^1, \ldots, X^m) = T(X_{t_1}^{i_1}, \ldots, X_{t_n}^{i_n}).$$

$ST^n(\mathbb{C})$ contains the multiplicative neutral element 1 for all $n \in \mathbb{N}$ and is equipped with the involution

$$\left(\overrightarrow{\prod_{1 \le i \le p}} (z_i - \sum_{k=1}^m \alpha_i^k X_k)^{-1} \right)^* = \overrightarrow{\prod_{p \le i \le 1}} (\bar{z}_i - \sum_{k=1}^m \alpha_i^k X_k)^{-1}$$

and so $\mathcal{F}$ is equipped with its natural extension.

We let $\mathcal{M}^P$ be the subset of linear forms τ on $\mathcal{F}$ such that

$$\tau(FG) = \tau(GF) \quad \tau(FF^*) \geq 0 \quad \tau(1) = 1.$$

We endow $\mathcal{M}^P$ with its weak topology. $\mathcal{M}^P$ is a metric space; a distance can for instance be given in the spirit of Levy's distance

$$d(\tau, \nu) = \sum_{k \geq 0} \frac{1}{2^k} |\tau(P_k) - \nu(P_k)|$$

with P_k a countable family of uniformly bounded functions dense in $\mathcal{F}$ (for instance by restricting the parameters α_i^j to take rational values, as well as the complex parameters).

Note that if we restrict $\tau \in \mathcal{M}^P$ to $F = (z - \sum_{i=1}^n \alpha_i X_{t_i}^{j_i})^{-1}$ with some fixed α_i but varying z, then this retriction is a linear form on the set of functionals $(z - .)^{-1}, z \in \mathbb{C}\backslash\mathbb{R}$. By the GNS construction, since τ is a tracial state, $Y = \sum_{i=1}^n \alpha_i X_{t_i}^{j_i}$ can be seen as the law of a self-adjoint operator on a C^*-algebra. Therefore, $\tau|_Y$ can be seen as a positive measure on the real line (by Riesz's theorem). Since we also have $\tau(1) = 1$, $\tau|_Y$ is a probability measure on the real line, it is the spectral measure of Y. Note in particular that for every $\tau \in \mathcal{M}^P$, any $z \in \mathbb{C}\backslash\mathbb{R}$ and $\alpha_i \in \mathbb{R}$

$$\left\| \left(z - \sum_{i=1}^n \alpha_i X_{t_i}^{j_i} \right)^{-1} \right\|_\infty^\tau$$

$$= \lim_{n \to \infty} \left(\tau\left(\left(\left(z - \sum_{i=1}^n \alpha_i X_{t_i}^{j_i} \right)^{-1} \left(\bar{z} - \sum_{i=1}^n \alpha_i X_{t_i}^{j_i} \right)^{-1} \right)^n \right) \right)^{\frac{1}{2n}}$$

$$\leq \frac{1}{|\Im(z)|}.$$

Hence, by non-commutative Hölder inequality Theorem 19.5, for all $\tau \in \mathcal{M}^P$, all $z_i \in \mathbb{C}\backslash\mathbb{R}$, all $\alpha_i^k \in \mathbb{R}$,

$$\left| \tau\left(\overrightarrow{\prod_{1 \leq i \leq p}} \left(z_i - \sum_{k=1}^m \alpha_i^k X_k \right)^{-1} \right) \right| \leq \prod_{1 \leq i \leq p} \frac{1}{|\Im(z_i)|}.$$

We let $\mathcal{M}_c^P$ be the set of laws of *continuous processes*, i.e., the set of $\tau \in \mathcal{M}^P$ such that for all, $n \in \mathbb{N}$, $i_1, \ldots, i_n \in \{1, \ldots, m\}^n$ and $T \in ST^n(\mathbb{C})$ such that

$$t_1, \ldots, t_n \in [0,1]^n \to \tau\left(T(X_{t_1}^{i_1}, \ldots, X_{t_n}^{i_n}) \right)$$

is continuous. Note that for all $n \in \mathbb{N}$, $ST^n(\mathbb{C})$ is countable, and therefore by the Arzela–Ascoli theorem, compact subsets of $\mathcal{M}_c^P$ are for example

$$\cap_{n\geq 0}\cap_{p\geq 0}\left\{\tau\in\mathcal{M}_c^P:\sup_{|t_i-s_i|\leq\eta_p^n}\max_{i_1,\ldots,i_n}|f_{t_1,\ldots,t_n}^{\tau,T_n}-f_{s_1,\ldots,s_n}^{\tau,T_n}|\leq\epsilon_n^p\right\}$$

with $f_{t_1,\ldots,t_n}^{\tau,T}=\tau\left(T_n(X_{t_1}^{i_1},\ldots,X_{t_n}^{i_n})\right)$, $\{T_n\}_{n\in\mathbb{N}}$ is some sequence that is dense in the countable space $\cup_{n\geq 0}ST^n(\mathbb{C})$, and ϵ_n^p goes to zero as p goes to infinity for all n while $\eta_p^n>0$.

We let $\mathbf{L}_N^P$ be the element of $\mathcal{M}^P$ given, for $F\in\mathcal{F}$, by

$$\mathbf{L}_N^P(F)=\frac{1}{N}\text{Tr}(F(H_N^1,\ldots,H_N^m))$$

with $H_N^1,\ldots,H_N^m$ m independent Hermitian Brownian motions (and denote $\mathbb{P}_N$ the law of one Hermitian Brownian motion).

2. *Exponential tightness.* We next prove:

Lemma 18.7. *For all $L\in R^+$, there exists a compact set $\mathcal{K}(L)$ of $\mathcal{M}_c^P$ such that*

$$\limsup_{N\to\infty}\frac{1}{N^2}\log\mathbb{P}_N^{\otimes m}\left(\mathbf{L}_N^P\in\mathcal{K}(L)^c\right)\leq -L.$$

Proof. Note that for all $T\in ST^n(\mathbb{C})$, all $\tau\in\mathcal{M}^P$,

$$\tau\left(T(X_{t_1}^{i_1},\ldots,X_{t_n}^{i_n})\right)-\tau\left(T(X_{s_1}^{i_1},\ldots,X_{s_n}^{i_n})\right)$$

$$=\sum_{j=1}^n\int_0^1 d\alpha\,\tau\left(\partial_i T(\alpha X_{t_1}^{i_1}+(1-\alpha)X_{s_1}^{i_1},\ldots,\right.$$

$$\left.\alpha X_{t_n}^{i_n}+(1-\alpha)X_{s_n}^{i_n})\sharp(X_{t_j}^{i_j}-X_{s_j}^{i_j})\right)d\alpha$$

$$=\sum_{j=1}^n\int_0^1\tau\left(\partial_j T(\alpha X_{t_1}^{i_1}+(1-\alpha)X_{s_1}^{i_1},\ldots,\right.$$

$$\left.\alpha X_{t_n}^{i_n}+(1-\alpha)X_{s_n}^{i_n})\sharp(X_{t_j}^{i_j}-X_{s_j}^{i_j})\right)d\alpha$$

$$=\sum_{j=1}^n\int_0^1\tau\left(D_j T(\alpha X_{t_1}^{i_1}+(1-\alpha)X_{s_1}^{i_1},\ldots,\right.$$

$$\left.\alpha X_{t_n}^{i_n}+(1-\alpha)X_{s_n}^{i_n})\times(X_{t_j}^{i_j}-X_{s_j}^{i_j})\right)d\alpha$$

with ∂_j (resp. D_j) the non-commutative derivative (resp. cyclic derivative) with respect to the jth variable (see Section 7.2.2). Now, $D_j T(\alpha X_{t_1}^{i_1}+(1-\alpha)X_{s_1}^{i_1},\ldots,\alpha X_{t_n}^{i_n}+(1-\alpha)X_{s_n}^{i_n}$ belongs to $\mathcal{F}$ for any $\alpha\in[0,1]$ and is therefore uniformly bounded (independently of α and τ). Thus, there exists a finite constant c, that depends only on T such that

$$|\tau\left(T(X_{t_1}^{i_1},\ldots,X_{t_n}^{i_n})\right)-\tau\left(T(X_{s_1}^{i_1},\ldots,X_{s_n}^{i_n})\right)|\leq c\sum_{i=1}^n\tau(|X_{t_j}^{i_j}-X_{s_j}^{i_j}|^2)^{\frac{1}{2}}.$$

By the characterization of the compact sets of $\mathcal{M}_c^P$, it is therefore enough to show that for all $L > 0$ and $\epsilon > 0$, there exists $\eta > 0$ such that

$$\mathbb{P}_N\left(\sup_{\substack{|t-s|\leq\eta\\0\leq s\leq t\leq1}}\frac{1}{N}\text{Tr}((H_N(t)-H_N(s))^2)\geq\epsilon\right)\leq e^{-N^2L}.$$

But

$$\mathbb{P}_N\left(\sup_{\substack{|t-s|\leq\eta\\0\leq s\leq t\leq1}}\frac{1}{N}\text{Tr}((H_N(t)-H_N(s))^2)\geq\epsilon\right)$$

$$\leq\mathcal{W}^{\otimes N^2}\left(\sup_{\substack{|t-s|\leq\eta\\0\leq s\leq t\leq1}}\frac{1}{N^2}\sum_{1\leq i,j\leq N}(B_{ij}(t)-B_{ij}(s))^2\geq\epsilon\right)$$

$$\leq e^{-N^2\epsilon\lambda}\mathcal{W}(e^{\lambda\sup_{\substack{|t-s|\leq\eta\\0\leq s\leq t\leq1}}(B_{ij}(t)-B_{ij}(s))^2})^{N^2}$$

where $\mathcal{W}$ is the Wiener law. Using the fact (see [137, theorem 4.1]) that there exists $\alpha > 0$ such that

$$\mathcal{W}(e^{\alpha\delta^{-\frac{1}{4}}\sup_{|t-s|\leq\delta}|B(t)-B(s)|})<\infty$$

we see that we can take above $\lambda=\alpha\eta^{-\frac{1}{4}}$ to conclude that if $\epsilon\eta^{-\frac{1}{4}}\geq L$,

$$\mathbb{P}_N\left(\sup_{\substack{|t-s|\leq\eta\\0\leq s\leq t\leq1}}\frac{1}{N}\text{Tr}((H_N(t)-H_N(s))^2)\geq\epsilon\right)\leq e^{-N^2(L+C)}$$

for some finite constant C.   □

3. *Statement of the large deviation result for processes.* According to Lemma 18.7, the rate function for a large deviation principle for the law of $\mathbf{L}_N^P$ under $P_N^{\otimes m}$ has to be infinite outside $\mathcal{M}_c^P$. For $\tau\in\mathcal{M}_c^P$, we let τ^t be the law of $(X_{s\wedge t}^i+S_{s-t\vee0}^i,i\in\{1,\ldots,m\},0\leq s\leq t)$ with X with law τ, and S an m-dimensional free Brownian motion, free with X. For $F(X^1,\ldots,X^m)=T(X_{t_1}-X_{t_0},\cdots X_{t_p}-X_{t_{p-1}})\in\mathcal{F}$ with $T\in ST^{pm}(\mathbb{C})$, $0\leq t_0\leq t_1\cdots\leq t_m\leq1$ and $s\in[0,1]$, we let

$$D_sF(X^1,\ldots,X^m)=\sum_{i=1}^p1_{s\in[t_{i-1},t_i]}D_iT$$

where D_i is the cyclic gradient with respect to the ith variable (it is m dimensional). Finally, for $\tau\in\mathcal{M}^P$, we let $\tau(.|\mathcal{B}_s)$ be the $L^2(\tau)$ projection over the algebra $\mathcal{B}_s$ generated by the set $\mathcal{F}_s$ of functions of $\mathcal{F}$ that only depend on $X_u,u\leq s$:

$$\tau(\tau(P|\mathcal{B}_s)Q)=\tau(PQ)\quad\forall P\in\mathcal{F},\forall Q\in\mathcal{F}_s.$$

Then we have

Theorem 18.8. *a) The law of* $\mathbf{L}_{\mathbf{N}}^{\mathbf{P}}$ *under* $\mathbb{P}_N^{\otimes m}$ *satisfies a large deviation upper bound for the weak* $\mathcal{F}$*-topology in the scale* N^2 *with good rate function*

$$I(\tau) = \sup_{t\in[0,1]} \sup_{F\in\mathcal{F}} \left\{ \tau^t(F) - \tau^0(F) - \frac{1}{2}\int_0^t \tau[|\tau^s(D_sF|\mathcal{B}_s)|^2]ds \right\}.$$

b) If $I(\tau) < \infty$*, there exists a map* $s \to K_s \in L^2(\mathcal{F}_s, \tau)^m$ *such that*
   i. $\inf_{F\in\mathcal{F}} \int_0^1 \tau[\|\tau^s(D_sF|\mathcal{B}_s) - K_s\|^2]ds = 0$.
   ii. For any $P \in \mathcal{F}$ *any* $t \in [0,1]$

$$\tau^t(F) = \tau^0(F) + \int_0^t \tau(\tau^s(D_sF|\mathcal{B}_s).K_s)ds. \tag{18.5}$$

Moreover, we then have

$$I(\tau) = \frac{1}{2}\int_0^1 \tau(\|K_s\|^2)ds.$$

c) When the infimum in b. i) above is achieved (i.e., there exists $F \in \mathcal{F}$ *such that* $K_s = \tau^s(D_sF|\mathcal{B}_s))$*,* τ *is uniquely determined by* (18.5) *and is the weak solution of the free stochastic differential equation*

$$dX_t = dS_t + K_t(X)dt.$$

d) If τ *is such that the infimum in b.i) above is achieved, the large deviation lower bound is also given by* $I(\tau)$*, i.e*

$$\liminf_{\epsilon\to 0} \liminf_{N\to\infty} \frac{1}{N^2} \log \mathbb{P}_N^{\otimes m}\left(d(\mathbf{L}_{\mathbf{N}}^{\mathbf{P}}, \tau) \le \epsilon\right) \ge -I(\tau)$$

where d *is a Levy distance compatible with the weak* $\mathcal{F}$*-topology.*

If we now use the contraction principle we deduce:

Corollary 18.9. *The law of* $\mathbf{L}^N = \mathbf{L}_{H_1^N(1),\cdots H_m^N(1)}$ *under* $\mathbb{P}_N^{\otimes m}$ *satisfies a large deviation upper bound with rate function*

$$\chi^*(\mu) = -\inf\{I(\tau); \tau|_{X_1^1,\dots,X_1^m} = \mu\}$$

and a large deviation lower bound with rate function

$$\chi(\mu) = -\inf\{I(\tau); \tau|_{X_1^1,\dots,X_1^m} = \mu,$$

$$\tau \text{ is such that the infimum in b. i) is achieved }\}$$

To complete the proof and identify χ^* with Voiculescu's original definition, one can use an abstract argument (cf. [56]) to see that the infimum has to be taken at the law of the free Brownian bridge:

$$dX_t = dS_t + \frac{X_t - X}{t - 1} dt.$$

Taking the previsible representation of the above process, we get that $K_t = \tau(\frac{X_t - X}{t-1} | X_t) = t^{-1} X_t - \mathcal{J}^{\tau_t}$ with $\mathcal{J}^{\tau_t}$ the conjuguate variable of the law of $X_t = tX + \sqrt{t(1-t)} S$. Plugging this result into the definition of I shows that χ^* is indeed the integral of the free Fisher information along free Brownian motion paths.

4. *Large deviation upper bound.* In the classical case where one considers large deviations for the empirical measure $N^{-1} \sum_{i=1}^{N} \delta_{B_t^i, t \in [0,1]}$, that are solved by Sanov's theorem, it can be seen that deviations (i.e., deviations with finite rate function) occur only along laws that are absolutely continuous with respect to Wiener law. By Girsanov's theorem, one knows that such laws are obtained as weak solutions of the SDE

$$dX_t = dB_t + b(t, (X_s)_{s \leq t}) dt$$

for some drift b. These heuristics will extend, as we shall see, to the non-commutative case (once one defines the right notion of weak solution to the SDE). We now compute a few Laplace transforms of quantities depending on a finite number of time marginals under $\mathbf{L}_N^P$. We shall give a pedestrian way to understand the Clark–Ocone formula used in [37].

• First, let $0 = t_0 \leq t_1 \cdots \leq t_{n-1} \leq t_n = 1$ and let us consider T_k^i self-adjoint elements of $ST^k(\mathbb{C})$ for $k \in \{0, \dots, n-1\}$, $i \in \{1, \dots, m\}$. Then, if we denote $\Delta_k t = t_{k+1} - t_k$, $\Delta_k H_N^i = H_N^i(t_{k+1}) - H_N^i(t_k)$ and $\Delta_k H_N = (\Delta_k H_N^i)_{1 \leq i \leq m}$,

$$\Lambda_N := \mathbb{E}\left[\exp\left\{N \sum_{k=0}^{n-1} \sum_{i=1}^{m}\right.\right.$$

$$\left.\left. \text{Tr}\left(T_k^i(\Delta_l H_N, l < k)\Delta_k H_N^i - \frac{1}{2}T_k^i(\Delta_l H_N, l < k)^2 \Delta_k t\right)\right\}\right]$$

$$= \int \prod \frac{d\Delta_k H_N^i}{\sqrt{2\pi\Delta_k t}^{N^2}}$$

$$\prod_{i,k} \exp\left\{-\frac{N}{2\Delta_k}\text{Tr}\left((\Delta_k H_N^i - T_k^i(\Delta_l H_N, l < k)\Delta_k t)^2\right)\right\}$$

$$= 1 \qquad\qquad\qquad\qquad\qquad\qquad\qquad\qquad (18.6)$$

where we have used that since the above centering of the Gaussian variables only depends on the past, it can be considered as constant and therefore does not change the integral. Putting $\Delta_k X^i = X^i(t_{k+1}) - X^i(t_k)$ and

$$f(\tau) = \sum_{k=0}^{n-1} \sum_{i=1}^{m} \left(\tau\left(T_k^i(\Delta_l X, l < k)\Delta_k X^i - \frac{1}{2}T_k^i(\Delta_l X, l < k)^2)\Delta_k t\right)\right)$$

we deduce that
$$\mathbb{E}[e^{N^2 f(\mathbf{L_N^P})}] = \Lambda_N = 1.$$

Hence, this simple computation already shows that considering processes allows us to compute the Laplace transforms $\mathbb{E}[e^{N^2 f(\mathbf{L_N^P})}]$ for all functions f as above.

However, the above computation is not sufficient to get a good upper bound since for instance it does not allow to compute the Laplace transform of $\mathbf{L_N^P}((\Delta_k X^i)^2)$. To do such a computation, we need infinitesimal calculus. Note that the upper bound that one would obtain by using only the previous functionals would allow to show already that the laws with finite entropy are such that there exists a drift K so that $dX_t - K(X_t)dt$ are the increments of a martingale. We could not, however, deduce that it has to be a free Brownian motion without using differential calculus (instead of finite variations).

• The idea to compute more general Laplace transforms is to generalize (18.6) by constructing more martingales. Indeed, the fact that $\Lambda_N(T,t)$ equals one can be seen as a consequence of the fact that

$$t \to \exp\Big\{ N^2 \sum_{k=0}^{n-1} \sum_{i=1}^{m} \Big(\mathbf{L_N^P}\Big(T_k^i(\Delta_l X, l < k)\Delta_k^t X^i \\ - \frac{1}{2} T_k^i(\Delta_l X, l < k)^2 \Delta_k^t t \Big) \Big) \Big\}$$

is a martingale for the canonical filtration of the underlying Brownian motions if $\Delta_k^t X^i := X_{t_{k+1} \wedge t}^i - X_{t_k \wedge t}^i$ and $\Delta_k^t t = t_{k+1} \wedge t - t_k \wedge t$. A simple way to construct martingales is simply to consider

$$M_t^F := \mathbb{E}[\mathrm{Tr}(F(H_1^N, \ldots, H_m^N)) | \mathcal{F}_t]$$

with $F \in \mathcal{F}$ (and $\mathcal{F}_t$ the canonical filtration of Brownian motion) and the associated exponential martingale

$$E_t^F = \exp\Big\{ N M_t^F - \frac{N^2}{2} \langle M^F \rangle_t \Big\}.$$

Then, since $F \in \mathcal{F}$ is uniformly bounded, E_t^F is a martingale implying that $\mathbb{E}[E_t^F] = \mathbb{E}[E_0^F] = 1$. Now, we can always write, with $H^N = (H_1^N, \ldots, H_m^N)$,

$$\mathbb{E}[\mathrm{Tr}(F(H^N)) | \mathcal{F}_t] = \mathbb{E}_{\tilde{H}}[\mathrm{Tr}(F(H^N(s \wedge t) + \tilde{H}^N(s - t))_{0 \le s \le 1})]$$

with $\tilde{H}^N$ a Hermitian Brownian motion, independent from H^N. Taking $F(X) = T(X_{t_1}^{i_1}, \ldots, X_{t_n}^{i_n})$ and applying Itô's calculus, one finds that

$$M_t^F = M_0^F + \int_0^t \mathrm{Tr}\Big(\mathbb{E}_{\tilde{H}}[D_s F(H^N(s \wedge t) + \tilde{H}^N(t - s))_{0 \le s \le 1}].dH_s^N \Big).$$

Indeed, when one performs Itô's calculus on $H^N(s) + \tilde{H}^N(t - s)$ the infinitesimal generator appears twice; once from $H^N(s)$ and once from $\tilde{H}^N(t - s)$ and the two contributions cancel. This shows that

$$\langle M^F \rangle_t = \frac{1}{N} \int_0^t \mathrm{Tr}\left(\|\mathbb{E}_{\tilde{H}}[D_s F(H^N(s \wedge t) + \tilde{H}^N(s - t))_{0 \leq s \leq 1}]\|^2 \right) ds.$$

It was proved in [37] that for all $t \in [0, 1]$,

$$G_F^t(\mathbf{L_N^P}) := \mathbb{E}[N^{-1}\mathrm{Tr}(F(H^N))|\mathcal{F}_t] - \mathbb{E}[N^{-1}\mathrm{Tr}(F(H^N))]$$

$$-\frac{1}{2} \int_0^t \mathrm{Tr}\left(\|\mathbb{E}_{\tilde{H}}[D_s F(H^N(s \wedge t) + \tilde{H}^N(s - t))_{0 \leq s \leq 1}]\|^2 \right) ds$$

is a continuous function of $\mathbf{L_N^P}$ (actually of its resctriction to $\mathcal{B}_t$ measurable functions). Furthermore, as $\mathbf{L_N^P}$ goes to τ, $G_F^t(\mathbf{L_N^P})$ goes to

$$G_F^t(\tau) = \tau^t(F) - \tau^0(F) - \frac{1}{2} \int_0^t \tau[|\tau^s(D_s F|\mathcal{B}_s)|^2]ds.$$

Therefore, since $E[e^{N^2 G_F^t(\mathbf{L_N^P})}] = 1$, we readily get the large deviation upper bound by taking a distance d compatible with our topology and picking an $\epsilon > 0$ to get

$$\mathbb{P}_N^{\otimes m}\left(d(\mathbf{L_N^P}, \tau) < \epsilon \right)$$

$$= \mathbb{P}_N^{\otimes m}\left(1_{d(\mathbf{L_N^P}, \tau) < \epsilon} e^{N^2(G_F^t(\mathbf{L_N^P}) - G_F^t(\mathbf{L_N^P}))} \right)$$

$$\leq e^{-N^2 G_F^t(\tau) + N^2 \kappa(\epsilon)} \mathbb{P}_N^{\otimes m}\left(1_{d(\mathbf{L_N^P}, \tau) < \epsilon} e^{N^2(G_F^t(\mathbf{L^N}))} \right)$$

$$\leq e^{-N^2 G_F^t(\tau) + N^2 \kappa(\epsilon)} \mathbb{P}_N^{\otimes m}\left(e^{N^2(G_F^t(\mathbf{L_N^P}))} \right)$$

$$= e^{-N^2 G_F^t(\tau) + N^2 \kappa(\epsilon)}$$

where $\kappa(\epsilon)$ goes to zero with ϵ. We finally can take the logarithm, divide by N^2, let N going to infinity, ϵ to zero and finally optimize over F to conclude.

● *Uniqueness of the solutions with smooth drift.* The second point of Theorem 18.8 is a consequence of Riesz's theorem. Moreover, taking $F = (X_t - X_s)G(X)$ with G $\mathcal{B}_s$ measurable, we deduce that

$$\tau\left(\left(X_t - X_s - \int_s^t K_u du \right) G(X) \right) = 0.$$

Hence

$$H_t = X_t - X_0 - \int_0^t K_u du$$

is a free martingale. To show that it is a free Brownian motion, we use a free version of Paul Lévy's well-known theorem on the characterization of Brownian motion as the unique martingale with continuous paths and square bracket equal to t, and that may be of independent interest (see [37, theorem 6.2])

Lemma 18.10. *Let* $(\mathcal{B}_s; s \in [0,1])$ *be an increasing family of von Neumann subalgebras, in a non-commutative probability space* $(\mathcal{A}, \tau)$, *and let*

$$(Z_s = (Z_s^1, \ldots, Z_s^m); s \in [0,1])$$

be an m-*tuple of self-adjoint processes adapted to* $(\mathcal{B}_s; s \in [0,1])$, *such that* Z *is bounded,* $Z_0 = 0$, *and for all* $s < t$ *one has:*
(a) $\tau(Z_t | \mathcal{B}_s) = Z_s$.
(b) $\tau(|Z_t - Z_s|^4) \leq K(t - s)^2$ *for some constant* K.
(c) *For any* $l, p \in \{1, \ldots, m\}$, *and all* $A, B \in \mathcal{B}_s$, *one has*

$$\tau(A Z_t^l B Z_t^p) = \tau(A Z_s^l B Z_s^p) + 1_{p=l}(t-s)\tau(A)\tau(B) + o(t-s),$$

then Z *is a free Brownian motion, i.e., for all* $s < t$ *the elements* $Z_t^l - Z_s^l; l \in \{1, \ldots, m\}$ *are free with* $\mathcal{B}_s$, *and have a semi-circular distribution of covariance* $(t-s)I_m$.

Proof. Because of the invariance of the conditions under time translation, it is enough to prove that $Z_t - Z_0$ is free with $\mathcal{B}_0$, and of semi-circular distribution with covariance $t I_m$. We can assume that $Z_0 = 0$, and one has for any $i_1, \ldots i_n \in \{1, \ldots, m\}$,

$$\tau(Z_t^{i_1} \ldots Z_t^{i_n}) = \tau((Z_s^{i_1} + (Z_t^{i_1} - Z_s^{i_1})) \ldots (Z_s^{i_n} + (Z_t^{i_n} - Z_s^{i_n}))).$$

From condition (a) we get $\tau(Z_t^l - Z_s^l | \mathcal{B}_s) = 0$, and expanding the above product using (b) and (c) gives

$$\tau(Z_t^{i_1} \ldots Z_t^{i_n}) - \tau(Z_s^{i_1} \ldots Z_s^{i_n})$$

$$= (t-s) \sum_{0 \leq k+p \leq n-2} \sum_{i_k = i_p} \tau(Z_s^{i_1} \ldots Z_s^{i_{k-1}} Z_s^{i_{k+p+1}} \ldots Z_s^{i_n})$$

$$\tau(Z_s^{i_{k+1}} \ldots Z_s^{i_{k+p-1}}) + o(t-s)$$

where we have used non-commutative Hölder's inequality in order to bound the terms containing at least three $(Z_t^l - Z_s^l)$ factors. It follows that the quantities $\tau(Z_t^{i_1} \ldots Z_t^{i_n})$ satisfy a system of differential equations whose initial conditions are known. It is easy to see that this system has a unique solution, resulting in the observation that there exists at most one process (in distribution) satisfying (a), (b) and (c).

Since the free m-dimensional Brownian motion also satisfies (a), (b) and (c), we conclude that $Z_t - Z_0$ is a free Brownian motion. For the freeness property with respect to $\mathcal{B}_0$, we consider a quantity of the form

$$\tau(A_1 Z_t^{i_1} A_2 Z_t^{i_2} \ldots A_n Z_t^{i_n})$$

that again satisfies the same differential equation as when Z_t is a free Brownian motion free with $\mathcal{B}_0$.   □

In order to apply Theorem 18.10 to the process Y we have to check the three conditions. First we apply (18.5) to $P = (X_t^l - X_s^l)Q_s$, where $Q_s \in \mathcal{B}_s \cap \mathcal{F}$. Although P does not belong to $\mathcal{F}$ one can again check that it is a limit of a sequence of P_n in $\mathcal{F}$, such that $\nabla_s P_n$ converges to $\nabla_s P$, so there is no problem in applying formula (18.5). One has $\nabla_u^k P = \delta_{kl} 1_{u \in [s,t]} Q_s + W$ where $\tau^u(W|\mathcal{B}_s) = 0$. We thus find that for all $Q_s \in \mathcal{B}_s$, one has

$$\tau((X_t^l - X_s^l)Q_s)$$
$$= \tau^0((X_t^l - X_s^l)Q_s) + \int_0^1 \tau(\tau^u(\nabla_u[(X_t^l - X_s^l)Q_s]|\mathcal{B}_u).K_u)du$$
$$= \tau\left(\int_s^t Q_s K_u^l du\right)$$

from which we get that condition (a) is satisfied by $X_t - \int_0^t K_s ds$. We now apply (18.5) to $P = (X_t^l - X_s^l)^4$ (the same remark as above applies). Since $\tau^0((X_t^l - X_s^l)^4) = 2(t-s)^2$, $\nabla_u^k[(X_t^l - X_s^l)^4] = 0$ for $u \notin [s,t]$ and $\nabla_u^k[(X_t^l - X_s^l)^4] = 4\delta_{kl}(X_t^l - X_s^l)^3$ for $u \in [s,t]$, one has

$$\tau((X_t^l - X_s^l)^4) = 2(t-s)^2 + \int_s^t \tau(\tau^u(4(X_t^l - X_s^l)^3|\mathcal{B}_u)K_u^{\tau,l})du$$

Since $K_u^{\tau,l}$ is uniformly bounded in norm, using Hölder's inequality and Gronwall lemma, we get the bound (b).
Condition (c) can be checked in a similar way as condition (a).
We conclude that X is solution to the stochastic differential equation

$$X_t = S_t + \int_0^t K_s ds$$

with $K_s = \tau(\nabla_s K|\mathcal{B}_s)$. Observing that for $K \in \mathcal{F}_{[0,1]}^m$, $X \to K_s(X)$ is uniformly Lipschitz, e.g., there exists a finite constant C such that for all $s \in [0,1]$,

$$\|K_s(X) - K_s(Y)\|_\infty \le C \sup_{u \le s} \|X_u - Y_u\|_\infty,$$

we can use the usual Gronwall argument to prove the uniqueness of the solution to this equation, establishing the uniqueness of τ.

• *Large deviation lower bound.* If τ is the law of the solution of

$$dX_t = dS_t + \tau^t(D_t K|\mathcal{B}_t)dt$$

for some $K \in \mathcal{F}$, we know that the unique strong solution of

$$dX_t^N = dH_t^N + \tau^t(D_t K|\mathcal{F}_t)(X^N)dt$$

will converge weakly to τ. Moreover, this law is absolutely continuous with respect to the law of the m-dimensional Hermitian Brownian motion H^N with density (see, e.g., (18.6))

$$d_N = \exp\left\{ N^2 \left(\mathbf{L_N^P}(K) - \sigma(K) - \frac{1}{2} \int_0^1 \mathbf{L_N^P}(D_t K | \mathcal{F}_t)^2 dt \right) \right\}.$$

Since $G_K^1(\mathbf{L_N^P}) = N^{-2} \log d_N$ is a continuous function of $\mathbf{L_N^P}$, we get the desired lower bound by the following chain of inequalities:

$$\mathbb{P}_N^{\otimes m}\left(d(\mathbf{L_N^P}, \tau) < \epsilon \right)$$

$$= \mathbb{P}_N^{\otimes m}\left(1_{d(\mathbf{L_N^P}, \tau) < \epsilon} e^{N^2 (G_K^1(\mathbf{L_N^P}) - G_K^1(\mathbf{L_N^P}))} \right)$$

$$\geq e^{-N^2 G_K^1(\tau) - N^2 \kappa(\epsilon)} \mathbb{P}_N^{\otimes m}\left(1_{d(\mathbf{L_N^P}, \tau) < \epsilon} e^{N^2 (G_K^1(\mathbf{L_N^P}))} \right)$$

$$= e^{-N^2 G_K^1(\tau) - N^2 \kappa(\epsilon)} \mathbb{P}_N^{\otimes m}\left(d(\mathbf{L_N^P}, \tau) < \epsilon \right)$$

$$= e^{-N^2 G_K^1(\tau) - N^2 \kappa(\epsilon)} \geq e^{-N^2 I(\tau) - N^2 \kappa(\epsilon)}$$

with $\kappa(\epsilon)$ going to zero as ϵ goes to zero.

Bibliographical Notes. A very nice introductory review on free entropy was written by Voiculescu [201]. The results of this section were proved in [37]. They were used in relation with the entropy dimension in [68, 156]. The problem of proving that in the definition of entropy one can replace the lim sup by a lim inf is still open, as well as the equality with the microstates-free free entropy.

Part VII

Appendix

19

Basics of Matrices

19.1 Weyl's and Lidskii's Inequalities

Theorem 19.1 (Weyl). *Denote* $\lambda_1(C) \leq \lambda_2(C) \leq \cdots \leq \lambda_N(C)$ *the (real) eigenvalues of an* $N \times N$ *Hermitian matrix* C. *Let* A, B *be* $N \times N$ *Hermitian matrices. Then, for any* $j \in \{1, \ldots, N\}$,

$$\lambda_j(A) + \lambda_1(B) \leq \lambda_j(A + B) \leq \lambda_j(A) + \lambda_N(B).$$

In particular,

$$|\lambda_j(A + B) - \lambda_j(A)| \leq \left(\mathrm{Tr}(B^2)\right)^{\frac{1}{2}}. \tag{19.1}$$

Theorem 19.2 (Courant–Fischer).

Let $A \in \mathcal{H}_N^{(2)}$ *with ordered eigenvalues* $\lambda_1(A) \leq \cdots \leq \lambda_N(A)$. *For* $k \in \{1, \ldots, N\}$,

$$\lambda_k(A) := \min_{w_1, \ldots, w_{N-k} \in \mathbb{C}^N} \max_{\substack{x \neq 0, x \in \mathbb{C}^N \\ x \perp w_1, \ldots, w_{N-k}}} \frac{x^* A x}{x^* x}.$$

Proof. We can without loss of generality assume that A is diagonal up to rotate the vectors $w_1, \ldots, w_{N-k}$ Then

$$\max_{\substack{x \neq 0, x \in \mathbb{C}^N \\ x \perp w_1, \ldots, w_{N-k}}} \frac{x^* A x}{x^* x} = \max_{\substack{\|x\|_2 = 1, x \in \mathbb{C}^N \\ x \perp w_1, \ldots, w_{N-k}}} \sum_{i=1}^N \lambda_i(A) |x_i|^2$$

$$\geq \max_{\substack{\|x\|_2 = 1, x \in \mathbb{C}^N, x_j = 0, j \leq k \\ x \perp w_1, \ldots, w_{N-k}}} \sum_{i=1}^N \lambda_i(A) |x_i|^2$$

$$\geq \lambda_k(A)$$

and equality holds when $w_i = u_{N-i+1}$ is the eigenvector corresponding to the eigenvalue $\lambda_{N-i+1}(A)$. Taking the minimum over the vectors w_i thus completes the proof. $\qquad\square$

A. Guionnet, *Large Random Matrices: Lectures on Macroscopic Asymptotics*, Lecture Notes in Mathematics 1957, DOI: 10.1007/978-3-540-69897-5_19, © 2009 Springer-Verlag Berlin Heidelberg, Reprint by Springer-Verlag Berlin Heidelberg 2012

Proof of Weyl's Inequalities Theorem 19.1. Let $(u_1, \ldots, u_{N-j})$ be the eigenvectors of the $N - j$ largest eigenvalues of A. Then, by Theorem 19.2,

$$
\begin{aligned}
\lambda_j(A+B) &= \min_{w_1,\ldots,w_{N-j}\in\mathbb{C}^N} \max_{\substack{x\neq 0,\, x\in\mathbb{C}^N \\ x\perp u_1,\ldots,u_{N-j}}} \frac{x^*(A+B)x}{x^*x} \\
&\leq \max_{\substack{x\neq 0,\, x\in\mathbb{C}^N \\ x\perp u_1,\ldots,u_{N-j}}} \frac{x^*(A+B)x}{x^*x} \\
&\leq \max_{\substack{x\neq 0,\, x\in\mathbb{C}^N \\ x\perp u_1,\ldots,u_{N-j}}} \frac{x^*Ax}{x^*x} + \max_{x\neq 0} \frac{x^*Bx}{x^*x} \\
&= \lambda_j(A) + \lambda_N(B).
\end{aligned}
$$

Replacing A, B by $-A, -B$ we obtain the second inequality.

$\square$

Theorem 19.3 (Lidskii). *Let $A \in \mathcal{H}_N^{(2)}$, $\eta \in \{+1, -1\}$ and $z \in \mathbb{C}^N$. We order the eigenvalues of $A + \eta z z^*$ in increasing order. Then*

$$
\lambda_k(A + \eta z z^*) \leq \lambda_{k+1}(A) \leq \lambda_{k+2}(A + \eta z z^*).
$$

Proof. Using the Courant–Fischer theorem one gets for $k \geq 2$,

$$
\begin{aligned}
\lambda_k(A + \eta z z^*) &:= \min_{w_1,\ldots,w_{N-k}\in\mathbb{C}^N} \max_{\substack{x\neq 0,\, x\in\mathbb{C}^N \\ x\perp w_1,\ldots,w_{N-k}}} \frac{x^*(A+\eta z z^*)x}{x^*x} \\
&\geq \min_{w_1,\ldots,w_{N-k}\in\mathbb{C}^N} \max_{\substack{x\neq 0,\, x\in\mathbb{C}^N \\ x\perp z,\, w_1,\ldots,w_{N-k}}} \frac{x^*Ax}{x^*x} \\
&\geq \min_{w_1,\ldots,w_{N-k+1}\in\mathbb{C}^N} \max_{\substack{x\neq 0,\, x\in\mathbb{C}^N \\ x\perp w_1,\ldots,w_{N-k+1}}} \frac{x^*Ax}{x^*x} \\
&= \lambda_{k-1}(A).
\end{aligned}
$$

Replacing $A' = A + \eta z z^*$, and η by $-\eta$ we also have proved $\lambda_k(A' - \eta z z^*) \geq \lambda_k(A')$, i.e., $\lambda_k(A) \geq \lambda_{k-1}(A + \eta z z^*)$. $\square$

One also has [33, Proposiion 28.2]:

Theorem 19.4 (Löwner). *Let $A, E \in \mathcal{H}_N^{(2)}$.*

$$
\sum_{k=1}^{N} |\lambda_k(A+E) - \lambda_k(A)|^2 \leq \sum_{k=1}^{N} \lambda_k(E)^2. \tag{19.2}
$$

19.2 Non-commutative Hölder Inequality

The following can be found in [158].

Theorem 19.5 (Nelson). *For any $P_1, P_2 \in \mathbb{C}\langle X_1, \cdots X_m \rangle$, any matrices* $\mathbf{A} = (A_1, \ldots, A_m) \in \mathcal{M}_N$ *and any $p_1, p_2 \in [0, 1]$ so that $p_1^{-1} + p_2^{-1} = 1$,*

$$|\mathrm{Tr}(P_1(\mathbf{A})P_2(\mathbf{A}))| \leq [\mathrm{Tr}(|P_1(\mathbf{A})|^{p_1})^{\frac{1}{p_1}}[\mathrm{Tr}(|P_2(\mathbf{A})|^{p_2})^{\frac{1}{p_2}}$$

with $|P| = \sqrt{PP^*}$. This non-commutative Hölder inequality extends when Tr *is replaced by any tracial state.*

20

Basics of Probability Theory

20.1 Basic Notions of Large Deviations

This appendix recalls basic definitions and main results of large deviations theory. We refer the reader to [75] and [74] for a full treatment.

In what follows, X will be assumed to be a Polish space (that is a complete separable metric space). We recall that a function $f : X \to \mathbb{R}$ is *lower semicontinuous* if the level sets $\{x : f(x) \le C\}$ are closed for any constant C.

Definition 20.1. *A sequence $(\mu_N)_{N \in \mathbb{N}}$ of probability measures on X satisfies a large deviation principle with speed a_N (going to infinity with N) and rate function I iff*

$$I : X \to [0, \infty] \text{ is lower semicontinuous.} \tag{20.1}$$

$$\text{For any open set } O \subset X, \ \liminf_{N \to \infty} \frac{1}{a_N} \log \mu_N(O) \ge -\inf_O I. \tag{20.2}$$

$$\text{For any closed set } F \subset X, \ \limsup_{N \to \infty} \frac{1}{a_N} \log \mu_N(F) \le -\inf_F I. \tag{20.3}$$

]When it is clear from the context, we omit the reference to the speed or rate function and simply say that the sequence $\{\mu_N\}$ satisfies the LDP. Also, if x_N are X-valued random variables distributed according to μ_N, we say that the sequence $\{x_N\}$ satisfies the LDP if the sequence $\{\mu_N\}$ satisfies the LDP.

Definition 20.2. *A sequence $(\mu_N)_{N \in \mathbb{N}}$ of probability measures on X satisfies a weak large deviation principle if (20.1) and (20.2) hold, and in addition (20.3) holds for all compact sets $F \subset X$.*

The proof of a large deviation principle often proceeds first by the proof of a weak large deviation principle, in conjuction with the so-called exponential tightness property.

Definition 20.3. *a. A sequence $(\mu_N)_{N \in \mathbb{N}}$ of probability measures on X is exponentially tight iff there exists a sequence $(K_L)_{L \in \mathbb{N}}$ of compact sets such that*

A. Guionnet, *Large Random Matrices: Lectures on Macroscopic Asymptotics*,   267
Lecture Notes in Mathematics 1957, DOI: 10.1007/978-3-540-69897-5_20,
© 2009 Springer-Verlag Berlin Heidelberg, Reprint by Springer-Verlag Berlin Heidelberg 2012

$$\limsup_{L\to\infty} \limsup_{N\to\infty} \frac{1}{a_N} \log \mu_N(K_L^c) = -\infty.$$

b. A rate function I is good if the level sets $\{x \in X : I(x) \le M\}$ are compact for all $M \ge 0$.

The interest in these concepts lies in the following:

Theorem 20.4. a. ([74, Lemma 1.2.18]) If $\{\mu_N\}$ satisfies the weak LDP and it is exponentially tight, then it satisfies the full LDP, and the rate function I is good.
b. ([74, Exercise 4.1.10]) If $\{\mu_N\}$ satisfies the upper bound (20.3) with a good rate function I, then it is exponentially tight.

A weak large deviation principle is itself equivalent to the estimation of the probability of deviations towards small balls:

Theorem 20.5. [74, Theorem 4.1.11] Let $\mathcal{A}$ be a base of the topology of X. For every $A \in \mathcal{A}$, define

$$\mathcal{L}_A = -\liminf_{N\to\infty} \frac{1}{a_N} \log \mu_N(A)$$

and

$$I(x) = \sup_{A\in\mathcal{A}:x\in A} \mathcal{L}_A.$$

Suppose that for all $x \in X$,

$$I(x) = \sup_{A\in\mathcal{A}:x\in A} \left\{ -\limsup_{N\to\infty} \frac{1}{a_N} \log \mu_N(A) \right\}.$$

Then, μ_N satisfies a weak large deviation principle with rate function I.

Let d be the metric in X, and set $B(x,\delta) = \{y \in X : d(y,x) < \delta\}$.

Corollary 20.6. Assume that for all $x \in X$

$$-I(x) := \limsup_{\delta\to 0} \limsup_{N\to\infty} \frac{1}{a_N} \log \mu_N(B(x,\delta))$$

$$= \liminf_{\delta\to 0} \liminf_{N\to\infty} \frac{1}{a_N} \log \mu_N(B(x,\delta)).$$

Then, μ_N satisfies a weak large deviation principles with rate function I.

From a given large deviation principle one can deduce large deviation principle for other sequences of probability measures by using either the so-called contraction principle or Laplace's method.

Theorem 20.7 (Contraction principle). [74, Theorem 4.2.1] Assume that the sequence of probability measures $(\mu_N)_{N\in\mathbb{N}}$ on X satisfies a large deviation principle with good rate function I. Then, for any function $F : X \to Y$ with

*values in a Polish space Y which is continuous, the image $(F\sharp\mu_N)_{N\in\mathbb{N}} \in$
$M_1(Y)^{\mathbb{N}}$ defined as $F\sharp\mu_N(A) = \mu_N\left(F^{-1}(A)\right)$ also satisfies a large deviation
principle with the same speed and rate function given for any $y \in Y$ by*

$$J(y) = \inf\{I(x) : F(x) = y\}.$$

Theorem 20.8 (Varadhan's lemma). *[74, Theorem 4.3.1] Assume that
$(\mu_N)_{N\in\mathbb{N}}$ satisfies a large deviation principle with good rate function I. Let
$F : X \to \mathbb{R}$ be a bounded continuous function. Then,*

$$\lim_{N\to\infty} \frac{1}{a_N} \log \int e^{a_N F(x)} d\mu_N(x) = \sup_{x\in X}\{F(x) - I(x)\}.$$

Moreover, the sequence

$$\nu_N(dx) = \frac{1}{\int e^{a_N F(y)} d\mu_N(y)} e^{a_N F(x)} d\mu_N(x) \in M_1(X)$$

satisfies a large deviation principle with good rate function

$$J(x) = I(x) - F(x) - \sup_{y\in X}\{F(y) - I(y)\}.$$

Large deviation principles are quite robust to exponential equivalence that we
now define.

Definition 20.9. *Let (X, d) be a metric space. Let $(\mu_N)_{N\in\mathbb{N}}$ and $(\tilde{\mu}_N)_{N\in\mathbb{N}}$ be
two sequences of probability measures on X. $(\mu_N)_{N\in\mathbb{N}}$ and $(\tilde{\mu}_N)_{N\in\mathbb{N}}$ are said
to be exponentially equivalent if there exists probability spaces $(\Omega, \mathcal{B}_N, P_N)$ and
two families of random variables $Z_N, \tilde{Z}_N$ on Ω with values in X with joint
distribution P_N and marginals μ_N and $\tilde{\mu}_N$ respectively so that for each $\delta > 0$*

$$\limsup_{N\to\infty} P_N\left(d(Z_N, \tilde{Z}_N) > \delta\right) = -\infty.$$

We then have:

Lemma 20.10. *[74, Theorem 4.2.13] If a large deviation principle for μ_N
holds with good rate function I and $\tilde{\mu}_N$ is exponentially equivalent to μ_N, then
a $\tilde{\mu}_N$ satisfies a large deviation principle with the same rate function I.*

$\mathcal{P}(\Sigma)$ possesses a useful criterion for compactness.

Theorem 20.11 (Prohorov). *Let Σ be Polish, and let $\Gamma \subset \mathcal{P}(\Sigma)$. Then $\overline{\Gamma}$
is compact iff Γ is tight.*

Since $\mathcal{P}(\Sigma)$ is Polish, convergence may be decided by sequences.

448

20.2 Basics of Stochastic Calculus

Definition 20.12 ([122], [168]). *Let $(\Omega, \mathcal{F})$ be a measurable space.*

- *A filtration $\mathcal{F}_t, t \geq 0$ is a non-decreasing family of sub-σ-fields of $\mathcal{F}$.*
- *A random time T is a stopping time of the filtration $\mathcal{F}_t, t \geq 0$ if the event $\{T \leq t\}$ belongs to the σ-field $\mathcal{F}_t$ for all $t \geq 0$.*
- *A process $X_t, t \geq 0$ is adapted to the filtration $\mathcal{F}_t, t \geq 0$ if for all $t \geq 0$ X_t is an $\mathcal{F}_t$-measurable random variable.*
- *Let $\{X_t, \mathcal{F}_t, t \geq 0\}$ be an adapted process so that $\mathbb{E}[\|X_t\|] < \infty$ for all $t \geq 0$. The process $\{X_t, \mathcal{F}_t, t \geq 0\}$ is said to be a martingale if for every $0 \leq s < t < \infty$,*

$$\mathbb{E}[X_t | \mathcal{F}_s] = X_s.$$

- *Let $\{X_t, \mathcal{F}_t, t \geq 0\}$ be a martingale so that $E[X_t^2] < \infty$ for all $t \geq 0$. The martingale bracket (or the quadratic variation) $\langle X \rangle$ of X is the unique adapted increasing process so that $X^2 - \langle X \rangle$ is a martingale for the filtration $\mathcal{F}$.*

Let $\{X_t, \mathcal{F}_t, t \geq 0\}$ be a real-valued adapted process and let B be a Brownian motion. Assume that $E[\int_0^T X_t^2 dt] < \infty$. Then,

$$\int_0^T X_t dB_t := \lim_{n \to \infty} \sum_{k=0}^{n-1} X_{\frac{Tk}{n}} \left(B_{\frac{T(k+1)}{n}} - B_{\frac{Tk}{n}} \right)$$

exists, the convergence holds in L^2 and the limit does not depend on the above choice of the discretization of $[0, T]$ (see [122, section 3]). The limit is called a stochastic integral.

One can therefore consider the problem of finding solutions to the integral equation

$$X_t = X_0 + \int_0^t \sigma(X_s) dB_s + \int_0^t b(X_s) ds \qquad (20.4)$$

with a given X_0, σ and b some functions on $\mathbb{R}^n$, and B a n-dimensional Brownian motion. This can be written under the differential form

$$dX_s = \sigma(X_s) dB_s + b(X_s) ds. \qquad (20.5)$$

There are at least two notions of solutions; the strong solutions and the weak solutions.

Definition 20.13. *[122, Definition 2.1] A strong solution of the stochastic differential equation (20.5) on the given probability space $(\Omega, \mathcal{F})$ and with respect to the fixed Brownian motion B and initial condition ξ is a process $\{X_t, t \geq 0\}$ with continuous sample paths so that*

1. *X is adapted to the filtration $\mathcal{F}$ given by*

$$\mathcal{G}_t = \sigma(B_s, s \leq t; X_0), \mathcal{N} = \{N \subset \Omega, \exists G \in \mathcal{G}_\infty \text{ with } N \subset G, P(G) = 0\},$$
$$\mathcal{F}_t = \sigma(\mathcal{G}_t \cup \mathcal{N}).$$

2. $P(X_0 = \xi) = 1$.
3. $P(\int_0^t (|b_i(X_s)| + |\sigma_{ij}(X_s)|^2) ds < \infty) = 1$ *for all* $i, j \leq n$.
4. *(20.4) holds almost surely.*

Definition 20.14. *[122, Definition 3.1] A weak solution of the stochastic differential equation (20.5) is a triple (X, B) and $(\Omega, \mathcal{F}, P)$ so that $(\Omega, \mathcal{F}, P)$ is a probability space equipped with a filtration $\mathcal{F}$, X is a continuous adapted process and B an n-dimensional Brownian motion. X satisfies (3) and (4) in Definition 20.13.*

There are also two notions of uniqueness:

Definition 20.15. *[122, Definition 3.4]*

- *We say that strong uniqueness holds if two solutions with common probability space, common Brownian motion B and common initial condition are almost surely equal at all times.*
- *We say that weak uniqueness, or uniqueness in the sense of probability, holds if any two weak solutions have the same law.*

Theorem 20.16. *[122, Theorems 2.5 and 2.9]*
Suppose that b and σ satisfy

$$\|b(t, x) - b(t, y)\| + \|\sigma(t, x) - \sigma(t, y)\| \leq K\|x - y\|,$$
$$\|b(t, x)\|^2 + \|\sigma(t, x)\|^2 \leq K^2(1 + \|x\|^2),$$

for some finite constant K independent of t and $\|.\|$ the Euclidean norm on $\mathbb{R}^n$, then there exists a unique strong solution to (20.5). Moreover, it satisfies

$$\mathbb{E}[\int_0^T \|b(t, X_t)\|^2 dt] < \infty$$

for all $T \geq 0$.

Theorem 20.17. *[122, Proposition 3.10]*
Any two weak solutions $(X^i, B^i, \Omega^i, \mathcal{F}^i, P^i)_{i=1,2}$ of (20.5) so that

$$\mathbb{E}[\int_0^T \|b(t, X_t^i)\|^2 dt] < \infty$$

for all $T < \infty$ and $i = 1, 2$ have the same law.

Theorem 20.18 (Itô (1944), Kunita–Watanabe (1967)). *[122, p. 149]*
Let $f : \mathbb{R} \to \mathbb{R}$ be a function of class C^2 and let $X = \{X_t, \mathcal{F}_t; 0 \le t < \infty\}$ be a continuous semi-martingale with decomposition

$$X_t = X_0 + M_t + A_t$$

where M is a local martingale and A the difference of continuous, adapted, non-decreasing processes. Then, almost surely,

$$f(X_t) = f(X_0) + \int_0^t f'(X_s)dM_s + \int_0^t f'(X_s)dA_s$$

$$+\frac{1}{2}\int_0^2 f''(X_s)d < M >_s, \quad 0 \le t < \infty.$$

We shall use the following well known results on martingales.

Theorem 20.19 (Burkholder–Davis–Gundy's inequality). *[122, p. 166]* Let $(M_t, t \ge 0)$ be a continuous local martingale with bracket $(A_t, t \ge 0)$. There exists universal constants λ_m, Λ_m so that for all $m \in \mathbb{N}$

$$\lambda_m E(A_T^m) \le E(\sup_{t \le T} M_t^{2m}) \le \Lambda_m E(A_T^m).$$

Theorem 20.20 (Novikov (1972)). *[122, p. 199]* Let $\{X_t, \mathcal{F}_t, t \ge 0\}$ be an adapted process with values in $\mathbb{R}^d$ such that

$$E[e^{\frac{1}{2}\int_0^T \sum_{i=1}^d (X_t^i)^2 dt}] < \infty$$

for all $T \in \mathbb{R}^+$. Then, if $\{W_t, \mathcal{F}_t, t \ge 0\}$ is a d dimensional Brownian motion,

$$M_t = \exp\left\{\int_0^t X_u.dW_u - \frac{1}{2}\int_0^t \sum_{i=1}^d (X_u^i)^2 du\right\}$$

is a $\mathcal{F}_t$-martingale.

Theorem 20.21 (Girsanov (1960)). *[122, p. 191]* Let $\{X_t, \mathcal{F}_t, t \ge 0\}$ be an adapted process with values in $\mathbb{R}^d$ such that

$$E[e^{\frac{1}{2}\int_0^T \sum_{i=1}^d (X_t^i)^2 dt}] < \infty.$$

Then, if $\{W_t, \mathcal{F}_t, P, 0 \le t \le T\}$ is a d dimensional Brownian motion,

$$\bar{W}_t^i = W_t^i - \int_0^t X_s^i ds, 0 \le t \le T$$

is a d-dimensional Brownian under the probability measure

$$\bar{P} = \exp\{\int_0^T X_u.dW_u - \frac{1}{2}\int_0^T \sum_{i=1}^d (X_u^i)^2 du\}P.$$

Theorem 20.22. *[122, p. 14] Let $\{X_t, \mathcal{F}_t, 0 \leq t < \infty\}$ be a submartingale whose every path is right-continuous. Then for any $\tau > 0$, for any $\lambda > 0$*

$$\lambda P(\sup_{0 \leq t \leq \tau} X_t \geq \lambda) \leq E[X_\tau^+].$$

We shall use the following consequence:

Corollary 20.23. *Let $\{X_t, \mathcal{F}_t, t \geq 0\}$ be an adapted process with values in $\mathbb{R}^d$ such that*

$$\int_0^T \|X_t\|^2 dt = \int_0^T \sum_{i=1}^d (X_t^i)^2 dt$$

is uniformly bounded by A_T. Let $\{W_t, \mathcal{F}_t, t \geq 0\}$ be a d-dimensional Brownian motion. Then for any $L > 0$,

$$P\left(\sup_{0 \leq t \leq T} \left| \int_0^t X_u.dW_u \right| \geq L\right) \leq 2e^{-\frac{L^2}{2A_T}}.$$

Proof. We denote in short $Y_t = \int_0^t X_u.dW_u$ and write for $\lambda > 0$,

$$P\left(\sup_{0 \leq t \leq T} |Y_t| \geq A\right) \leq P\left(\sup_{0 \leq t \leq T} e^{\lambda Y_t} \geq e^{\lambda A}\right) + P\left(\sup_{0 \leq t \leq T} e^{-\lambda Y_t} \geq e^{\lambda A}\right)$$

$$\leq P\left(\sup_{0 \leq t \leq T} e^{\lambda Y_t - \frac{\lambda^2}{2}\int_0^t \|X_u\|^2 du} \geq e^{\lambda A - \frac{\lambda^2 A_T}{2}}\right)$$

$$+ P\left(\sup_{0 \leq t \leq T} e^{-\lambda Y_t - \frac{\lambda^2}{2}\int_0^t \|X_u\|^2 du} \geq e^{\lambda A - \frac{\lambda^2 A_T}{2}}\right).$$

By Theorem 20.20, $M_t = e^{-\lambda Y_t - \frac{\lambda^2}{2}\int_0^t \|X_u\|^2 du}$ is a non-negative martingale. Thus, By Chebychev's inequality and Doob's inequality

$$P\left(\sup_{0 \leq t \leq T} M_t \geq e^{\lambda A - \frac{\lambda^2 A_T}{2}}\right) \leq e^{-\lambda A + \frac{\lambda^2 A_T}{2}} E[M_T]$$

$$= e^{-\lambda A + \frac{\lambda^2 A_T}{2}}$$

Optimizing with respect to λ completes the proof. $\qquad\square$

Theorem 20.24 (Rebolledo's Theorem). *Let $n \in \mathbb{N}$, and let M_N be a sequence of continuous centered martingales with values in $\mathbb{R}^n$ with bracket $\langle M_N \rangle$ converging pointwise (i.e., for all $t \geq 0$) in L^1 towards a continuous deterministic function $\phi(t)$. Then, for any $T > 0$, $(M_N(t), t \in [0, T])$ converges in law as a continuous process from $[0, T]$ into $\mathbb{R}^n$ towards a Gaussian process G with covariance*

$$E[G(s)G(t)] = \phi(t \wedge s).$$

20.3 Proof of (2.3)

Put

$$V(\mathbf{i}^1,\ldots,\mathbf{i}^l) = [[i_n^j]_{n=1}^k]_{j=1}^l\,, \; I = \bigcup_{j=1}^{l}\{j\} \times \{1,\ldots,k\}\,, \; A = [\{i_n^j, i_{n+1}^j\}]_{(i,n)\in I}.$$

We visualize A as a left-justified table of l rows. Let $G' = (V', E')$ be any spanning forest in $G(\mathbf{i}^1,\ldots,\mathbf{i}^l)$, with c connected components. Since every connected component of G' is a tree, we have

$$|V| = |V'| = c + |E'|. \tag{20.6}$$

Now let $X = \{X_{in}\}_{(i,n)\in I}$ be a table of the same "shape" as A, but with all entries equal either to 0 or 1. We call X an *edge-bounding table* under the following conditions:

- For all $(i,n) \in I$, if $X_{in} = 1$, then $A_{in} \in E'$.
- For each $e \in E'$ there exists distinct $(i_1, n_1), (i_2, n_2) \in I$ such that $X_{i_1 n_1} = X_{i_2 n_2} = 1$ and $A_{i_1 n_1} = A_{i_2 n_2} = e$.
- For each $e \in E'$ and index $i \in \{1,\ldots,j\}$, if e appears in the ith row of A then there exists $(i,n) \in I$ such that $A_{in} = e$ and $X_{in} = 1$.

For any edge-bounding table X the corresponding quantity $\frac{1}{2}\sum_{(i,n)\in I} X_{in}$ bounds $|E'|$ by the second required property. At least one edge-bounding table exists, namely the table with a 1 in position (i, n) for each $(i, n) \in I$ such that $A_{in} \in E'$ and 0's elsewhere. Now let X be an edge-bounding table such that for some index i_0 all the entries of X in the i_0th row are equal to 1. Then the graph $G(\mathbf{i}_0)$ is a tree (since all edges of $G(\mathbf{i}_0)$ could be kept in G'), and hence every entry in the i_0th row of A appears there an even number of times and *a fortiori* at least twice. Now choose $(i_0, n_0) \in I$ such that $A_{i_0 n_0} \in E'$ appears in another row than i_0. Let Y be the table obtained by replacing the entry 1 of X in position (i_0, n_0) by the entry 0. Then Y is again an edge-bounding table. Proceeding in this way we can find an edge-bounding table with 0 appearing at least once in every row, and hence we have $|E'| \leq \lceil\frac{|I|-l}{2}\rceil = \frac{kl-l}{2}$. Together with (20.6) and the definition of I, this completes the proof.

References

1. AIDA, S., AND STROOCK, D. Moment estimates derived from Poincaré and logarithmic Sobolev inequalities. *Math. Res. Lett. 1*, 1 (1994), 75–86.

2. ALBEVERIO, S., PASTUR, L., AND SHCHERBINA, M. On the $1/n$ expansion for some unitary invariant ensembles of random matrices. *Comm. Math. Phys. 224*, 1 (2001), 271–305. Dedicated to Joel L. Lebowitz.

3. ALON, N., KRIVELEVICH, M., AND VU, V. H. On the concentration of eigenvalues of random symmetric matrices. *Israel J. Math. 131* (2002), 259–267.

4. AMBJØRN, J., CHEKHOV, L., KRISTJANSEN, C. F., AND MAKEENKO, Y. Matrix model calculations beyond the spherical limit. *Nuclear Physics B 404* (1993), 127–172.

5. AMBROSIO, L., AND SANTAMBROGIO, F. Necessary optimality conditions for geodesics in weighted Wasserstein spaces. *Atti Accad. Naz. Lincei Cl. Sci. Fis. Mat. Natur. Rend. Lincei (9) Mat. Appl. 18*, 1 (2007), 23–37.

6. ANDERSON, G., GUIONNET, A., AND ZEITOUNI, O. *An introduction to random matrices*. Work in progress.

7. ANDERSON, G. W., AND ZEITOUNI, O. A CLT for a band matrix model. *Probab. Theory Related Fields 134*, 2 (2006), 283–338.

8. ANÉ, C., BLACHÈRE, S., CHAFAÏ, D., FOUGÈRES, P., GENTIL, I., MALRIEU, F., ROBERTO, C., AND SCHEFFER, G. *Sur les inégalités de Sobolev logarithmiques*, vol. 10 of *Panoramas et Synthèses [Panoramas and Syntheses]*. Société Mathématique de France, Paris, 2000. With a preface by Dominique Bakry and Michel Ledoux.

9. AUFFINGER, A., BEN AROUS, G., AND PÉCHÉ, S. Poisson convergence for the largest eigenvalues of heavy tailed random matrices. *http://front. math.ucdavis.edu/0710.3132* (2007).

10. BAI, Z., AND SILVERSTEIN, J. *Spectral analysis of large dimensional random matrices*. Science Press, Beijing. 2006.

11. BAI, Z. D. Convergence rate of expected spectral distributions of large random matrices. I. Wigner matrices. *Ann. Probab. 21*, 2 (1993), 625–648.

12. BAI, Z. D. Circular law. *Ann. Probab. 25*, 1 (1997), 494–529.

13. BAI, Z. D. Methodologies in spectral analysis of large-dimensional random matrices, a review. *Statist. Sinica 9*, 3 (1999), 611–677. With comments by G. J. Rodgers and Jack W. Silverstein; and a rejoinder by the author.

276   References

14. BAI, Z. D., AND SILVERSTEIN, J. W. No eigenvalues outside the support of the limiting spectral distribution of large-dimensional sample covariance matrices. *Ann. Probab. 26*, 1 (1998), 316–345.

15. BAI, Z. D., AND YIN, Y. Q. Necessary and sufficient conditions for almost sure convergence of the largest eigenvalue of a Wigner matrix. *Ann. Probab. 16*, 4 (1988), 1729–1741.

16. BAIK, J., BEN AROUS, G., AND PÉCHÉ, S. Phase transition of the largest eigenvalue for nonnull complex sample covariance matrices. *Ann. Probab. 33*, 5 (2005), 1643–1697.

17. BAIK, J., DEIFT, P., AND JOHANSSON, K. On the distribution of the length of the longest increasing subsequence of random permutations. *J. Amer. Math. Soc. 12*, 4 (1999), 1119–1178.

18. BAKRY, D., AND ÉMERY, M. Diffusions hypercontractives. In *Séminaire de probabilités, XIX, 1983/84*, vol. 1123 of *Lecture Notes in Math.* Springer, Berlin, 1985, pp. 177–206.

19. BAKRY, D., AND ÉMERY, M. Inégalités de Sobolev pour un semi-groupe symétrique. *C. R. Acad. Sci. Paris Sér. I Math. 301*, 8 (1985), 411–413.

20. BARNDORFF-NIELSEN, O. E., AND THORBJØRNSEN, S. A connection between free and classical infinite divisibility. *Infin. Dimens. Anal. Quantum Probab. Relat. Top. 7*, 4 (2004), 573–590.

21. BARTHE, F. Levels of concentration between exponential and Gaussian. *Ann. Fac. Sci. Toulouse Math. (6) 10*, 3 (2001), 393–404.

22. BARTHE, F., CATTIAUX, P., AND ROBERTO, C. Concentration for independent random variables with heavy tails. *AMRX Appl. Math. Res. Express*, 2 (2005), 39–60.

23. BELINSCHI, S. T., AND BERCOVICI, H. A property of free entropy. *Pacific J. Math. 211*, 1 (2003), 35–40.

24. BEN AROUS, G., DEMBO, A., AND GUIONNET, A. Aging of spherical spin glasses. *Probab. Theory Related Fields 120*, 1 (2001), 1–67.

25. BEN AROUS, G., AND GUIONNET, A. Large deviations for Wigner's law and Voiculescu's non-commutative entropy. *Probab. Theory Related Fields 108*, 4 (1997), 517–542.

26. BEN AROUS, G., AND GUIONNET, A. The spectrum of heavy tailed random matrices. *Comm. Math. Phys. 278* (2008), 715–751.

27. BEN AROUS, G., AND PÉCHÉ, S. Universality of local eigenvalue statistics for some sample covariance matrices. *Comm. Pure Appl. Math. 58*, 10 (2005), 1316–1357.

28. BEN AROUS, G., AND ZEITOUNI, O. Large deviations from the circular law. *ESAIM Probab. Statist. 2* (1998), 123–134 (electronic).

29. BENDER, E. A., AND CANFIELD, E. R. The number of degree-restricted rooted maps on the sphere. *SIAM J. Discrete Math. 7*, 1 (1994), 9–15.

30. BERCOVICI, H., AND VOICULESCU, D. Free convolution of measures with unbounded support. *Indiana Univ. Math. J. 42*, 3 (1993), 733–773.

31. BERTOLA, M., EYNARD, B., AND HARNAD. Duality, bimaster loop equations, free energy and correlations for the chain of matrices. *J. High Energy Phys.*, 11 (2003), 018, 45 pp. (electronic).

32. BESSIS, D., ITZYKSON, C., AND ZUBER, J. B. Quantum field theory techniques in graphical enumeration. *Adv. in Appl. Math. 1*, 2 (1980), 109–157.

33. BHATIA, R. *Perturbation bounds for matrix eigenvalues*, vol. 53 of *Classics in Applied Mathematics*. Society for Industrial and Applied Mathematics (SIAM), Philadelphia, PA, 2007. Reprint of the 1987 original.

34. BIANE, P. On the free convolution with a semi-circular distribution. *Indiana Univ. Math. J. 46*, 3 (1997), 705–718.

35. BIANE, P. Processes with free increments. *Math. Z. 227*, 1 (1998), 143–174.

36. BIANE, P., BOUGEROL, P., AND O'CONNELL, N. Littelmann paths and Brownian paths. *Duke Math. J. 130*, 1 (2005), 127–167.

37. BIANE, P., CAPITAINE, M., AND GUIONNET, A. Large deviation bounds for matrix Brownian motion. *Invent. Math. 152*, 2 (2003), 433–459.

38. BOBKOV, S. G., AND GÖTZE, F. Exponential integrability and transportation cost related to logarithmic Sobolev inequalities. *J. Funct. Anal. 163*, 1 (1999), 1–28.

39. BORODIN, A. Biorthogonal ensembles. *Nuclear Phys. B 536*, 3 (1999), 704–732.

40. BOUGEROL, P., AND JEULIN, T. Paths in Weyl chambers and random matrices. *Probab. Theory Related Fields 124*, 4 (2002), 517–543.

41. BOULATOV, D., AND KAZAKOV, V. One-dimensional string theory with vortices as the upside-down matrix oscillator. *Internat. J. Modern Phys. A 8*, 5 (1993), 809–851.

42. BOULATOV, D. V., AND KAZAKOV, V. A. The Ising model on a random planar lattice: the structure of the phase transition and the exact critical exponents. *Phys. Lett. B 186*, 3-4 (1987), 379–384.

43. BOULATOV, D. V., AND KAZAKOV, V. A. The Ising model on a random planar lattice: the structure of the phase transition and the exact critical exponents. *Phys. Lett. B 186*, 3-4 (1987), 379–384.

44. BOUSQUET-MELOU, M., AND SCHAEFFER, G. The degree distribution in bipartite planar maps: applications to the Ising model. *arXiv:math.CO/0211070* (2002).

45. BOUTET DE MONVEL, A., KHORUNZHY, A., AND VASILCHUK, V. Limiting eigenvalue distribution of random matrices with correlated entries. *Markov Process. Related Fields 2*, 4 (1996), 607–636.

46. BOUTTIER, J., DI FRANCESCO, P., AND GUITTER, E. Census of planar maps: from the one-matrix model solution to a combinatorial proof. *Nuclear Phys. B 645*, 3 (2002), 477–499.

47. BRASCAMP, H. J., AND LIEB, E. H. On extensions of the Brunn-Minkowski and Prékopa-Leindler theorems, including inequalities for log concave functions, and with an application to the diffusion equation. *J. Functional Analysis 22*, 4 (1976), 366–389.

48. BRENIER, Y. Polar factorization and monotone rearrangement of vector-valued functions. *Comm. Pure Appl. Math. 44*, 4 (1991), 375–417.

49. BRÉZIN, E., AND HIKAMI, S. An extension of the HarishChandra-Itzykson-Zuber integral. *Comm. Math. Phys. 235*, 1 (2003), 125–137.

50. BRÉZIN, E., ITZYKSON, C., PARISI, G., AND ZUBER, J. B. Planar diagrams. *Comm. Math. Phys. 59*, 1 (1978), 35–51.

51. BRÉZIN, É., AND WADIA, S. R., Eds. *The large N expansion in quantum field theory and statistical physics*. World Scientific Publishing Co. Inc., River Edge, NJ, 1993. From spin systems to 2-dimensional gravity.

52. BRU, M.-F. Wishart processes. *J. Theoret. Probab. 4*, 4 (1991), 725–751.

53. BRYC, W., DEMBO, A., AND JIANG, T. Spectral measure of large random Hankel, Markov and Toeplitz matrices. *Ann. Probab. 34*, 1 (2006), 1–38.

278   References

54. CABANAL-DUVILLARD, T. Fluctuations de la loi empirique de grandes matrices aléatoires. *Ann. Inst. H. Poincaré Probab. Statist. 37*, 3 (2001), 373–402.

55. CABANAL DUVILLARD, T., AND GUIONNET, A. Large deviations upper bounds for the laws of matrix-valued processes and non-communicative entropies. *Ann. Probab. 29*, 3 (2001), 1205–1261.

56. CABANAL-DUVILLARD, T., AND GUIONNET, A. Discussions around Voiculescu's free entropies. *Adv. Math. 174*, 2 (2003), 167–226.

57. CAFFARELLI, L. A. Boundary regularity of maps with convex potentials. *Comm. Pure Appl. Math. 45*, 9 (1992), 1141–1151.

58. CAFFARELLI, L. A. The regularity of mappings with a convex potential. *J. Amer. Math. Soc. 5*, 1 (1992), 99–104.

59. CAFFARELLI, L. A. Monotonicity properties of optimal transportation and the FKG and related inequalities. *Comm. Math. Phys. 214*, 3 (2000), 547–563.

60. CHADHA, S., MAHOUX, G., AND MEHTA, M. L. A method of integration over matrix variables. II. *J. Phys. A 14*, 3 (1981), 579–586.

61. CHAN, T. The Wigner semi-circle law and eigenvalues of matrix-valued diffusions. *Probab. Theory Related Fields 93*, 2 (1992), 249–272.

62. CHASSAING, P., AND SCHAEFFER, G. Random planar lattices and integrated superBrownian excursion. *Probab. Theory Related Fields 128*, 2 (2004), 161–212.

63. CHATTERJEE, S. Concentration of Haar measures, with an application to random matrices. *J. Funct. Anal. 245*, 2 (2007), 379–389.

64. CIZEAU, P., AND BOUCHAUD, J.-P. Theory of lévy matrices. *Physical Review E 50*, 3 (1994), 1810–1822.

65. COLLINS, B. Moments and cumulants of polynomial random variables on unitary groups, the Itzykson-Zuber integral, and free probability. *Int. Math. Res. Not.*, 17 (2003), 953–982.

66. COLLINS, B., MAUREL-SEGALA, E., AND GUIONNET, A. Asymptotics of unitary matrix integrals. *http://front.math.ucdavis.edu/0608.5193* (2006).

67. COLLINS, B., AND ŚNIADY, P. New scaling of Itzykson-Zuber integrals. *Ann. Inst. H. Poincaré Probab. Statist. 43*, 2 (2007), 139–146.

68. CONNES, A., AND SHLYAKHTENKO, D. L^2-homology for von Neumann algebras. *J. Reine Angew. Math. 586* (2005), 125–168.

69. CORI, R., AND VAUQUELIN, B. Planar maps are well labeled trees. *Canad. J. Math. 33*, 5 (1981), 1023–1042.

70. GROSS, D., PIRAN, T., AND WEINBERG, S. Two dimensional quantum gravity and random surfaces. *Jerusalem winter school, World Scientific* (1991).

71. DEIFT, P., AND GIOEV, D. Universality at the edge of the spectrum for unitary, orthogonal, and symplectic ensembles of random matrices. *Comm. Pure Appl. Math. 60*, 6 (2007), 867–910.

72. DEIFT, P., KRIECHERBAUER, T., AND MCLAUGHLIN, K. T.-R. New results on the equilibrium measure for logarithmic potentials in the presence of an external field. *J. Approx. Theory 95*, 3 (1998), 388–475.

73. DEIFT, P. A. *Orthogonal polynomials and random matrices: a Riemann-Hilbert approach*, vol. 3 of *Courant Lecture Notes in Mathematics*. New York University Courant Institute of Mathematical Sciences, New York, 1999.

74. DEMBO, A., AND ZEITOUNI, O. Large deviations and applications. In *Handbook of stochastic analysis and applications*, vol. 163 of *Statist. Textbooks Monogr.* Dekker, New York, 2002, pp. 361–416.

75. DEUSCHEL, J.-D., AND STROOCK, D. W. *Large deviations*, vol. 137 of *Pure and Applied Mathematics*. Academic Press Inc., Boston, MA, 1989.

76. DI FRANCESCO, P. Exact asymptotics of meander numbers. In *Formal power series and algebraic combinatorics (Moscow, 2000)*. Springer, Berlin, 2000, pp. 3–14.

77. DI FRANCESCO, P., AND ITZYKSON, C. A generating function for fatgraphs. *Ann. Inst. H. Poincaré Phys. Théor. 59*, 2 (1993), 117–139.

78. DI FRANCESCO, P., GINSPARG, P., AND ZINN-JUSTIN, J. 2d gravity and random matrices. *Phys. Rep.*, 254 (1995).

79. DIJKGRAAF, R., AND VAFA, C. On geometry and matrix models. *Nuclear Phys. B 644*, 1-2 (2002), 21–39.

80. DOLL, J.-M. Q-states potts model on a random planar lattice. *http://arxiv.org/abs/hep-th/9502014* (1995).

81. DUMITRIU, I., AND EDELMAN, A. Matrix models for beta ensembles. *J. Math. Phys. 43*, 11 (2002), 5830–5847.

82. DUMITRIU, I., AND EDELMAN, A. Global spectrum fluctuations for the β-Hermite and β-Laguerre ensembles via matrix models. *J. Math. Phys. 47*, 6 (2006), 063302, 36.

83. DUNFORD, N., AND SCHWARTZ, J. T. *Linear operators, Part I*. Interscience Publishers Inc., New York, 1958.

84. DYKEMA, K. On certain free product factors via an extended matrix model. *J. Funct. Anal. 112*, 1 (1993), 31–60.

85. DYSON, F. J. A Brownian-motion model for the eigenvalues of a random matrix. *J. Mathematical Phys. 3* (1962), 1191–1198.

86. ERCOLANI, N. M., AND MCLAUGHLIN, K. D. T.-R. Asymptotics of the partition function for random matrices via Riemann-Hilbert techniques and applications to graphical enumeration. *Int. Math. Res. Not.*, 14 (2003), 755–820.

87. EYNARD, B. Master loop equations, free energy and correlations for the chain of matrices. *J. High Energy Phys.*, 11 (2003), 018, 45 pp. (electronic).

88. EYNARD, B., KOKOTOV, A., AND KOROTKIN, D. $1/N^2$-correction to free energy in Hermitian two-matrix model. *Lett. Math. Phys. 71*, 3 (2005), 199–207.

89. FERRER, A. P., EYNARD, B., DI FRANCESCO, P., AND ZUBER, J.-B. Correlation functions of Harish-Chandra integrals over the orthogonal and the symplectic groups. *J. Stat. Phys. 129*, 5-6 (2007), 885–935.

90. FÖLLMER, H. An entropy approach to the time reversal of diffusion processes. In *Stochastic differential systems (Marseille-Luminy, 1984)*, vol. 69 of *Lecture Notes in Control and Inform. Sci.* Springer, Berlin, 1985, pp. 156–163.

91. FONTBONA, J. Uniqueness for a weak nonlinear evolution equation and large deviations for diffusing particles with electrostatic repulsion. *Stochastic Process. Appl. 112*, 1 (2004), 119–144.

92. FRIEDLAND, S., RIDER, B., AND ZEITOUNI, O. Concentration of permanent estimators for certain large matrices. *Ann. Appl. Probab. 14*, 3 (2004), 1559–1576.

93. FÜREDI, Z., AND KOMLÓS, J. The eigenvalues of random symmetric matrices. *Combinatorica 1*, 3 (1981), 233–241.

94. GENTIL, I., GUILLIN, A., AND MICLO, L. Modified logarithmic Sobolev inequalities and transportation inequalities. *Probab. Theory Related Fields 133*, 3 (2005), 409–436.

95. GINIBRE, J. Statistical ensembles of complex, quaternion, and real matrices. *J. Mathematical Phys. 6* (1965), 440–449.

280   References

96. GÖTZE, F., AND TIKHOMIROV, A. The rate of convergence for spectra of GUE and LUE matrix ensembles. *Cent. Eur. J. Math. 3*, 4 (2005), 666–704 (electronic).
97. GOULDEN, I. P., AND JACKSON, D. M. *Combinatorial enumeration*. A Wiley-Interscience Publication. John Wiley & Sons Inc., New York, 1983. With a foreword by Gian-Carlo Rota, Wiley-Interscience Series in Discrete Mathematics.
98. GROSS, D. J., AND MATYTSIN, A. Some properties of large-*N* two-dimensional Yang-Mills theory. *Nuclear Phys. B 437*, 3 (1995), 541–584.
99. GROSS, D. J., PIRAN, T., AND WEINBERG, S., Eds. *Two-dimensional quantum gravity and random surfaces* (River Edge, NJ, 1992), vol. 8 of *Jerusalem Winter School for Theoretical Physics*, World Scientific Publishing Co. Inc.
100. GUIONNET, A. Large deviations upper bounds and central limit theorems for non-commutative functionals of Gaussian large random matrices. *Ann. Inst. H. Poincaré Probab. Statist. 38*, 3 (2002), 341–384.
101. GUIONNET, A. First order asymptotics of matrix integrals; a rigorous approach towards the understanding of matrix models. *Comm. Math. Phys. 244*, 3 (2004), 527–569.
102. GUIONNET, A., AND MAÏDA, M. Character expansion method for the first order asymptotics of a matrix integral. *Probab. Theory Related Fields 132*, 4 (2005), 539–578.
103. GUIONNET, A., AND MAÏDA, M. A Fourier view on the *R*-transform and related asymptotics of spherical integrals. *J. Funct. Anal. 222*, 2 (2005), 435–490.
104. GUIONNET, A., AND MAUREL-SEGALA, E. Combinatorial aspects of matrix models. *ALEA Lat. Am. J. Probab. Math. Stat. 1* (2006), 241–279 (electronic).
105. GUIONNET, A., AND MAUREL-SEGALA, E. Second order asymptotics for matrix models. *Ann. Probab. 35*, 6 (2007), 2160–2212.
106. GUIONNET, A., AND SHLYAKHTENKO, D. Free diffusions and matrix models with strictly convex interaction. *http://arxiv.org/abs/math/0701787* (2007).
107. GUIONNET, A., AND ZEGARLINSKI, B. Lectures on logarithmic Sobolev inequalities. In *Séminaire de Probabilités, XXXVI*, vol. 1801 of *Lecture Notes in Math*. Springer, Berlin, 2003, pp. 1–134.
108. GUIONNET, A., AND ZEITOUNI, O. Concentration of the spectral measure for large matrices. *Electron. Comm. Probab. 5* (2000), 119–136 (electronic).
109. GUIONNET, A., AND ZEITOUNI, O. Large deviations asymptotics for spherical integrals. *J. Funct. Anal. 188*, 2 (2002), 461–515.
110. GUIONNET, A., AND ZEITOUNI, O. Addendum to: "Large deviations asymptotics for spherical integrals" [J. Funct. Anal. **188** (2002), no. 2, 461–515; mr 1883414]. *J. Funct. Anal. 216*, 1 (2004), 230–241.
111. HAAGERUP, U., AND THORBJØRNSEN, S. A new application of random matrices: $\mathrm{Ext}(C^*_{\mathrm{red}}(F_2))$ is not a group. *Ann. of Math. (2) 162*, 2 (2005), 711–775.
112. HAMMOND, C., AND MILLER, S. J. Distribution of eigenvalues for the ensemble of real symmetric Toeplitz matrices. *J. Theoret. Probab. 18*, 3 (2005), 537–566.
113. HARER, J., AND ZAGIER, D. The Euler characteristic of the moduli space of curves. *Invent. Math. 85*, 3 (1986), 457–485.
114. HARGÉ, G. A convex/log-concave correlation inequality for Gaussian measure and an application to abstract Wiener spaces. *Probab. Theory Related Fields 130*, 3 (2004), 415–440.
115. HARISH-CHANDRA. Fourier transforms on a semisimple Lie algebra. I. *Amer. J. Math. 79* (1957), 193–257.

116. HARISH-CHANDRA. Fourier transforms on a semisimple Lie algebra. II. *Amer. J. Math. 79* (1957), 653–686.

117. HIAI, F., AND PETZ, D. *The semicircle law, free random variables and entropy*, vol. 77 of *Mathematical Surveys and Monographs*. American Mathematical Society, Providence, RI, 2000.

118. HIAI, F., PETZ, D., AND UEDA, Y. Free transportation cost inequalities via random matrix approximation. *Probab. Theory Related Fields 130*, 2 (2004), 199–221.

119. HOUGH, J. B., KRISHNAPUR, M., PERES, Y., AND VIRÁG, B. Determinantal processes and independence. *Probab. Surv. 3* (2006), 206–229 (electronic).

120. JOHANSSON, K. On fluctuations of eigenvalues of random Hermitian matrices. *Duke Math. J. 91*, 1 (1998), 151–204.

121. JOHANSSON, K. Universality of the local spacing distribution in certain ensembles of Hermitian Wigner matrices. *Comm. Math. Phys. 215*, 3 (2001), 683–705.

122. KARATZAS, I., AND SHREVE, S. E. *Brownian motion and stochastic calculus*, second ed., vol. 113 of *Graduate Texts in Mathematics*. Springer-Verlag, New York, 1991.

123. KAZAKOV, V. Solvable matrix models. In *Random matrix models and their applications*, vol. 40 of *Math. Sci. Res. Inst. Publ.* Cambridge Univ. Press, Cambridge, 2001, pp. 271–283.

124. KAZAKOV, V. A., STAUDACHER, M., AND WYNTER, T. Character expansion methods for matrix models of dually weighted graphs. *Comm. Math. Phys. 177*, 2 (1996), 451–468.

125. KAZAKOV, V. A., AND ZINN-JUSTIN, P. Two-matrix model with $ABAB$ interaction. *Nuclear Phys. B 546*, 3 (1999), 647–668.

126. KEATING, J. P. Random matrices and number theory. In *Applications of random matrices in physics*, vol. 221 of *NATO Sci. Ser. II Math. Phys. Chem.* Springer, Dordrecht, 2006, pp. 1–32.

127. KENYON, R., AND OKOUNKOV, A. Limit shapes and the complex Burgers equation. *Acta Math. 199*, 2 (2007), 263–302.

128. KENYON, R., OKOUNKOV, A., AND SHEFFIELD, S. Dimers and amoebae. *Ann. of Math. (2) 163*, 3 (2006), 1019–1056.

129. KIPNIS, C., AND LANDIM, C. *Scaling limits of interacting particle systems*, vol. 320 of *Grundlehren der Mathematischen Wissenschaften [Fundamental Principles of Mathematical Sciences]*. Springer-Verlag, Berlin, 1999.

130. KIPNIS, C., AND OLLA, S. Large deviations from the hydrodynamical limit for a system of independent Brownian particles. *Stochastics Stochastics Rep. 33*, 1-2 (1990), 17–25.

131. KIPNIS, C., OLLA, S., AND VARADHAN, S. R. S. Hydrodynamics and large deviation for simple exclusion processes. *Comm. Pure Appl. Math. 42*, 2 (1989), 115–137.

132. KONTSEVICH, M. Intersection theory on the moduli space of curves and the matrix Airy function. *Comm. Math. Phys. 147*, 1 (1992), 1–23.

133. LANDO, S. K., AND ZVONKIN, A. K. *Graphs on surfaces and their applications*, vol. 141 of *Encyclopaedia of Mathematical Sciences*. Springer-Verlag, Berlin, 2004. With an appendix by Don B. Zagier, Low-Dimensional Topology, II.

134. LASRY, J.-M., AND LIONS, P.-L. Mean field games. *Jpn. J. Math. 2*, 1 (2007), 229–260.

460

282   References

135. LE GALL, J.-F. The topological structure of scaling limits of large planar maps. *Invent. Math. 169*, 3 (2007), 621–670.
136. LEDOUX, M. Remarks on logarithmic Sobolev constants, exponential integrability and bounds on the diameter. *J. Math. Kyoto Univ. 35*, 2 (1995), 211–220.
137. LEDOUX, M. Isoperimetry and Gaussian analysis. In *Lectures on probability theory and statistics (Saint-Flour, 1994)*, vol. 1648 of *Lecture Notes in Math.* Springer, Berlin, 1996, pp. 165–294.
138. LEDOUX, M. *The concentration of measure phenomenon*, vol. 89 of *Mathematical Surveys and Monographs*. American Mathematical Society, Providence, RI, 2001.
139. LEDOUX, M. A remark on hypercontractivity and tail inequalities for the largest eigenvalues of random matrices. In *Séminaire de Probabilités XXXVII*, vol. 1832 of *Lecture Notes in Math.* Springer, Berlin, 2003, pp. 360–369.
140. LEDOUX, M. Spectral gap, logarithmic Sobolev constant, and geometric bounds. In *Surveys in differential geometry. Vol. IX*, Surv. Differ. Geom., IX. Int. Press, Somerville, MA, 2004, pp. 219–240.
141. LENOBLE, O., AND PASTUR, L. On the asymptotic behaviour of correlators of multi-cut matrix models. *J. Phys. A 34*, 30 (2001), L409–L415.
142. LOEPER, G. The reconstruction problem for the Euler-Poisson system in cosmology. *Arch. Ration. Mech. Anal. 179*, 2 (2006), 153–216.
143. MAÏDA, M. Large deviations for the largest eigenvalue of rank one deformations of Gaussian ensembles. *Electron. J. Probab. 12* (2007), 1131–1150 (electronic).
144. MARCKERT, J.-F., AND MIERMONT, G. Invariance principles for random bipartite planar maps. *Ann. Probab. 35*, 5 (2007), 1642–1705.
145. MARˇ CENKO, V. A., AND PASTUR, L. A. Distribution for some sets of random matrices. *Math. USSR-Sb.* (1967), 457–483.
146. MATYTSIN, A. On the large-N limit of the Itzykson-Zuber integral. *Nuclear Phys. B 411*, 2-3 (1994), 805–820.
147. MATYTSIN, A., AND ZAUGG, P. Kosterlitz-Thouless phase transitions on discretized random surfaces. *Nuclear Phys. B 497*, 3 (1997), 658–698.
148. MAUREL-SEGALA, E. High order expansion for matrix models. *http://front. math.ucdavis.edu/0608.5192* (2006).
149. MCCANN, R. J. Existence and uniqueness of monotone measure-preserving maps. *Duke Math. J. 80*, 2 (1995), 309–323.
150. MCKEAN, JR., H. P. *Stochastic integrals*. Probability and Mathematical Statistics, No. 5. Academic Press, New York, 1969.
151. MECKES, M. W. Concentration of norms and eigenvalues of random matrices. *J. Funct. Anal. 211*, 2 (2004), 508–524.
152. MEHTA, M. L. A method of integration over matrix variables. *Comm. Math. Phys. 79*, 3 (1981), 327–340.
153. MEHTA, M. L. *Random matrices*, third ed., vol. 142 of *Pure and Applied Mathematics (Amsterdam)*. Elsevier/Academic Press, Amsterdam, 2004.
154. MEHTA, M. L., AND MAHOUX, G. A method of integration over matrix variables. III. *Indian J. Pure Appl. Math. 22*, 7 (1991), 531–546.
155. MILMAN, V. D., AND SCHECHTMAN, G. *Asymptotic theory of finite-dimensional normed spaces*, vol. 1200 of *Lecture Notes in Mathematics*. Springer-Verlag, Berlin, 1986. With an appendix by M. Gromov.
156. MINEYEV, I., AND SHLYAKHTENKO, D. Non-microstates free entropy dimension for groups. *Geom. Funct. Anal. 15*, 2 (2005), 476–490.

157. MINGO, J. A., AND SPEICHER, R. Second order freeness and fluctuations of random matrices. I. Gaussian and Wishart matrices and cyclic Fock spaces. *J. Funct. Anal. 235*, 1 (2006), 226–270.

158. NELSON, E. Notes on non-commutative integration. *J. Funct. Anal. 15* (1974), 103–116.

159. NICA, A., AND SPEICHER, R. *Lectures on the combinatorics of free probability*, vol. 335 of *London Mathematical Society Lecture Note Series*. Cambridge University Press, Cambridge, 2006.

160. NORRIS, J. R., ROGERS, L. C. G., AND WILLIAMS, D. Brownian motions of ellipsoids. *Trans. Amer. Math. Soc. 294*, 2 (1986), 757–765.

161. OKOUNKOV, A. Infinite wedge and random partitions. *Selecta Math. (N.S.) 7*, 1 (2001), 57–81.

162. PASTUR, L. Limiting laws of linear eigenvalue statistics for Hermitian matrix models. *J. Math. Phys. 47*, 10 (2006), 103303, 22.

163. PASTUR, L., AND SHCHERBINA, M. On the edge universality of the local eigenvalue statistics of matrix models. *Mat. Fiz. Anal. Geom. 10*, 3 (2003), 335–365.

164. PASTUR, L. A. Spectral and probabilistic aspects of matrix models. In *Algebraic and geometric methods in mathematical physics (Kaciveli, 1993)*, vol. 19 of *Math. Phys. Stud.* Kluwer Acad. Publ., Dordrecht, 1996, pp. 207–242.

165. PAUWELS, E. J., AND ROGERS, L. C. G. Skew-product decompositions of Brownian motions. In *Geometry of random motion (Ithaca, N.Y., 1987)*, vol. 73 of *Contemp. Math.* Amer. Math. Soc., Providence, RI, 1988, pp. 237–262.

166. PÉCHÉ, S., AND SOSHNIKOV, A. Wigner random matrices with non-symmetrically distributed entries. *J. Stat. Phys. 129*, 5-6 (2007), 857–884.

167. PEDERSEN, G. K. C^*-algebras and their automorphism groups, vol. 14 of *London Mathematical Society Monographs*. Academic Press Inc. [Harcourt Brace Jovanovich Publishers], London, 1979.

168. REVUZ, D., AND YOR, M. *Continuous martingales and Brownian motion*, third ed., vol. 293 of *Grundlehren der Mathematischen Wissenschaften [Fundamental Principles of Mathematical Sciences]*. Springer-Verlag, Berlin, 1999.

169. RIDER, B., AND SILVERSTEIN, J. W. Gaussian fluctuations for non-Hermitian random matrix ensembles. *Ann. Probab. 34*, 6 (2006), 2118–2143.

170. ROGERS, L. C. G., AND SHI, Z. Interacting Brownian particles and the Wigner law. *Probab. Theory Related Fields 95*, 4 (1993), 555–570.

171. ROYER, G. *An initiation to logarithmic Sobolev inequalities*, vol. 14 of *SMF/AMS Texts and Monographs*. American Mathematical Society, Providence, RI, 2007. Translated from the 1999 French original by Donald Babbitt.

172. RUDIN, W. *Functional analysis*, second ed. International Series in Pure and Applied Mathematics. McGraw-Hill Inc., New York, 1991.

173. RUELLE, D. *Statistical mechanics: rigorous results*. Benjamin, Amsterdam, 1969.

174. RUZMAIKINA, A. Universality of the edge distribution of eigenvalues of Wigner random matrices with polynomially decaying distributions of entries. *Comm. Math. Phys. 261*, 2 (2006), 277–296.

175. SAGAN, B. E. *The symmetric group*, second ed., vol. 203 of *Graduate Texts in Mathematics*. Springer-Verlag, New York, 2001. Representations, combinatorial algorithms, and symmetric functions.

176. SCHAEFFER, G. Bijective census and random generation of Eulerian planar maps with prescribed vertex degrees. *Electron. J. Combin. 4*, 1 (1997), Research Paper 20, 14 pp. (electronic).

284   References

177. SCHENKER, J., AND SCHULZ-BALDES, H. Gaussian fluctuations for random matrices with correlated entries. *Int. Math. Res. Not. IMRN*, 15 (2007), Art. ID rnm047, 36.

178. SHIRYAEV, A. N. *Probability*, second ed., vol. 95 of *Graduate Texts in Mathematics*. Springer-Verlag, New York, 1996. Translated from the first (1980) Russian edition by R. P. Boas.

179. SINAI, Y., AND SOSHNIKOV, A. Central limit theorem for traces of large random symmetric matrices with independent matrix elements. *Bol. Soc. Brasil. Mat. (N.S.) 29*, 1 (1998), 1–24.

180. SOSHNIKOV, A. Universality at the edge of the spectrum in Wigner random matrices. *Comm. Math. Phys. 207*, 3 (1999), 697–733.

181. SOSHNIKOV, A. Universality at the edge of the spectrum in Wigner random matrices. *Comm. Math. Phys. 207*, 3 (1999), 697–733.

182. SOSHNIKOV, A. Determinantal random point fields. *Uspekhi Mat. Nauk 55*, 5(335) (2000), 107–160.

183. SOSHNIKOV, A. A note on universality of the distribution of the largest eigenvalues in certain sample covariance matrices. *J. Statist. Phys. 108*, 5-6 (2002), 1033–1056. Dedicated to David Ruelle and Yasha Sinai on the occasion of their 65th birthdays.

184. SOSHNIKOV, A. Poisson statistics for the largest eigenvalues in random matrix ensembles. In *Mathematical physics of quantum mechanics*, vol. 690 of *Lecture Notes in Phys.* Springer, Berlin, 2006, pp. 351–364.

185. SPEICHER, R. Free calculus. In *Quantum probability communications, Vol. XII (Grenoble, 1998)*, QP-PQ, XII. World Sci. Publishing, River Edge, NJ, 2003, pp. 209–235.

186. SUNDER, V. S. *An invitation to von Neumann algebras.* Universitext. Springer-Verlag, New York, 1987.

187. 'T HOOFT, G. A planar diagram theory for strong interactions. *Nuclear Physics B 72*, 3 (1974), 461–473.

188. TALAGRAND, M. New concentration inequalities in product spaces. *Invent. Math. 126*, 3 (1996), 505–563.

189. TALAGRAND, M. A new look at independence. *Ann. Probab. 24*, 1 (1996), 1–34.

190. TAO, T., AND VU, V. Random matrices: the circular law. *http://front.math. ucdavis.edu/0708.2895* (2007).

191. TRACY, C. A., AND WIDOM, H. The distribution of the largest eigenvalue in the Gaussian ensembles: $\beta = 1, 2, 4$. In *Calogero-Moser-Sutherland models (Montréal, QC, 1997)*, CRM Ser. Math. Phys. Springer, New York, 2000, pp. 461–472.

192. TUTTE, W. T. A census of planar triangulations. *Canad. J. Math. 14* (1962), 21–38.

193. TUTTE, W. T. A new branch of enumerative graph theory. *Bull. Amer. Math. Soc. 68* (1962), 500–504.

194. TUTTE, W. T. On the enumeration of planar maps. *Bull. Amer. Math. Soc. 74* (1968), 64–74.

195. VILLANI, C. *Topics in optimal transportation*, vol. 58 of *Graduate Studies in Mathematics*. American Mathematical Society, Providence, RI, 2003.

196. VILLANI, C. *Optimal transport, old and new*. Lecture Notes for the 2005 Saint-Flour Probability Summer School. www.umpa.ens-lyon.fr/ cvillani/ StFlour/oldnew-9.ps, 2007.

197. VOICULESCU, D. Limit laws for random matrices and free products. *Invent. Math. 104*, 1 (1991), 201–220.

198. VOICULESCU, D. The analogues of entropy and of Fisher's information measure in free probability theory. III. The absence of Cartan subalgebras. *Geom. Funct. Anal. 6*, 1 (1996), 172–199.

199. VOICULESCU, D. The analogues of entropy and of Fisher's information measure in free probability theory. VI. Liberation and mutual free information. *Adv. Math. 146*, 2 (1999), 101–166.

200. VOICULESCU, D. Lectures on free probability theory. In *Lectures on probability theory and statistics (Saint-Flour, 1998)*, vol. 1738 of *Lecture Notes in Math.* Springer, Berlin, 2000, pp. 279–349.

201. VOICULESCU, D. Free entropy. *Bull. London Math. Soc. 34*, 3 (2002), 257–278.

202. VOICULESCU, D. Aspects of free probability. In *XIVth International Congress on Mathematical Physics.* World Sci. Publ., Hackensack, NJ, 2005, pp. 145–157.

203. VOICULESCU, D. V., DYKEMA, K. J., AND NICA, A. *Free random variables*, vol. 1 of *CRM Monograph Series.* American Mathematical Society, Providence, RI, 1992. A noncommutative probability approach to free products with applications to random matrices, operator algebras and harmonic analysis on free groups.

204. WEYL, H. *The Classical Groups. Their Invariants and Representations.* Princeton University Press, Princeton, N.J., 1939.

205. WIGNER, E. P. On the distribution of the roots of certain symmetric matrices. *Ann. of Math. (2) 67* (1958), 325–327.

206. WISHART, J. The generalized product moment distribution in samples from a normal multivariate population. *Biometrika 20* (1928), 35–52.

207. WITTEN, E. Two-dimensional gravity and intersection theory on moduli space. In *Surveys in differential geometry (Cambridge, MA, 1990)*. Lehigh Univ., Bethlehem, PA, 1991, pp. 243–310.

208. ZAKHAREVICH, I. A generalization of Wigner's law. *Comm. Math. Phys. 268*, 2 (2006), 403–414.

209. ZINN JUSTIN, P. The dilute potts model on random surfaces. *J. Stat. Phys. 98* (2000), 245–264.

210. ZINN-JUSTIN, P., AND ZUBER, J.-B. On some integrals over the U(N) unitary group and their large N limit. *J. Phys. A 36*, 12 (2003), 3173–3193. Random matrix theory.

211. ZVONKIN, A. Matrix integrals and map enumeration: an accessible introduction. *Math. Comput. Modelling 26*, 8-10 (1997), 281–304. Combinatorics and physics (Marseilles, 1995).

Index

Adapted process, 270
Affiliated operator, 227
Algebra
  C^*-algebra, 227
  Unital algebra, 227
  von Neumann algebra, 227

Bakry–Emery criterion, 54
Bakry–Emery theorem, 55
Brascamp–Lieb inequalities, 77
Brownian motion
  Hermitian, 165
  Symmetric, 165
  Dyson's Brownian motion, 167
Burkholder–Davis–Gundy inequality,
  272

c-convexity, 94
Catalan numbers, 9
Closed path, 35
Complex Burgers equation, 177, 186
Convergence in moments, 96
Convex potentials, 94
Courant–Fischer formula, 263
Cyclic derivative, 95

Dick path, 9
Dudley's distance, XI

Empirical distribution of matrices, 230
Empirical measure, 175
Exponential equivalence, 269
Exponential tightness, 267

Filtration, 270
Formal expansion, 105
Free additive convolution, 241
Free cumulants, 237
Free energy, 118
Free Fisher information, 248
Freeness
  Asymptotic Freeness, 232
  Definition, 231

Gaussian ensembles, 147
Gaussian matrix moments, 102

Harish–Chandra–Itzykson–Zuber
  integral, 211
Herbst's lemma, 50

Itô's calculus, 272

Klein's lemma, 68

Large deviation principle, 267
Largest eigenvalue, 33
Legendre transform, 246
Leibniz rule, 95
Lidskii's theorem, 264
Logarithmic Sobolev inequality, 49
Lower semi-continuous function, 267

Map
  Colored maps, 100
  Colored planar maps, 42
  Genus of a map, 13
  One-color map, 13

288   Index

Maps number $\mathcal{M}_k^g(P)$, 97
Martingale, 270
Martingale bracket, 270
Matrix integral, 89
Matrix models, 89
Models on random graphs
  Ising model, 217
  Potts model, 217

Non-commutative derivative ∂_i, 95
Non-commutative entropy
  Microstates entropy, 245
  Microstates-free entropy, 248
  Non-commutative entropy of one
    variable, 150
Non-commutative Hölder inequality,
  264
Non-commutative laws, 95, 228
Non-commutative polynomials
  $\mathbb{C}\langle X_1, \ldots, X_m \rangle$, 93
Non-commutative probability space,
  228
Novikov's criteria, 272

Opérateurs carré du champ, 54
Opérateurs carré du champ itéré, 54

Partition
  Block of a partition, 11
  Colored non-crossing pair partition,
    41
  Definition, 10
  Non crossing partition, 11
  Pair partition, 11
  Partition, 10
Poincaré's inequality, 59
Prohorov's theorem, 269

R-transform, 241
Rate function, 267
Ricci tensor, 62
Rooted tree, 8

Schur polynomial, 213
Schwinger–Dyson equations
  Finite dimensional, 109
  Limiting equation, 111
  SD[V], 111
Self-adjoint operator, 227
Self-adjoint polynomials, 94
Semicircle law, 7
Space of non-commutative laws
  $\mathbb{C}\langle X_1, \ldots, X_m \rangle^{\mathcal{D}}$, 96
Star
  Colored star, 42
  One-color star, 12
State
  Definition, 228
  Tracial states, 228
Stochastic Differential equation, 271
Stochastic integral, 270
Stopping time, 270
Strong solution, 270
Strong uniqueness, 271

Talagrand's concentration inequality, 60
Topological expansion, 105
Tree, 8
Tutte, 93

Varadhan's lemma, 269
Voiculescu's theorem, 42

Weak solution, 271
Weak topology, XI
Weak uniqueness, 271
Weyl's inequality, 263
Wick's formula, 99
Wigner's theorem, 16

Young shape, 212

List of Participants of the Summer School

Lecturers

Maury BRAMSON	University of Minnesota, Minneapolis, USA
Alice GUIONNET	École Normale Supérieure de Lyon, France
Steffen LAURITZEN	University of Oxford, UK

Participants

Marie ALBENQUE	Université Denis Diderot, Paris, France
Louis-Pierre ARGUIN	Princeton University, USA
Sylvain ARLOT	Université Paris-Sud, Orsay, France
Claudio ASCI	Università La Sapienza, Roma, Italy
Jean-Yves AUDIBERT	École Nationale des Ponts et Chaussées, Marne-la-Vallée, France
Wlodzimierz BRYC	University of Cincinnati, USA
Thierry CABANAL-DUVILLARD	Université Paris 5, France
Alain CAMANES	Université de Nantes, France
Mireille CAPITAINE	Université Paul Sabatier, Toulouse, France
Muriel CASALIS	Université Paul Sabatier, Toulouse, France
François CHAPON	Université Pierre et Marie Curie, Paris, France
Adriana CLIMESCU-HAULICA	CNRS, Marseille, France
Marek CZYSTOLOWSKI	Wroclaw University, Poland

468

Manon DEFOSSEUX	Université Pierre et Marie Curie, Paris, France
Catherine DONATI-MARTIN	Université Pierre et Marie Curie, Paris, France
Coralie EYRAUD-DUBOIS	Université Claude Bernard, Lyon, France
Delphine FERAL	Université Paul Sabatier, Toulouse, France
Mathieu GOURCY	Université Blaise Pascal, Clermont-Ferrand, France
Mihai GRADINARU	Université Henri Poincaré, Nancy, France
Benjamin GRAHAM	University of Cambridge, UK
Katrin HOFMANN-CREDNER	Ruhr Universität, Bochum, Germany
Manuela HUMMEL	LMU, München, Germany
Jérémie JAKUBOWICZ	École Normale Supérieure de Cachan, France
Abdeldjebbar KANDOUCI	Université de Rouen, France
Achim KLENKE	Universität Mainz, Germany
Krzysztof LATUSZYNSKI	Warsaw School od Economics, Poland
Liangzhen LEI	Université Blaise Pascal, Clermont-Ferrand, France
Manuel LLADSER	University of Colorado, Boulder, USA
Dhafer MALOUCHE	École Polytechnique de Tunisie, Tunisia
Hélène MASSAM	York University, Toronto, Canada
Robert PHILIPOWSKI	Universität Bonn, Germany
Jean PICARD	Université Blaise Pascal, Clermont-Ferrand, France
Júlia RÉFFY	Budapest University of Technology and Economics, Hungary
Anthony REVEILLAC	Université de La Rochelle, France
Alain ROUAULT	Université de Versailles, France
Markus RUSCHHAUPT	German Cancer Research Center, Heidelberg, Germany
Erwan SAINT LOUBERT BIÉ	Université Blaise Pascal, Clermont-Ferrand, France
Pauline SCULLI	London School of Economics, UK
Sylvie SEVESTRE-GHALILA	Université Paris 5, France

Frederic UTZET   Universitat Autonoma de Barcelona,
                                    Spain
Nicolas VERZELEN   Université Paris-Sud, Orsay, France
Yvon VIGNAUD   Centre de Physique Théorique,
                                    Marseille, France
Pompiliu Manuel ZAMFIR   Stanford University, USA

List of Short Lectures Given at the Summer School

Louis-Pierre Arguin	Spin glass systems and Ruelle's probability cascades
Sylvain Arlot	Model selection by resampling in statistical learning
Claudio Asci	Generalized Beta distributions and random continued fractions
Wlodek Bryc	Families of probability measures generated by the Cauchy kernel
Thierry Cabanal-Duvillard	Lévy unitary ensemble and free Poisson point processes
Manon Defosseux	Random words and the eigenvalues of the minors of random matrices
Delphine Féral	The largest eigenvalue of rank one deformation of large Wigner matrices
Mathieu Gourcy	A large deviation principle for 2D stochastic Navier-Stokes equations
Mihaï Gradinaru	On the stochastic heat equation
Katrin Hofmann-Credner	Limiting laws for non-classical random matrices
Jérémie Jakubowicz	Detecting segments in digital images
Krzysztof Latuszyński	$(\varepsilon - \alpha)$-MCMC approximation under drift condition
Liangzhen Lei	Large deviations of kernel density estimator
Gérard Letac	Pavage par séparateurs minimaux d'un arbre de jonction et applications aux modèles graphiques

294   List of Short Lectures Given at the Summer School

Hélène Massam	Discrete graphical models Markov w.r.t. an undirected graph: the conjugate prior and its normalizing constant
Manuel Lladser	Asymptotics for the coefficients of mixed-powers generating functions
Robert Philipowski	Approximation de l'équation des milieux poreux visqueuse par des équations différentielles stochastiques non linéaires
Anthony Réveillac	Stein estimation on the Wiener space
Alain Rouault	Asymptotic properties of determinants of some random matrices
Markus Ruschhaupt	Simulation of matrices with constraints by using moralised graphs
Pauline Sculli	Counterparty default risk in affine processes with jump decay
Frederic Utzet	On the orthogonal polynomials associated to a Lévy process
Yvon Vignaud	Rigid interfaces for some lattice models